*Statistics: A New Approach*

# STATISTICS
## A NEW APPROACH

by W. Allen Wallis

PROFESSOR OF STATISTICS AND ECONOMICS, UNIVERSITY OF CHICAGO

and Harry V. Roberts

ASSOCIATE PROFESSOR OF STATISTICS, UNIVERSITY OF CHICAGO

The Free Press, Glencoe, Illinois

First Printing April 1956
Second Printing August 1956
Third Printing October 1956
Fourth Printing November 1956
Fifth Printing September 1957
Sixth Printing September 1958
Seventh Printing January 1959

LIBRARY OF CONGRESS CATALOG CARD NO. 56-8453

*TO  GARFIELD  V.  COX*

# Preface

Statistics is a lively and a fascinating subject; but studying it is too often excruciatingly dull. In this new approach, we have tried to bring out its liveliness by lavish use of real examples from a wide variety of fields, choosing the examples for their intrinsic interest, their closeness to everyday experience, or the significance of the information contained in them. We have tried to bring out the fascination of statistics as a subject in its own right by emphasizing its fundamental ideas and the principles and criteria involved in applying them, and by keeping technical details from dominating the scene.

Statistical reasoning, like mathematical reasoning, legal reasoning, or any other form of reasoning, is essentially independent of its content. As with mathematical or other forms of reasoning, illustrations of the content are indispensable for the beginner, both as motivation and as a means of learning. The illustrations should appeal to the student's experience, interest, and comprehension, but they need not come predominantly from any one field. (Only a minority will pursue careers within their fields of college specialization.) There are, indeed, two strong reasons for *not* concentrating the illustrations in any one field. The first is that the unity of statistical methods, and even their very nature, may be obscured if the methods are always presented in conjunction with a specific subject matter. The second is that statistics ought to be part of a "general" or "liberal" education, in the sense of an introduction to the problems, materials, and methods of the major arts and sciences and the development of competence to exercise constructive, critical judgment. As a subsidiary contribution to this purpose, the illustrations used in teaching statistics ought to convey information that itself adds to a general education.

It is not possible, in an introductory course, to teach a student all he is likely to need to know about statistics, even for a single field. This is a compelling reason for developing adaptability and flexibility by emphasizing ideas, principles, criteria, and methods, rather than the fullest assortment of techniques and details. Techniques and details, beyond a comparatively small range of fairly basic methods, are likely to do more harm than good in the hands of beginners. One of our chief objectives is to make our readers understand that this is so, and why it is so, and at the same time to put them in a position to read understandingly material based on statistical methods beyond their own competence to execute. Indeed, this seems to us the appropriate function of an introduction to statistics: neither a diluted version of an advanced course nor a how-to-do-it manual, but a treatment specifically for readers rather than writers of statistics. Methods of organizing and presenting data, and the "common sense" interpretation of statistics, must be emphasized equally with, and treated integrally with, modern methods of statistical inference.

This orientation toward those who will be scientists, business men, professional people, administrators, or intelligent housewives and mothers, rather than specialists in statistics, actually proves a boon, we find, to potential students of intermediate or advanced statistics. It means that subsequent courses can build on the introductory course, without having to rework it. More important, it gives budding statisticians an early glimpse of an important field, with its own significant ideas and principles—a field which frequently seems unattractive because too often viewed first through a haze of details, technicalities, and clerical work.

As for mathematics, this book is virtually devoid of it. None beyond the high school level is used. The only mathematics likely to be unfamiliar is the $\sum$ notation, and that is amply explained. Avoidance of mathematics is, of course, almost a necessity with beginning students. At the stage when students usually study statistics first, even those who will specialize (or are specializing) in mathematics or the physical sciences seldom know their mathematics well enough to use it as a medium for learning another subject—just as students specializing in a foreign language do not ordinarily know the language well enough to learn statistics effectively in that language.

Avoidance of mathematics is not, however, merely a necessity in introductory statistics; it is, we feel, a real virtue. Elementary statistics courses that draw freely on, say, first year college mathematics, unavoidably teach mathematics at the expense of statistics, or sometimes fail to teach either. The great ideas of statistics are lost in a sea of algebra. If the introductory statistics course emphasizes statistical concepts and principles, those students who are well enough grounded in mathematics will find little difficulty in introducing mathematical formulations themselves, and perhaps deriving some results; and they will even find their comprehension of mathematics heightened by an independent grasp of the statistical ideas they are trying to encompass in their mathematical formulations. We have used the material of this book with some undergraduates specializing in mathematics and the physical sciences, and with groups holding degrees in engineering sciences, and they have found it distinctly more profitable than material showing, for example, the algebraic derivation of the binomial distribution and its parameters. As a matter of fact, most of them find the nature of the binomial distribution and its parameters better illuminated by this approach than by an algebraic approach.

As the preceding paragraphs imply, the material of this book has been used already for a number of years. It has been evolving steadily through a continuous series of revisions for nine years now. Perhaps three thousand students have used it, coming from nearly all fields (though mostly the social sciences, business, and economics), all levels of academic advancement (though mostly sophomores and juniors), and all levels of mathematical training (though mostly so little as to be effectively none). Some of the eight or nine preliminary versions of the book were tailored to the needs and capabilities of special groups, until experience convinced us that a single approach was best for all groups. About twenty different teachers have used the material, again representing a wide range of specialties and levels of statistical training, and most of them have given us thoughtful critiques which have influenced subsequent revisions. Whatever merits the book has must be credited in large measure to the assistance we have had from these teachers and students of earlier versions.

While we have found a single approach best for all kinds of students, we have not covered the same material or placed the same emphases for all students. Though all of the material in the book has been taught to some students, and much of it has been taught to all students, no students have yet covered all of it. The material as presented here offers considerable flexibility with regard to selection and sequence of topics. Chap. 3, for example, on misuses of statistics, can well be presented before Chap. 2, on effective uses. We have usually done that. The advantage is that Chap. 3 always proves one of the most intriguing, and serves not only to allay students' apprehensions about the dullness of statistics, but actually to create a positive interest. On the other hand, Chap. 2 is nearly as intriguing, and has the advantage of putting statistics' best foot forward. Chap. 2 is really two chapters in one, and the last section, Sec. 2.8, may have to be omitted in a short course. Furthermore, Sec. 2.8 can well be introduced just after Chap. 15, on the planning of research, for it is as useful near the end of the course, where it serves to tie together many separate topics, as near the beginning, where it serves as a preview.

We expect to prepare almost immediately a manual for teachers which will, among other things, indicate more fully the possibilities for picking and choosing among the topics and rearranging their sequence to meet differences in length of course, background of students, or objectives of the teacher. Suffice it to add here that many chapters are so constructed that parts can be omitted—for example, the last parts of Chaps. 2, 5, and 9; and many chapters can be omitted. Thus, the book is adaptable to courses ranging from less than a quarter to a full year.

One topic whose omission will be noted is index numbers. Our preliminary editions (except the last one) covered this, including descriptions of two important indexes, the Bureau of Labor Statistics' Index of Consumer Prices, and the Federal Reserve Board's Index of Industrial Production. But the detailed methods of compiling these indexes have changed so frequently that we have had a hard time keeping dittoed material up to date, and decided that it would be folly to go into print. Besides, the statistical basis of index numbers, weighted and standardized means, is covered in Chaps. 7 and 9, and examples are included which illustrate some of the special problems of index numbers. Finally, space given to particular data because of their importance in public affairs might just as well be given to, say, measures of unemployment—the methods for which have, however, also changed recently and presumably may be changed again. It seems likely that the more important the data, the more frequently will methods of compilation be improved. In short, basic statistical methods are the subject of this book, and for explanations of specific data it is better to rely on supplementary readings.

Our treatment of Student's *t* distribution will shock some mathematical statisticians. For perhaps a quarter of a century, this distribution, discovered in 1908, has been regarded by the cognoscenti as the very hall-mark of statistical sophistication. This is not the place to argue our position, but we suggest that, far-reaching as have been the consequences of the *t* distribution for technical statistics, in elementary applications it does not differ enough from the normal distribution and does not introduce enough of a new principle, to justify giving beginners this added complexity in lieu of some other topic. We have, therefore, confined our discussion of it to one paragraph in Chap. 13, technical notes to Chaps. 13 and 14, and a footnote in

Chap. 17. We venture to suggest that, in an elementary book, this treatment may be one stage more sophisticated than the usual display of sophistication with respect to Student's $t$.

The claim of novelty implied by our subtitle is nebulous and ambiguous, and we are perfectly willing not to press it. We had in mind three possible interpretations: that statistics represents a new approach to problems of scientific knowledge and practical action; that this book is a newcomer to the list of books by which a beginner may approach statistics; and that there are some novel features to our approach. Of the last, we may mention the following, without claiming that any one of them alone is unique:

(1) Statistics is treated here as a cohesive and important body of knowledge, worthy of attention and interesting in its own right, apart from its contributions to other fields.

(2) The universality of statistical methods is emphasized by choosing illustrations from a wide variety of fields.

(3) The illustrations have been chosen, in the main, as vehicles for conveying significant or interesting information, as well as illustrating statistical methods or achievements.

(4) An integrated treatment is accorded descriptive and analytical statistics. The collection of data is always aimed at drawing conclusions, and the soundness of conclusions depends equally on the meaning of the individual observations and their interrelations, and on allowance for sampling error.

(5) The materials are closely articulated, in the sense that the same examples are viewed from different standpoints, that the relation of topics in one chapter to those in others is indicated, that the *Do-It-Yourself* examples sometimes pick up points from earlier chapters and lay the groundwork for points in later chapters, and so on. To overcome the resistance of students to turning to other parts of the book, all tables, charts, and examples have been numbered to correspond with the pages on which they appear; thus "Table 61" means the table on page 61, and "Example 83C" means the third example on page 83.

(6) Although Student's $t$ and the $F$ ratio are explained so that the student should be able to take them in his stride when he encounters them in reading, he is advised not ordinarily to use them himself but to use the shortcut methods of Chap. 19. These, being nonparametric and involving simpler computations, are more nearly foolproof in the hands of the beginner—and, ordinarily, only a little less powerful. This is one of the ways we have eliminated technical detail without sacrificing ideas, principles, or accuracy, and without making the

book incomplete. Should the student actually need to make a $t$ or an $F$ test, he can do so by using formulas presented in technical notes to Chaps. 13 and 14, which adapt $t$, $F$, and $\chi^2$ to the normal distribution.

(7) New methods are introduced, but only conservatively, where they are clearly essential and clearly consonant with tried and proven methods and principles, as in the case of the new measures of association in Chap. 9 and the new shortcuts in Chap. 19. On the other hand, traditional material has been omitted where it involves technical details that might clutter up the presentation, or where experience has shown it to be sterile, as in the case of the traditional resolution of time series into independent secular, cyclical, seasonal, and random components. Thus, what is presented is coherent, sound, and useful, not faddish, tentative, or untried; in fact, we are confident that it will remain sound and useful half a century hence, whatever new developments may by then merit higher priority or suggest a different approach.

(8) The history of statistics and some of the illustrious figures who have contributed to it are mentioned occasionally to convey an appreciation of its continuity and permanence. Similarly, matters now subject to research, and viewpoints currently evolving, are mentioned from time to time, without, however, any pretense that these have yet attained permanence or applicability. These glimpses of the intellectual quality and challenge of statistics may, we hope, attract an occasional reader to a field in which opportunities for interesting, useful, and remunerative careers are exceptional, but usually unknown until too late in a student's education.

A book so long in the making naturally incorporates the work of many people besides the authors. Our greatest obligation, as we have indicated, is to the thousands of students who have studied the preliminary versions, literally hundreds of whom have made useful contributions, and to the score of teachers who have taught it, virtually all of whom have helped us. Among faculty members, H. Gregg Lewis, Josephine J. Williams, and Edgar Z. Friedenberg must be singled out for special thanks.

We are deeply indebted to Leonard J. Savage for a remarkably thorough and penetrating critique, both microscopic and macroscopic, of the last pre-publication edition; innumerable suggestions of his have been used throughout, and there can be no doubt that the book has benefited immeasurably by them. Frederick Mosteller and William H. Kruskal were also exceptionally generous, going over the manuscript line by line, discussing substantive and expository issues

*Preface*

at length, and making a great number of invaluable contributions. Only someone who has grappled with the problem of trying to be intelligible and interesting to elementary students, yet avoid violence to the standards of his colleagues and profession, can begin to appreciate the depth and breadth of our debt to Professors Savage, Mosteller, and Kruskal. The book would be improved had we been able to execute more adequately all of their suggestions.

We have been unusually fortunate, also, in receiving many useful suggestions from K. Alexander Brownlee, David Carr, Leo A. Goodman, Ruth Sawtell Wallis, and Wilson D. Wallis.

One of the early versions benefited by the capable assistance of Zenon S. Malinowski, two other versions by similar assistance from Lyle R. Johnson, one by the work of Margaret A. Labadie, and the earliest version by the work of Raymond Charles. To Winifred Ver Nooy, the University of Chicago Reference Librarian, we are indebted for a quantity and quality of assistance far beyond the call of duty.

To Louise Forsyth and Naomi Shoop we are indebted for providing a wide variety and large amount of services in the preparation of the manuscript. They have found people to do typing, proofreading, computing, and checking, and, with Elaine S. Smith, have provided most efficient and expeditious coordination of such activities, meeting reasonable and unreasonable demands with equal promptness and pleasantness. Among a number of competent and conscientious typists of the manuscript, Alease Hargis and Walter R. Paichel deserve especial thanks. Dean Haskin has done outstanding work in editing the final manuscript and supervising the proofreading and preparation of the front-matter and index; it may seem a reflection on us that the manuscript afforded opportunities for as much improvement as she accomplished, but is actually a measure of her capacities and diligence and of our confidence in her. In the proofreading she had excellent assistance from Margaret A. Labadie and Barsha Sprague. In re-checking the computations at the galley proof stage, valuable aid was given by Stanley Kupferberg, Margaret A. Labadie, and Albert Madansky. The charts, which speak for themselves, are the work of Sue Allen and Mary Jane Owen.

To those who know Garfield V. Cox, or know of him, the dedication calls for no comment; to those who do not, no comment would be adequate.

W. ALLEN WALLIS
HARRY V. ROBERTS

*The University of Chicago*
*29 February 1956*

# Contents

Contents

# Part II.  *Statistical Description*

## Part III.   Statistical Inference

Contents

# Part IV.   Special Topics

Contents

*Contents*

Contents

## *Appendix*

# List of Examples*

---

*Only examples with numbers and titles are included in this list. The number of an example is the same as its page number.

*List of Examples*

## List of Examples

# List of Tables*

*Only numbered tables are included in this list. The number of a table is the same as its page.

## List of Tables

## List of Tables

## List of Tables

# List of Figures*

---

*Only titled figures are included in this list. The number of a figure is the same as its page.

## List of Figures

# PART I
# THE NATURE OF STATISTICS

# Chapter 1

# *The Field*
# *of Statistics*

## 1.1
## WHAT IS STATISTICS?

Statistics is a body of methods for making wise decisions in the face of uncertainty.

This modern conception of the subject is a far cry from that usually held by laymen. Indeed, even the pioneers in statistical research have adopted it only within the past decade or so.

To the layman, the term "statistics" usually carries only the nebulous—and, too often, distasteful—connotation of "figures." He may even be vague about the distinction between mathematics, accounting, and statistics. In this sense, statistics are numerical descriptions of the quantitative aspects of things, and they take the form of counts or measurements. Statistics on the membership of a certain club might, for example, include a count of the number of members, and separate counts of the numbers of members of various kinds, as male and female, or over and under 21 years of age. They might include such measurements as the weights and heights of the members, or the lengths of time they can hold their breaths. Further, they might include numbers computed from such counts or measurements as those already mentioned, for example, the proportion of members who are married, the average height, or the ratios between weights and heights (that is, pounds of weight per inch of height). In this sense, the *Statistical Abstract of the United States* is a typical—and excellent—collection of statistics.

3

But in addition to meaning numerical facts, "statistics" refers to a subject, just as "mathematics" refers to a subject as well as to symbols, formulas, and theorems, and "accounting" refers to principles and methods as well as to accounts, balance sheets, and income statements. The subject, in this sense of statistics, is a body of methods of obtaining and analyzing data in order to base decisions on them. It is a branch of scientific method, used in dealing with phenomena that can be described numerically, either by counts or by measurements. It is in this sense that the word "statistics" is used in this book, except in the few places where the context makes it quite clear that the facts-and-figures sense is intended, for example, in the phrase "statistical data."

The purposes for which statistical data are collected can be grouped into two broad categories, which may be described loosely as practical action and scientific knowledge. Practical action here includes not only such actions by administrators as setting a bus schedule or admitting a student to school, but also such acts by individuals as having the oil changed in a car or carrying an umbrella. Scientific knowledge here includes not only knowledge gained by scientists through research, such as experiments with serums to relieve colds or analyses of records of business cycles, but also conclusions by an individual on such questions as whether coffee keeps him awake or whether his colds recur at regular intervals.

These two purposes, practical action and scientific knowledge, are by no means sharply distinct, since knowledge becomes the basis of action. For statistics, the important difference between the two purposes is that in practical action the alternatives being considered can be listed and, in principle at least, the consequences of taking each can be evaluated for each possible set of subsequent developments; whereas scientific knowledge may be employed by persons unknown for decisions not anticipated by the scientist. Thus, the consequences of error—obviously an important consideration in reaching a decision—can be taken into account more explicitly in the case of decisions for the specific "rifle-shot" purposes of practical action than in the case of decisions for the unspecified "shot-gun" purposes of scientific knowledge. The difference is, however, one of degree rather than of kind.

Statistical data, then, are collected to help decide questions of practical action or questions in scientific research. A decision about the allocation of military manpower or about a physical theory, for example, requires that the right kind of information be obtained. Statistics helps decide what kind of information is needed and how much. It then participates in the collection, tabulation, and interpretation of the data.

It is in developing methods for finding out what data mean that statisticians have evolved the present broad concept of their field. In most problems concerning the administration of business, governmental, or personal affairs, or in the search for scientific generalizations, complete information cannot be obtained; hence incomplete information must be used. Statistics provides rational principles and techniques that tell when and how judgments can be made on the basis of this partial information, and what partial information is most worth seeking. In short, statistics has come to be regarded, as we said in the first sentence, as a method of making wise decisions in the face of uncertainty.

## 1.2
## STATISTICS AND SCIENTIFIC METHOD

Statistics is not a body of substantive knowledge, but a body of methods for obtaining knowledge. As such it should be viewed against the background of general methods of obtaining knowledge—of general scientific method, in short.

There is no such thing as *the* scientific method. That is, there are no procedures, formal or informal, which tell a scientist how to start, what to do next, or what conclusions to reach. Scientists rely on the same everyday methods of reasoning that are common to all intelligent problem solving. "The scientific method, as far as it is a method, is nothing more than doing one's damnedest with one's mind, no holds barred."[1]

It is enlightening, nevertheless, to recognize four stages which recur in intelligent problem-solving, or scientific method.

### 1.2.1   Four Stages in Scientific Inquiry

(1) *Observation.* The scientist observes what happens; he collects and studies facts relevant to his problem.

(2) *Hypothesis.* To explain the facts observed, he formulates his "hunches" into a hypothesis, or theory, expressing the patterns he thinks he has detected in the data.

---

1. P. W. Bridgman, "The Prospect for Intelligence," *Yale Review*, Vol. 34 (1945), pp. 444–461; quoted in James B. Conant, *On Understanding Science* (New Haven: Yale University Press), 1947, p. 115. Warren Weaver, in his presidential address to the American Association for the Advancement of Science, put the same point this way: ". . . the impressive methods that science has developed . . . involve only improvement—great, to be sure—of procedures of observation and analysis that the human race has always used. . . . In short, every man is to some degree a scientist." (*Science*, Vol. 122, 1955, p. 1258.)

(3) *Prediction*. From the hypothesis or theory, he makes deductions. These, if the theory is satisfactory, constitute new knowledge, not known empirically, but deduced from the theory. If the theory is to be of value, it must make possible such new knowledge. These new facts are usually called "predictions," not in the sense of foretelling history, but rather of anticipating what will be seen if certain observations, not yet made, are made.

(4) *Verification*. He collects new facts to test the predictions made from the theory. With this step the cycle starts all over again. If the theory is substantiated, it is put to more severe tests by making more specific or more far-reaching predictions from it and testing them, until ultimately some deviation is found requiring modification of the theory. If the theory is contradicted, a new hypothesis consistent with the larger number of facts now available is formulated and then tested by steps (3) and (4); and so on. There is no final truth in science, for although failure to refute a hypothesis may increase confidence in it, no amount of testing can literally "prove" that it will always hold.

In actual scientific work these four stages are so intertwined that it would be hard to fit the history of any particular scientific investigation into such a rigid scheme. Sometimes the different stages are merged or blurred, and frequently they do not occur in the sequence listed. To know what facts to collect, one must already have some hypothesis about what facts are relevant to the problem, but such a hypothesis in turn presupposes some factual knowledge; and so forth. Nonetheless, the four stages help to focus discussion of scientific method.

Statistics is pertinent chiefly at the first and fourth stages, observation and verification, and to some extent at the second stage, formulating a hypothesis. The methods most important at the second stage, however, are primarily those of intuition, insight, imagination, and ingenuity. Very little can be said about them formally; perhaps they can be learned, but they cannot be taught. As someone has said, referring to an apocryphal story, many men noticed falling apples before Sir Isaac Newton, yet no interpretations of comparable interest were recorded by these earlier observers. The methods used at the third stage, prediction, are those of pure logic, utilizing sufficient knowledge of the field to provide those premises not given by the theory under test. The role of statistics at the first, second, and fourth stages deserves a little fuller consideration.

Statistics is helpful in the first stage, observation, because it suggests what can most advantageously be observed, and how the resulting observations can be interpreted. Not everything can be observed; it is necessary to be selective. The statistician visualizes in detail the analysis that will be made of the observations, and the interpretation that might result from these observations. In connection with the interpretation he especially emphasizes the degree of confidence in the conclusion and the necessary allowance for error. Then he compares the different kinds and quantities of observations that could be made with the resources available, and recommends making those observations that will effect a good compromise between the conflicting goals of high confidence in the conclusions and small allowances for error.

At the second stage, statistics helps to classify, summarize, and present the results of observation in forms that are comprehensible and likely to be suggestive of fruitful hypotheses. The branch of statistics dealing with methods for doing this is called *descriptive statistics*, in contrast to *analytical statistics*, the branch dealing with methods of planning the observation of, analyzing, and basing decisions on, the data so summarized. Often, of course, summarization of important observations must necessarily be "impressionistic" or "literary" rather than numerical; this is true, for example, of anthropological studies of the character and values of cultures, or of art criticism. The statistical approach is limited to those aspects of things that can be described and summarized numerically. This limitation is not, however, as confining as it may at first appear. Many things that are "qualitative" or "subjective" nevertheless have a quantitative aspect; for example, an important aspect of a certain organic disease may be the number of times it occurs. Many subjective or qualitative impressions can be sharpened or corrected by statistical study of subsidiary details, as when the impression that racial discrimination is decreasing is checked against the number of occurrences of certain specific kinds of incident. Even though at the stage of deriving new hypotheses such extra-statistical considerations as knowledge of and intuition for the subject matter may predominate, skillful statistical organization of the materials still plays a significant role.

At the fourth stage of scientific method, hypotheses are considered verified to the extent that predictions deduced from them are borne out by later events. Sometimes, especially in the natural sciences, it is possible to speed up the testing of predictions by experimentation. Frequently, however, a prediction can be tested only by waiting to see whether it comes true; for example, some astronomical predictions forecast the course of events (history), and some medical predictions indicate what would happen to human beings under circumstances that can come about only through accident. Statistics is relevant in either situation, for the essential problem is to determine whether or not the new data observed are concordant with the prediction.

In checking a prediction with new numerical data, it is crucial to realize that the data and the prediction can seldom be expected to agree exactly, even if the theory is correct. Discrepancies may arise simply because of chance circumstances ("experimental error") that are not inconsistent with the theory. Furthermore, many important theories of modern science are probabilistic or *stochastic* rather than deterministic, in that they do not predict precisely how each observation will turn out, but only what proportion of the observations will in the long run turn out in each of a number of possible ways. Genetic theories, for example, do not in general specify the characteristics of each individual offspring of a given parentage, but only the proportions in which certain different kinds of offspring will appear. Such theories, furthermore, do not specify the proportions for any one set of observations, but only the "long-run" proportions or probabilities. In comparing a set of observations with theory, the question to be considered is, therefore, "Is the discrepancy reasonably attributable to chance?" If the discrepancy can reasonably be attributed to chance, the theory is not contradicted, and there is no adequate reason to seek special "causes" to explain the discrepancy. If the discrepancy cannot reasonably be attributed to chance, it is appropriate to look for causes—that is, to modify the theory.

Modern statistical reasoning has given a definite meaning to the verification of a hypothesis. A hypothesis is verified—"tested" is perhaps a better word—to the extent that the influence of chance in the evidence has been correctly interpreted. Statistical procedures have been evolved for measuring the risk of incorrect interpretation objectively, in terms of numerical probabilities; or, to put it differently, for measuring the risks of erroneous conclusions.

## 1.2.2 Concrete Examples of the Four Stages

Illustrations of the process just described are found in everyday experience as well as in scientific inquiries.

EXAMPLE 8    OVERHEATED CAR

(1) *Observation.* The driver of a car notices that the engine temperature is too high. (This observation might be made to verify a theory. For example, he might have observed something that made him suspect—formulate the theory—that his engine was overheated.)

(2) *Hypothesis.* He formulates the hypothesis that the fan belt is broken, and that the fan and water pump, which he knows to be driven by the fan belt, are not working for this reason.

(3) *Prediction.* From this hypothesis he deduces that the generator will not be working, since it is also driven by the fan belt, and that the ammeter will, therefore, show a zero or negative rate of charge.

(4) *Verification.* He observes the ammeter. If it shows no charging, this strengthens his confidence in the hypothesis that the fan belt is broken. It does not *prove*, however, that the fan belt is broken. Many other hypotheses are consistent with the observed data, for example, that the battery is fully charged and a regulator has stopped the charging, that something has put all the instruments out of order, and so forth.

EXAMPLE 9   THEFT OF FINISHED PRODUCT

(1) *Observation.* A certain business enterprise has to have a great deal of waste material hauled away. The net weights of four truckloads chosen at random ranged between 14,200 and 14,500 pounds.

(2) *Hypothesis.* The variation from truckload to truckload is random, in accordance with certain statistical principles (normal distribution) that we will study later.

(3) *Prediction.* Practically all future truckloads will fall between 13,900 and 14,800 pounds. If this is true, it may result in a decision to dispense with regular weighings and pay a flat rate per truckload.

(1a) *Observation.* Several truckloads are found to weigh 16,000 pounds. This contradicts the initial prediction and demands a new hypothesis.

(2a) *Hypothesis.* The unusually heavy truckloads may be related to trucks or drivers.

(1b) *Observation.* The heavy loads do coincide with a particular driver.

(2b) *Hypothesis.* The fact that one driver is consistently taking out unusually heavy loads, together with the already known facts that there have been shortages of the firm's finished product and that the finished product is substantially denser than the waste, suggests the hypothesis that the driver may be smuggling out finished product at the bottom of his load. More facts are required.

This example was carried out by a student during a statistics course. He got no farther with his investigation before the course ended, and we do not know what happened next. But even this much illustrates the point that actual problems go through fairly definite stages on the way to their solutions. It also illustrates an experience which is common and important: that a study started for one objective (in this example, to eliminate a work operation) may contribute to unforeseen objectives (in this case, detection of theft). Serendipity—the knack of spotting and exploiting good things encountered accidentally while searching for something else—is as valuable in statistics as it is in other arts.

It would be wrong to leave the impression that people think of the four stages as they solve real-life problems, or that it would help them much if they did. But analyzing the process this way in retrospect is helpful in understanding how an inquiry progresses, and at what points statistics fits into it.

## 1.3
## APPLICATIONS OF STATISTICS

So far we have discussed statistics at a general level. Now we pause to consider some of the kinds of practical and scientific problems to which statistics is applied.

Statistical methods have been increasingly used in business. One element common to all problems faced by business managers is the need to make decisions in the face of uncertainty; and, as we have seen, the essence of modern statistics lies in the development of general principles for dealing wisely with uncertainty. It is not surprising, then, that statistical methods are widely applicable in nearly all areas of managerial decision. Applications are made in market and product research, investment policies, quality control of manufactured products, selection of personnel, the design of industrial experiments, economic forecasting, auditing, the selection of credit risks, and many others. The scientific management movement of this century has especially emphasized the need for collecting facts and interpreting them carefully, as has its currently popular offspring "operations research."

Governments have long collected and interpreted data concerning the State; for example, data about population, taxes, wealth, and foreign trade. In fact, the word *statistics* is derived from *state*. The first article of the Constitution of the United States provides that the government shall collect statistics—a decennial census to serve as a basis for representation of the states in Congress. Perusal of the *Statistical Abstract of the United States* will give an idea of the breadth and detail of statistical data currently compiled by the government: area and population; education; law enforcement; climate; labor force, employment, and earnings; elections; foreign commerce; transportation; and comparative international statistics are a few of its 34 major headings. The Department of Commerce has been a particularly important compiler of business statistics since World War I, and since World War II the President's Council of Economic Advisers has published data from a variety of sources on general economic conditions.[2]

Investigations in the social sciences have relied increasingly on statistical methods. The sample survey has supplied information at

2. An excellent guide to government statistics is Philip M. Hauser and William R. Leonard (eds.), *Government Statistics for Business Use* (revised ed.; New York: John Wiley and Sons, Inc., 1956).

moderate cost on many topics, including incomes and savings; consumer anticipations about future expenditures; attitudes toward atomic energy, civil defense, public libraries, and international relations; voting, actual and intended; unemployment; and the effect of television on family life. Understanding of personality has been gained by statistical analysis of psychological tests and experiments. Attempts have been made to measure statistically the extent of monopoly in business at different times and thus the extent to which monopoly is increasing or decreasing. Archaeologists have used statistics in drawing inferences from excavated potsherds. The increasing use of mathematical "models" (that is, theories formulated in mathematical symbols) which attempt to explain social behavior has brought an increasing interest in statistical techniques by which the validity of these models can be tested.

The demands of research in certain biological sciences, notably anthropometry, agronomy, and genetics, brought forth a rebirth of statistics at the beginning of the twentieth century, and the use of statistical methods in this area continues to grow. The development of genetics, especially, has been intimately related to the development of statistics. Experiments about crop yields with different fertilizers and types of soil, or the growth of animals under different diets and environments, are frequently designed and analyzed according to statistical principles. Statistical methods also affect research in medicine and public health. The first large-scale, statistically well-designed medical experiment in the United States was done in 1952 to test the efficacy of gamma globulin as protection against poliomyelitis, though a statistically comparable experiment had been done in England in 1946 to test the efficacy of streptomycin in the treatment of tuberculosis.

The physical sciences, especially astronomy, geology, and physics, were among the fields in which statistical methods were first developed and applied (as early as the beginning of the nineteenth century), but until recently these sciences have not shared the twentieth century developments of statistics to the same extent as the biological and social sciences. Currently, however, the physical sciences seem to be making increasing use of statistics, especially in astronomy, chemistry, engineering, geology, meteorology, and certain branches of physics.

In the humanities—history, linguistics, literature, music, and philosophy, for example—the use of statistical tools is not common; but even in these fields statistics finds an increasing number of significant applications. A modern historian, for example, can use the evidence of attitude studies as well as more impressionistic data to characterize public opinion on, say, the question of isolationism in the United States just before World War II. An important historical question on which statistical evidence, even though fragmentary, has helped to give an answer, is whether the welfare of the working classes in England rose or fell during the industrial revolution of the late eighteenth and early nineteenth centuries. The power of statistics in resolving such an issue is illustrated by the fact that two authors who have done much to disseminate the view that the position of the working class greatly deteriorated during the early nineteenth century, admitted candidly toward the end of their lives that

> statisticians tell us that when they have put in order such data as they can find, they are satisfied that earnings increased and that men and women were less poor when this discontent was loud and active than they were when the eighteenth century was beginning to grow old in a silence like that of autumn. The evidence, of course, is scanty, and its interpretation not too simple, but this view is probably more or less correct.[3]

These allusions to statistical applications are not intended to be exhaustive, but simply suggestive of the diversity of applications of the underlying methods and ideas of statistics. Many more concrete illustrations will be given in later chapters. Statistics is a tool which can be used in attacking problems that arise in almost every field of empirical inquiry. While the details of the appropriate statistical techniques vary from one field to another and from one problem to another, it is important to recognize the basic similarity of approach. We hope to bring out this similarity and give insight into the scope and applicability of statistics by drawing illustrations from many fields.

But there is a deeper reason for stressing a broad range of applications. It is that the statistical approach, though universal in its underlying ideas, must be tailored to fit the peculiarities of each concrete problem to which it is applied. It is dangerous to apply statistics in cookbook style, using the same recipes over and over, without careful study of the ingredients of each new problem. A wide range of illustrations will, we hope, emphasize the need to begin from basic principles in attacking each new problem.

Our interest lies in statistical method. It is important to recognize, however, that statistics cannot be used to full advantage in the absence of good understanding of the subject to which it is applied. The statistician working in meteorology, for example, without a good understanding of meteorology is likely to produce technically compe-

---

3. J. L. and Barbara Hammond, *The Bleak Age* (revised ed.; London: Pelican Books, 1947), p. 15.

tent trivia that contribute little to meteorology. Conversely, the meteorologist without a good understanding of statistics is likely to get bogged down in awkward, inefficient, and misdirected attempts to obtain evidence on important meteorological problems, and he is liable to fall into erroneous conclusions. The skill and knowledge of statistician and meteorologist must be blended. Sometimes the two abilities are combined in the same person, but more often the meteorologist consults with a statistician. Such collaboration relieves neither partner of the need to understand something of the other's field, but it does relieve each of the necessity of qualifying as an expert in two fields.

## 1.4
## FACTORS RELATED TO THE GROWTH OF STATISTICS

The great and continuing growth in the use of statistics can be explained by the economist's rubrics of demand and supply: The demand for statistics has increased, and so has the supply. Either increase, in the absence of a compensating decrease in the other, would bring about an increase in the use of statistics. The two increases together have magnified each other's effects.

### 1.4.1 Increased Demand for Statistics

The areas in which statistics is applied most are, as we have just seen, business, government, and science. The extraordinary growth of all three of these is one of the most distinctive features of the present century.

A striking, though indirect, reflection of the increasing importance of business is the fact that from 1910 to 1952, while the total civilian population of the continental United States was increasing by two-thirds (from 92 to 153 million), the farm population declined by more than a fifth (from 32 to 25 million); correspondingly, the percentage of the population classified as urban rose by nearly one-third (from 46 percent in 1910 to 59 percent in 1950)[4] and the percentage of

---

4. *Statistical Abstract: 1955*, Tables 8, 9, and 16, pp. 13 and 24. The urban population from 1910 to 1950 was defined as "all persons living in incorporated places of 2,500 inhabitants or more and in other areas classified as urban under special rules relating to population size and density" (p. 2). Beginning with 1950, however, a new definition is used, differing chiefly by adding the residents of unincorporated places of 2,500 or more and of "the densely settled urban fringe . . . around cities of 50,000 or more." By this new definition, 64 percent of the population was urban in 1950

workers engaged in nonfarm occupations rose by more than one-fourth (from 69 to 88).[5] Similarly, from 1929 to 1953, the number of business firms in operation increased half as much again as did the civilian population of the continental United States (by 39 and 26 percent, respectively).[6] The increased magnitude of business would alone account for a considerable increase in the need for statistics, but the need has been still further augmented by the increasing complexity of business, as firms have become larger (in manufacturing the average number of employees per firm increased by more than a fourth between 1929 and 1953 [from 41 to 52]),[7] as government regulations and taxes have become more pervasive and complicated, as labor relations have become more involved, and as technology has advanced.

The increase in the magnitude of government is so often commented on that it will suffice here simply to cite two statistical facts: First, there were nearly two and one-half times as many government employees in 1953 as in 1919.[8] Second, the total expenditures of the federal government were about 150 times as great in 1953 as in 1900.[9] The increased complexity of government operations is illustrated by the facts that in 1910 there was no federal income tax and no social security program. Thus, government activities, even more than business activities, have increased in size and in complexity, thereby greatly increasing the demand for statistics.

The growth of scientific research has been equally dramatic. By 1954, funds used for research and development were three times as much as they had been only ten years earlier; universities were using more than five times as much, industry more than three times as much, and governments about twice as much.[10] Science too has become more complex, and this has resulted in a large increase in the demand for statistics in research.

---

5. *Statistical Abstract: 1955*, Table 218, p. 185.

6. *Statistical Abstract: 1955*, Table 577, p. 488, and Table 8, p. 13. The number of business firms in operation on June 30, 1953, was 4.2 million.

7. *Statistical Abstract: 1954*, Tables 226 and 577, pp. 191 and 488.

8. The numbers were 2.7 and 6.6 million, respectively. These include state and local, as well as federal employees, but not the armed forces, which numbered 3.6 million in 1953. *Statistical Abstract: 1955*, Tables 226 and 264, pp. 191 and 226.

9. *Statistical Abstract: 1955*, Table 407, p. 349.

10. Governments in 1954 were *providing*, in contrast with using, two and one-half times as much money for research as in 1944, industry was providing four times as much, and universities four times as much. *Statistical Abstract: 1955*, Table 593, p. 499. These figures somewhat exaggerate the growth of research activities, since they partly reflect price increases.

## 1.4.2 Decreasing Costs of Statistics

The cost and the time required for summarizing and analyzing masses of data put a limit on the use of statistics. This limit has become progressively less restrictive because of technological improvements in processing numerical data. The development of tabulating and computing machines has resulted in great savings of money and time, and, consequently, a marked impetus to the use of statistics. Recent developments, such as electronic calculators, have been spectacular. More conventional devices, such as desk calculators and card sorting and tabulating machines, have made it easy for scientists and administrators to complete statistical work that would have been too expensive and slow to undertake fifty years ago.

The development of statistical theory has also had the effect of reducing the costs of compilation of statistical data, especially by making it possible to base reliable conclusions on samples. At the beginning of the century, the idea of "taking a sample" had scant theoretical basis to serve as a guide to practice or to give confidence in the results. Great advances in the theory of sampling have occurred since that time; now procedures can be guided by these tried-and-proven theoretical developments. Two striking, though specialized, examples will illustrate the swiftness of these developments: (1) Sound techniques of *sampling human populations* when complete listings of individuals are not available have been developed almost entirely since 1935. As a result, estimates of unemployment are now prepared monthly with errors that almost surely are under 20 percent. Similarly, sampling methods make available important information from censuses long before the complete tabulations can be prepared and published. These are only two of the practical applications these sampling techniques have found; empirical research both in the social sciences and business is making increasing use of them in cases where, if it were necessary to have a complete listing of the individuals under study before selecting a sample, the cost and delay would be prohibitive. (2) Equally revolutionary developments have occurred, almost entirely since 1935, in the collection and analysis of many other kinds of data, especially through that branch of statistics known as the *design of experiments*.

Of great interest even to the nonspecialist in statistics is the fact that much of the basic progress in statistical theory of the past few decades can be attributed directly to a single individual, Sir Ronald Fisher (born 1890). As one writer puts it, "Fisher is the real giant in the development of the theory of statistics. His first paper was published in 1912, and his work continues unabated today. Although hundreds of scholars have contributed to the science of statistics, this one man must be credited with at least half of the essential and important developments as the theory now stands."[11] Fisher is not only the greatest figure in the history of statistics, but one of the greatest figures in the history of scientific method generally.

As rapid as the recent development of statistical theory has been, it would be wrong to give the impression that the current body of theory is complete or final. In spite of the rapid developments we have been outlining, the list of unsolved statistical problems is long, and statistical research today is more vigorous than ever before.

## 1.5
## CONCLUSION

*Statistics are* numerical facts, but *statistics is* a body of methods for making decisions when there is uncertainty arising from the incompleteness or the instability of the information available. The decisions may be made either for the practical purpose of selecting a course of action or for the scientific purpose of gaining general knowledge.

Intelligent problem-solving, or scientific method, involves the observation of facts, the formulation of hypotheses describing the relations among the facts, the deduction from the hypotheses of things that must be true if the hypotheses are true, and the verification of these deductions by observing more facts. Statistics assists in planning the initial observations, in organizing them and formulating hypotheses from them, and in judging whether the new observations agree sufficiently well with the predictions from the hypotheses.

For the past two decades there has been a remarkable and sustained growth in the use of statistics. Partly, this is because business, government, and science, the three fields in which applications of statistics are most numerous and diverse, are growing in volume and complexity, both absolute and relative to other activities. Partly, too, it is because of a technological revolution in data handling, affecting especially computing and tabulating equipment, and a scientific revolution in statistical theories and techniques.

---

11. Alexander McFarlane Mood, *Introduction to the Theory of Statistics* (New York: McGraw-Hill Book Company, Inc., 1950), p. 282.

# Chapter 2

# *Effective Uses*
# *of Statistics*

## 2.1
## COMMON SENSE AND STATISTICS

Most of us pass through two stages in our attitudes toward statistical conclusions. At first we tend to accept them, and the interpretations placed on them, uncritically. In discussion or argument, we wilt the first time somebody quotes statistics, or even asserts that he has seen some. But then we are misled so often by skillful talkers and writers who deceive us with correct facts that we come to distrust statistics entirely, and assert that "statistics can prove anything"—implying, of course, that statistics can prove nothing.

He who accepts statistics indiscriminately will often be duped unnecessarily. But he who distrusts statistics indiscriminately will often be ignorant unnecessarily. A main objective of this book is to show that there is an accessible alternative between blind gullibility and blind distrust. It is possible to interpret statistics skillfully. In fact, you can do it yourself. The art of interpretation need not be monopolized by statisticians, though, of course, technical statistical knowledge helps. This book represents an attempt to illustrate the fact that many important ideas of technical statistics can be conveyed to the nonstatistician without distortion or dilution.

Statistical interpretation depends not only on statistical ideas, but also on "ordinary" clear thinking. Clear thinking is not only indispensable in interpreting statistics, but is often sufficient even in the absence of specific statistical knowledge. Before we turn to the main stream of our exposition of statistical ideas, we shall devote this chapter and the next one to a series of statistical examples which can be interpreted reasonably well without any statistical background.

In the next chapter we will consider misuses of statistics, but in this one we will consider effective uses. First we will give quick sketches of successful applications of statistics in World War II, in business, in the social sciences, in the biological sciences, in the physical sciences, and in the humanities. Then we will take a closer, more detailed look at three examples, one each from the social, biological, and physical sciences.

One warning is needed before the examples are discussed. A receptive yet critical mind is essential. The rewards of open-minded skepticism are great, yet such skepticism is harder to apply than to advocate, especially when the problem in which statistical methods have been used is interesting. If one is interested in race relations in a community, in the effectiveness of an advertising campaign, or in the sexual habits of the population, it may seem tedious and pedantic to be critical about statistical methods. Statisticians are not much more immune to this attitude than anyone else, although they may be more consciously aware of it. One of the authors once recorded this reaction to an interesting book:

> When I first examined the volume, paying attention mostly to its fascinating substantive findings and scarcely at all to its methods, I was very favorably impressed indeed. When I diverted my attention to the general methods I began to note shortcomings; but I felt that these were technicalities—mere blemishes on the surface of the monument, which might modify some of the findings in detail but surely would not affect the broad conclusions. After all, many of [the] figures would still be important and interesting even if we had to allow for an error factor as large as two or even three. But when I spent some time studying the statistical methods in detail, I realized that my confidence in the basic significance of the findings cannot be securely buttressed by factual material included in the volume. In fact, it now seems to me that the inadequacies in the statistics are such that it is impossible to say that the book has much value beyond its role in opening a broad and important field.[1]

Even in the successful examples that follow, one should note potential flaws, and consider what effect they might have on special applications of the findings.

---

1. W. Allen Wallis, "The Statistics of the Kinsey Report," *Journal of the American Statistical Association*, Vol. 44 (1949), p. 466.

## 2.2
## SOME USES OF STATISTICS IN WORLD WAR II

EXAMPLE 19A   AIRCRAFT LOSSES IN RELATION
TO TIME SINCE OVERHAUL

In order to minimize flying time lost for overhauling engines, and at the same time avoid plane losses that overhauling could have prevented, a study was made of the relation between aircraft losses and flying time since the last overhaul. Contrary to expectation, it was found that the number of planes lost decreased as the time since overhaul increased; that is, the risk of failure was greatest right after overhaul, and declined steadily until the next overhaul, when it increased again. This result led to a great extension of the amount of flying time between overhauls. It also led to an investigation and reorganization of the overhauling system, so that overhauling made the planes less rather than more likely to fail. This improvement in the overhauling system illustrates again the point about serendipity that we made in Example 9 (Theft of Finished Product), that unanticipated by-products of a systematic statistical study may be at least as useful as the specific objectives. This important study required little more than intelligent and careful collection of data, and their proper organization and interpretation.

EXAMPLE 19B   MERCHANT SHIP LOSSES
IN RELATION TO CONVOY SIZES

In 1942–1943, large merchant ship losses by submarine attack led to a study of the relation between the number of ships lost and the size of the convoy. Data on losses for various sizes of convoy revealed that there was no tendency for the number of ships lost to vary with the size of the convoy, though, of course, there was considerable variation in losses even for convoys of a given size. Since the *number* of ships lost did not increase with the size of the convoy, the size of the convoys was increased to reduce the *percentage* loss. One possible explanation of this independence of loss and convoy size is a constant attack potential of a submarine group.

EXAMPLE 19C   ARMY USE OF SAMPLING INSPECTION
OF MASS-PRODUCED ITEMS

Because of the tremendous amount of work required for complete inspection of mass-produced items, the Army, with the guidance of Bell Telephone System statisticians, introduced sampling inspection plans. Under such plans, only a small part of an entire lot (perhaps only 100 out of 5,000), is inspected in order to determine whether the entire lot should be accepted or rejected. Unless such sampling is statistically sound it may be worse than useless, but if properly done it is usually superior to inspecting each item, because the few items can be inspected more accurately; and, of course, it is far cheaper. This is a subject that we shall discuss further, especially in Chap. 16.

#### EXAMPLE 20A   OPA SAMPLE STUDIES OF TIRE INVENTORIES

During World War II, the Office of Price Administration attempted to take complete inventories of tires in the hands of dealers. Later, a number of dealers were selected on a statistical basis and the complete inventory was estimated on the basis of this relatively small group of dealers. Not only was a nuisance eliminated for many dealers, but the figures proved more accurate than the complete inventory previously attempted. The increase in accuracy was due to the fact that there was a huge number of nonresponses in the "complete" counts, which were made by mailed questionnaires, but in the sample it was possible by energetically following up to keep nonresponses at a low level. Those who failed to respond without follow-up proved to be quite different from those who responded readily.

#### EXAMPLE 20B   ESTIMATES OF ENEMY OUTPUT

During the war, German industrial output and capacity were estimated by British and American statisticians from the manufacturing serial numbers on captured equipment. According to checks after the war, many of these estimates were quite as good as the estimates made by the Germans themselves. They were, moreover, available substantially sooner than the estimates of the Germans, since the Germans waited for complete coverage, whereas the British and Americans were forced to rely on sampling methods. The Germans never did know their total production figures for V-2 missiles, most of which were produced towards the end of the war, while the British and American estimates subsequent to the firing of the first missile, were found by special studies after the war to have been quite accurate.

#### EXAMPLE 20C   RELATION BETWEEN TRAINING AND BOMBING ACCURACY

The question of the most fruitful division of flying time between training and bombing missions in the case of B-29 airplanes operating from the Marianas was resolved by a statistical study of the relation between training time and accuracy of bombing. It was found that with an increase of training time from four percent to 10 percent of the available flying time, the number of bombs on the target doubled.

### 2.3
## SOME USES OF STATISTICS IN BUSINESS

#### EXAMPLE 20D   FITTING A NEW PRODUCT TO CONSUMER TASTES

A flour manufacturer wanted to bring out a new pie crust mix. The proposed mix was put in plain packages with only an identifying letter, and the same was done with the pie crust mix of the leading competitor. Each of 75 families was given a package of each of the two mixes and asked to fill out a questionnaire comparing the two brands for taste, texture, and other qualities. The competitor's product was preferred by most of the 75 families. The flour company therefore revised the formula of its mix. The revised formula was tested on another group of 75 families and found still to be less preferred than the competitor's mix. The same thing was repeated several more times. Meanwhile, another competitor introduced a new mix which gained preference over all others. The company made similar tests of its proposed formulas against this new competitor. Finally, a mix was developed that seemed to be preferred to all competitive products. When it was marketed it proved highly successful.

EXAMPLE 21A   ESTIMATING SALES BY DEALERS TO CONSUMERS

A manufacturer of household appliances wanted to know the current rate of sales of his appliances to consumers. The only information readily available was the rate of factory shipments to distributors and dealers. There were more than ten thousand distributors and dealers, however, and only vague, qualitative reports were available on the rate at which dealers sold appliances to the public. It may take as long as two or three months before changes in dealers' sales are reflected in changes in orders to the factory. A group of about 100 dealers was chosen, and each was asked to fill in a monthly report of his sales. Changes in the rate of sales to consumers were estimated from these reports. These data proved extremely valuable later when orders to the factory declined sharply. It was found from the reports of the sample dealers that the decline was due to reductions of inventories by dealers, rather than to any marked drop in retail sales.

EXAMPLE 21B   VALUATION OF PLANT AND EQUIPMENT

A telephone company wished to determine the value of its capital equipment, such as poles, cables, batteries, tools, buildings, and so on. It had lists of these things, and knew the original and the replacement costs, but the question was what percent of their useful life still remained. For many items this is costly to estimate, perhaps because the item is in a remote area or an inaccessible position, or because the evaluation requires highly paid experts or requires dismantling equipment and disrupting service. Furthermore, the number of items was large. Instead of examining every piece of equipment, which would have been prohibitively slow and costly, the company chose a relatively small sample of each kind of equipment according to statistical principles. The average condition of each kind of item was then estimated from the samples, and an allowance was made for possible error due to sampling. The results were as useful as if a full examination had been made.

EXAMPLE 21C   QUALITY ASSURANCE

A manufacturer of electric fuzes wished to guarantee that his product met specified standards of quality. Quality was defined in terms of the time required for the fuze to break the circuit, when a certain current was applied. To find out whether a fuze met the standard of quality, the item had to be destroyed. A statistically designed sampling plan, similar to those mentioned in Example 19C (Army Use of Sampling Inspection), enabled the manufacturer to draw reliable inferences about the quality of an entire lot from the information obtained by testing relatively few items.

EXAMPLE 22A   EXPERIMENTING ON A MANUFACTURING
PROCESS

A manufacturer made a very expensive product which, like the one in the preceding example, could be tested only by destroying the product. His engineers wanted to know the effect on the final product of changes in six variable components of the manufacturing process, such as amounts and kinds of various ingredients, and the length of time certain operations were continued. The traditional approach would have been to vary one component at a time and observe the effect. In this case, however, that would have been too costly, for it was necessary that each component be tried at five different "levels"; for example, each of five different concentrations of one of the ingredients had to be tried. If, say, five units of the product were to be made and tested at each of five levels of one of the components, 25 units of the final product would have had to be specially made and tested, and for six components a total of 150 units would have been required. A statistician, however, devised a plan for the experiment which permitted the simultaneous evaluation of the effects of all six components on the basis of testing only 25 units of the final product. It was possible to determine whether each component had an effect on quality beyond that which might be reasonably attributed to chance or random variations in the product. For those components which had a significant effect, it was possible to estimate what the effect was, so that an approximation to the "optimum" level of the component could be used in the manufacturing process. This is not to say that everything that could have been learned from a full 150 tests was learned from 25; but the specific questions of most importance were answered reliably enough from 25 tests.

EXAMPLE 22B   ESTIMATING SALES OF DIFFERENT STYLES

In order to place orders with manufacturers, a mail order company wanted to predict how total sales of a certain garment would be divided among individual styles. To a randomly selected sample of catalog receivers, the company mailed, in advance, a special booklet made up of the catalog pages describing the various styles of the garment. Actual orders from recipients of the special booklet were then tabulated, and predictions were made on the basis of these orders. These predictions were found, at the end of the sales season, to have been substantially more accurate than the predictions of experienced buyers.

*2.4 Statistics in the Social Sciences*

EXAMPLE 23A   SEASONAL PATTERNS OF ACCIDENT RISKS

A personnel manager wanted to find out the times during the year when the largest number of accidents occurred in his plant. With this information he hoped to be able to give safety instruction when the need for it was greatest. A statistical study of the accident records for this plant showed that, while there were variations among the months in the number of accidents, these variations were no greater than might reasonably be expected by chance alone. Thus, there was no best season for safety indoctrination, and the decision could be based on other grounds.

EXAMPLE 23B   USE OF RESERVED FACILITIES

A large firm arranges frequent educational programs for its 600 supervisory employees from foremen up. Each program includes discussion, and 30 is considered the best number of participants. Each program is therefore given 20 times, twice each morning and afternoon, Monday through Friday. Each employee is assigned to a session, but is free to come to any other session instead of the assigned one, if he feels that he ought not to leave his work at the time assigned. Originally, 30 employees were assigned to each session. Records of actual attendance showed fewer than 10 at some sessions and more than 60 at others. They also showed considerable uniformity for corresponding sessions of different weeks. The number of assignments was therefore varied, 90 being assigned to a session that had averaged 10 in attendance on the assumption that one-third of those assigned to that session would attend it, 15 to a session that had averaged 60, and so forth. Thereafter, actual attendance was seldom less than 25 or more than 33.

Hotels, airlines, physicians, restaurants, and others who make reservations that are subject to cancellation by clients sometimes use a variant of this system, but the problem is more complicated when there is an inflexible upper limit on the number who can be accommodated. Another variant was used with some success during World War II by the cafeterias in the Pentagon. They posted charts showing the average length of the line at various times. The charts revealed considerable variation in the wait encountered at times fairly close together. After the charts were posted these variations were appreciably reduced, as some people with control of their own lunch periods avoided the times of longest delay.

Recently, problems of this general kind have been considered extensively under the name "queuing theory." (In England, a waiting line is called a queue.)

## 2.4
# SOME USES OF STATISTICS IN THE SOCIAL SCIENCES

EXAMPLE 23C   CONTENT ANALYSIS

A political scientist studied British attitudes toward the United States during the period 1946–1950, insofar as these attitudes were expressed in newspapers and records of parliamentary debates. By a technique known as "content analysis," which consists of finding the relative frequency of appearance of different "themes," he hoped to describe British attitudes and to detect changes through time. A sampling scheme was devised whereby instead of reading all of the issues of the leading British papers during the period of interest, he read only a selected sample.

EXAMPLE 24A    CONSUMER FINANCES

For several years the Federal Reserve Board has sponsored at least one survey a year in order to determine, among other things, basic facts about income, savings, and holdings of liquid assets by individuals. A sample consisting of about 3,000 families has been used in this study. One interesting by-product of this work has been some indication that consumer plans for purchasing durable goods may be helpful in forecasting general business conditions.

EXAMPLE 24B    SUCCESS IN COLLEGE

Numerous studies have demonstrated considerable correlation between scholastic performance in high school and performance in college. There is also some, but less, correlation with performance on entrance (or "aptitude") examinations. Findings from such studies have been used to predict "success" of students applying for entrance to college, on the basis of their high school performance and entrance examinations.

EXAMPLE 24C    PUBLIC OPINION

Impressions about public attitudes on important issues are often ambiguous and unreliable. Carefully designed statistical studies can usually give more accurate pictures. An interesting example is the results of the following two questions asked in Belgium in 1948.[2]

|  | Yes | No (in percent) | No Opinion |
|---|---|---|---|
| Do you believe that the American government sincerely wants peace? | 68.5 | 13.9 | 17.6 |
| Do you believe that the government of the U.S.S.R. sincerely wants peace? | 18.0 | 60.2 | 21.8 |

EXAMPLE 24D    HOUSING SUPPLY

The following use of statistical reasoning appeared as part of a discussion of rent controls and the postwar housing shortage:

The present housing shortage appears so acute, in the light of the moderate increase in population and the actual increase in housing facilities since 1940, that most people are at a loss for a general explanation. Rather they refer to the

2. Institut Universitaire d'Information sociale et économique (Centre belge pour l'étude de l'opinion et des marchés), *Cinq Années de Sondages* (Brussels: 1950), p. 62. The wording of the question is here translated from the French.

rapid growth of some cities—but all cities have serious shortages. Or they refer to many marriages and the rise of birth rates—but these numbers are rarely measured, or compared with housing facilities.

Actually the supply of housing has about kept pace with the growth of the civilian nonfarm population, as the following estimates based on government data show:

| Date | NONFARM | | |
| --- | --- | --- | --- |
| | Occupied Dwelling Units | Civilian Population | Persons per Occupied Dwelling Unit |
| June 30, 1940 | 27.9 million | 101 million | 3.6 |
| June 30, 1944 | 30.6 million | 101 million | 3.3 |
| End of Demobilization (Spring 1946) | More than 31.3 million | About 111 million | Less than 3.6 |

Certain areas will be more crowded in a physical sense than in 1940, and others less crowded, but the broad fact stands out that the number of people to be housed and the number of families have increased by about 10 percent, and the number of dwelling units has also increased about 10 percent.[3]

Thus, the authors found that an explanation of the unavailability of housing had to be sought elsewhere than in a physical shortage.

## 2.5
## SOME USES OF STATISTICS IN THE BIOLOGICAL SCIENCES

### EXAMPLE 25A  HEIGHTS OF PARENTS AND CHILDREN

By recording the heights of parents and children and grouping these data by the height of one parent, it has been found that for every inch by which the parent's height exceeds (or falls below) the average for adults of the same sex and generation, the average of the children's heights, when grown, exceeds (or falls short of) the average for their sex and generation by about half as many inches. If data are grouped by the heights of both parents, the children's average is about four-fifths as far from the general mean, and in the same direction, as the parents'.

### EXAMPLE 25B  MENDELIAN HEREDITY

Gregor Mendel discovered the foundations of the modern science of genetics about a century ago, by methods that were essentially statistical.

---

3. Milton Friedman and George J. Stigler, *Roofs or Ceilings? The Current Housing Problem* (Irvington-on-Hudson, New York: Foundation for Economic Education, Inc., 1946), pp. 17–18.

Mendel, working with garden peas, noted the characteristics of the parents and counted the number of offspring having various characteristics. The regularities he observed led to the formulation of his theories.

EXAMPLE 26A   ANIMAL POPULATIONS

To determine the number of mice in a field or the number of fish in a lake, biologists catch a sample, count them, mark them (often with metal tags), and release them. Later they catch another sample. If, say, ten percent of those in the second sample are marked, they can infer that the total population is about ten times as large as the first (tagged) sample. Various elaborations are necessary to allow for special circumstances (such as that in a large woods or lake, a mouse or fish is likely to remain in a certain general area), to improve the estimates, and to calculate an allowance for error in the estimate.

## 2.6

# SOME USES OF STATISTICS IN THE PHYSICAL SCIENCES

EXAMPLE 26B   DIVISION OF THE TERTIARY ROCKS

Charles Lyell, the geologist, published the three volumes of his celebrated *Principles of Geology* in 1830, 1832, and 1833.

Geologists prior to Lyell had recognized the sequences of strata which we know as Primary and Secondary, using in the first place the regularity of order of superposition in the same locality. They observed, too, that particular components of these formations could be recognized, though far apart, by their characteristic fossils. They could not by these means recognize or establish the order among Tertiary rocks, for, in the part of the world then accessible, these occur in patches, and not over wide areas overlying one another. Lyell determined the order and assigned to the successive rock masses the names they now bear by a purely statistical argument. A rich group of strata might yield so many as 1,000 recognizable fossil species, mostly marine molluscs. A certain number of these might be still living in the seas of some part of the world, or at least be morphologically indistinguishable from such a living species. . . .

With the aid of the eminent French conchologist M. Deshayes, Lyell proceeded to list the identified fossils occurring in one or more strata, and to ascertain the proportions now living. To a Sicilian group with 96 percent surviving he gave, later, the name of Pleistocene (mostly recent). Some sub-appenine Italian rocks, and the English Crag with about 40 percent of survivors, were called Pliocene (majority recent). Forty percent may seem to be a poor sort of majority, but no doubt scrutiny of the identifications continued after the name was first bestowed, and the separation of the Pleistocene must have further lowered the proportion of the remainder. The Miocene, meaning "minority recent," had 18 percent, and the Eocene, "the dawn of the recent," only 3 or 4 percent of living species. Not only did Lyell immortalize these statistical estimates in the names still used for the great divisions of the Tertiary Series, but in an Appendix in his

### 2.7 *Statistics in the Humanities*

third volume he occupies no less than 56 pages with details of the classification of each particular form, and of the calculations based on the numbers counted. There can be no doubt that, at the time, the whole process, and its results, gave to Lyell the keenest intellectual satisfaction.[4]

### EXAMPLE 27A   RADIOCARBON DATING

Radiocarbon, or Carbon 14, is present in all living things. While things are living, the quantity of radiocarbon is proportional to the quantity of nonradiocarbon, or Carbon 12. After death, the quantity of nonradiocarbon remains stable, but the radiocarbon disintegrates. The ratio of radio- to nonradiocarbon indicates, therefore, how long a specimen has been dead.

There are standard chemical methods of determining the amount of nonradiocarbon in a specimen. The radiocarbon emits small particles that can be detected by special counting devices, such as the Geiger counters and scintillometers used in uranium prospecting or in measuring the radioactive fallout after atomic explosions. The average rate at which these particles are emitted is proportional to the amount of radiocarbon present, and hence provides a means of measuring the amount of radiocarbon present, but the actual emission at any given time is a matter of chance. From counts of the number of particles emitted in a given period it is possible, by statistical methods, to determine the average rate, and to determine the necessary allowance for uncertainty in this average due to the chance character of the emissions. The average rate of emission indicates the amount of radiocarbon present, and the ratio of this to the amount of nonradiocarbon indicates the age of the specimen. Allowances for uncertainty in the age are calculated from the corresponding allowances for uncertainty about the average rate of emission of particles from the radiocarbon.

Radiocarbon dating has become a standard means of dating ancient materials such as textiles, leather, and wood charcoal from campfires; it has revolutionized the dating of archaeological objects.

## 2.7
## SOME USES OF STATISTICS IN THE HUMANITIES

### EXAMPLE 27B   LINGUISTIC DATING

A method statistically similar to Carbon 14 dating, and in fact suggested by it, has been used in linguistics. In place of Carbon 14, it uses a list of two to three hundred concepts for which there are words in virtually all languages. By studying languages which are known to be descendants of a common language, for which the date of separation is known, and for which there are writings at various known dates following the separation, it has been found that after separation the number of common words tends to

4. Sir Ronald Fisher, "The Expansion of Statistics" (Inaugural Address as President of the Royal Statistical Society), *Journal of the Royal Statistical Society, Series A (General)*, Vol. 96 (1953), pp. 1–6.

diminish by about 20 percent per 500 years. Thus, after 500 years, about 80 percent of the words are still the same, after 1,000 years, about 64 percent, and so forth. Knowing this, it is possible to calculate, from the number of words they have in common, when two related languages separated. This method, however, is not so firmly established and widely used as radiocarbon dating.

EXAMPLE 28   LITERARY STYLE

Statistical studies of the length of sentences, the relative frequency of various parts of speech, the frequency of use of individual words, and the frequency of various word sequences have been used to help answer such disputed questions as whether a given author wrote a certain work, whether a work came early or late in an author's career, and what portions of joint works were written by the respective authors.

## 2.8
## THREE DETAILED EXAMPLES

### 2.8.1   Nature and Purpose of the Examples

The remainder of this chapter is devoted to rather detailed examinations of three successful statistical studies, one each in the social, biological, and physical sciences. The first example, on long-term trends in the frequency of mental disease, involves a historical study in which the investigators had to rely on existing data and records, whereas the second example, on the effect of vitamins B and C on human endurance under severe physical stress in extreme cold, and the third, on making rain by "seeding" clouds, involve experiments arranged by the investigators for their specific purposes. The second and third examples, though completely different in subject matter, are in many respects similar statistically.

The purposes of presenting these three examples are: (1) to dispel any aura of magic that may have resulted from the brief summaries in the earlier part of the chapter; (2) to present a glimpse of the inner "works" of a statistical investigation; (3) to impart a feel for the necessity of caution, judgment, and detailed information in drawing conclusions from even the best research, and (4) to indicate the extent to which the over-all soundness of an investigation depends on care and skill with a large number of details. In these examples, therefore, instead of omitting details and focusing on the major methods and findings, we shall give particular attention to details, though it will be impractical to recapitulate the original studies in full detail.

### 2.8.2 *Mental Disease*

It is not essential to study these examples intensively now; indeed, in a quick reading they could, if necessary, be omitted altogether. There are occasional references to them later, especially in Chap. 15, but nothing in later chapters depends on familiarity with the details of these examples.

## 2.8.2 Long-Term Trends in the Frequency of Mental Disease

2.8.2.1 *Purpose of Study.* Mental health is a matter of growing concern. War, depression, urbanization, industrialization, competition, and the breakdown of family and community ties: these and many other aspects of the increased complexity and insecurity of modern civilization are said to have aggravated the problem. To determine how much increase there has been in the frequency of mental disease over the past century was the purpose of a careful investigation by Herbert Goldhamer, a sociologist, and Andrew W. Marshall, a statistician.[5]

Such a purpose is too broad for a single inquiry and too vague for a systematic one. Goldhamer and Marshall, therefore, proceeded to narrow and to define their objectives. The result is a study which leads to reliable and specific conclusions about a single facet of the broader problem. Studies of other facets, made equally carefully but perhaps by other investigators in other times or countries, each building on its predecessors, will eventually cumulate into an understanding of the broader problem. Even this broader problem is, as Goldhamer and Marshall emphasize by their title, only a facet of the still more fundamental question of the relation between the psychogenic psychoses (serious mental disorders which apparently result from mental influences, rather than from physiological influences, injuries, or other causes) and the characteristics of contemporary social existence, particularly the characteristics associated with the growth of "civilization," especially its increased personal responsibility and freedom.

First, Goldhamer and Marshall specified more explicitly what they mean by *frequency*. Obviously they are not interested in the total number of cases, since this would mainly reflect the fact that the population of the United States is about six times as large as it was a century ago. Nor are they interested in the number of cases of mental disease per capita of population. True, this would allow for changes in the total population, but it would not allow for the fact that the

---

5. Herbert Goldhamer and Andrew W. Marshall, *Psychosis and Civilization: Two Studies in the Frequency of Mental Disease* (Glencoe, Illinois: Free Press, 1953).

frequency of mental disease varies with age, and now a larger proportion of the population is at the ages most susceptible to mental disease. This change in the age distribution of the population would cause an increase in the per capita rate of mental disease even if the rate of onset for each specific age group were unchanged. What they decided to use was not a single frequency figure but a set of *age-specific rates*. These show, separately for each age group, the number of cases of mental disease per capita of population of that age group.

Furthermore, mental disease is too broad and vague a concept, so the authors narrowed it to include only the *major psychoses*. These are characterized by "behavior, such as extreme agitation, excitement, deep depression, delusions, hallucinations, suicidal and homicidal acts" that is "sufficiently recognizable as insanity, irrespective of the classificatory terminology used" so that we may be confident that such cases would be diagnosed as mentally diseased either a century ago or today. Some of the lesser forms of disturbance, such as neuroses, "nervous breakdowns," and "maladjusted personalities," are not clearly classifiable as mental disease, have been considered such only comparatively recently, and may be classified differently by different diagnosticians even contemporaneously. Thus, data that included them probably could not be found at all, and if found would be virtually worthless for comparisons among widely separated times.

In order to obtain data on the number of cases of psychosis, it was necessary to narrow the study still further, covering *hospital admissions*. Not all admissions to mental hospitals, but only *first admissions*, are relevant, for the authors wish their figures to show the rate of onset of mental disease.

Whether hospital first admissions for psychosis are an adequate index of the number of cases of psychosis in the population requires consideration. It may be that the relation between hospital first admissions and the actual frequency of mental disease has changed in the course of time. It seems plausible, for example, that the proportion of the afflicted population hospitalized is larger now than a century ago and that hospitalization follows sooner after the onset of the disease. These differences seem plausible because there has been an increase in the number of beds available, relative to the need for them, and also because the proportion of cases hospitalized is greater for cases near hospitals and, now that the population is more urban (the proportion urban rose from 15 to 60 percent between 1850 and 1950), more people are close to hospitals. This matter of the relation between hospital first admissions and the total

### 2.8.2 Mental Disease

incidence of mental disease is one that, as we shall see, is kept constantly in mind by the authors in their analysis and interpretation.

Narrowing the study to hospital first admissions is dictated not by the desire for definiteness and precision, but by the availability of data. A shift has been made from what would be studied ideally, namely, rates of first onset for the whole population, to what can be studied practically, namely, rates of first admission to hospitals. Such shifts are commonly necessary in research. They require especially good judgment, for neither a precise study of irrelevant trivialities nor a meaningless study of the central issue is of any value. Good researchers must balance tenacious adherence to strategic objectives against attacks on targets of opportunity. As a matter of fact, this whole investigation is an excellent example of the role of serendipity, the art of successfully exploiting good things encountered accidentally while searching for something else, in scientific progress. Goldhamer and Marshall intended to study variations in the frequency of mental disease among different groups of the contemporary population. While searching in the Library of Congress for pertinent data, they came across data which, they recognized, made possible a study of long-term trends, something they had presumed would be impossible.

A final step in defining the specific objectives, also dictated by expediency, was to confine the study to the state of Massachusetts:

> Massachusetts was chosen as the state of inquiry because its facilities for the care of the mentally ill during the last half of the 19th century were, despite their obvious limitations, more advanced than those of most other states. The relatively small size of Massachusetts is also favorable to our inquiry, since it diminishes the mean distance of the population from a mental hospital. It has been well known for some time now that the tendency, especially in the past, to hospitalize the mentally ill is inverse to their distance from a mental hospital. Massachusetts further recommends itself for study during the latter half of the 19th century because of the work of such leaders in institutional psychiatry as Dr. Edward Jarvis and Dr. Pliny Earle. Their studies and reports, together with the documents and reports of official state agencies and hospitals, made feasible an investigation which one might have supposed quite impossible at this late date.[6]

Clearly the authors would not be greatly interested in a single state unless they felt that conclusions might be drawn which would apply reasonably well to some larger area, such as the entire United States. In short, they were faced with the problem of drawing conclusions about a *population* or *universe* (here the entire United States) on the basis of a *sample* from that universe (here Massachusetts).

---

6. Goldhamer and Marshall, *op. cit.*, pp. 25–26. Supporting footnote omitted.

Sampling concepts pervade modern statistics and will be discussed at length in this book. We will see that when the element of *randomness* is introduced in the selection of the sample, effective techniques are available for drawing inferences about the population from which the sample was selected. In the present example, however, randomization was not possible, since Massachusetts was the only state for which accurate and meaningful information was available. It was necessary, therefore, to rely on expert judgment to decide how far the results of the sample might be generalized—to decide, that is, how far what is true of Massachusetts is approximately true of the entire United States or some other large region of importance. Goldhamer and Marshall feel that their findings do apply more widely, and they present some supporting evidence to be described later in this chapter. In so deciding, they were acting very much like a surgeon at a private clinic who decides that a new operative technique is successful even though he knows that the patients on whom he has tried it are not randomly selected but are definitely atypical in many respects, for example, income. He may judge, however, that in respect to the operation, his patients are similar to patients of the same age and sex of other income groups. He would want, of course, to examine every bit of evidence that bears on his judgment. While Goldhamer and Marshall have reason to believe that Massachusetts fairly reflects major trends in the incidence of the psychoses, they undoubtedly would have preferred more evidence on this point than they had.

The authors summarize their purpose, then, as follows:

> The immediate aim of this report is to establish acceptable estimates of age-specific first admission rates to institutions caring for the mentally ill in Massachusetts for the years 1840 to 1885 in order to compare these rates with those of the contemporary period. No antiquarian zeal or historical interest has moved us to engage in this laborious task.* Our interest is in providing a more adequate test than is now available of contending views concerning the incidence of the major mental disorders in our own day and in an earlier period. We assume that a more adequate test of these beliefs will throw light on the validity of contentions concerning the psychologically pathic effects of contemporary social existence. . . .

2.8.2.2 *What Was Known Already?* Goldhamer and Marshall mention and discuss briefly eight previous studies related to their

---

*Nor were we concerned to show that a judicious use of documents usually reserved for historical study can extend the horizons of comparative statistical social analysis to phenomena other than those of a demographic and economic character. Yet the present study does show that the past is not always as irrecoverable, statistically speaking, as is sometimes assumed. [Footnote in original. Goldhamer and Marshall, *op. cit.*, p. 21.]

### 2.8.2 Mental Disease

problem. Most of these suggest that the incidence of mental illness has *not* increased; only one seems to conclude than an increase has taken place. Goldhamer and Marshall did not consider these previous studies to be decisive, chiefly because: (1) the time periods investigated did not extend very far into the past, some covering periods as short as a decade and only one extending back into the 19th century, and (2) most of the studies (including the one covering the longest period of time) did not make adequate adjustments for the changing age composition of the population.

In addition, the Goldhamer-Marshall volume is influenced by their preliminary study of their own data. This suggested strongly that there has in fact been no increase—that the 19th century rates of mental disease were fully as high as those of today, quite contrary to what we suggested in introducing this example. This tentative finding influenced their analysis, in that wherever questions of judgment, of estimating, or of selection among different figures arose they were careful that if they erred it would be in the direction of overstating the amount of increase. If, despite such bending over backward, it appeared that there had been no increase, they could be fairly confident that this conclusion was justified. Of course, if there had actually been a decrease they might fail to establish it this way; but for their purpose of determining whether contemporary social existence has caused an increase in the psychoses it is sufficient if they show either that there has been an increase or that there has not. Put another way, they were trying to decide between two actions, (1) to investigate further the notion that the pace of modern life is a cause of psychosis, and (2) to drop this notion and seek other causes. The first decision would be right if there has been an increase in the psychoses, the second if there has been no change or if there has been a decrease. In statistical terms that we shall introduce later (Chap. 12), but which some may find self-explanatory in this context, the null hypothesis was that there has been no change, and it was tested against the one-sided alternative hypothesis that there has been an increase.

2.8.2.3 *How the Data Were Obtained.* Though the basic records were surprisingly good for so early a period, it was nevertheless a lengthy and demanding task to derive from them the age-specific rates of first admission. In the authors' own words:

It is not to be denied that even the presentation of *first* admission rates for so early a period is in itself a novelty and that the attempt to carry this to a further stage of refinement by calculating *age-specific* first admission rates may seem extremely hazardous. However, we shall present quite fully the sources of our data and all assumptions involved in their utilization. The reader will be in a position to judge for himself whether any of our assumptions were or were not warranted. . . .[7]

While it is desirable to limit statistical investigations to things that can actually be observed, it is sometimes necessary to introduce assumptions which cannot themselves be tested completely. An author's willingness to tell exactly what he did, without vague general assertions as to the high quality of the work, is presumptive evidence of the soundness of the work and of the integrity of the author.[8]

Goldhamer and Marshall present eleven pages in which the data, method of collection, and assumptions are discussed in detail. A few highpoints of this discussion will be summarized. Before you read this summary, you might well pause to think of possible questions and objections. Then, as you read, notice how many of your questions are answered by the authors and how many unanswered, and how many questions the authors thought of that did not occur to you.

(a) Almost complete records had survived for all the mental hospitals and town almshouses in Massachusetts for the period 1840 to 1885.

> As for the general accuracy of the admission bookkeeping and reporting of the hospital and state reports of this period, we can only say that intensive reading, analysis and cross-checking of them have given us the highest esteem for the thoroughness and integrity of the hospital admission data they present. The system of financial accounting for the support of state and town paupers and for receipts from privately paying patients made accurate records imperative. In the earlier years especially, the hospital "paper work" was looked after by the medical superintendent himself. One is impressed by the almost obsessive detail and thoroughness of the statistical summaries of some of the reporters. In a number of respects the early Massachusetts hospital reports are more illuminating than those produced today.[9]

-----

7. Goldhamer and Marshall, *op. cit.*, p. 26.

8. One recent investigator asserts boldly in the introductory chapter to his book that "It is a fact that no number reported in this study is exact." After a discussion of the sources of error and the measures taken to cope with them, he goes so far as to extend an invitation to bona fide scholars or journalists to come to the office where the documents are filed and to read some or all of them at their leisure. Samuel A. Stouffer, *Communism, Conformity, and Civil Liberties: A Cross-section of the Nation Speaks Its Mind* (Garden City, N. Y.: Doubleday and Company, Inc., 1955), pp. 23–25.

This attitude is in sharp contrast to that of a research organization which refuses to release its original data on the ground that they might be misinterpreted. They might. There will be somebody to misuse virtually any set of data; in fact, wherever there is freedom of speech there will be (as advocates of freedom recognize) erroneous and mendacious statements. There will be much more misinterpretation, error, and prevarication, however, and it will gain wider currency and survive longer, when there is protection from independent inquiry and analysis.

9. Goldhamer and Marshall, *op. cit.*, pp. 31–32.

*2.8.2 Mental Disease*

(b) There were some gaps in the hospital records. The authors present a detailed table which shows the "Number of first admissions to institutions caring for the insane, Massachusetts, by 5-year periods, 1840 to 1884, and 1885."[10] (The quotation marks enclose the title of the table itself; the clarity, completeness, and brevity of this title provide a good model for statistical practice.) The table is accompanied by a page of footnotes which, together with textual discussion, describes exactly how the data were obtained for each hospital and how allowances were made for incomplete information. For example, for McLean Hospital, a private mental hospital in Boston, the exact number of first admissions was available from reports for the years 1868 to 1885, but for the years 1840 to 1867 only total admissions were available. The ratio of first to total admissions during 1868–85 was 0.70. It was assumed that the same ratio prevailed in the earlier period, and an estimate was made on the basis of this assumption. For some of the other hospitals, data on first admissions were available for the entire period 1840–85.

Another example (the comment in brackets is ours):

> Beginning with 1870 our table shows a small number of admissions from other private hospitals. [The number of patients admitted to these hospitals in 1870–74 was only 114, about three-tenths of one percent of the total.] We know that there were two or three private mental "hospitals" in earlier years, but the number of patients (about 25) that they housed at that time is negligible. Since we were unable to secure further data, we have no entries prior to 1870 for this class of small private hospitals. . . .[11]

This is one of the points where the authors used judgment in such a way that the error, if any, would tend to favor the hypothesis that mental illness had really increased.

(c) The care needed to obtain meaningful numbers is illustrated in the derivation of the age distribution of first admissions (comments in brackets are ours):

> In order to arrive at age-specific first admission rates we required the age distribution of patients admitted for the first time to an institution. For the entire period, 1840 to 1885, wherever first admission data were not available it was possible to secure from the official reports of the state mental hospitals the age distribution of their *total* admissions. For the South Boston Hospital [a private hospital] we were able to secure the age distribution of total admissions only for the years 1850–54 and 1860–64. The age distribution for these two periods was very close to that provided by the state hospital reports for the corresponding years; we have assumed that the age distribution of the South Boston admissions for the remaining years in our series is likewise similar to that of the other hospi-

---

10. *Ibid.*, pp. 28–29.
11. *Ibid.*, p. 27.

tals. [Here a footnote account is given of a technical statistical analysis which showed that the differences between the age distributions are easily explainable by random or "chance" variation.] For the McLean Hospital we were not able to secure the age distribution of total admissions until 1876. This age distribution was virtually identical with that of the rest of the state and hence we have used the total state age distribution for McLean in the earlier years as well.

What we required, of course, was the age distribution of *first* admissions rather than total admissions. Here, however, we were able to secure only scattered evidence primarily supplied by Dr. Earle in his Northampton reports. They provided, for several years, the age distribution of first admissions. This differed so little from the age distribution of total admissions that the use of the total admissions age distribution gave us a fully satisfactory basis for the calculation of age-specific first admission rates.* It should also be pointed out that in the period we are considering 65 to 75 percent of total admissions were first admissions. Consequently the age distribution of total admissions is in any case considerably weighted by first admissions. The final piece of evidence bearing on the use of the total admission age distribution for first admissions is provided by direct data on the age distribution of first admissions for 1880–85. This period in Table [37A] is based on the exact reporting of the age distribution of first admissions. We found that the assumptions we had used in the earlier years gave us an extremely striking continuity with the age distribution in this last period where direct evidence is available. Our method of estimation is therefore such that had we applied it to the period 1880–85 we would have come out with an age distribution that is virtually identical with that provided by the official reports.[12]

(d) Another question is the extent to which the figures on numbers admitted to hospitals reflect out-of-state admissions. It was possible to adjust quite accurately for this,[13] but we will not discuss it.

(e) A rate of incidence is a special kind of fraction. Therefore it is necessary to know not only the numerator (that is, the number of first admissions) but also the denominator (the total number of people in the relevant age group). The discussion up to this point has referred only to the numerator. The sizes of the total population and of the various age groups were obtained from Federal and state censuses.

Although population enumeration was probably less accurate in the 19th century than today, the difference in the amount of error can hardly be such as

---

*It might be supposed that the average age of readmissions would normally be higher than the average age of first admissions. This is not, in fact, the case. Thus, during the contemporary period, both Illinois and Massachusetts show a *lower* age for readmissions than for first admissions. This is because patients with the psychoses of the senium [that is, old age] are much less likely to be discharged and hence to be readmitted than are patients who enter a hospital at an earlier age. In the 19th century, readmissions do not show a younger age of admission than first admissions because in this period (as we shall show shortly) admissions in the older age groups form a very much lower proportion of admissions than is the case today. [Footnote in original.]

12. Goldhamer and Marshall, *op. cit.*, pp. 32–34.
13. *Ibid.*, p. 34.

### 2.8.2 Mental Disease

to affect, to any appreciable degree, comparisons between the earlier and contemporary period.[14]

2.8.2.4 *Analysis.* One of the first findings was that the age pattern of first admissions was considerably different in the 19th century than today. That is, even if the average number of first admissions per 100,000 people in the total population had been the same at both times, and the age composition of the populations had been the same, the rates for specific age groups would have differed. In the 19th century, more of the first admissions occurred in the middle age groups, especially in the years 20 through 49, and fewer in the under-20 and over-60 age groups.

Next the authors present the following two tables:

TABLE 37A

AGE-SPECIFIC FIRST-ADMISSION RATES FOR MAJOR PSYCHOSES, MASSACHUSETTS, BY 5-YEAR PERIODS, 1840–84 AND 1885[15]

| Age | 1840–1844 | 1845–1849 | 1850–1854 | 1855–1859 | 1860–1864 | 1865–1869 | 1870–1874 | 1875–1879 | 1880–1884 | 1885 |
|---|---|---|---|---|---|---|---|---|---|---|
| 10–19 | 12.2 | 13.4 | 14.4 | 16.4 | 15.4 | 14.5 | 15.8 | 19.1 | 17.2 | 18.6 |
| 20–29 | 50.1 | 51.8 | 52.5 | 62.6 | 59.0 | 62.7 | 70.0 | 74.0 | 76.5 | 84.6 |
| 30–39 | 71.7 | 73.8 | 69.8 | 82.0 | 78.0 | 75.3 | 101.0 | 104.1 | 99.5 | 109.2 |
| 40–49 | 80.5 | 80.6 | 83.6 | 92.3 | 85.0 | 71.5 | 97.0 | 99.4 | 99.7 | 109.0 |
| 50–59 | 77.5 | 85.5 | 61.7 | 63.5 | 62.9 | 72.7 | 77.4 | 78.9 | 83.4 | 90.0 |
| 60– | 50.1 | 59.9 | 44.5 | 48.0 | 68.0 | 60.5 | 66.8 | 68.0 | 80.2 | 67.8 |
| Total | 39.4 | 41.1 | 39.0 | 44.6 | 43.1 | 43.2 | 51.9 | 56.3 | 57.8 | 62.2 |

TABLE 37B

AGE-SPECIFIC FIRST-ADMISSION RATES FOR MAJOR PSYCHOSES, BY SEX, MASSACHUSETTS, 1880–84 AND 1885[16]

| Age | 1880–1884 | | 1885 | |
|---|---|---|---|---|
| | Male | Female | Male | Female |
| 10–19 | 19.3 | 14.7 | 22.0 | 15.0 |
| 20–29 | 87.9 | 66.8 | 96.4 | 75.0 |
| 30–39 | 103.6 | 95.6 | 111.0 | 107.9 |
| 40–49 | 104.7 | 95.2 | 110.0 | 108.1 |
| 50–59 | 88.5 | 78.5 | 102.9 | 78.8 |
| 60– | 84.8 | 74.5 | 70.4 | 65.5 |

14. *Ibid.*, pp. 34–35.

15. *Ibid.*, p. 49. Detailed explanatory notes in the original are omitted here. Notice that the figures in the row labeled "Total" are rates for all ages combined from 10 up. These are not simple averages of the rates for separate ages, but weighted averages, which will be discussed in Chap. 7.

16. *Ibid.*, p. 50.

Many readers will be reading these lines after only a cursory glance at the tables. Actually, you can easily learn to read tables accurately and quickly, once you overcome this tendency to skim over or skip them entirely. It is unwise to be dependent upon someone else's interpretation of tables, just as it is unfortunate to be completely dependent upon an interpreter in dealing with a foreigner; and fortunately it is easier to learn to read tables than to learn a foreign language. There is danger that the main facts of the table may be obscured, if for no other reason than an author's desire to use variety in his wording when he is putting the facts into prose. It is not uncommon, either, for an author to misinterpret his own tables, or to overlook important matters shown by them. Moreover, tables can usually be read more quickly than verbal descriptions of them. The only explanation that should be needed for an interpretation of Tables 37A and 37B is that the rates are given on a base of 100,000 people. If, for example, there were 200 first admissions in a group of 200,000 people, the rate per 100,000 would be 100, that is, $(200/200,000) \times 100,000 = 100$. Note that the second table presents rates that are specific for both age and sex. The method by which this information on the sex distribution was obtained is discussed fully in the original report.[17]

Now we let the authors resume the story:

We wish to test the hypothesis that in the central age groups the incidence of the major mental disorders has not increased over the last two to three generations. We bring to bear on this problem first admission rates for Massachusetts in the 19th century, and the question now arises: What rates from the contemporary period should be compared with them? The most immediate comparison that suggests itself is, of course, with the contemporary Massachusetts age-specific rates. This, however, is not necessarily the most desirable choice of comparative data. Hospital admission rates are a function (a) of the actual incidence of mental disease and (b) of factors that influence the proportion and type of mentally ill persons who are hospitalized. Our comparisons should, therefore, strive to ensure as much comparability as possible with respect to these latter factors, and where strict comparability cannot be attained, we must at least take them into account in our interpretations of the 19th century and contemporary rates.

The more important extraneous factors that need to be considered in testing the hypothesis are (a) level of hospital facilities relative to demand as measured, for example, by marked differences in the ratio of admissible patients who are rejected for lack of accommodations to the total number of admissions, or as measured by the sudden rise in admission rates resulting from the opening of new hospitals in the areas most immediately accessible to them; (b) accessibility to the institutions as defined, for example, by the relation of admission rate to

---

17. *Ibid.*, p. 48.

### 2.8.2 Mental Disease

distance from a hospital (where other factors have been held constant); (c) motivation to use facilities for a given level of facilities available and accessible; (d) range or type of patients admitted, in terms of diagnostic class, degree of severity of the mental illness required to secure admission, and (partly related to this) whether admissions are for relatively long periods or just for a few days to permit observation or temporary care; (e) composition of the population with respect to factors (other than age) that influence admission rates both in terms of their relation to the foregoing factors and to the true incidence of mental disease, e.g., proportion of foreign-born and urban dwellers in the population. The large foreign-born (especially Irish) immigration of the mid-19th century renders it imperative to ensure that our 19th century rates, relative to those of today, are not rendered incomparable by differing proportions of the foreign-born population and different relations of foreign-born and native-born rates. . . .

Since the factors that influence the choice of a standard of comparison from the contemporary period were not constant through the period 1840–85, it follows that the rates chosen from the contemporary period for comparison with those of 1885 are not necessarily the appropriate ones to use in a comparison with 1860 or 1840. Consequently, in what follows we provide a variety of contemporary rates with which we compare the rates of different parts of our 19th century series; and in each case we indicate why these particular contemporary rates have been chosen for comparative purposes. The attempt to choose contemporary rates that provide the greatest constancy of the conditions (a) to (e) discussed above, must, to a considerable extent, be impressionistic. Sufficiently exact data on the factors involved, and on the weighting to be assigned to each, to permit the construction of a single quantitative measure, are not available. We have, however, in all cases chosen contemporary rates that we believe provide a severe test of the hypothesis under study—that is, we have chosen contemporary rates in which the operation of factors (a) to (e) are on the whole prejudicial to the hypothesis.[18]

In short, rates of first admissions to mental hospitals, even when given for specific age and sex groups, are not sufficient for making comparisons of the true incidence of mental disease if the *other factors* enumerated above are not comparable. In later chapters we shall show some of the techniques by which it is sometimes possible to make allowance for "other factors" when more complete evidence is available than Goldhamer and Marshall were able to obtain. The method pursued in the present study was dictated by the incompleteness of the evidence available as to the effect of the other factors on mental illness or on hospitalization for mental illness. The authors made a series of comparisons in which selected 19th-century rates were compared with selected 20th-century rates, the selections being made in such a way that the rates are affected in much the same way by the extraneous factors. We shall summarize the main comparisons actually made.[19]

---

18. *Ibid.*, pp. 50–52. Footnote omitted.
19. *Ibid.*, pp. 53–76. The italicized headings are direct quotations.

(a) *Comparison of 1885 and Contemporary Massachusetts Rates.* In the first comparison the other factors were not really comparable, but were rather strongly "loaded" against the hypothesis the authors were establishing. Present-day Massachusetts first admission rates are higher than either the national average or the average for the New England States; a large part of contemporary admissions is for observation and temporary care; the percentage of urbanization is higher than in the 1880's; a larger proportion of current admissions are for nonpsychotic conditions. All these factors might be expected to exaggerate the magnitude of mental illness today by comparison with the 19th century. As might be expected, it turns out that the 19th-century rates are lower for most age groups, but, surprisingly enough, the rates for women in the 19th century were as high for ages 30–49 as they are today.

(b) *Comparison of 1885 Rates with Contemporary Massachusetts Rates for Admissions with Mental Disorder.* By using only those admitted "with mental disorders" in the calculation of current rates, it is possible to eliminate some of the artificial excess of current rates and hence make a fairer comparison. When this is done, ". . . the 1885 female rates exceed the contemporary rates for the entire age range 20–50. The 1885 male rates show substantial agreement with the contemporary rates for ages up to 40; the contemporary rate for the age group 40–50 is 13 percent in excess of the corresponding rate for 1885."[20]

(c) *Comparison of 1885 Massachusetts Rates with Contemporary Massachusetts Rates for Court and Voluntary Admissions.*

In this comparison we exclude the observation and temporary care first admissions, but include all regular admissions to public and private hospitals both with and without mental disorder [that is, in the contemporary figures]. . . . This comparison . . . reveals that the male 1885 rate for the combined age group 20–40 slightly exceeds that of the contemporary period and that the 1885 female rates for ages 20–50 exceed the corresponding 1930 figures. . . .[21]

(d) *Comparison of 1885 Massachusetts Rates with Contemporary Rates for Northeastern United States.*

The comparison of Massachusetts late 19th century rates with those for contemporary Massachusetts imposes a quite severe test of our hypothesis. Nonetheless, for the central age groups, the hypothesis has stood up to the test applied. A further comparison that suggests itself is to juxtapose our late 19th century rates for Massachusetts with first admission rates for the combined Northeastern states (New England and Middle Atlantic states). The two pre-

20. *Ibid.*, pp. 56–57.
21. *Ibid.*, p. 58.

## 2.8.2 Mental Disease

ceding comparisons have provided somewhat greater comparability with respect to the classes of patients received in the two periods. In terms of comparability of level of facilities available, a better comparison can probably be achieved by using rates for a larger area in which the facilities may be presumed to deviate less from those of our late 19th century period. It would be desirable to provide a comparison which simultaneously attempts to equate level of facilities and class of patients, but the contemporary data do not permit this very readily. Our inability to make such a comparison means, of course, an increase in the severity of the test to which our hypothesis is subjected. In choosing the Northeastern states for our next comparison, we have by no means selected a low-rate area. These states have a first admission rate that is 20 percent above the average for the country as a whole. They are, taken together, highly urbanized and indus-trialized states and the great proportion of their population and admissions come from states that have well-developed mental hospital systems. Further, in making this comparison we include all admissions, both with and without mental dis-order, to state, county, and city mental hospitals, Veterans Administration hospitals, and private mental hospitals. . . . Here again we find that the female rates for 1885 show complete parity with those of 1940 for the age groups 20–50. The male rates for 1885 show complete parity for the age groups 20–40; in the age group 40–50 the contemporary rate exceeds the 1885 rate by 17 percent.[22]

(e) *Comparison of Massachusetts 1855–59 Rates with United States 1940 Rates for First Admissions with Psychosis.* After a careful analysis the authors concluded:

> . . . the various conditions inhibiting admissions to mental hospitals were at least no less in Massachusetts of 1855–59 than they are currently in the United States as a whole. Probably there is no state in the Union today in which the restrictions on admission to institutions for the mentally ill approach those that existed in Massachusetts in 1855–59.[23]

The conclusion of this comparison was similar to the earlier ones.

(f) *Comparison of Massachusetts 1840–44 Rates with the Contemporary Period.* In the early 1840's, the Massachusetts rates of first admission undoubtedly understated greatly the true rates of mental illness because facilities were so extremely limited. "Some conception of the limited facilities available at this time is perhaps conveyed by the fact that during the first six years of its operation as the first state mental hospital, Worcester received additions of four wings which were no sooner completed than they were immediately 'filled to the overflowing.'"[24] The authors decide not to make precise tabular com-parisons in this case, but by an "impressionistic" comparison with selected states which had low admission rates in 1940, they conclude that for women at least, "Given the restrictions on admissions in this earliest period it is quite impossible to suppose that this difference

22. *Ibid.*, pp. 59–61. Footnotes omitted.
23. *Ibid.*, p. 64. Footnote omitted.
24. *Ibid.*, pp. 66–67.

reflects a real increase in the incidence of mental disorders among women in these age groups in the intervening 100 years."[25]

(g) *Comparison of Suffolk County, Massachusetts, 19th Century Rates with Contemporary Rates for New York City.* Next, the authors turned to a slightly different kind of comparison:

> So far we have dealt in our analysis with Massachusetts as a single unit. There are, however, several reasons why a special analysis of the Boston area recommends itself. In the first place, some interest is attached to the question whether . . . the disparity usually found today between rates for large metropolitan centers and for smaller towns and country areas existed in the earlier period. Secondly, and this is more important for our present purposes, we may presume that comparisons of 19th century and contemporary metropolitan rates provide a somewhat greater constancy of conditions than is feasible when state rates as a whole are compared. Large urban centers probably have social characteristics that are more continuous over time. Perhaps more important is the fact that Boston residents, throughout our entire period, had two hospitals locally available . . . [these hospitals] did counteract to some extent the operation of the "law of distance."[26]
>
> . . . As we are principally interested in the central age groups 20–50, our rates for these ages may . . . be taken as quite conservative estimates. We emphasize this because the rates we are about to present may astonish the reader and we wish to assure him that he is not dealing with inflated rates in the central age groups.
>
> The only large urban center for which we were able to find contemporary first admission age-specific rates, including admissions to both state and private institutions, is New York City. . . .
>
> . . . in 1840–44 Suffolk County had higher rates than contemporary New York City in the age group 40–60 and, by the mid-19th century period, almost uniformly higher rates except in the oldest age group. The reader may at this point feel that we have proved too much. We must confess that when these results became evident we ourselves felt intimidated by them. However a thorough reexamination of our data and our procedures has convinced us that these findings must stand.[27]

The final comparison, which we shall not discuss in detail, shows that the surprisingly high 19th-century rates are not unique to Massachusetts, although the best and most complete data are available for that state.

The remainder of the study may be described briefly with much less attention to the detailed evidence. The comparisons described above were concerned chiefly with the middle age groups, especially the ages 20–49. By contrast, the rates for older people were much

25. *Ibid.*, p. 68.

26. By the "law of distance" the authors are referring to the tendency for the rate of hospitalization for mental disease to be greater for people who live closer to mental hospitals.

27. Goldhamer and Marshall, *op. cit.*, pp. 68–73. Footnotes omitted.

### 2.8.2 Mental Disease

higher today than in the 19th century. Yet the authors show that even this differential may possibly be due to "other factors" rather than to a "real" increase. They first point out that the preponderance of admissions of old people is due to senile and arteriosclerotic psychoses, "diseases of the senium." Therefore,

> Three major possibilities present themselves: (1) that there has been a true increase in the incidence of the arteriosclerotic psychoses and that consequently the admission rates for the older age groups have risen as a result of this; (2) that there has not been a true increase in the incidence of such mental diseases, but that the tendency to hospitalize such cases has increased; (3) that the increase in rates is a result of a combination of the two foregoing factors. . . . [28]

Medical research suggests that it is possible that a part of the increase in the senile psychoses is "real," but the evidence is not very conclusive. But there is fairly good evidence, statistical and impressionistic, that a much higher proportion of old people with psychoses are hospitalized today than in the 19th century. Hence the authors conclude:

> While not excluding the very real possibility that part of the increase in admissions in the oldest age groups is due to a true increase in arteriosclerosis, the foregoing considerations strongly suggest that a major share of the increase in the age-specific rates for arteriosclerosis is due to the different hospitalization patterns for the older age groups in the 19th and 20th century periods. . . . [29]

The discrepancy between 19th- and 20th-century rates for those under 20 also turns out to be illusory. Contemporary admissions for this group include a larger proportion admitted for mental deficiency, as opposed to psychosis, than was true in the 19th century. "We conclude, therefore, that there is no evidence of an increase during the last century in the incidence of psychoses among persons under the age of 20, and that consequently the findings for ages 20–50 can now be stated to be true of all ages under 50."[30]

The authors also analyze the effect of the foreign-born population and conclude that this factor does not affect the conclusions already reached.[31]

We have frequently quoted the procedures used in making comparisons in order to convey without oversimplification the method used by the authors. Even so, we have omitted much of the careful discussion of scources of data and the footnote documentation for assertions made in the text. Such thorough attention to detail is certainly onerous and at times may seem overly pedantic. Yet the

28. *Ibid.*, p. 77.
29. *Ibid.*, p. 81.
30. *Ibid.*, p. 83.
31. *Ibid.*, pp. 83–89.

temptation to overlook details and to fill in with unsubstantiated assertions can easily lead to erroneous conclusions.

2.8.2.5 *Conclusions.* We have now described the analysis of the evidence. The main findings are summarized by the authors in five short paragraphs:

1. When appropriate comparisons are made which equate the class of patients received and the conditions affecting hospitalization of the mentally ill, age-specific first admission rates for ages under 50 are revealed to be just as high during the last half of the 19th century as they are today.

2. There has been a very marked increase in the age-specific admission rates in the older age groups. The greater part of this increase seems almost certainly to be due to an increased tendency to hospitalize persons suffering from the mental diseases of the senium. However, there is a possibility that some of the increase may be due to an actual increase in the incidence of arteriosclerosis.

3. The 19th and 20th century distributions of age-specific rates, that is, the distributions of admissions by age independent of changes in the age structure of the population, are radically different. In the 19th century there was relatively a much higher concentration of admissions in the age group 20–50; and today there is relatively a high concentration in the ages over 50 and more particularly over 60. This, of course, in no way affects the results summarized in paragraph (1) above.

4. Nineteenth century admissions to mental hospitals contain a larger proportion of psychotic cases and of severe derangement than do contemporary admissions. This is in part due to the more limited facilities of that period which tended to restrict admissions to the severer cases, and to the different distribution of age-specific rates.

5. Male and female age-specific rates show a greater degree of equality in the 19th century than today. This is largely due to the differences discussed in paragraphs (3) and (4) above.[32]

Each of these conclusions is supported by the statistical evidence obtained. It is never safe to assume that conclusions are supported by evidence unless one actually examines carefully the evidence adduced. An author may state some evidence and then a conclusion without showing any connection.

As we indicated in discussing the purpose of the research, Goldhamer and Marshall are interested *in much more* than these findings. In their final section they discuss carefully the possible significance of their findings for problems other than those they attacked directly. In doing so they carefully distinguish between what they say on the basis of the evidence of this study and what they say on other grounds. We select a few sentences from this excellent discussion:

---

32. *Ibid.*, pp. 91–92. In our description of the study, we have not brought out the evidence supporting the fifth conclusion.

### 2.8.2 Mental Disease

In addition to random fluctuations, admission data do show short-term changes that coincide with marked social changes such as those incident to wars and depressions. . . . our findings concerning the stability of secular trends for the psychoses [are] not intended in any way to minimize the importance of possible shorter-term fluctuations that have occurred in the past or may occur in the future.

The necessary restriction of the findings to the psychoses, and more especially to those of a psychogenic or functional character, raises the question whether, nonetheless, the findings have any presumptive value for statements about long-term trends in the incidence of neuroses, psychoneuroses, and character disorders. . . . the implications of the present report, for views on long-term trends in neurosis rates, depend on the theoretical orientation to the neuroses that the reader favors.

. . . A single study, such as the one reported here, can help to sharpen the formulation of alternatives, narrow the range of possible solutions to the theoretical problems at issue, and indicate promising directions for further research. Since the secular trend of admission rates has remained constant over the past 100 years, intensive research on short-term fluctuations is especially indicated. This research will first need to determine whether these fluctuations represent true changes in incidence. If this is found to be so, it should then be possible to relate these rate changes to the specific alterations in life circumstances associated with the periods of changing rates. This would remove analysis from the level of the rather vague ascription of causation to broad social developments associated with the "growth of civilization" and lead to the analysis of the more concrete changes in social life that characterize the short-term periods under study. Only the combined and continuing research of laboratory, clinical, and social psychiatry can eventually enable us to discard those views that are inconsistent with observed fact. To this process the present report contributes the finding that, whatever may be the causal agents of the functional psychoses, they will almost certainly have to be sought for among those life conditions that are equally common to American life of a hundred years ago and today.[33]

This particular study has been described in such detail for several reasons. First, it is a competent, objective, and thorough investigation, and illustrates well the nature of such an inquiry. Second, the subject studied is important, but especially susceptible to erroneous impressions, hunches, and intuitions. Third, this investigation serves as an introduction to many of the basic problems encountered in statistical practice. Fourth, the statistical techniques used by Goldhamer and Marshall are elementary enough to be understood even before reading the rest of this book. Fifth, the study emphasizes the vital need for close integration of knowledge of subject matter with knowledge of statistical method, and for broad perspective on the general problem but meticulous treatment of the minutiae of the specific inquiry.

---

33. *Ibid.*, pp. 92–97.

## 2.8.3 Vitamins and Endurance [34]

2.8.3.1 *The Problem.* By 1952 there was a good deal of evidence that extremely large doses of certain vitamins might enable animals and possibly humans better to withstand severe physical and psychological stresses that exist under conditions of extreme cold. It had been reported, for example, that the ability of rats to continue swimming in very cold water (48° F.) was enhanced by vitamin supplementation of the diet. There was related, but less conclusive, evidence for human beings. On the basis of such evidence, the Canadian Army had decided upon vitamin supplementation for certain combat and survival rations. Supplementation on the scale needed (many times the normal requirements for the vitamins in question) was somewhat expensive, however, so the United States Army decided to conduct a special experiment involving simulation of battle conditions before supplementing its own combat rations. This experiment, which we shall describe in some detail, illustrates the care, persistence, and ingenuity needed to answer what at first appears to be a simple question, even though the investigators could specify what evidence they wanted, instead of being limited, as were Goldhamer and Marshall, to data that happened to have survived.

The objective of the experiment was "to determine the effect of supplementation with large amounts of ascorbic acid [vitamin C] and B-complex vitamins on the physical performance of soldiers engaging in a high-activity program in a cold environment, with and without caloric restriction." This objective is narrower than the original objective in two interesting and important ways. Originally, the intent was to ascertain the effect of supplementation on the physical performance of soldiers engaging in combat-type activity. The change to "a high-activity program" was necessary because of the near-impossibility of simulating combat conditions effectively; as we shall see later, this change was of considerable importance in interpreting the results of the experiment. Originally, also, the intent was to ascertain the separate effects of ascorbic acid (vitamin C) and of B complex. The decision to narrow the objective to the study of the combined effect was dictated primarily by statistical considerations relating to the comparative smallness of the number of men available. We shall describe these considerations later.

---

34. This section is based upon Staff of Army Medical Nutrition Laboratory, *The Effect of Vitamin Supplementation on Physical Performance of Soldiers Residing in a Cold Environment* (Report No. 115 of Medical Nutrition Laboratory, Office of Surgeon General, United States Army, 15 September 1953). We have also drawn on unpublished statistical memoranda prepared during the investigation.

### 2.8.3 Vitamins

The statistical problems of this investigation were anticipated in advance, when the study design could be molded to meet them. The scientific staff, though mostly M.D.'s and Ph.D.'s with specialization in physiology, were more conversant with statistical principles than are most research workers. Moreover, they worked closely with professional statisticians from the planning stage to the analysis and interpretation of the study.

2.8.3.2 *Statistical Planning*. Initially, it appeared that about 100 soldiers would be available, all volunteers and almost all from a Medical Corps establishment in Texas. Ideally, as we shall see in succeeding chapters, a random sample drawn from all combat soldiers in the Army would have given a better basis for generalizations beyond the group participating in the experiment. Such a random sample was, however, impossible for administrative reasons. The scientists had to make an extra-statistical decision, namely, that results for the population sampled would apply to the target population of interest—that is, that the sampling process actually used was satisfactory for studying the physiological response to vitamin supplementation.

All the men were to be housed in relatively insubstantial barracks, unheated during the night, in a cold and lonely spot in Wyoming, called Pole Mountain, at an elevation of 8,310 feet. Their clothing would be inadequate except when they were quite active. For most of ten weeks in January, February, and March 1953, they were to engage in strenuous outdoor activities: marches, forced marches, calisthenics, and sports. There were to be no leaves or passes. The diet was designed for monotony; the caloric total was ample at the start—4,100 calories—but a three-week period of short rations was scheduled for the end of the experiment, with only about 2,100 to 2,500 calories per day—about enough to maintain a stenographer's weight. It was anticipated that many of the men would collapse under the combination of strenuous activity and restricted diet, and the experiment would have to be terminated before the three-week period was over. Throughout the experiment there were to be periodic measurements of physical condition and performance, and also of psychological attitudes and aptitudes.

So much for the bleak regimen in store for the 100 volunteers. What about the statistical design? Your first reaction might be simply to give the vitamin supplements to everyone and see what happened. But when experiments are done that way, their findings are nearly valueless. The fatal defect is that no one knows how the men would have performed under these conditions in the absence of vitamin supplementation. The only way to find out what would happen without supplementation was to withhold the vitamins from some of the 100 soldiers. Then a comparison of performance could be made between those who did and those who did not receive the supplementation, the latter being called the "control group."

But which group of men should not receive the supplementation? The control and experimental groups should be such that, chance factors aside, both groups would react the same if treated the same. Then, if the supplemented group did better than the unsupplemented group even after allowance for chance factors, a decision could be made in favor of the vitamin supplementation. There is, basically, only one method of separating the men into the two groups so that the experimenter can draw valid conclusions about the effect of the supplementation: the separation should make proper use of *random sampling*. For example, the names of the 100 men could be put on slips of paper, the slips shuffled thoroughly, and the names for the control group selected by a blindfolded person. Random selection of the control group has two advantages. First, it protects against any bias of selection, conscious or unconscious, that might tend to make the control group systematically different from the other group. Second, only when the selection is essentially random is it possible to measure the influence of chance on the differences between the two groups, and so decide whether or not the actual difference exceeds that which would be expected from chance alone. These advantages will be explained in later chapters.

An important technical contribution of the statisticians to the design came about in the following way. The scientists suggested dividing the men randomly into four groups of 25 each. One group was to receive both vitamins C and B. The second was to receive vitamin C but not vitamin B, the third was to receive vitamin B but not vitamin C, and the fourth group was to receive neither vitamin. An alternative suggestion was to divide the 100 men at random into two groups of 50 each, one receiving *both* C and B, the other receiving *neither*. The statisticians recommended the second suggestion. With this two-group design, a more adequate evaluation could be made of the combined effect of vitamins C and B, though at the cost of not learning about the separate effects. With the four-group design, it was more likely that important *true* effects of vitamin supplementation would be obscured by chance factors, in which case a promising line of experimentation would be wrongly abandoned. If a significant effect for vitamins B and C together were detected by the experiment, further experiments to refine the findings by isolat-

### 2.8.3 Vitamins

ing individual effects would be inevitable. This reasoning was reinforced by the arbitrariness of the designation "vitamin B complex," which includes many distinct elements, each potentially as important as vitamin C. Hence even the original four-group experiment could not show which specific B-complex component was responsible for any effects of vitamin B complex. Moreover, in terms of the immediate military problem, the cost of multiple-ingredient vitamin capsules for front-line troops would not be much greater than that of capsules which contained only the effective ingredient or ingredients, since the cost of distribution from manufacturer to the Army and then to the troops represented the bulk of the total cost. Thus, if the combined B-complex and C supplementation proved effective, interim action could be taken, and later experiments could investigate more carefully the specific source of the benefits.

There was one major qualification in the recommendation of the statisticians that the vitamins be studied only in combination. This would be disastrous if, in truth, vitamin C and vitamin B complex each had beneficial effects, but the two together tended to cancel each other. The extent of this danger had to be evaluated by the scientists on the basis of their knowledge of the physiological effects of vitamins. They decided that the danger was remote, and adopted the two-group design.

An important question was whether 100 men were enough to make the experiment worth performing at all. The basic approach to this question was as follows: Suppose that the vitamins really have an effect which, if it could be detected despite inevitable chance variations, would be worth knowing about. What would be the probability of detecting such an effect in an experiment with 100 men? To answer the question, the statisticians needed to know (1) how big a difference would be "worth knowing about," and (2) how great the chance variation among men treated alike was apt to be. Both of these questions were studied in terms of one of the proposed measures of physical performance, the Army Physical Fitness Test, which consisted of five exercises: pull-ups, squat-jumps, push-ups, sit-ups, and squat-thrusts. This test had been used in the Army for several years with a standardized scoring system. It was known that an average improvement of about 20 points on this test might be expected during six weeks of the basic training period. If the same amount of improvement could be achieved merely by vitamin supplementation, supplementation would seem worthwhile. Next, records of past performance on the test were procured for a group of soldiers at an eastern camp. These records gave some idea about variation of scores on this test among individuals, and also about the variation of scores for the same individual on different occasions. This information made possible an assessment of the probabilities that such variation would obscure a true average effectiveness of supplementation of 20 units in an experiment based on 80, 100, or indeed any number of men that might be contemplated. It turned out that 100 men was about the smallest number for which an experiment would probably give valuable information.

Between the completion of the plans and the start of the experiment, the number of soldiers available was reduced to 87. This jeopardized the success of the experiment, but it was felt still to be worth doing. The previous decision to use two groups rather than four now seemed especially desirable.

In the planning period, several other issues were discussed by the statisticians and scientists. Some of them may not at first appear statistical, but all were relevant to the design of the experiment and to subsequent analysis of its results.

(1) It was essential to the success of the experiment that the soldiers themselves not learn who was receiving the supplements. Such knowledge might influence performance by its effect on the morale of the participants. It had been decided, therefore, to give capsules to everyone. The capsules for the control group had no nutritive value except for a trivial amount of vitamin C—just enough for normal requirements in the low-calory phase of the experiment. All capsules appeared identical in every respect but one: for ease of administration, capsules given to the supplemented group were colored orange while the others were colored brown. Thus, it would be known who was receiving the same treatment, although it would not be known what the treatments were. Even this knowledge could affect the experiment adversely. For example, one of the measures of physical performance was to be the ability to complete forced marches. Suppose that the first two men who fell out of a forced march turned out to be members of the group receiving capsules of the same color. Other men receiving capsules of this color might then suspect that they were not getting the superior treatment, and that they might soon have to drop out. Since, as the subsequent experiment confirmed, physical performance is very much influenced by attitude, falling out might become epidemic among those receiving capsules of the same color. Thus, what was really a matter of a few men being unable to continue would be exaggerated in the data because of psychological contagion. Moreover, all subsequent performance for

### 2.8.3 *Vitamins*

the duration of the experiment might be strongly influenced by the memory of this one unhappy forced march. Unfortunately, it was not possible to correct this situation by making all capsules the same color. It was hoped that this defect in the design could be compensated by fostering strong inter-platoon competition in all the performance tests. Since each platoon contained men receiving capsules of both colors, men might identify themselves primarily with their own platoons rather than with their capsule colors.

(2) At the recommendation of the statisticians, all performance measurements were made once for each man before vitamin supplementation started. This permitted a more powerful analytical technique, based on the amount of improvement (or deterioration) of each man during the course of the experiment, rather than on his final performance alone. The importance of initial measurements of physical performance before supplementation was stressed by one of the statisticians in these words: ". . . failure to do so would be tantamount to removal of more than three quarters of the men from the experiment."

(3) The general strategy of the two-group design was modified to take into account the fact that the men were organized into four platoons, and that every effort would be made to foster inter-platoon competition. Instead of subdividing the entire group of men randomly into a supplemented and control group, a random subdivision was made within each platoon. This was as if four small experiments were performed instead of one big one, and there was reason to believe that the four small experiments combined would yield more reliable results than one big one.

(4) A still finer subdivision of the experiments by squads within platoons was considered and rejected.

2.8.3.3 *Execution of the Experiment.* Complex administrative problems arose in carrying out the experiment. A staff of 47 people—officers, enlisted men, and civilians—was needed, even though the subjects themselves handled the camp chores. The following jobs, among others, had to be done:

(1) Menus had to be devised to give the desired caloric values along with as much monotony as could be injected without causing excessive rejection of the food.

(2) The food had to be prepared with more than usual care in order that the theoretical caloric levels could actually be offered.

(3) All food not eaten by each subject had to be sorted and weighed in order to estimate his caloric intake, both in total and for protein, fat, and carbohydrate separately.

(4) The capsules had to be given to the right men, and it was necessary to be sure they were actually swallowed.

(5) Uniforms and barrack temperatures had to be adjusted to the weather.

(6) All activities and work details had to be scheduled properly.

(7) All performance tests had to be carefully supervised and recorded. For example, records had to be kept of the time and distance at which each man fell out on a forced march. Alertness was needed to notice such things as that fewer men fell out on forced marches if they were picked up and brought home in an open truck rather than a heated ambulance, and that still fewer gave up if they had to walk home anyway at their own pace. Total physical exhaustion was rarely encountered. "Experience with the forced march as a measure of performance, and specifically endurance, demonstrated that the usual cause of dropping out was loss of the will to proceed. It is not proper to call a man a quitter if he stops after marching 20 miles uphill into a fierce wind, yet in only rare instances did men apparently reach the limit of their capacity to march."

(8) Many special records had to be kept. For example, one Army enlisted man, trained as a meteorologist, kept detailed records of the weather.

(9) Twelve technicians in the laboratory section were needed to make the various physical and biochemical determinations, such as blood pressure, body weight, skinfold thickness, blood glucose, blood and urinary ascorbic acid, hemoglobin, and the like.

2.8.3.4 *Analysis of the Findings.* As the experiment drew to a close, attention was focused more closely on the details of the analysis. The general nature of the analysis had been determined before the experiment had even started, but there were many detailed questions to be answered. There were also innovations and improvisations in the experiment itself that had not been anticipated.

As data were collected in the field, rough analyses had been made by the supervising scientists, partly out of curiosity to see if the answer was going to be obvious. The most striking finding to emerge from these rough analyses was that the average physical performance for the entire group, supplemented and controls combined, had improved steadily throughout the experiment. In the last three weeks, when the 2,100–2,500 calory diet had been expected to cause the experiment to terminate, the men not only carried on but continued to show improvement on the physical tests. When they departed on their "convalescent" furloughs, they were actually in better physical condition than at the start of the experiment. The unanticipated improvement of the men during the entire experiment, and especially that during the short-ration period, might have been attributed to the vitamins had there not been a control group which showed similar improvement. This outcome of the experiment thus underscores our earlier comments about the need for a control group.[35]

The answer to the basic question, then, was not obvious from the rough analysis. It would have been obvious only if the effect of the vitamin supplementation had been large and consistent. The actual differences were relatively small. Careful analysis was needed to decide whether the supplemented and control groups differed more than could reasonably be ascribed to chance.

As we have seen, there were many measures of physical status and performance. One of the most important was the Army Physical Fitness Test, described earlier. Initially, the combined fitness score—the sum of scores on the five components—was the focus for analysis. Before actual numerical work could begin, certain decisions about treatment of the data had to be made. The fitness test had been administered weekly during the experiment. A major problem arose because some of the subjects had missed an occasional test on account of injury or illness, or had participated when their physical conditions were below par for one of these reasons. When the latter occurred, a decision was typically made by the medical officers *before* the actual test whether or not the man's score would be included. However, six subjects presented more serious problems, and these were not finally resolved until the analysis was about to begin. To illustrate, we quote the description of two of these cases.

*Test Subject No.* 311: A thin, slight man of 22 developed an upper respiratory infection during the second week of capsule administration. . . . Soon thereafter, following vigorous physical exercise he developed a large hematoma in the right thigh. A pneumonitis ensued with fever, anorexia, vomiting, and 7½ pounds weight loss. He was at bed rest and light activity for approximately one month, a week of which was spent in the F. E. Warren AFB Station Hospital. During this time he missed four consecutive weeks of physical and metabolic tests. Following this illness his performance was generally poor and he continued to lose weight on the restricted caloric diet. It was decided to eliminate all of his data from the experiment. [This was the only subject for whom all data were discarded.]
*Test Subject No.* 432: This 30 year old platoon sergeant was granted emergency leave during the third week of the test . . . because of acute illness of several members of his family. He was absent from the test site for 10 days during which

---

35. The need for controls is also illustrated by the experience of an elderly man who, having difficulty in hearing conversation, placed in his ear a plastic button with a cord long enough to run under his collar. Thereafter, he had no difficulty in hearing. People mistook the button and cord for a hearing aid, and talked louder. Had this man had a real hearing aid, he might have attributed all of the improvement in his hearing to the aid.

he administered nursing care to his family and continued to take capsules at the usual rate. No significant change of weight occurred during his absence, and tests of physical performance after his return showed no deterioration. It was decided to include all of the data collected from this man.

The final analysis, you will recall, was to be based essentially on improvement between the initial and final fitness test scores, and other performance measures. There were 44 men in the supplemented group and 40 in the control group for whom usable data were available for the first and last fitness tests. The results are shown in Table 54. The average score for the supplemented group was lower at the beginning and higher at the end; the average improvement was therefore greater for the supplemented group than for the controls.

On first glance, then, the supplementation appears effective. Actually, however, Table 54 shows only the over-all average for these two groups of 40 and 44 for the particular time period of the experiment. The table does not by itself tell whether these findings apply more generally. This question is what we had in mind earlier in our allusions to the effects of chance and the problem of allowing for those effects in interpreting the data. It is possible to analyze the original data from which Table 54 was computed in order to reach a decision as to whether the greater improvement shown for the supplemented group is more than we would ordinarily expect by chance alone. The analysis used, though not the idea underlying it, is too technical for this book. The conclusion was that differences at least as great as the ones observed in Table 54 would arise purely by chance about 17 times in 100, *even if the supplementation had no effect.* The italicized clause, to use again the technical terminology first introduced in our discussion of the study of the incidence of major psychoses, expressed the null hypothesis. The evidence of the experiment is not strong enough to warrant discarding the null hypothesis, at least so far as this analysis was concerned.

TABLE 54

MEAN PHYSICAL PERFORMANCE SCORES OF SOLDIERS, INITIAL AND FINAL TESTS, VITAMIN-SUPPLEMENTED AND CONTROL GROUPS

| Group | Initial Test | Final Test |
|-------|------|------|
| Control | 175.33 | 330.33 |
| Supplemented | 164.50 | 340.07 |

Several other analyses of the same type were made for other aspects of the fitness test data. For example, a separate analysis of each

### 2.8.3 Vitamins

of the five component tests was made. In addition, analyses were devised which utilized not only the beginning and ending scores, but also the intermediate scores. None of these analyses provided convincing evidence against the null hypothesis.

The same analytical procedure was applied to several of the other physical and psychological tests. For some of these measures, the control, and for others the treated, group appeared slightly better. For the most part, the differences were readily ascribable to chance. On one type of test, however, the supplemented group appeared superior by a margin exceeding what would be expected by chance alone. The average drop in body temperature after periods of passive exposure to cold, both indoors and out, was less for the treated than for the control group. On the other hand, the loss of body weight during the experiment appeared significantly greater for the treated group.

Some of the measurements, such as performance on the forced marches, could not be analyzed by the approach just described because the data were qualitative (for example, a man did or did not fall out on a forced march) rather than quantitative (for example, scores on the fitness test). There was a variety of minor problems of analysis, but we shall report only the main conclusion: no convincing evidence in favor of supplementation.

This account may make the analysis sound tedious and complicated. It was. Moreover, many key questions arising during the analysis had to be handled by relatively crude statistical methods because more refined methods were not possible, given the then current state of statistical knowledge. There were a few interesting methodological by-products, statistical and medical, such as a better method for scoring the physical performance tests. Much was learned that would enable future experiments of this type to be more effectively conducted, and this was thoroughly discussed in the final report.

The most important criticism of the experiment was not a statistical one, but the problem of the meaning of *cold stress*. One crucial element of combat stress was missing: long, anxious, sleepless waiting in the cold. As the final report stated,

> The type and degree of cold stress should be precisely defined prior to the study and adhered to throughout. . . . continued high energy activity is not compatible with body cooling despite the wearing of minimal uniforms. On the other hand, prolonged inactivity in the cold (simulating the fixed battlefield condition) is not compatible with high energy output. . . .

Our description of the experiment has necessarily neglected many important phases, but perhaps we have gone far enough to give you an appreciation of what underlay the brief statement of conclusions and recommendations, which we quote in full:

> Under the conditions of this experiment, supplementation of an adequate diet with large amounts of ascorbic acid and B-complex vitamins in men subjected to the stresses of high physical activity, residence in a cold environment and, during the later part of the experiment, caloric deficit, did not result in significantly better physical performance than that of unsupplemented men.
>
> Vitamin supplementation of the type used in this study resulted in a reduction in the fall in rectal temperature on exposure to cold.
>
> A caloric deficit of 1,200 calories per day for 22 days did not lead to detectable impairment of physical performance.
>
> The present study indicates that the current army minimal allowances of water soluble vitamins are capable of supporting good physical performance under the conditions of this study.
>
> An ascorbic acid intake of about 60 mg per day (control group) resulted in whole blood ascorbic acid levels of 0.3 to 1.2 mg % with a mean value of 0.7 mg %.
>
> RECOMMENDATIONS
>
> 1. That Army rations to be used in cold weather not be supplemented with ascorbic acid and B-complex vitamins. This recommendation is subject to change if further studies should reveal benefits not detected in the present study.
>
> 2. That further studies be made on the effect of vitamin supplementation on the physiological and pathological response of human subjects to cold exposure while at rest.

## 2.8.4 Artificial Rain-Making[36]

The next study to be considered in detail concerns a problem far removed from the evaluation of the effects of vitamin supplementation, but many of the statistical problems are surprisingly similar. The over-all objective of the research was "to obtain a more complete understanding of the fundamental physical processes which govern the formation of precipitation." The specific purpose of the part of the work reported here was to find out whether "seeding" cumulus clouds—that is, injecting appropriate materials, in this case water—causes the clouds to rain.

When the study began, the state of knowledge about cloud-seeding was not unlike that about vitamin supplementation when that study

---

36. This section is based on Roscoe R. Braham, Jr., Louis J. Battan, and Horace R. Byers, *Artificial Nucleation of Cumulus Clouds* (Chicago: Report No. 24, Cloud Physics Project, Department of Meteorology, University of Chicago, March 31, 1955). We have also drawn on unpublished statistical memoranda prepared in connection with this study.

### 2.8.4 Artificial Rain-Making

began, except that rain-making had attracted more public attention and claims of success were more numerous. Most of the evidence was sketchy and inconclusive, for a reason which may not be hard to guess if you think back to the vitamin study. Many clouds had been treated in many ways, but relatively little was known about what would have happened to them without treatment. It was almost as if the vitamin supplementation experiment had been performed without a control group and the inference drawn that supplementation was responsible for the physical performance observed.

In the present experiment, clouds were seeded by airplanes specially equipped with radar and photographic equipment, meteorological instruments, recording systems, and so on. Elaborate instrumentation was needed to obtain detailed information on all phases of cloud behavior. Moreover, the best way to detect the occurrence of rain was by the appearance of an "echo" on a radar screen. While radar equipment can tell whether or not a cloud produces precipitation, it does not tell how much precipitation is released or whether any of it reaches the ground. Thus, even if it could be shown that seeding initiated precipitation, it would not be known whether the seeding had simply altered the timing of precipitation that would have occurred later anyway, or whether it had increased the total amount of precipitation. Almost all attention in this experiment was focused on the simple question of initiation of precipitation, with a view to further study if initiation of precipitation were demonstrated.

The main statistical problem was to devise a method of deciding whether precipitation would have occurred in the absence of seeding. As in the experiment with vitamin supplementation, the need for a control group—unseeded clouds—was apparent to the scientists from the start. The precise way in which the control group was to be selected and the method by which the resulting data were to be analyzed were evolved during the course of consultation with several statisticians.

For reasons mentioned in the discussion of the vitamin experiment, it was essential that random selection be used in deciding which clouds were to be seeded and which were not. The main problem was to decide whether the proportion of seeded clouds producing rain differed from the proportion of unseeded clouds producing rain by a larger amount than could be ascribed reasonably to chance factors. Statistically, then, the basic problem in this experiment was similar to that of the vitamin study. One difference was that many of the measurements in the vitamin study were quantitative, for example, scores on the Army Physical Fitness Test, while in this study the basic measurements were qualitative, that is, rain or no rain.

In the vitamin experiment we saw that four small two-group experiments, one for each of four platoons, were preferable to one large two-group experiment. This was because the performance of soldiers within platoons was likely to be more homogeneous than the performance of all the soldiers in the experiment. Similarly, in the cloud-seeding study it seemed desirable to make each pair of clouds a separate small experiment. One cloud of each pair, selected at random, would be seeded. The two clouds in each pair, having been chosen at nearly the same time from the same part of the sky, would tend to be more like each other in respect to the probability of rain in the absence of seeding than would clouds selected at different times or different places. The analysis of an experiment performed in this way is simple. In fact, the method used is presented in this book (Sec. 13.3.2.2), and later in this section we shall outline the idea briefly. Before coming to this, however, we must examine the remaining problems in the design stage.

The chief problem of execution in the paired-cloud design was to keep bias, conscious or unconscious, from entering into the selection of the cloud to be seeded in each pair. The following quotation from the report describes the procedure used:

> After a cloud had been selected for study, the senior scientist, acting as flight controller, instructed the meteorological-instrument engineer to release the treating reagent on the next pass. The cloud was treated or *not* treated depending upon further instructions available only to the meteorological-instrument engineer. The senior scientist who selected the clouds for study was physically isolated from other scientists and had no knowledge of which clouds were treated until after each mission was completed. . . . Whether or not the cloud was treated, observations and measurements continued until it had dissipated, developed into a well-defined rainstorm, or lost its identity by merging with other clouds. . . .

Thus, the scientist who selected the clouds for study did not himself know which cloud in each pair had been treated. After the scientist ordered release of the reagent (water, in this experiment), the meteorological-instrument engineer opened a sealed envelope and read instructions which told him whether or not to execute the order. These envelopes had been prepared earlier by the statisticians, who assured random selection by using a method equivalent to tossing a fair coin. It was essential that the man selecting the second cloud in each pair not know whether or not the first cloud had been treated. Had he known, for example, that the first cloud had been treated, he might unconsciously have tended to pick a less (or more) juicy-looking cloud for the next pass. A systematic factor would then have

### 2.8.4 Artificial Rain-Making

worked for or against the treated clouds; this would have invalidated subsequent analysis based on the assumption that treated and un-treated clouds differed only by chance.

> The only conceivable way in which the controller might have known whether or not a cloud was treated was through the possible effect of the treatment on the behavior of the airplane. . . . When treating with water from the large valve, . . . the decrease in load on the airplane caused the plane to rise about 40 ft. in 14 seconds. . . . From the controller's position the change of altitude was detectable in clear still air, but totally undetectable in cloud or during air-plane maneuvers. The reason for this lay in the natural turbulence found in the cloud and in the preoccupation of the controller with other duties.
>
> As further substantiation of the fact that the effect of the release was not detectable during normal flight conditions, consider the outcome of flight 199, October 25, 1954. On this date, the crew obtained what they thought was a valid cloud pair, that is, the clouds met the eligibility requirements, echoes did not form in the interval between the inspection and treatment passes, and obser-vation of the clouds continued for a satisfactory length of time. After landing, the ground crew started to refill the water tank and it was discovered that the valve had not opened. . . .
>
> On the basis of all the evidence at hand, it is the unqualified judgment of the experimenters that the *controller* had no knowledge as to which cloud of any pair was treated until the information was revealed by the meteorological-instrument engineer after the test was completed.

Another problem of planning was to determine the number of cloud pairs needed in the experiment. Again, there is a close parallel with the vitamin experiment. The basic question, for any proposed number of cloud pairs, was the probability of detecting any given true effect of seeding in the midst of the effects of chance variation. This question could be handled less satisfactorily in the cloud-seeding experiment than in the vitamin experiment, because fewer perform-ance data were available on unseeded clouds than on unsupplemented soldiers. Again, however, it appeared that an experiment of the scale permitted by available resources was large enough to give a reason-able prospect of informative results.

The most important results of the experiment are given in the following quotation:

> . . . flight operations in the Caribbean area were carried out during two periods, October 1953 to February 1954 and October 1954 to November 1954. During the first period, 32 valid pairs of clouds were studied using the small water valve. The results are shown in Table [60].
>
> Each unit in the table represents a pair of clouds, e.g., the number "4" in this table represents four pairs of clouds in each of which the untreated cloud produced an echo and the treated cloud did not. In the initial analysis, only the data in the lower left and upper right entries were considered. The numbers in

TABLE 60

RESULTS OF TREATING CLOUDS WITH WATER USING SMALL VALVE
Total number of pairs—32

| Untreated cloud of pair | Treated cloud of pair | |
|---|---|---|
| | Echo | No Echo |
| Echo | 3 | 4 |
| No Echo | 3 | 22 |

the other diagonal represent those pairs of clouds of which both members behaved the same, and thus do not contribute to a test of the hypothesis that treatment makes no difference on the average.

From the fact that in three pairs the treated cloud developed an echo whereas the untreated did not, and in four pairs the reverse was true, it is obvious that the experiment does not support the efficacy of the treatment. A chance division of the 7 pairs could not be more even than 3 and 4.

The meteorologists suspected that the reason for the lack of effect might be the smallness of the amount of water being released. A larger valve was therefore installed, increasing the amount of water released. The large valve was used for the remainder of the operations. The results of precipitation initiation tests using the large valve are represented in Table 61.

The probabilities of obtaining the results in the lower left-upper right diagonal (or more unusual results) under the hypothesis of no treatment effects from the data in Table [61] (a), (b), and (c) are 0.11, 0.072 and 0.017 respectively. These probabilities were calculated under the assumption that large valve water treatment does not affect the average probability of precipitation initiation and are relevant if the alternative assumption under test is that water treatment increases the average probability of precipitation initiation. In designing this experiment, it had been decided that in problems of this type, 0.05 would be an acceptable level of significance. On this basis, it must be concluded from Table [61] (c) that the null hypothesis, i.e., that treatment had no effect, is false and that treated clouds had a higher probability of precipitation than untreated clouds. The probability of 0.017 given by the composite table must be viewed with some caution because it was calculated on the assumption that the total sample size was decided in advance, or at least decided on issues totally independent of what the initial results happened to be, whereas in fact this was not the case. It was decided to return to the Caribbean area after the data in Table [61] (a) were obtained. It is our opinion (and only an opinion since rigorous statistical techniques for this situation do not exist and since the rule for continuing experimentation was not made formal) that this will have little effect on the calculated probability. The experimental procedure and the types of clouds selected were the same during both seasons of operation.

**2.8.4** *Artificial Rain-Making*

TABLE 61

RESULTS OF TESTS FOR PRECIPITATION INITIATION IN
TROPICAL CUMULUS CLOUDS TREATED WITH WATER USING LARGE VALVE

(a) January-February 1954; total number of pairs—15

| Untreated cloud of pair | Treated cloud of pair | |
|---|---|---|
| | Echo | No Echo |
| Echo | 3 | 1 |
| No Echo | 5 | 6 |

(b) October-November 1954; total number of pairs—31

| Untreated cloud of pair | Treated cloud of pair | |
|---|---|---|
| | Echo | No Echo |
| Echo | 2 | 5 |
| No Echo | 12 | 12 |

(c) All data; total number of pairs—46

| Untreated cloud of pair | Treated cloud of pair | |
|---|---|---|
| | Echo | No Echo |
| Echo | 5 | 6 |
| No Echo | 17 | 18 |

While the logic of the basic statistical analysis is clear in this quotation, it may be well to amplify it. In a group of paired tests, those pairs in which both clouds performed identically tell us nothing about whether treated or untreated clouds are more likely to rain. We are interested, therefore, only in pairs in which one cloud rained and the other did not. In the experiment summarized in Table 61(a), for example, there were six such pairs. In five of these pairs, the seeded cloud rained; in the sixth, the unseeded cloud rained. Now if seeding had no effect whatever, we would, except for chance, expect a 3–3 division rather than a 5–1 division. The situation is precisely analogous to an experiment you might do yourself with a coin. If you toss a fair coin six times, the most probable outcome is three heads and three tails. But chance alone will fairly often give you 4–2 or 2–4. How unusual, then, is the 5–1 result or one more extreme, that is, 6–0, on the basis of chance alone? If you have a fair coin and a little time, you can try this out and see how often you get either all heads or all but one head in six tosses. If you do this enough times, and your coin is really fair, you will come very close to 0.11 or 11 percent, the figure given in the quotation. The necessary principles for calculating such probabilities are given in Chap. 10 of this book. The testing procedure is discussed more fully in Sec. 13.3.2.2.

Since 0.11 is a moderately small probability, we might suspect that 5 or more heads in 6 coin tosses was due not to chance alone, but to some inherent tendency for the coin to turn up heads more often than tails. Similarly, when the scientists in the experiment were confronted with the data of Table 61(a), they suspected that the treatment was actually having some effect. The suspicion was not very convincing, but it did lead to a renewal of experimentation in the fall of 1954. That time the division was 12–5, somewhat more convincing, since the probability is only 0.072 by chance alone. The combined evidence of the two experiments suggests quite strongly that seeding was effective.

We have quoted also some of the qualifications necessary in interpreting the final probability of 0.017 that a result at least this favorable to the cloud seeding procedure could arise by chance alone. This basic qualification is that the result of Table 61(a), the 5–1 split, was the first evidence of effectiveness given by the experiment. Earlier seeding efforts had given no hint whatever of success. Even though the experimental procedure had been modified prior to the 5–1 split, the experimenters could hardly put much confidence in this result alone. It is the evidence of Table 61(b) that is most convincing. Table 61(a) is not irrelevant to the final conclusion, but after all an event of probability 0.11 may well occur among a set of events, just as 5 or more heads in 6 coin tosses is fairly likely to occur if a number of such tosses are made.

A closely related point is that probably the experiment would not have been continued, so would have had no opportunity to produce Table 61(b), if just *one* cloud pair in Table 61(a) had been switched. The fact that the second phase of the experiment was run at all, then, partly reflects good luck in the first phase.

In this brief description, we have necessarily left out a great deal, even of the statistical problems. We have gone far enough, however, to illustrate the point made in the abstract in Chap. 1, namely, that the same statistical ideas are often applicable in problems that appear at first to be very different.

## 2.9
## CONCLUSION

Statistical methods are used effectively in the most diverse subjects, ranging from minor business and personal decisions to abstruse questions of pure science and scholarship.

### 2.9 Conclusion

Brief illustrations serve to indicate the range of applicability of statistics, but they can give only the barest hints about the way statistics enters into these applications. Statistics, when used effectively, becomes so intertwined in the whole fabric of the subject to which it is applied as to be an integral part of it. Full appreciation of the ways in which statistics enters into an investigation requires, therefore, a detailed analysis of the subject matter and of all the methods brought to bear on it.

# Chapter 3

# *Misuses*
# *of Statistics*

## 3.1
## THE INTERPRETATION OF STATISTICS

The most important thing to know about the interpretation of statistical data is that they do have to be interpreted. They seldom if ever "speak for themselves." Statistical data in the raw simply furnish facts for someone to reason from. They can be extremely useful when carefully collected and critically interpreted. But unless handled with care, skill, and above all, objectivity, statistical data may seem to prove things which are not at all true.

"In earlier times," Stephen Leacock wrote, "they had no statistics, and so they had to fall back on lies. Hence the huge exaggerations of primitive literature—giants or miracles or wonders! They did it with lies and we do it with statistics; but it is all the same." Disraeli averred that there are three kinds of lies: lies, damned lies, and statistics. It is sometimes said that statistics are used the way a drunk uses a lamp post: for support, rather than for light. A famous statement about history has been paraphrased to say that the unsupported declaration "statistics prove" should be read "I choose to assert without evidence," or even "I choose to assert, contrary to the evidence." The view that statistical conclusions are usually wrong is often supplemented by the view that when they are not wrong they are self-evident and trivial: "A statistician is a person who draws a mathematically precise line from an unwarranted assumption to a foregone conclusion."

**64**

*3.2 Shifting Definitions*

Misuses, unfortunately, are probably as common as valid uses of statistics. The ability to discriminate between a valid and an invalid use of statistics is more important for most people than knowing how themselves to make effective use of statistics. No one—administrator, executive, scientist, or responsible citizen in general—can afford to be misled by bad statistics; and everyone needs knowledge that can be gained only through the effective use of statistics.

Unfortunately, emphasis on misuses may give the mistaken impression that statistics are seldom or never reliable. Notice, however, that most misuses represent potentially good uses of statistics. "We share with Socrates the pious hope that men avoid mistakes once they are aware of them. But we are frivolous enough to suppose that men do this out of a spirit of pure contrariness, and hence are more affected by the sight of a horrible example than by a good precept."[1]

The examples which follow are divided into categories for purposes of discussion; the classifications are not to be taken very seriously, however, for many of the examples fall equally well into several categories.[2]

## 3.2
## MISUSES DUE TO SHIFTING DEFINITIONS

EXAMPLE 65A  UNEMPLOYMENT IN DIFFERENT COUNTRIES

In a comparison of unemployment in Germany and in the United States during the 1930's, those working on government projects were considered employed in Germany but unemployed in the United States. Which is right depends upon the purpose of the study, but for comparisons the definitions of unemployment must be the same in both countries.

EXAMPLE 65B  EMPLOYMENT, UNEMPLOYMENT,
AND PARTIAL EMPLOYMENT

Late in 1949, Georgi Malenkov, then a member of the Soviet Politburo, asserted that there were 14 million unemployed in the United States, and that this showed that the United States was in a serious depression. (Actually, this figure would exceed by more than 1 million the U. S. Bureau of Labor Statistics' estimate of the average number unemployed in 1933, the highest

---

1. Ernst Wagemann, *Narrenspiegel der Statistik, Die Umrisse eines statistischen Weltbildes* (A Fool's Mirror of Statistics, the Outline of a Statistical View of the World) (3d ed.; Bern: A. Francke Ag. Verlag, 1950). We are indebted to Norma E. Kruskal for the translation from which the quotation is taken.

2. Several of the examples which follow are taken from, or based on, those given by Jerome B. Cohen, "The Misuse of Statistics," *Journal of the American Statistical Association*, Vol. 33 (1938), pp. 657–674. Other excellent presentations of statistical fallacies are given by Darrell Huff, *How to Lie with Statistics* (New York: W. W. Norton and Company, Inc., 1954), and by Wagemann in the book referred to in the preceding footnote.

in history, and it would fall only $2\frac{1}{2}$ percent below the peak percentage—25, also in 1933.) Malenkov based his estimate on American data of good accuracy, but he defined "unemployed" to include all members of the labor force who worked less than full time. According to the definitions used by the U. S. Bureau of the Census, unemployment at this time was about 4 million. The basic difficulty here is that "employed" and "unemployed" do not cover all cases; they are extremes between which there is a wide range of possibilities. Furthermore, if there are many part-time workers, this may reflect either such scarcity of work that workers cannot find enough full-time employment, or such scarcity of labor that employers cannot find enough full-time workers. Full-time employment is itself hard to define, for hours of work that would now be considered full-time in America (for example, 35 to 40 hours per week) would be considered part-time in other countries (certainly in Mr. Malenkov's) or at other times.

### Example 66A   Car Registrations

Automobile registration figures are not an entirely satisfactory measure of the number of automobiles in the hands of the public, for three reasons. First, some states issue a new registration upon sale of a car, and some states transfer the old registration to the seller's new car, if any. Second, station wagons, sedan-type delivery cars, taxis, jeeps, and certain other types are classified as passenger cars in some states and not in others. Third, some cars are registered by dealers before they are sold to consumers. This last factor became important when the 1954 registration figures were scrutinized by two manufacturers each hoping to claim sales leadership. Enough unsold cars of both makes had been registered by dealers at the end of the year, to make it difficult to tell which make had led in sales.

### Example 66B   Overhead Cost

In studies of overhead and variable cost, confusion sometimes occurs between the economist's and the accountant's definitions of overhead cost. In economic analysis, overhead costs are those that do not change with the volume of output; accountants, however, sometimes allocate fixed costs among different years or different products in proportion to the volume of production.

### Example 66C   Personal Income

Questionnaire studies of personal income, whether based on censuses or samples, usually suggest an aggregate personal income for the country as a whole which is at least 5 percent below the aggregate actually believed to be correct by economists. While the reasons for the understatement are complex, one basic cause is that people tend to think of income as wages and salaries only, rather than as their income from all sources.

### 3.2 Shifting Definitions

EXAMPLE 67A  INDUSTRIAL CONCENTRATION

In measuring the extent of industrial concentration or "monopoly," the percentage of the total sales of an industry made by the four leading companies is often used. One difficulty with this measure, called the "concentration ratio," is that it may be affected greatly by the definition of "industry." If industries are defined relatively narrowly—for example, the "household electric toaster industry,"—the concentration ratio tends to be higher than if broader definitions are used—for example, the "household appliance industry," or the "electrical equipment industry."

EXAMPLE 67B  WAGE RATES AND WAGES EARNED

Although wages, in the sense of hourly wage rates, in a certain industry were up 10 percent, wages in the sense of average weekly earnings, had gone down, due to a reduction in hours of work. Weekly earnings depend not only on average hourly earnings but on the number of hours worked. Wages may thus appear to have gone either up or down, or even both at once, if in defining wages no distinction is made between wage rates and wage payments.

EXAMPLE 67C  HOURLY WAGE RATES

In a labor dispute, the union presented figures showing that during a period when prices had risen, average hourly wage rates had decreased. The management presented figures showing that this average had increased. The management had averaged the straight-time rates of the individual workers (that is, rates for regular working hours rather than for overtime, holidays, nights, and so forth) and this average had increased. The union had gotten an hourly rate for each employee by dividing his earnings by the number of hours he worked. The decrease in the union's average represented a decrease in the proportion of the work that was done at overtime rates. Again, either conclusion may be correct, depending on which definition of hourly wages is appropriate. The union's definition will often be appropriate when wages are viewed as income of workers, the management's when they are viewed as costs of production. Which way they should be viewed depends upon the problem.

EXAMPLE 67D  SEVERITY OF DISEASE

The stages of severity of a disease may be defined differently from one hospital to the next, and comparisons between hospitals are thereby made difficult.

EXAMPLE 67E  DURATION OF LABOR

In a study designed to find factors related to the difficulty of labor in childbirth, "length of labor" was used as a measure of difficulty. One shortcoming of this definition is that the beginning of labor is sometimes not clearly defined.

EXAMPLE 68A   LONDON VS. NEW YORK

Whether the city with the world's greatest population is New York or London depends on what areas are referred to by "New York" and "London." The City of London proper had a population in 1955 of only about 5,200, and New York County, or Manhattan, one of the five boroughs of New York City, had 1,910,000. The analogous political units, however, are the City of New York, with a population of 8,050,000 in 1955, and the County of London, 3,325,000 in 1955. Each of these is a municipality made up of boroughs, 29 in London and 5 in New York. A comparison often made (though inaccurately) is that between Greater London and the City of New York—probably because of the coincidence that the City of New York, when it was formed by consolidation of New York, Brooklyn, and other areas in 1898, was referred to as "Greater New York." "Greater London," with a 1955 population of 8,315,000, is defined as the area within 15 miles of the center of the City of London. It has been estimated that the area within 15 miles of the center of New York has a population of 10,350,000. The "New York Standard Metropolitan Area," however, had a 1955 population of 13,630,000. (A Standard Metropolitan Area is defined by the U. S. Bureau of the Census as a county or group of counties containing at least one city of 50,000 or more, plus such contiguous counties as are metropolitan in character and integrated with the central city by certain specified criteria.) A metropolitan area defined for London on a basis similar to that used for New York would have a population of approximately 10,000,000.[3]

## 3.3
## MISUSES DUE TO INACCURATE MEASUREMENT OR CLASSIFICATION OF CASES

When confronting statistical data, one useful question is, "How could they know?" Another is, "Who says so, and does he have a personal interest in the data being the way he reports them?" The answers to these questions do not settle anything about the quality of the data, for sometimes it is possible to find out about things for which the question, "How could they know?" suggests the answer, "They couldn't." (Would you have thought of the method described in Example 26A for finding out how many fish there are in a lake?) Moreover, some people are capable of great objectivity even when their own interests are involved. Nevertheless, the answers to these questions may properly stimulate skepticism.

EXAMPLE 68B   CRIME RECORDS

Thomas F. Murphy, while Commissioner of Police in New York, issued a report that showed an increase of 34.8 percent in felonies in 1950 over 1949.

---

3. This example was prepared for us by the Map Publications Editorial Department, Rand McNally and Company, Chicago, for which we are indebted to Duncan M. Fitchet.

### 3.3 Inaccurate Measurement

This rise was one of the largest recorded in recent years. It was attributed by sources close to the commissioner's office . . . [to] a recent overhauling of police customs in recording crimes and not . . . [to] any appreciable increase in crime itself in the city. . . .

At police headquarters it was declared that the introduction of Mr. Murphy's system virtually eliminated the traditional "burying" of uninvestigated complaints or of unsolved crimes. . . . [4]

EXAMPLE 69A   INSPECTION ERRORS

In the inspection of manufactured products, sometimes every item is inspected and it is reported that all defective items have been eliminated. Actually, few inspections are completely accurate. Even with several inspectors each inspecting every item, some defective items are missed. More generally, measuring every one of a large group of things does not insure complete accuracy because of errors in the individual measurements.

EXAMPLE 69B   INFANT SEX RATIO

The sex ratio in live births is about 105 or 106 males to 100 females. Experts argue that this ratio may be a trifle high. Errors in reporting, editing, tabulating, and transcribing, though rare, tend to run predominantly in the direction of reporting girls as boys, or of omitting girls more often than boys. This effect is negligible for nearly all purposes, but the point is that the errors do not necessarily cancel out.

EXAMPLE 69C   LANGUAGES OF THE WORLD

Consider the difficulty of obtaining information on the numbers speaking the various languages of the world. The following are the figures given by the 1950 editions of two well-known almanacs:[5]

WHAT ALMANAC D'YA READ?

| [From the World Almanac, 1950] *Tabulation of Those Who Speak the Chief Languages* | | [From the Information Please Almanac, 1950] *Languages of the World* | |
|---|---|---|---|
| Arabic | 29,000,000 | Arabic | 58,000,000 |
| Chinese | 488,573,000 | Chinese | 450,000,000 |
| Czech | 7,500,000 | Czech | 8,000,000 |
| Dutch | 16,548,500 | Dutch | 10,000,000 |
| German | 78,947,000 | German | 100,000,000 |
| Hungarian | 8,001,112 | Hungarian | 13,000,000 |
| Italian | 43,700,000 | Italian | 50,000,000 |
| Japanese | 97,700,000 | Japanese | 80,000,000 |
| Portuguese | 48,800,000 | Portuguese | 60,000,000 |
| Rumanian | 19,400,000 | Rumanian | 16,000,000 |
| Spanish | 80,000,000 | Spanish | 110,000,000 |
| Swedish | 6,266,000 | Swedish | 7,000,000 |
| Turkish | 16,160,000 | Turkish | 18,000,000 |

4. *New York Times*, June 7, 1951, p. 1.
5. *New Yorker*, Vol. 26 (September 23, 1950), p. 80.

*Misuses of Statistics*

EXAMPLE 70A   INTERVIEWER EFFECT

The person who collects the data may consciously or unconsciously affect the response. For example, "When Negroes were asked if the army is unfair to Negroes, 35 percent said yes to Negro interviewers; only 11 percent said yes to white interviewers."[6]

EXAMPLE 70B   DESTRUCTION OF PLANES

Accurate statistics about military operations are particularly difficult to obtain, even when great efforts are made. During the Battle of Britain in 1940, for example, the British estimated that the ratio of German to British air losses was 3 to 1. An American general was so impressed by the thoroughness of British methods that he believed the British claims of German losses were conservative. Yet a postwar check of German records showed that the correct ratio was 2 to 1.[7]

EXAMPLE 70C   DESTRUCTION BY PLANES

A similar example is given by the following quotation:

Air attack by a single combat plane is a fleeting thing, and the results achieved do not always conform to first estimates. Air reports of destroyed vehicles, particularly armored vehicles, were always optimistic by far. This was not the fault of pilots. Each fighter-bomber airplane was equipped with a movie camera which automatically recorded the apparent results of every attack. The films were examined at bases and became the basis of "Air Claims," but we found that this method provided no accurate estimate of the damage actually inflicted. Exact appraisal could be made only after the area was captured by the ground troops.[8]

## 3.4
## MISUSES DUE TO METHODS OF SELECTING CASES

EXAMPLE 70D   BRITISH TEXTILE UNEMPLOYMENT

During the early part of 1952, there was a slump in the textile industry of Lancashire. The extent of the decline was the subject of some controversy:

Unemployment figures issued by the Ministry of Labor are misleading. They are based on counts of workers made on Mondays. But nearly all workers now employed only three days a week work on Monday and therefore are not included in the official short-time count. For example, the official count for mid-February shows 18,400 operatives on short time. However, the official "estimate," which

---

6. Based on work done at the National Opinion Research Center, reported in the *University of Chicago Magazine*, April, 1952, p. 10.

7. Winston S. Churchill, *Their Finest Hour* (Boston: Houghton Mifflin Company, 1949), pp. 337–339.

8. Dwight D. Eisenhower, *Crusade in Europe* (Garden City, New York: Doubleday and Company, Inc., 1948), p. 324.

### 3.4 Methods of Selecting Cases

is acknowledged to be correct by those in the know but that is not made public, gives the number of unemployed as 24,000 for the same period.[9]

Notice that whereas Example 65B involved matters of definition and interpretation—whether part-time workers are employed or unemployed, and whether they are symptomatic of prosperity or depression—this example hinges on the method of determining the number of part-time workers.

EXAMPLE 71A  CENSUS UNDERENUMERATION

In China, one census taken for military and taxation purposes showed a total population of only 28 million; but a few years later a census of the same territory for the purpose of famine relief showed 105 million. Such an increase could not possibly have actually occurred. People evade the census taker if taxes and military service are involved, but seek him out when it is a question of receiving aid. In general, census-taking is more difficult and the results less accurate than people commonly suppose. Even the United States census of 1950, for example, is reliably estimated to have understated the total population by 3.6 percent and to have erred by much larger percentages in its counts of some groups within the population.[10] In the capital city of one important Latin American country, the only census in recent years was abandoned after only two districts of the city had been canvassed, and the total population of the city is known only through intelligent "guesstimates."

EXAMPLE 71B  MOVIE CENSORSHIP

The Chicago Police Department in 1952 prohibited the showing of the Italian film, "The Miracle." An interested organization reported the following investigation:

> In the past few months the Chicago Division has shown "The Miracle" at several private meetings. Of those filling out questionnaires after seeing the film, less than 1 percent felt it should be banned. "It thus seems," said Sanford I. Wolff, Chairman of the Chicago Division's Censorship Committee and Edward H. Meyerding, the Chicago ACLU's Executive Director, "that the five members of the Censorship Board do not represent the thinking of the majority of Chicago citizens."[11]

The statement quoted seems to be based on the assumption that those who saw the film at the private showings and filled out questionnaires represent the majority of citizens. Actually, it would be unwarranted to assume that the replies to the questionnaire represent the opinions of those who attended, to assume that those who attended were representative of those invited, or to assume that those invited were representative of the majority of citizens.

---

9. *New York Times*, March 24, 1952.

10. Ansley J. Coale, "The Population of the United States in 1950 Classified by Age, Sex, and Color—A Revision of Census Figures," *Journal of the American Statistical Association*, Vol. 50 (1955), pp. 16–54.

11. *Civil Liberties* (published by the American Civil Liberties Union), December, 1952.

Many people find it hard to analyze separately the various elements in a complex and emotion-laden issue like this. For example, many who object to police censorship of moving pictures will resent our pointing out the statistical fallacies in this or any other attack on it, while many who approve the action will interpret our criticism of the attack as support of the action.

#### EXAMPLE 72A  MENTAL DISEASE IN MEN AND WOMEN

The incidence of mental and nervous diseases appears to be higher among men than among women. A difficulty with the figures, however, is that men are more likely to be detected and institutionalized, since they are more likely to earn their livings in ways for which these disorders incapacitate them, and they are less likely to be supported by some other member of the family if unable to support themselves.

#### EXAMPLE 72B  SCHOOL CHILDREN PER FAMILY

In a certain city, the average number of school-age children per family having school-age children was estimated by questioning a sample of children in schools. The figure obtained was much too high, because a greater proportion of large families than of small families was covered by the data. Consider two families, for example, one with a single school-age child and the other with six. The average number per family is seven divided by two, or $3\frac{1}{2}$. But if each of the seven children were asked the number of school-age children in his family, the total of the seven replies would be thirty-seven and the average $5\frac{2}{7}$. This example is discussed a little further in Chap. 4.

#### EXAMPLE 72C  FAMILIES SELECTED THROUGH WAGE EARNERS

An error similar to that in the preceding example was made in estimating family earnings by sampling wage earners listed in employers' records. Those families with more than one wage earner had a greater probability of being included in the sample. Multiple-earner families tend to have higher incomes than single-earner families, not only because of the multiplicity of earners but because the heads of the families tend to be at the ages where earnings are highest, that is, at the ages where their children are old enough to earn money but have not yet left home.

#### EXAMPLE 72D  ERRORS OF EXECUTIVES

Consider the following statement by the head of a market research company:

> A "box score" which we have kept for a number of years shows that executives are right, or substantially right, only about 58 percent of the time in their decisions on questions of marketing policy and strategy. . . . [12]

12. Arthur C. Nielsen, "Evolution of Factual Techniques in Marketing Research," in Nugent Wedding (ed.), *Marketing Research and Business Management, University of Illinois Bulletin*, Vol. 49 (1952), pp. 52–53.

### 3.4 Methods of Selecting Cases

The impression may be conveyed by this statement that executives arrive at the wrong conclusions 42 percent of the time when they solve their own problems. But the executives presumably bring to an outside consultant only the problems they consider beyond their own capacities. Thus, the market research company probably has a very biased sample from which to estimate the proportion of cases in which executives are right about marketing problems. In fact, the figures might even be interpreted as showing that the executives are wrong 58 percent of the time when they think they need outside advice on marketing problems! Moreover, there seems to be an assumption implicit in the quotation that the market research firm's answers are invariably correct, and this can hardly be quite true.

EXAMPLE 73  AGES OF EXECUTIVES

This quotation is from a publication issued by a management consulting firm:

> The managements of a representative 65 companies are today, on the average, seven years older than were the managements of these same companies 20 years ago. Here is what we found in a recent survey:
>
> |  | 1929 | 1949 |
> |---|---|---|
> | Average age of all officers (excluding chairmen of boards) | 47 years | 54 years |
> | Average age of presidents | 53 years | 59 years |
>
> In about 80 percent of these companies, those holding top management positions were older in 1949 than were their counterparts in 1929.
>
> Taking the senior officers alone, i.e., presidents, vice-presidents, treasurers, controllers and secretaries, we found that they averaged 48 years of age in 1929 and 55 in 1949. More significantly the junior officers, who are normally regarded as replacements for the senior group, are not much younger than their superiors. They now average 52, and their advance in years since 1929 has followed the same upward trend as that of the senior officers in 65 companies studied.
>
> With the average age of presidents today at 59 and that of all senior officers at 55, it is apparent that replacements will have to be made in the next five to ten years at a more rapid rate than has been the case in the past.[13]

The investigators are no doubt correct with respect to the particular companies studied (although note that the figures in the text and the table refer to different groups). These data do not, however, constitute evidence for or against the proposition that the average age of *all* executives has increased in the last twenty years. The proposition may be true, but these data do not show it. The fallacy is rooted in the fact that the same 65 companies were used in getting the average ages for 1929 and 1949. This means that the sample is limited to companies which have been in business for at least twenty years. Any generalizations made from the sample must there-

---

13. Booz, Allen and Hamilton, *Management Personnel: Is Your Company Building and Protecting its Most Valuable Asset?* (1949). Leaflet.

fore be restricted to companies which have been in business for twenty years. There is reason to presume, at least in the absence of evidence to the contrary, that the average age of executives in firms at least 20 years old is higher than in firms under 20 years old. Thus the method of selecting the sample is correlated with the very characteristic (age of executives) that is being studied. To study the change in average age one should select for 1929 a sample of corporations then in business, and then select for 1949 a different sample of the corporations then in business. Note that if a similar study had been made of ages of heads of families in 1929 and 1949, using the same procedures used in this study of ages of executives, similar but even more striking results would have been obtained.

EXAMPLE 74A   LITERARY DIGEST

In 1936, the *Literary Digest*, a magazine that ceased publication in 1937, mailed 10,000,000 ballots on the presidential election. It received 2,300,000 returns, on the basis of which it confidently predicted that Alfred M. Landon would be elected. Actually, Franklin D. Roosevelt received 60 percent of the votes cast, one of the largest majorities in American presidential history. One difficulty was that those to whom the *Literary Digest's* ballots were mailed were not properly selected. They over-represented people with high incomes, and in the 1936 election there was a strong relation between income and party preference. In the preceding four elections, ballots obtained in the same way had correctly predicted the winners, but in those elections there was much less relation between income and party preference.

## 3.5
## MISUSES DUE TO INAPPROPRIATE COMPARISONS

This classification is closely related to misuses due to shifting definitions, due to shifting composition of groups, and due to misinterpretation of correlation and association.

EXAMPLE 74B   POWER OUTPUT

On the same day, two New York papers published exactly contrary headlines. One stated that electric power output had gone up; the other that it had gone down. The first was comparing the power output of the current week to that of the preceding week; the second was comparing it to the corresponding week a year earlier.

EXAMPLE 74C   EARNINGS AND RECEIPTS

In March, 1947, one New York paper headed an article "Douglas Aircraft Clears $2,000,000." Another paper carried the headline: "Douglas Aircraft Loses $2,000,000." The company had lost $2 million on its current

### 3.5 Inappropriate Comparisons

operations, but had received a $4 million refund on taxes paid in previous years.

#### EXAMPLE 75A   NUMBERS OR PROPORTIONS OF ILLITERATES

The true statement that there are more illiterates in New York than in California requires further examination. There are many more people in New York, and on a percentage basis, we get the opposite conclusion. Whether to use absolute figures or percentages depends upon the particular purpose.

#### EXAMPLE 75B   NUMBERS OR PROPORTIONS KILLED

During World War II, about 375 thousand people were killed in the United States by accidents and about 408 thousand were killed in the armed forces. From these figures, it has been argued that it was not much more dangerous to be overseas in the armed forces than to be at home. A more meaningful comparison, however, would consider rates, not numbers, of deaths, and would also consider the same age groups. This comparison would reflect adversely on the safety of the armed forces during the war— in fact, the armed forces death rate (about 12 per thousand men per year) was 15 to 20 times as high, per person per year, as the over-all civilian death rate from accidents (about 0.7 per thousand per year). Peacetime versions of the same fallacy are also common: "Homes are more dangerous than places of work, since more accidents occur at home." "Beds are the most dangerous things in the world, because more people die in bed than anywhere else." "Sick people are more likely to die when cared for in hospitals than when cared for at home."

#### EXAMPLE 75C   PLEASANT AND UNPLEASANT WORDS

A psychologist found that a group of young children used "pleasant" words much more frequently than "unpleasant" words. From this finding it was concluded that children learn pleasant words more easily and rapidly than unpleasant words. It would be better to define "ease of learning" by the ratio of "pleasant" or "unpleasant" words actually learned to the total number of "pleasant" or "unpleasant" words which the children had equal opportunities to learn.

#### EXAMPLE 75D   HEREDITY VS. ENVIRONMENT

Sometimes people try to make quantitative comparisons that cannot possibly be meaningful. One illustration is the heredity-versus-environment controversy. The following quotation shows how a logically meaningless comparison can give rise to a meaningful question:

> Hence it is really not legitimate to ask: What is the relative importance of heredity and environment? This question belongs in the scrap basket with the type of general conclusions in some of the studies quoted: "It appears that heredity is twice as important as environment in determining intelligence." The new

approach would be: Given a stated environment, how much variation will heredity permit for such and such a characteristic (among so and so individuals)? Or, given a stated heredity, how much variation could a given range of environment introduce for such and such a character.[14]

EXAMPLE 76A   DIVORCE RATES

After the 1930 census, it was stated on the basis of tabulations of one-half of the states, that the divorce rate had apparently fallen from 1920 to 1930. When all the results were tabulated, it was found that the divorce rate had not changed. The error was due to the fact that the first states reporting were the less populous, agricultural, lower-divorce-rate states. These states should have been compared with the same states in 1920, instead of with the whole country in 1920, though even then the result could have been interpreted only as applying to these states, not the whole country.

EXAMPLE 76B   INCOMES AND PRICES

In the 1936 election, the Democrats claimed that employment and production had risen greatly while the cost of living had not gone up at all. Their bases for comparison were 1933 for employment and production and 1925–29 for cost of living. The Republicans, on the other hand, claimed that the cost of living had gone up, but not employment. They compared cost of living with 1933 and employment with the 1925–29 average.

EXAMPLE 76C   POSTWAR JAPANESE PRODUCTION

In 1949, an article in *Fortune* criticized the American regime in Japan for its handling of economic problems. It claimed that industry in Japan was stagnant by comparison with prewar production. The reply was made that Japan had made a greater improvement since 1946 than any other country in the world. The disagreement about the economic status of Japan turned chiefly on the base date for comparison.

EXAMPLE 76D   RUSSIAN DOCTORS

"According to Yaroslavsky, the number of doctors in Russia had increased from 1,380 in 1897 to 12,000 in 1935."[15] In studying the effectiveness of the Soviet regime in expanding the number of doctors, it would be desirable to compare the increase between, say, 1897 and 1916, with that between 1917 and 1935, since the regime took over in 1917.

EXAMPLE 76E   PRICES DURING AND AFTER CONTROL

The Office of Price Administration based its claims of effectiveness in holding down prices on the Bureau of Labor Statistics' Cost of Living Index (now called the Consumer Price Index); but after OPA was discontinued in

---

14. Gladys C. Schwesinger, *Heredity and Environment: Studies in the Genesis of Psychological Characteristics* (New York: Macmillan Company, 1933), p. 459.
15. Bernard Pares, *Russia* (New York: Mentor Books, 1949), p. 137.

**3.6 Shifting Composition of Groups**

July, 1946, some of its supporters showed alarming price rises on the basis
of the same Bureau's Index of Spot Primary Market Prices of 22 Commodities. A wholesale price index generally fluctuates more than a consumer
price index, and this specific index, based on daily quotations of 22 raw materials, fluctuates more than the Bureau of Labor Statistics' Wholesale Price
Index, which is based on weekly and monthly average prices of 2,000 commodities (only 900 at the time the OPA ended) of all kinds. At this particular
time, moreover, most of the rise in the spot prices index was due to one
transaction (and that by the government) in one commodity, silk.

EXAMPLE 77A   PROPORTION OF CHINA LOST

In mid-1949 the following argument was adduced to support the position that the Chinese Nationalist government was not yet defeated in the
war against the Communists on the Chinese mainland: "The Nationalists
retain control of about half the area that is China. The Communists hold
no more territory than the Japanese held at the height of the occupation."
The mere fact that the Nationalists held about 50 percent of the territory
of China tells nothing at all about the proportion they held of population,
important cities, resources, or transportation facilities.

EXAMPLE 77B   PROPORTION OF UNITED STATES VULNERABLE

In 1954, a statistically similar example appeared in a Chicago newspaper
article, which attributed the following statement to a leading civil defense
official:

> Even if Russian planes destroy the 70 largest industrial groups of cities, it
> would knock out only 3 percent of the nation's real estate. Ninety-seven percent
> would still be ready for business.[16]

# 3.6
# MISUSES DUE TO SHIFTING COMPOSITION OF GROUPS

This category is closely related to inappropriate comparisons and
to misinterpretation of correlation or association.

EXAMPLE 77C   GROUP AVERAGE DOWN,
EACH INDIVIDUAL UP (OR OUT)

A manufacturing plant found that the average monthly earnings of its
employees had fallen 8 percent during a certain period. This might seem to
"prove" that earnings had gone down. As a matter of fact, however, the
earnings of every single employee were exactly 10 percent higher than at
the beginning of the period. The reason the average earnings fell despite
this increase was that many of the higher-paid employees were dropped at

---

16. *Chicago Daily News*, September 28, 1954.

the time the increase was made, so that the new average included only the lower-paid workers.

### EXAMPLE 78A OLD GRADS

The alumni of a certain class of a college had an average age of 87 one year and 85 the next year. The explanation is not that they were literally getting younger every year, but that the oldest members had died during the year.

### EXAMPLE 78B ARIZONA TUBERCULOSIS DEATH RATES

The death rate from tuberculosis is far higher in Arizona than in any other state. This might seem to indicate that Arizona has a bad climate for tuberculosis. Actually, it reflects the fact that its climate is considered beneficial by many people who have tuberculosis, who therefore go there. For causes of death such as heart disease, cancer, and cerebral hemorrhage (which are, in order, the three leading causes of death in the United States) Arizona has extremely low death rates—largely because it has a relatively young population.

### EXAMPLE 78C REGIONAL DIFFERENCES IN INCOME

Regional comparisons of income reveal a difference between average income in the North and in the South. An analysis of the North alone, however, reveals differences in average income between white and colored and between urban and rural areas. An analysis of the South alone reveals similar intraregional variation. Since the proportions of the population colored and white or urban and rural are not the same in the North and South, the average incomes of the two regions would be different even if the incomes of corresponding groups were the same in the North and the South. Part of the difference between the two regions is thus due to factors that operate within regions; this part simply reflects differences in racial composition and in urbanization, rather than differences in income of corresponding groups in the two regions.

## 3.7
## MISUSES DUE TO MISINTERPRETATION OF ASSOCIATION OR CORRELATION

This kind of misuse is really a special case of inappropriate comparisons. It exemplifies the familiar but often ignored fact that correlation or association does not necessarily indicate causation.

### EXAMPLE 78D FEET AND HANDWRITING

In a study of schoolboys, an educator discovered a correlation between size of feet and quality of handwriting. The boys with larger feet were, on the average, older.

### 3.7 Misinterpretation of Association

EXAMPLE 79A STORKS' NESTS

There is reported to be a positive correlation between the number of storks' nests and the number of births in northwestern Europe. Only the most romantic would contend that this indicates that the stork legend is true. A more prosaic interpretation is this: as population and hence the number of buildings increases, the number of places for storks to nest increases.

EXAMPLE 79B PROPAGANDA LEAFLETS

During the Italian campaign of World War II, it was found that there was a positive correlation between the number of propaganda leaflets dropped on the Germans and the amount of territory captured from them. While this is consistent with the hypothesis that the leaflets were effective, it is also consistent with other hypotheses, for example, that leaflets were dropped when major offensives were about to begin.

EXAMPLE 79C BUSINESS SCHOOL ALUMNI

A study of the alumni of a certain university's graduate school of business showed that students whose grades had been about average had higher incomes, on the average, than either the poor or the excellent students who graduated at the same time. Before drawing conclusions from this, it would be necessary to investigate the possibility that a higher proportion of the excellent than of the average have gone into teaching, where earnings are less than in business.

EXAMPLE 79D KENNY TREATMENT

Sister Elizabeth Kenny, originator of a method of treating poliomyelitis, declared yesterday that if a true knowledge of her method were available to the medical world, recoveries from the disease would be increased at least 10 percent.

The Australian asserted that when she arrived in this country in 1940, the percentage of recoveries was about 15 percent and that since, the recovery rate had risen to about 75 percent.

"This could be increased to 85 percent if the medical profession had the full knowledge of the Kenny treatment," she declared. "I'm not saying this . . . statistics already published prove it."[17]

From the newspaper article it is impossible to be sure what statistics Sister Kenny was talking about. It appears, however, that she was attributing much of the improvement after 1940 to her method of therapy. It is entirely possible that other things which happened after 1940 might account for some or all of the increase in recovery rate. For one thing, the diagnosis of polio improved so much that it became possible to detect many more mild cases. This improvement in diagnosis had the effect of increasing by equal amounts both the numerator and denominator of the ratio of recoveries to total cases. Hence, the recovery rate seemed to be increased.

---

17. *New York Times*, August 26, 1949, p. 12.

## 3.8
## MISUSES DUE TO DISREGARD OF DISPERSION

EXAMPLE 80A  CALIFORNIA WEATHER

In California, the weather is usually called unusual, as though the average were the usual value. Actually, substantial departures from the average, especially in winter rainfall, are typical of California. In some cases, it may actually be impossible for any single observation to be equal to the average, as when the average number of persons in a certain category is 2.38.

EXAMPLE 80B  WADING IN THE TOMBIGBEE RIVER

Congressman John Jennings, Jr., of Tennessee, in the U. S. House of Representatives on June 6, 1946, stated that in dry weather the average depth of the Tombigbee River is only one foot. "In other words," he said, "you can wade up it from its mouth to the spring branch in which it originates." While such a wading trip may be possible—we are assured by a native of the region that it is possible, although he has not himself made the trip—it does not follow from the statement about average depth. Something would have to be known about dispersion.

EXAMPLE 80C  MINIMUM SALARY SCALE

The President of an institution proposed to set $12,000 as the minimum annual salary for a certain class of employees. He asked an assistant to calculate the addition to the payroll that would result. The assistant found that there were 250 such employees and that their average salary was $11,000. He therefore reported that the cost would be 250 times $1,000, or $250,000 per year. Actually, the cost turned out to be $450,000. To see the point (if it is not obvious) suppose there were 50 employees at each of the following salaries: $7,000, $9,000, $11,000, $13,000, and $15,000. The average is $11,000. The increase in the payroll caused by each group would be $250,000, $150,000, $50,000, $0, and $0 respectively. The figure of $250,000 that the assistant calculated is what it would cost to raise the *average* to $12,000.

EXAMPLE 80D  SASKATCHEWAN WHEAT

One farmer [in Saskatchewan] has reported harvesting 104 bushels of wheat from a two-acre strip. Thirty-five bushels an acre is considered well above the average.[18]

Even if 35 bushels is a high average, it does not follow that 52 bushels per acre is unusually high for the best two-acre strip in a large area. It might even be unusually low. Averages based on large numbers vary much less than individual measurements, or than averages of very few measure-

---

18. *New York Times*, September 1, 1955, p. 1.

ments. Especially when an observation is selected as being the most unusual, the average is almost useless as a good bench mark against which to judge its unusualness.

The fact that John Adams lived nearly 30 years after his inauguration as President (1797) does not by itself permit the inference that Presidents of the United States have unusual longevity. Similarly, the claim that Brand A gives "up to" $2\frac{1}{2}$ times as much wear as the average of three leading competitors does not mean that Brand A gives more wear on the average.

## 3.9
## MISUSES DUE TO TECHNICAL ERRORS

The preceding statistical misuses, though frequent, are relatively obvious—at least after they have been pointed out. The errors are errors of common sense or of logic more than of statistics in any very technical sense. It is their frequency in statistical applications that justifies their emphasis here. There are, however, misuses that arise from more technical statistical deficiencies. Misuses of this kind will be discussed from time to time in later chapters. For the present, we give some illustrations that show the ever present danger of the most prosaic but most common technical error, a mistake in calculation.

EXAMPLE 81A   ERRORS IN COMPUTING STANDARD ERRORS

Ericksen's failure to find statistically significant differences between groups was due to erroneous computation of the standard errors of the differences between means. Instead of using the standard errors of the separate means, he used the standard deviations of the score distributions; hence, all of his 9 reported critical ratios tend to be quite low, 0.32 or less. With the use of the proper formula, some of the logically expected differences are statistically significant. . . . [19]

EXAMPLE 81B   ERRORS IN COMPUTING
AVERAGE PERCENTAGE

A firm manufacturing complicated electrical devices had found that it had to expect about 5 percent of the units made to be defective, but that the rate need never exceed 10 percent with good materials, machines adjusted properly, and skillful workmanship. One week, more than 10 percent defective units were reported, so special care was given the next week's production. Nevertheless, 16.4 percent defective units were reported. One of the engineers called in to make an intensive survey of the production line looked at the inspector's records and found:

---

19. Quinn McNemar, "Opinion-Attitude Methodology," *Psychological Bulletin*, Vol. III (1946), p. 304.

| Day | Number Inspected | Number Defective | Percent Defective |
|---|---|---|---|
| Monday | 70 | 0 | 0 |
| Tuesday | 68 | 2 | 3.0 |
| Wednesday | 68 | 3 | 4.4 |
| Thursday | 70 | 1 | 1.4 |
| Friday | 72 | 4 | 5.5 |
| Saturday | 32 | 1 | 3.1 |
| Total | 380 | 11 | 16.4 |

The correct percent defective was therefore, 11/380 or 2.9 percent. The inspector had used a wrong method of calculating—adding the daily percentages—and furthermore had added wrong and made two small errors in calculating the daily percentages!

EXAMPLE 82A   ERRORS IN UNITS OF MEASUREMENT

During World War II, military and scientific people developing a promising new bombing device were disheartened when a statistician's calculations showed that the device would have virtually no chance of hitting its targets. Those responsible for the project were hastily gathered together from all parts of the country to consider this bombshell. Another statistician noticed that it would be physically impossible for a bomb to get as far from the target as the average error shown by the calculations. Hurried long-distance phoning and frantic all-night checking revealed that in the computations angular errors had been measured in degrees, but interpreted as if they were in radians (a radian is 57.3 degrees).

A frequent error of the same kind is to confuse two kinds of logarithms, "natural" and "common," the former being 2.3 times the latter.

## 3.10
## MISUSES DUE TO MISLEADING STATEMENTS

EXAMPLE 82B   CO-EDS MARRYING FACULTY

The statement, "One-third of the women students at Johns Hopkins University during its first year married faculty members," creates an impression unwarranted by the facts. There were only three women students. Similarly: "Thirty-three percent of the women married two percent of the men."

EXAMPLE 82C   CRAZY RADAR MECHANICS

The preceding example led a student to tell us of an effective use he had made of the same form of statement. During World War II, he was responsible for airborne radar in the Mediterranean Theater. He was able to obtain only seven radar mechanics for the Troop Carrier Command, which was authorized to have, and badly needed, forty to fifty. Repeated requests and complaints submitted through normal channels accomplished

### 3.10 Misleading Statements

nothing. "One month I was informed," the student told us, "that one of the seven mechanics had suffered a mental breakdown precipitated by overwork. In my next monthly report, under the heading 'Troop Carrier Command—Personnel' all I wrote was 'Over fourteen percent (14%) of the radar mechanics went crazy last month due to overwork.' Almost immediately after the report reached Washington, thirty-five additional radar mechanics were sent us by high priority air."

### EXAMPLE 83A  PALO ALTO SUMMER RAIN

At Palo Alto, California on July 25, 1946, nineteen times as much rain fell between 6 A.M. and noon as during all the preceding Julies since the weather station opened in 1910. That is, in six hours 19 times as much rain fell as in a 26,784-hour period, a rate during those six hours about 85,000 times "normal." Actually, this "deluge" consisted of only 0.19 inches; the only measurable rain in all the thirty-six previous Julies was on one occasion when 0.01 inches fell.

### EXAMPLE 83B  GROWTH OF CHILDREN

> PALO ALTO, Calif., Aug. 30 (Science Service)—How tall a growing child will be when he is grown up is now being predicted to within a quarter of an inch by scientists at the Leland Stanford University here. . . . [The scientists] report eight cases in which the adult heights came to within one-quarter of an inch of the heights predicted while the subjects were children. . . . [20]

The first sentence of this quotation leads the reader to believe the implausible proposition that the scientists can predict *any* child's height to within one-quarter of an inch. But the second sentence suggests that they may be able to do no more than anyone else, namely, be right occasionally—say 8 times in several hundred.

### EXAMPLE 83C  PAJAMA SALES

> Pulling its drawstring tighter, the men's pajama industry dolefully reported some raw facts last week: Men were buying only one-third of a pair of pajamas each a year. . . . [21]

The statistical finding, of course, is that an average of one-third of a pair of pajamas per man was sold. While hardly anyone would be misled by this example, which is attributable to journalistic flippancy, statements of this kind often convey the impression that statisticians are chiefly concerned with quaint curiosities, and warrant jibes like the following: "The average statistician is married to $\frac{7}{10}$ of a wife, who tries her level best to drag him out of the house $2\frac{1}{4}$ nights a week with only 50 percent success . . . [etc., etc., etc.]."

---

20. *New York Times*, August 31, 1949, p. 20.
21. *Newsweek*, August 1, 1949, pp. 50–51.

## EXAMPLE 84A   WORLD HEALTH ORGANIZATION

Each American citizen contributes slightly more than 2 ct/yr toward support of the World Health Organization.[22]

This is a way of making a sum apparently in excess of $3 million per year sound small. It would have sounded still smaller if it had been expressed as a twenty-fifth of a cent per week or a two-hundredth of a cent per day. Literally, of course, no individual American citizen contributes anything toward support of the World Health Organization. Whatever amount the American government contributes is derived from a number of sources, including taxes levied on citizens and others.

## EXAMPLE 84B   1948 GALLUP POLL

In 1948 the American Institute of Public Opinion (usually called the Gallup poll) predicted confidently, on the basis of a series of polls culminating in one involving about 3,000 interviews with what it called a "scientifically" selected sample of voters, that Thomas E. Dewey would be elected president by a substantial margin. Other polling organizations made the same prediction, some even more confidently. Actually, Harry S. Truman was elected by a small margin. Many factors combined to produce this failure, a major one being statements and interpretations based as much on assumptions as on the data. A confident prediction in favor of either candidate was not justified by the data, but the prediction was based on a number of assumptions, including important ones concerning the voting of the substantial group who had not made up their minds when interviewed. In the preceding three elections, similar methods had successfully predicted the winner, though there had been large errors in the margins predicted, and this may have been conducive to carelessness in statements made in 1948. Incidentally, the upshot of the 1948 failure was that in 1952 Gallup presented his data without predicting a winner. The actual winner was Dwight D. Eisenhower, and his margin, 10.5 percent of the 62 million votes cast, was substantial enough—it was exceeded in only 7 of the 17 elections from 1888 to 1952—so that, to be useful, a method of forecasting presidential elections would have to be able to detect it.

## EXAMPLE 84C   OMISSIONS

Forecasters of business (and other phenomena) sometimes relate successes which, if they were the only forecasts made, would be strong evidence of prognosticating ability, but which in fact are selected from a long list of forecasts that, in the aggregate, are unimpressive despite occasional successes. In general, conclusions that would be correct if the figures cited were the only ones relevant, may be seriously qualified or even reversed if the complete data are taken into account.

---

22. *Science*, Vol. 120 (1954), p. 935.

*3.11 Misleading Charts*

The principle involved can be brought out by a simple example. Suppose you were told that a certain coin had been tossed 10 times and had showed heads each time. You would be fairly confident that the coin would continue to show more heads than tails, for the probability is a trifle less than 1 in 500 that a fair coin would show the same side on 10 tosses. But now suppose the fact was that 1,000 coins had each been tossed 10 times, and this one selected afterwards because it had shown heads all 10 times. If you had known about the other 999 coins, and the method of selecting the one about which you were told, you would not have felt much confidence that the coin would continue to show more heads than tails; for six times out of seven, on the average, sets of 1,000 fair coins tossed 10 times each, would have one coin that falls the same way all 10 times.

As the courts have recognized since time immemorial, it is not sufficient to draw conclusions from the truth and nothing but the truth; it must also be the whole truth.

## 3.11
## MISUSES DUE TO MISLEADING CHARTS

Example 85   Details Magnified Out of Context

Fig. 85A shows the course of consumer prices from 1933 to 1953. The rise at the end of the chart can be made quite startling on a chart confined only to it, as in Fig. 85B. The omission of the zero line leaves room to magnify the vertical scale, in this illustration by a factor of 10. The horizontal scale has also been magnified, but only by a factor of 3, so the rise is $3\frac{1}{3}$ times as steep in Fig. 85B as in Fig. 85A.

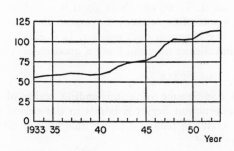

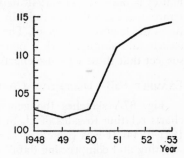

FIG. 85A.  Consumer  price  index,  1933–53.  (1947–49 = 100.)

FIG. 85B.   Detail of Fig. 85A.

Source: U. S. Bureau of Labor Statistics, *Monthly Labor Review.*

**Misuses of Statistics**

EXAMPLE 86A    PERSPECTIVE

Perspective diagrams are hard to interpret. Fig. 86 is supposed to depict the change in the national debt from about 1860 to the present time.[23] This presentation grossly distorts the amplitude of the recent fluctuations. The visual impression is that the debt in 1948 is about $10\frac{1}{2}$ times the debt

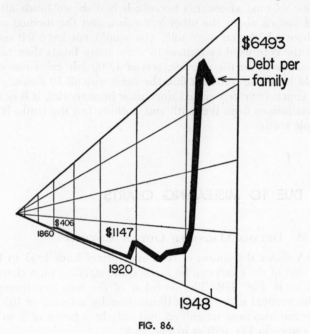

$6493
Debt per
← family

$406
1860

$1147

1920

1948

FIG. 86.

of 1920, but the ratio between 1948 and 1920 computed from the debt figures is only $5\frac{2}{3}$. The 1948 figure appears to be about 63 times the 1860 figure, but actually was only 16 times it. Thus, the chart gives two to four times the legitimate impact. The purpose of any chart is to present the facts clearly and simply. Such a perspective diagram does neither. It is easy to suspect that those who use charts that distort may not have a good case.

EXAMPLE 86B    DECEPTIVE CHANGES OF SCALE

Fig. 87A sketches the general appearance of a misleading series of charts relating to sales of U. S. Government Series E bonds in the period 1941–1944. It was presented as a model of what "a lively imagination in selecting and compressing data" can do.[24]

23. This is the cover design used by the Committee on Public Debt Policy for its *National Debt Series*, issued between World War II and the Korean War.

24. J. A. Livingston, "Charts Should Tell A Story," *Journal of the American Statistical Association*, Vol. 40 (1945), pp. 342–350.

### 3.11 Misleading Charts

A quick glance at the curves and the titles raises the question of how the volume of outstanding bonds can be gaining at all if, as is suggested by the first two diagrams, redemptions are running as high as sales. Examination of the scales, which are shown on the original charts, reveals, however, that the redemption scale is more than three times as large as the sales scale. The

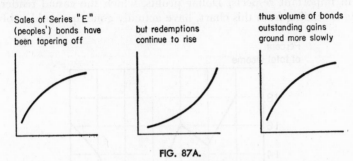

Sales of Series "E" (peoples') bonds have been tapering off

but redemptions continue to rise

thus volume of bonds outstanding gains ground more slowly

**FIG. 87A.**

curves all end at about the same level, but this represents 1,000 million dollars per month on the sales scale and only 300 million dollars per month on the redemptions scale. And the third scale is totally different, a logarithmic scale. If the data on volume outstanding are plotted on an arithmetic scale the resulting curve looks like Fig. 87B, the final level being about 25 billion

**FIG. 87B.**

dollars. The absolute amount added to the volume of outstanding bonds is greater each month than the month before. As a *percentage* of the total outstanding in the previous month, however, the increase in volume of outstanding bonds is decreasing. The logarithmic scale shows the latter—the percentage rate of growth. On a logarithmic scale, equal rises on the vertical axis represent equal *percentage* increases.

EXAMPLE 87  CARELESS SCALES AND LABELS

The following graph was intended to show that the steel industry was in much worse financial shape than was generally realized.[25] The point may or may not be well taken, but the chart exaggerates the effect of taxes upon profits.

---

25. *Steelways* (published by the American Iron and Steel Institute), March, 1948, pp. 6–7.

In the first place, the base line is not zero, so a false impression is made of the amount of the profits, especially for 1945 when profits fall to the base line.

The title of the chart is "Profits after taxes," but the vertical axis measures profits after taxes not directly, but as a percent of total income, which differs in important respects. Dollar profits, which the casual reader may think he is reading from this chart, have actually gone up considerably.

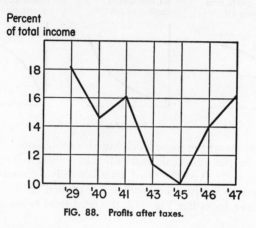

FIG. 88.   Profits after taxes.

A final criticism is the manner in which the years are marked off on the horizontal axis. Certain years are simply omitted. The same distance is used to represent a one-year interval in three cases, a two-year interval in two other cases, and an eleven-year interval in one case. If the years 1929 to 1940 had been included year by year, an entirely different impression might have been created because of the tremendous rise in profits since the early thirties.

# 3.12
# CONCLUSION

Our examples illustrate errors in the use of the basic definitions underlying an investigation, in the application of those definitions in the measurement or classification of individual people or objects, and in the selection of individuals for measurement. They also illustrate errors in the use of the resulting data, by making comparisons improperly, by failing to allow for such indirect causes of differences as heterogeneity of groups with regard to important variables, by disregarding the variability that is usually present even under apparently constant conditions, by technical errors, and by misleading verbal or graphical presentations.

The misinterpretations involved in most of the examples are simple and obvious. Some readers find them amusing; many are. Some find them distressing; many are that, too. Some find them irritating. Others find us irritating, for what they consider a negativistic, quibbling, and pettifogging attitude. Some become negativistic, quibbling pettifoggers themselves, conjuring up imaginary fallacies or exaggerating the consequences of real ones. Some readers are dismayed at the lack of systematic criteria which will automatically and authoritatively classify any particular use of statistics as sound or unsound.

Some readers, and we hope you are among them, recognize that each illustration is an example of a general type of fallacy. They realize that, easy as it may be to recognize and cope with such fallacies when they are exhibited caged here, it is not always easy to recognize them or cope with them in their native habitats. These people ponder each example carefully, however amusing, distressing, or irritating they may find it. They appreciate, too, that all these misuses, far from showing statistics to be useless, exemplify its usefulness; for each misuse represents an opportunity to find a sound basis for practical action or to obtain valid general knowledge.

For the statistician, not only death and taxes but also statistical fallacies are unavoidable. With skill, common sense, patience, and above all (as we said at the beginning of this chapter) objectivity, their frequency can be reduced and their effects minimized. But eternal vigilance is the price of freedom from serious statistical blunders.

## DO IT YOURSELF

The examples discussed in this chapter and the preceding one illustrate good and bad uses of statistics. It would be useful to analyze some of these in more detail. Our comments have frequently emphasized only one or two things when much more might profitably have been said. Similar examples are easy to find in newspapers and magazines (especially letters to the editor), advertisements, speeches, and technical, scientific, and scholarly literature. Often the most interesting illustrations are buried in what appears to be nonstatistical writing, and it takes practice even to recognize that these pertain to statistics. It is especially instructive to look for such examples and to analyze them.

In this section we present additional examples. Most, but *not all*, involve some misuse of statistics, and all are worth analyzing. You should write out your conclusions more fully than we have done for the illustrations in this chapter.

### EXAMPLE 90A

JOBS TO BE FEWER FOR 1950 SENIORS. June graduates of the nation's colleges and universities are faced with a sharp decline in employment prospects with large corporations, Dr. Frank S. Endicott, director of Northwestern University's Bureau of Placement has disclosed.

Reporting the results of a survey of 169 well-known companies that regularly contact the country's colleges and universities for graduating seniors, Dr. Endicott announced a decrease of about 25 percent in personnel requirements for new college graduates for 1950.

The large industrial concerns reporting, which in 1949 hired 8,321 college men and women, expect to take only 6,270 graduates in 1950. . . . [26]

### EXAMPLE 90B

Mr. Bennett: The number of men in prison today, if you put together all the state prisons and all the federal prisons, is less than it was just prior to the war. That means, it seems to me, that we are doing a better job of law enforcement all the way through.

Mr. Levi: Would you say that that means that crime is on the decrease?

Mr. Bennett: If you can measure crime in terms of the men who go to prison, I think it is. . . . [27]

### EXAMPLE 90C

The dental profession is understaffed; only 22 percent of the population receives dental care within a year.

### EXAMPLE 90D

In spite of all our effort at education, the American people are becoming more ignorant. . . . Women college graduates, aged 45–49, have had barely half enough children to replace their parents; high school graduates, same age, four-fifths enough children for replacement . . . BUT women, same age, with *fourth grade education or less, have had nearly twice the number necessary to replace the parents*. Fourth-graders are practically doubling their numbers every generation; college women are dying out 50 percent every generation.

### EXAMPLE 90E

Brilliant students more frequently become leaders, if we can accept listing in *Who's Who* as evidence of leadership. One out of 6 college students who won Phi Beta Kappa scholarship and leadership honors is now listed in *Who's Who*, about 50 times as high a rate as for non-honor students.

### EXAMPLE 90F

There were more civilian than military amputees during the war. During

---

26. *New York Times*, January 8, 1950.
27. *University of Chicago Roundtable*, No. 586, June 12, 1949, pp. 1–2.

the period of the war, 120,000 civilians suffered amputations, but only 18,000 military personnel.

EXAMPLE 91A

Persons involved in illicit (sexual) activities, each performance of which is punishable as a crime under the law . . . constitute more than 95 percent of the total male population. . . . [Therefore] only a relatively small proportion of the males who are sent to penal institutions for sex offenses have been involved in behavior which is materially different from the behavior of most of the males in the population.

EXAMPLE 91B

. . . it is the farmer, not the importuning salesman, who leads in the consumption of alcohol. Results of a survey of drinkers classed by occupation, which was published last week by the Keeley Institute of Dwight, Ill. . . . show that of 13,471 patients treated in this well-known rehabilitation center from 1930 through 1948, a total of 1,553 (11.5 percent) were farmers. Next in line came salesmen, merchants, mechanics, clerks, lawyers, foremen and managers, railroad men, doctors and manufacturers.

Since Keeley is located in the heart of the farm belt, it might be expected that its proportion of farmer patients would be unusually large. This is not the case, according to James H. Oughton, institute director. The patients in the survey were drawn from all over the world.[28]

EXAMPLE 91C

Grim Statistics. — If the Korean conflict continues at its present pace, an impressive statistic will be made some time this spring. One day an American soldier will fall in battle. He will be the 1,000,000th to die in our wars since the nation was born.

A few months later, another grim milestone will be reached. At that time, the Association of Casualty and Surety Companies estimates, the 1,000,000th American — motorist or pedestrian — will perish in a modern highway traffic accident. Our wars go back to 1776. The traffic death figure starts with 1900.

Neither figure makes pleasant reading. But war has been a dangerous business since nations first became civilized enough to have armies. Getting from one place to another, except for an occasional Columbus or Lindbergh, is not supposed to be a risk of life and limb.

The surprising parallel of these statistics points up vividly something we already know. Highway safety is a seven-day-a-week job, 52 weeks a year. It is a job for every motorist and pedestrian as well as for the officials who regulate traffic in these days of multiple horsepower.[29]

EXAMPLE 91D

The age-specific death rates [that is, death rates for specified age groups, obtained by dividing the number of deaths by the number of women in the age group] for breast cancer have remained constant for 35 years.

28. *Newsweek*, May 2, 1949, pp. 47–48.
29. *Chicago Daily News*, February 2, 1951, p. 12.

**Misuses of Statistics**

It is argued that this constancy shows that no advances have occurred in methods of treatment for this kind of cancer during this period.

EXAMPLE 92A

It is three times as dangerous to be a pedestrian while intoxicated as to be a driver. This is shown by the fact that last year 13,943 intoxicated pedestrians were injured and only 4,399 intoxicated drivers.

EXAMPLE 92B

The following data refer to residential rent changes after the removal of rent control in seven cities. They comprise all rental dwellings, including those under lease and those removed from control prior to the ending of all rent control in the city. There was a controversy between real estate men and the rent administrator as to which set of figures best represented the changes in rent following the removal of rent control. The rent administrator liked the percentage increase for units reporting increases, while the real estate men advocated the percentage increase for all rental units.

| City | Survey Period (1949) | Percent Increase for All Rental Units | Units Reporting Increases | | |
|------|---------------------|------------------------------------|---------------------------|---|---|
| | | | Percent of all units in city | Average dollar amount | Average percent increase |
| Dallas | Apr 15–Nov 15 | 16.7 | 59 | 13.96 | 36.1 |
| Knoxville | May 15–Nov 15 | 13.9 | 57 | 6.83 | 25.7 |
| Jacksonville | Jun 15–Nov 15 | 10.9 | 52 | 6.59 | 25.8 |
| Houston | Aug 15–Nov 15 | 10.5 | 33 | 12.03 | 40.0 |
| Topeka | Jul 15–Nov 15 | 9.0 | 36 | 9.08 | 30.2 |
| Spokane | Jul 15–Nov 15 | 8.2 | 46 | 5.71 | 19.0 |
| Salt Lake City | Jun 15–Nov 15 | 6.6 | 44 | 6.46 | 16.0 |

EXAMPLE 92C

An automobile manufacturer asserted in an advertisement that 50 percent of all the cars it had ever made were still in existence, and claimed this as evidence of durability.

EXAMPLE 92D

I have always been helplessly captivated by research organizations. They can prove almost any proposition you care to present them.

That two and two equal four, for instance. Or that two and two doesn't equal four. Or, if pressed hard, that there are no such figures.

Lately, I've been bewitched by a couple of full-page ads run in newspapers on successive days by NBC and CBS, each proving that it has the largest audience in radio. NBC—according to NBC—has 3,000,000 more families listening to it than the second network in the daytime, 4,870,000 more families at night.

My own rather suspect arithmetic figures this as roughly 12 percent more listeners than CBS daytime, 14 percent more at night.

**Do It Yourself**

The day after this claim, CBS, trumpeting like an enraged elephant, charged forth with its figures. CBS — says CBS — has an audience 29 percent higher than the "second-place network" in the daytime, 32 percent higher at night.

NBC quotes Broadcast Measurement Bureau as the source for its figures; CBS doesn't quote any source but, since it is one of the most research-happy organizations in America, I'm sure the network is up to its hips in surveys, polls, and statistics — all of them unassailable.[30]

EXAMPLE 93A

The frequency of divorce for couples with children is only about one half of that for childless couples.

The more children you have, the less likelihood there is that you will ever be divorced.

This is an idea that many have suspected to be true. But it remained for the Metropolitan Life Insurance Company to prove it with the statistics printed below. The possibility that a childless couple will be divorced is about twice as great as it would be if they had two or more dependent children. This holds true not only for young couples but for those married 20 years.

Divorce in families where there are children is, however, more common than is generally realized, according to the Metropolitan's statisticians. For, of the 421,000 absolute decrees granted in a recent year's time, 42 percent were to couples with children.

| | |
|---|---|
| Marriages with no children have | 15.3 divorces per 1,000 marriages |
| Marriages with 1 child have | 11.6 divorces per 1,000 marriages |
| Marriages with 2 children have | 7.6 divorces per 1,000 marriages |
| Marriages with 3 children have | 6.5 divorces per 1,000 marriages |
| Marriages with 4 or more children have | 4.6 divorces per 1,000 marriages[31] |

EXAMPLE 93B

. . . in spite of the higher charges made necessary by more expensive facilities and rising labor and maintenance costs, it is cheaper, on the average, says Mr. Hatfield, to go to a hospital today than it was in the 1920's. This is because improved diagnostic and treatment techniques, including prompter surgery and the germ-killing "miracle drugs," are getting people out of bed and back to work sooner. The average patient in a general hospital now stays only seven and eight-tenths days, as compared to twenty or twenty-five days thirty years ago. The charge per hospital day has gone up, but the charge per average hospital stay has gone down.[32]

EXAMPLE 93C

During the period of rent control the following information was submitted by a landlord in support of a petition for a rent increase in a six-family apartment building.

---

30. John Crosby, *Chicago Daily News*, April 28, 1950.
31. "More Children Equal Less Divorce," *Look*, February 13, 1951, p. 80.
32. Steven M. Spencer, "They Didn't Know What They Started," *Saturday Evening Post*, February 24, 1951.

INCOME AT 1275 LEXERD ROAD
Comparative Figures for June 30, 1942 and June 30, 1950

|  | June 30, 1942 | June 30, 1950 | Percentage Increase |
|---|---|---|---|
| Receipts | $3,956.76 | $4,715.76 | 12 percent |

EXPENSE INFORMATION AT 1275 LEXERD ROAD
Comparative Figures for June 30, 1942 and June 30, 1950

|  | June 30, 1942 | June 30, 1950 | Percentage Increase |
|---|---|---|---|
| Janitor's Wages and Materials | $363.62 | $507.91 | 40 percent |
| Fuel | 461.96 | 796.05 | 73 percent |
| Cost of Ash Hauling | 27.00 | 35.00 | 29 percent |
| Legal Fees and Management | 96.36 | 189.92 | 97 percent |
| Insurances | 36.34 | 71.74 | 97 percent |
| Taxes | 560.03 | 696.99 | 24 percent |

Because the expenses of this building have increased more than double the percentage of the rentals, we request that the increase shown on the face of this petition be allowed to reduce the present inequitable ratio of expense and income.

EXAMPLE 94A

A common medical belief states that after a certain operation it is dangerous for women to have children. An interested doctor studied the records for a group of women who had been through this operation. For each woman the records showed the number of children borne and the number of years the woman had lived after the operation. The doctor discovered that the more children a woman bore after the operation the longer, on the average, she survived the operation. He concluded that the traditional medical belief was the reverse of the true situation.

EXAMPLE 94B

After the New Hampshire preferential primary in 1952 it was reported that Senator Taft had received a slightly higher percentage of the total vote in a group of 17 cities in which he had not campaigned personally than in a group of 15 cities in which he had. One newspaper writer jibed that "Senator Taft should have stayed at home."

EXAMPLE 94C

One of the simplest and most practical systems of managing investment funds and building capital is dollar averaging! It is a systematic savings plan, ideally suited to the individual with a steady income who can see his way clear to investing a fixed amount regularly over a period of years. Strictly an unemotional approach, it avoids the wide margin for error involved in trying to gear purchases and sales to the swings of the general market.

**Do It Yourself**

Dollar averaging merely involves the buying of equal dollar amounts of stock at regular intervals. . . . The simple arithmetic principle involved is that the same amount of money will buy more shares when the price is low than when the price is high. As a consequence, the average cost of purchases will be lower than the average price no matter what the various prices may be, although this does not rule out the possibility of a paper loss at some intermediate phase of the program. . . .

Only sound stocks with reasonably assured prospects of long-term progress should be chosen. . . . Dollar averaging can be started with one stock but the long-range objective should be to obtain adequate diversification. While the emphasis is on long-term holding, stocks acquired under the program should be carefully watched. If there is any evidence of a weakness in fundamental position, a switch of holdings should be undertaken.[33]

EXAMPLE 95A

CONFIRMATION: Periscope, July 25: "Best-informed sources expect the Communists to poll a maximum 6 percent vote at the August election in the Western zone of Germany." Actual result, August 14: The Communists received 5.7 percent.[34]

EXAMPLE 95B

In an attempt to minimize the seriousness of the lynching problem, it was recently claimed that more whites than Negroes have been lynched regularly over the years in the United States. (The claim, incidentally, is actually false, but the argument is worth examining anyway.)

EXAMPLE 95C

In a study of unemployment it was found that in 1930 and 1940 there were more unemployed male workers in the age group 20–24 than in any five-year age group between 40 and 59. It was then concluded that older workers are not strongly discriminated against in present day industry.

EXAMPLE 95D

U. S. STATISTICS PROVE IT . . . . . MARRY AND LIVE LONGER: There's no question about it: Married people live longer than single people; and people who were once married live longer than people who were never married.

The Public Health Service's National Office of Vital Statistics has proved this to be a fact. Basing his figures on the 1950 census and mortality rates for 1949, 1950 and 1951, statistician Dewey Shurtleff showed that deaths among bachelors were almost two-thirds greater than among husbands.

Among divorced men and widowers, the rate was half again more.[35]

---

33. Quoted from *Standard and Poor's* by Alger Perrill and Company of Chicago, January 14, 1952.

34. *Newsweek*, August 29, 1949, p. 5.

35. *Chicago Daily News*, April 8, 1955.

EXAMPLE 96A

In an article in the *New York Herald Tribune* in May, 1954, the Alsops said that if the probability of knocking down an attacking airplane were 0.15 at each of five defense barriers (radar plus fighter planes), and if an attacking plane had to pass all five barriers to get to the target, then the probability of knocking down the plane before it passed all five barriers would be 0.75.

EXAMPLE 96B

Experience is not necessarily the best teacher for making safe drivers, a traffic expert said here Tuesday.

In fact, a Minnesota Study has shown that 61 percent of those involved in accidents have spent more than 10 years behind the wheel.

The figures were cited by Spaulding Southhall, of the National Chicago Safety Conference.

The study also showed that 21 percent of those involved in accidents had six to ten years' driving experience, and 17 percent, one to five.

"Apparently drivers become more complacent about their driving as the years go by," Southhall said.

"As a consequence their records become worse."

He said that teen-agers continue to have the worst accident record, but that it was becoming obvious experience alone would not improve their driving.[36]

EXAMPLE 96C

Most doctors know that visitors often do more to stir up hospital patients than to soothe them. But the doctor's own ward rounds can have the same effect, sometimes with fatal results, reported Finnish Doctor Klaus A. J. Jarvinen in the *British Medical Journal*.

Studying the histories of 39 Helsinki hospital patients who died of coronary occlusion after stays of seven to 42 days, Dr. Jarvinen discovered that six of them, subject to severe emotional stress, had died during or after a physician's visit. Among the cases:

An accountant, 58, came to the hospital 21 days after an attack of angina pectoris. He seemed in satisfactory condition until the 16th day in the hospital. The head physician was making his round; as the doctor drew closer, the patient became nauseated, suffered a severe attack and died within two hours.

After suffering chest pains during a tantrum, a female post-office clerk, 68, was admitted for treatment. In the ward, she grew excited over trivialities. After nine days, when the doctor approached, she became restless. Asked how she felt, she tried to answer, and died on the spot. . . .[37]

EXAMPLE 96D

The following quotation compares fire losses in Miami before and after the introduction of the Layman method (based on spray of a fine mist) of fire fighting.

---

36. *Chicago Daily News*, June 2, 1953.
37. *Time*, February 21, 1955, p. 37.

*Do It Yourself*

Miami's experience, which is impressive, involves a slum area containing scores of one-story tinderbox dwellings, ten feet apart or less. Beginning in 1948, all fires in this district were handled by the Layman method. In 1946 and 1947 this district had 99 fires, with a total fire loss of $474,093. During 1948 and 1949, it had 107 fires with loss of only $67,737.[38]

## EXAMPLE 97A

A firm constructing capital equipment compared costs estimated in advance with those realized. They separated jobs done by firm bid from those on cost-plus, or for their own account to be sold later. The estimated cost as a percentage of actual cost was lower, on the average, on fixed-price than on the other jobs. They concluded that the fixed price leads to poorer efficiency.

## EXAMPLE 97B

We understand that much concern is felt in academic circles in the United States over the fact that many American universities find themselves obliged for financial reasons to devote an undue share of their research activities to the fulfillment of government contracts. In this country, the danger seems not to be immediate. . . . The proportion of the government expenditure on scientific research which goes to the universities is relatively small. . . . In the years ended 31st July, 1950 and 1951 the sums so received (including a certain amount of capital) amount to £1.6m. and £1.7m. respectively. These sums are a relatively small proportion of the total expenditure of the universities in these years, £22m. and £24m. respectively.[39]

## EXAMPLE 97C

AVERAGE TEACHER GETTING YOUNGER. Grand Rapids, Mich. (UP) — The average Grand Rapids school teacher is getting younger, a survey showed today. The board of education said the average Grand Rapids teacher was 42 years old this year compared to a 43 year average in 1952.[40]

## EXAMPLE 97D

Under the 1943 amendment of the Indeterminate Sentence Law, judges are permitted to set maximum and minimum sentences within the statutory minimum and maximum sentences. The result has been that many judges give sentences that do not provide sufficient spread between the minimum and the maximum to permit a parole period. Consequently, over half the men are released from the penitentiary without parole supervision. A study made by the parole board revealed that the commission of new crimes is more frequent among those released without than among those with parole supervision, which suggests either the value of parole supervision or the poor judgment of judges in selecting offenders whom they decide either do not need, or will not benefit from, parole

---

38. *National Safety News,* reprinted in *Reader's Digest,* April, 1954.
39. University Grants Committee, *University Development: Report on the Years 1947 to 1952* (London: Her Majesty's Stationery Office, July, 1953), Cmd. 8875, p. 43.
40. *Manistee [Michigan] News-Advocate,* July 7, 1955.

supervision. A partial test of the alternative explanations can be made by utilizing data available in the records of the parole board for a comparison of the probabilities of committing new crimes (according to expectancy tables) of the men discharged into the community without parole and of those with parole.

EXAMPLE 98A

The following notice was enclosed with an appeal for funds:

Your gift is a voluntary, personal matter. No one should tell you how much to give. But most donors, desiring to give a fair share, want to know, "What are others giving?" Below is a factual answer secured through a survey of gifts last year in relation to income. With the help of numerous payroll departments, it is possible to show *Actual Gifts* without revealing names or salaries. Here is what the most generous 40% of donors in various income groups actually give to the UNITED CRUSADE.

TOP FORTY PERCENT OF DONORS (Average Gift)

| Annual Salary or Wages | Weekly[a] | Monthly[a] | Total Gift (for Year) |
|---|---|---|---|
| $ 3,000 | $ .20 | $ .95 | $ 11.52 |
| 4,000 | .30 | 1.25 | 14.95 |
| 5,000 | .40 | 1.65 | 19.82 |
| 6,000 | .70 | 3.00 | 35.90 |
| 9,000 | .90 | 3.90 | 46.99 |
| 12,000 | 1.80 | 7.80 | 93.86 |
| 15,000 | 2.20 | 9.50 | 114.22 |
| 20,000 | 5.30 | 23.00 | 276.08 |

[a] Rounded to nearest five cents. Those with larger incomes give proportionately higher amounts.

EXAMPLE 98B

PER CAPITA INCOME IN U. S. UP 1% IN 1954. Washington, Sept. 25 (AP) — Total national personal income in the United States rose by 1 percent last year, the Commerce Department announced today.

In its annual study of American incomes, the department reported that in 1954 personal income nationally rose by $2,000,000,000 to a new high of $285,368,000,000.

But the per capita income declined slightly because the population increase — 2,823,000 persons for the year — was at a faster rate than the jump in spending money in people's hands. . . . [41]

EXAMPLE 98C

Patents are of little value since the Supreme Court invalidates most of the patents that come before it.

EXAMPLE 98D

Is it safe to take the children swimming? Last summer New York City investigators queried all polio victims on their activities for the month preceding

---

41. *San Francisco Chronicle*, September 26, 1955, p. 12.

their illness. These were the results: Fifty-five per cent had not been swimming at all. Twenty-one per cent had bathed in safe and approved city waters. Twenty-three per cent had bathed outside city limits, but mostly at beaches with high sanitary standards. Less than one per cent of the polio victims had bathed in polluted waters.[42]

42. *This Week*, June 14, 1950.

# Chapter 4

# *Basic Ideas*

## 4.1
## INTRODUCTION

The following four examples illustrate a problem of interpretation which was not explicitly brought out in Chaps. 2 and 3.

EXAMPLE 100A  FAMILY INCOME, 1952

The Bureau of the Census reported that in 1952 one-third of the families in the nation had money incomes of less than $3,000, and one-third had money incomes above $4,500. A critic of the report wrote:

> Only 25,000 families were interviewed. . . . The main fault of the report is the limited scope of its available figures. 25,000 families are not a sound representation for the 46,000,000 families in the U. S. . . .

EXAMPLE 100B  RETAIL DRUG STORES

A student wrote as follows about a statistical study of retail druggists:

> It is based upon voluntary financial returns of 1,378 individual drug stores. The smallness of this number in comparison to the 55,796 drug stores shown in the 1948 Census of Business seems to invalidate this survey. To make broad generalizations about 56 thousand stores on the basis of data for only about 14 hundred stores is quite unjustifiable.

EXAMPLE 100C  CIGARETTE SALES

A restaurant attempted to evaluate the effect on its cigarette business of changing from sales by the cashier to sales by a coin-operated vending machine. The number of packages sold during the first month of machine sales was 51 percent less than during the last month of cashier sales. As a basis of comparison, sales for the same two months in two comparable restaurants were used. These showed decreases of 15 percent and 3 percent, respectively. As a result of this comparison, it was concluded that the "installation of the cigarette vending machine was detrimental to sales."[1]

---

1. Richard R. Still, "The Effect of an Automatic Vending Machine Installation on Cigarette Sales," *Journal of Marketing*, Vol. 17 (1952), pp. 61–63.

**100**

Example 101   Railroad Telegrapher

During the presidential campaign of 1952, a prominent speaker said that since the period before World War II the cost of living for a particular railroad telegrapher had increased more than his income. It was implied that the same was true for the country as a whole.

In the first two examples, samples of 25,000 and 1,378 were considered too small for "broad generalizations." In the second two examples, samples of 3 and 1 were considered large enough. Which are right? This problem cannot be analyzed as readily by common sense as could the examples of Chaps. 2 and 3. It is true that a sample of 25,000 might be inadequate for some purposes. It is also true that very small samples might be adequate for some purposes—a single case showed quite conclusively that atomic bombs will sometimes explode. But common sense is a poor guide in determining how much evidence should be assembled to answer a given question, or whether the evidence actually presented is enough to justify conclusions someone has drawn.

The problem thus posed is a purely statistical one, the problem of *sampling*. A large part of this book will be concerned with this problem and its innumerable ramifications. In this chapter we shall introduce a few ideas that are basic to what comes later. At the risk of oversimplifying the idea of statistical sampling, we shall deliberately ignore almost all complications and ramifications that arise in practice, and concentrate on the heart of the matter. Later chapters will introduce some variations and qualifications, but they will only elaborate, not revise, the ideas of this chapter.

## 4.2
## SAMPLES AND POPULATIONS

The two most fundamental concepts of statistics are those of a *sample* and a *population*.

A *sample* is often referred to as "the data" or "the observations": numbers that have been observed. The *population*, on the other hand, is the totality of all possible observations of the same kind. In Example 100A, the sample consisted of income figures for 25,000 families. The population consisted of all the numbers that would be obtained by selecting indefinitely many families and measuring their incomes by the same methods that were used to select these 25,000 families and measure their incomes. Since in 1952 there were 46 million families in the country, it might seem that the population contains 46 million numbers. Actually, it contains many more, for the process by which a number is obtained for a given family might have yielded different numbers for that same family, depending on the interviewer used, the member of the family interviewed, the time of the interview, and so forth. The population in this example can be regarded as infinite.

A single population can give rise to many different samples; thus, the population is stable but samples vary. A central problem in statistics is to determine what generalizations about the population can be drawn from the one particular sample which is actually available in a practical problem.

We shall try to bring out some of these basic ideas by describing a sampling demonstration in which many samples were drawn from a single population.

## 4.3
## SAMPLING DEMONSTRATIONS

### 4.3.1 Apparatus and Method

A closed box containing an unknown number of red and green beads was used. The bottom of the box is a sliding panel with 20 depressions, into each of which one bead falls. This panel can be slid out of the box by pushing it with another panel which takes its place and keeps the remaining beads from escaping from the box. In this way, a sample of 20 beads is obtained each time a panel is removed.

The procedure in the demonstration was to mix the beads by shaking the box well, take a sample of 20 by sliding out the panel, record the number of red beads, and put the sample beads back into the box. This procedure was repeated 49 times, producing 50 samples in all.

The purpose of the demonstration was to illustrate the kind of results to be expected if repeated samples were drawn with the same initial conditions. Since the beads from one sample were returned to the box before another sample was drawn, the population remained the same. It was as if one person studied the population of beads by drawing a single sample of 20 and counting the number of red beads, a second person studied the same population, and altogether 50 people studied it independently by exactly the same methods of selection and measurement ("measurement" in this example being simply a matter of classifying each bead as red or not red and then counting the number of red beads in the panel). Had we not replaced the beads after each sample of 20, the contents of the box would have changed every time a sample was drawn. We wanted each sample to arise in the same way it would have in a practical situation in which we were taking only one sample.

The problem of the beads in the box is statistically equivalent to innumerable practical problems. For example, suppose there is a group of 700 income tax returns of which an unknown proportion are incorrect. From this population we might take a sample of 75 returns, audit them, count the number that are incorrect by more than a specified amount, compute the proportion incorrect in the sample, and apply this figure to the whole 700. Or suppose there is a population of 1,000 hospital patients with the same illness, of which an unknown proportion will respond favorably to a certain treatment. Or, again, suppose a meteorologist interested in rain-making wants to estimate the proportion of clouds of a given type which, under specified conditions, would produce rain within 20 minutes after seeding by a certain method. All of these situations, and many more, reduce to this: From a population in which an unknown proportion of the items have a characteristic in which we are interested, a sample is drawn and the proportion of the sample having that characteristic is determined. On the basis of what has been learned from the sample, a decision is made concerning the entire population. The consequences of this decision depend on what proportion of the population has the characteristic whose frequency in the sample was computed. The consequence of treating or not treating patients in a certain way, for example, depends on the proportion in the population of patients who will respond favorably to the treatment.

Let us introduce some symbols to represent the various quantities described in the preceding paragraph. This will make our further discussion briefer, more precise, and more general. It will also serve as an introduction—a gentle one, we promise—to the use of simple mathematical notation for statistical ideas.

The number of individuals in the population may be represented by $N$. For the case of the box of beads, $N$ is unknown, but much larger than 20; for the income tax example $N = 700$; for the medical example $N = 1,000$; and for the cloud-seeding example $N = \infty$. This last, read "$N$ equals infinity," means that no definite number, however large, will be as large as the number of possible reactions to seeding of all the clouds of the given type that ever have existed or ever will exist. The proportion of the items or individuals in the population that have the characteristic in which we are interested—redness for the beads, errors for the tax returns, and so on—is an unknown number between 0 and 1, inclusive; we may denote it by $P$. If the numerical value of $P$ were known, it would determine the proper decision in the matter at hand and there would be no statistical problem and no need for a sample. Since $P$ is not known, we draw a sample and base our decision on that. Let $n$ represent the size of the sample, that is, the number of items from the population which are actually observed. For the beads, $n = 20$; for the tax returns, $n = 75$; and for the other examples the value of $n$ was not stated, but would be known. Let $X$ represent the number of items in the sample which have the characteristic in which we are interested. Finally, let $p$ represent the proportion of the individuals in the sample having the characteristic in which we are interested; that is, $p = X/n$. For example, if 19 red beads are found in a sample of 20 from a population of 3,000, then $N = 3{,}000$, $P$ is unknown, $n = 20$, $X = 19$, and $p = 19/20 = 0.95$.

*Always distinguish clearly between P, the proportion in the population, and p, the proportion in the sample.* The consequences of the decision depend on $P$, and if it were known we would know what decision to make. When $P$ is unknown, we face the statistical problem of basing our decision on the incomplete information about $P$ given by $p$.

For demonstration purposes, the box has several advantages over most practical situations. It is compact, and many samples can be taken in a short time. Furthermore, there is (except for the color blind) no trouble in distinguishing between a red bead and a green bead, but it may be difficult or time-consuming to distinguish a correct tax return from an incorrect one, a patient who responds favorably from one who does not, or even a cloud that rains from one that does not.

The demonstration which we will describe shows that samples vary, but it also shows that random samples vary according to a regular pattern. It would take many more than 50 samples to disclose the pattern perfectly, but 50 will suffice to show the general outline.

### 4.3.2 Preliminary Sample

Typically, in practical work we have only one sample. So we first examined one sample of 20 beads and considered it by itself, as if it were to be the only sample. In this preliminary sample, the number of red beads, $X$, was 13. Thus, the proportion $p$ in the sample was $13/20 = 0.65$; that is, the sample contained 65 percent red beads.

### 4.3 Sampling Demonstrations

With only this information available about the contents of the box, some might conclude that there are exactly 65 percent red beads in the whole box. Others might say that a sample of 20 is too small to tell anything about the proportion in the box. Both of these extreme positions are wrong. Clearly there is no basis for thinking that the sample is a perfect miniature replica of the population. If the true proportion, $P$, were 0.628, for example, 0.65 would be as near as the sample proportion, $p$, could possibly come in a sample of 20; but chance could lead to such results as 0.55, 0.70, and so on, even if $P$ were 0.628. But, just as clearly, *something* has been learned from the sample. The possibilities that $P = 0$ or $P = 1$ have been ruled out completely. The possibility that $P = 0.1$ is not ruled out completely, but probably no one will take that possibility seriously, for if $P$ were really as small as 0.1, the probability of getting as many as 13 red beads in a sample of 20 would be extremely small—it would happen, on the average, less than once in a hundred million times. Clearly, by this type of reasoning, some ideas about $P$ can be obtained from $p$.

### 4.3.3 Fifty Samples from Population I

The preliminary sample of 20 contained 13 red beads. But as we have seen, this is partly accidental. We have to judge it against the variation in the samples that would occur if the same conditions were

TABLE 105

RESULTS OF 50 SAMPLES FROM POPULATION I
(20 Beads per Sample; Total of Sample Sizes, 1,000)

| Sample Number | Red Beads | Sample Number | Red Beads | Sample Number | Red Beads |
|---|---|---|---|---|---|
| 1 | 12 | 18 | 12 | 35 | 12 |
| 2 | 12 | 19 | 8 | 36 | 11 |
| 3 | 9 | 20 | 12 | 37 | 10 |
| 4 | 12 | 21 | 9 | 38 | 10 |
| 5 | 9 | 22 | 11 | 39 | 9 |
| 6 | 9 | 23 | 9 | 40 | 10 |
| 7 | 11 | 24 | 10 | 41 | 11 |
| 8 | 8 | 25 | 14 | 42 | 9 |
| 9 | 11 | 26 | 11 | 43 | 11 |
| 10 | 8 | 27 | 12 | 44 | 12 |
| 11 | 11 | 28 | 14 | 45 | 12 |
| 12 | 10 | 29 | 10 | 46 | 12 |
| 13 | 7 | 30 | 14 | 47 | 13 |
| 14 | 8 | 31 | 10 | 48 | 10 |
| 15 | 13 | 32 | 18 | 49 | 10 |
| 16 | 13 | 33 | 8 | 50 | 13 |
| 17 | 13 | 34 | 15 | Total | 548 |

repeated many times. Since this is a demonstration, rather than a real problem, we have an opportunity to find out something about "what might have been" if we "had it all to do over again." We did it all over again—not once, but 50 times. The results are shown in Table 105.

If these results are grouped in a *frequency distribution*, as in Table 106, there begins to emerge a general pattern of variation for samples of 20 from this particular population. The bulk of the samples had 9, 10, 11, or 12 red beads. The preliminary sample we talked of earlier, with 13 red beads, is now seen not to have been one of the most usual results, though it is not particularly unusual either. At any rate, we would have been foolish to draw any conclusions from that preliminary sample without taking into account "what might have been"—that is, without taking into account the pattern of variation among samples which is revealed (partially) by Table 106.

TABLE 106

FREQUENCY DISTRIBUTION OF SAMPLE
RESULTS, POPULATION I

| Number of Red Beads | Number of Samples |
|---|---|
| Less than 7 | 0 |
| 7 | 1 |
| 8 | 5 |
| 9 | 7 |
| 10 | 9 |
| 11 | 8 |
| 12 | 10 |
| 13 | 5 |
| 14 | 3 |
| 15 | 1 |
| 16 | 0 |
| 17 | 0 |
| 18 | 1 |
| More than 18 | 0 |
| Total | 50 |

*Source:* Table 105.

## 4.3.4 Fifty Samples from Population II

What appears in a sample depends on chance, as this demonstration reveals. But the pattern of chance variation in repeated samples depends on the population. This was illustrated by changing the contents of the box, thus forming a new population. The pattern of variability changes when the contents are changed. This fact, that the pattern of variability depends on the population, provides the

### 4.3 Sampling Demonstrations

possibility of making generalizations about a population from a random sample. The problem is one of stripping away the effects of chance, as far as possible, to see the reflection of the population in the sample.

The new set of beads was called Population II. The results of the 50 samples from it are given in Table 107A and the frequency distribution is shown in Table 107B.

TABLE 107A

FIFTY SAMPLES FROM POPULATION II

(20 Beads per Sample; Total of Sample Sizes, 1,000)

| Sample Number | Red Beads | Sample Number | Red Beads | Sample Number | Red Beads |
|---|---|---|---|---|---|
| 1 | 3 | 18 | 4 | 35 | 2 |
| 2 | 4 | 19 | 4 | 36 | 5 |
| 3 | 1 | 20 | 4 | 37 | 4 |
| 4 | 2 | 21 | 5 | 38 | 3 |
| 5 | 1 | 22 | 3 | 39 | 4 |
| 6 | 2 | 23 | 5 | 40 | 4 |
| 7 | 3 | 24 | 2 | 41 | 3 |
| 8 | 1 | 25 | 2 | 42 | 2 |
| 9 | 5 | 26 | 2 | 43 | 3 |
| 10 | 1 | 27 | 2 | 44 | 4 |
| 11 | 3 | 28 | 2 | 45 | 2 |
| 12 | 3 | 29 | 2 | 46 | 3 |
| 13 | 4 | 30 | 4 | 47 | 7 |
| 14 | 3 | 31 | 2 | 48 | 3 |
| 15 | 3 | 32 | 2 | 49 | 2 |
| 16 | 3 | 33 | 4 | 50 | 4 |
| 17 | 3 | 34 | 3 | Total | 152 |

TABLE 107B

FREQUENCY DISTRIBUTION OF SAMPLE RESULTS, POPULATION II

| Red Beads in Sample | Number of Samples |
|---|---|
| 0 | 0 |
| 1 | 4 |
| 2 | 14 |
| 3 | 15 |
| 4 | 12 |
| 5 | 4 |
| 6 | 0 |
| 7 | 1 |
| More than 7 | 0 |
| Total | 50 |

*Source:* Table 107A.

The pattern of variability for the samples differs between Populations I and II in several respects, of which two are conspicuous: the patterns cluster around different values, and the tightness of the clustering differs. The samples from Population I cluster mainly from 9 to 12 beads, with approximately equal numbers above and below 11; those from Population II cluster mainly from 2 to 4, with approximately equal numbers above and below 3. A span of 12 (7 through 18) is necessary to encompass all samples from Population I, while a span of 7 (1 through 7) is sufficient for Population II. These two differences, in the location and in the dispersion of the patterns, can be seen in Fig. 109 which displays the results of all the samples from both populations.

A chart like Fig. 109 is an excellent means of recording data as they occur. It shows vividly the variation from sample to sample, yet it also indicates the general level about which the variation occurs. A frequency distribution like Tables 106 and 107B can be prepared easily from a chart like Fig. 109; for example, looking across at a certain horizontal level, say the one representing samples with 10 red beads, we see quickly that there were 9 such samples from Population I. Furthermore, if there were relations among the successive samples, such as trends or cycles, these would be apt to catch the eye. Our method of sampling in this experiment included precautions to preclude such relations, except such as are due to chance —for example, the run of 6 consecutive twos in the Population II (samples 24–29). We shall have more to say about the practical uses of such charts in Sec. 4.8 and especially in Chap. 16.

### 4.3.5  Conclusions from the Demonstrations

From the results of the sampling demonstration, two important conclusions stand out:

(1) Sample results vary by chance.
(2) The pattern of chance variation depends on the population.

These two facts correspond with the basic problem and the basic tool of statistical analysis. The basic problem is that a sample is not a miniature replica of a population, so when decisions about a population are based on a sample it is necessary to make allowance for the role of chance in determining the characteristics of the particular sample that is available. The basic tool is knowledge of the patterns of sampling variability that result from various populations, and therefore of the probability of getting the observed sample from any

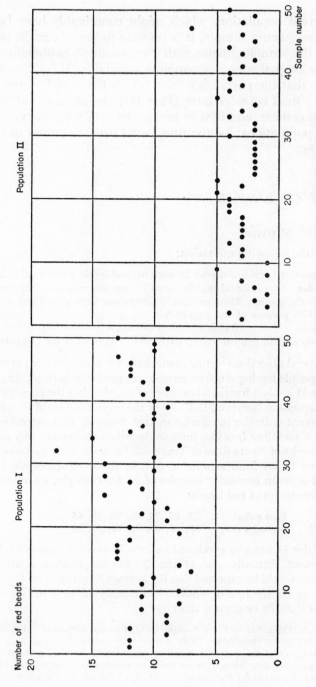

FIG. 109. Number of red beads per sample in 100 samples of 20 beads each. Source: Tables 105 and 107A.

of the different populations which might conceivably have been its source. Thus, given a sample, it is possible to say of certain populations that the sample might with "reasonable" probability have arisen in the normal course of sampling variability. It is "reasonable" to conclude that the population from which the sample came is one of these. We shall consider later (Part III) the meaning of "reasonable" in this context, and how to find patterns of sampling variability for various populations; for the time being we are content to convey general ideas.

## 4.4
## VARIABILITY OF SAMPLES

Example 110  Motives

Consider the following quotation:

> Interviews were performed with 20 persons, and motives extracted from the experience data. It was found that between 75 and 80 percent of these persons had a particular motive. Then another 20 interviews were added and still approximately 75 percent of the people had this particular motive. Additional interviews were added in blocks of 20 until 140 interviews had been taken. This particular motive still applied to approximately 75 percent of the 140 persons.[2]

There are several objections to this quotation—for example, the percent of a group of 20 people having a characteristic can not be in between 75 and 80, since this would imply a fractional number of people—but the point pertinent here is the impression conveyed that *each* of the seven samples of 20 contained about 75 percent with the particular motive. Suppose that three-fourths of the beads in a sampling box like the one described earlier are red, and that successive samples of 20 are drawn. There will be much more variation in the percent of red beads from sample to sample than the quotation implies. In two series of seven successive samples of 20, for example, we obtained the following percentages of red beads:

First series:    75, 85, 75, 60, 70, 70, 65
Second series:   55, 65, 85, 70, 60, 90, 80

Only two of our 14 samples produced exactly 75 percent, and only six from 70 to 80 percent. Actually, our 14 samples happen to show a little more variability than would be expected "on the average," but even on the average only about one sample in five would yield exactly 75 percent. Nearly half would be less than 70 or greater than 80.[3]

2. William A. Yoell, "How Big a Sample in Qualitative Research?" *Advertising Age and Advertising and Selling*, September, 1950.

3. Another aspect of the quotation deserves comment. Cumulating the samples, as the author apparently did, introduces an artificial appearance of stability. Thus, our first series shows 75 percent for the first sample of 20, 80 percent for the first two samples,

### 4.4 Variability of Samples

It is a fundamental fact that different results are obtained under apparently fixed conditions. Thus, holes drilled with a given drill will all have different measurements, even with the same operator and the same material. The holes may be similar in the sense that any differences among their dimensions are of no practical importance. If fine enough measurements are taken, however, they will always show some variation. Part of the difference in dimensions will be "real," and part may be due to inaccuracies in measurement. If the same hole is measured repeatedly by the same person or different persons, there will be some variation in the recorded measurements if the gradations are fine enough. Similarly, a litter of rats will exhibit varying individual growth despite the most carefully controlled heredity, environment, and measurement. We can regard this as a controlled "process," much as we regard the machine process that drills repeated holes. By a "controlled" or "apparently fixed" process we mean one in which the individual differences cannot be associated with identifiable or "assignable" causes, but rather are such as we observed in sampling from Populations I and II, which we ascribe to "chance."

Thus, the items in a population almost always vary among themselves. Even if there were no "real" differences, the process of measurement might introduce variability. It is this variability among the items of the population that leads to variation among samples, for the different samples include different sets of the population items. The greater the variation among the population items (that is, the less homogeneous the population), the greater, other things equal, will be the variability among samples. In the sampling box, for example, when one of the two colors predominates there is, in a sense, less variability in the population than if the two colors are equally represented, hence less variability from sample to sample. This is illustrated by the fact that the probability that both beads of a pair

---

78 percent for the first three, and so on. Cumulated in this way, our series (rounded to two figures) become:

First series:    75, 80, 78, 74, 73, 72, 71

Second series:   55, 60, 68, 69, 67, 71, 72

There is an appearance of stability here, at least in comparison with the original series, but it results from the arithmetic of cumulating, not the stability of successive samples. For example, even if an eighth sample were to show 100 percent, either series would be raised only by 4. As the number of samples already taken becomes greater, the effect of the next sample on the over-all average becomes less. (To be specific, sample number $k$ will change the previous average by one $k$th as much as it differs from the previous average; thus, in our first series, sample number 3 with 75 percent is 5 below the previous average of 80, so it lowers the average by $\frac{5}{3}$ to $78\frac{1}{3}$, which is rounded off to 78.)

will be the same color is higher the more one color predominates in the population. At one extreme, when there is no variability in the population—that is, the beads are completely homogeneous in color—there is no variation among the samples. At the other extreme, when the population is evenly divided, its variability is at a maximum and the variation from sample to sample is also at a maximum. That is why the samples from Population I were more variable than those from Population II.

In analyzing samples we will repeatedly fall back on the question, "Can the observed differences reasonably be explained by chance?" Only when the answer is "no" does the evidence imply assignable causes.

## 4.5
## REASONS FOR USING SAMPLES

It would have been possible to determine the proportion of red beads in the sampling box by counting the whole population, but in many situations this would either be impossible or impractical. It would be most impractical, for example, for a mail-order company to open every outgoing package to classify it as satisfactory or unsatisfactory, and thereby determine the proportion of its orders being filled correctly. In general, when observing an item destroys it, as in this case and in measurements of durability or breaking point and in tests of functioning on such items as fuzes or matches, inspection of the whole population is out of the question.

Even where complete inspection is possible, sampling may have economic advantages. Resources—materials, time, personnel, and equipment—constitute a limitation in any investigation, and it is necessary to balance the information obtained against the expenditure. It may be that measuring only a sample instead of the entire population results in a margin of potential error, known as *sampling error* (referring to error that in all probability *might* occur because of sampling, not to error that necessarily *has* occurred or will occur because of sampling), small enough for practical purposes—that is, small enough so that a reduction in this risk of error would not be worth the cost of achieving it by further observations. For example, if measuring the useful life remaining in telephone equipment on only three percent of the units gives a sampling error for the whole plant of less than 0.5 percent (as was true in one actual case), it would not ordinarily be worthwhile to measure more items unless the measurements were virtually costless—which, of course, they are not.

### 4.5 Reasons for Using Samples

The concept of sampling error will require further clarification later; its nature is implied, however, by the following (true) statement about sampling error: In situations like that of the sampling box, involving determination of the proportion of a population having a specified characteristic, the sample proportion will be within 0.05 of the population proportion for at least 95 percent, on the average, of samples of size 404 or more.[4] (The principle underlying this calculation is explained in Sec. 14.5.2.)

Two reasons have been given so far for using samples instead of complete surveys: that sometimes the measuring process destroys the items, and that the gain in accuracy from a complete survey may not be worth the cost. A third reason is that the individual measurements may not be as accurate for a complete survey as for a sample. A large number of measurements made hurriedly or superficially may not represent as much true information as a small number made carefully. In extreme cases, poor data can be so misleading as to be worse than no information at all. A rather paradoxical example of the effective use of samples is the Bureau of the Census' use of them to check on the accuracy of the census. Although sampling error is almost absent from the census, the nonsampling errors may be considerable—that is, such errors as those arising from failure to make questions clearly understood, from misrecording replies, from faulty tabulation, from omitting people who should have been interviewed. In the sample census, however, these nonsampling errors may be reduced enough to offset the sampling error, for it is cheaper and easier to select, train, and supervise a few hundred well-qualified interviewers to conduct a few thousand careful interviews than it is to select, train, and supervise 150,000 interviewers to conduct a com-

---

4. This assertion would be expressed symbolically as follows:

$$Pr(P - 0.05 \leq p \leq P + 0.05 \mid n \geq 404) \geq 0.95.$$

This would be read: "The probability that small $p$ is at least capital $P - 0.05$ but no more than capital $P + 0.05$, given that $n$ is at least 404, is at least 0.95." Actually, $\leq$ reads "is less than or equal to," $\geq$ reads "is greater than or equal to," and $\mid$ reads "under the condition that," so a more literal translation is: "The probability that capital $P - 0.05$ is less than or equal to small $p$ and small $p$ is less than or equal to capital $P + 0.05$, under the condition that $n$ is greater than or equal to 404, is greater than or equal to 0.95."

If you have had the perseverance to master the first paragraph of this note you will see that reading and understanding the meaning of such a "mathematical" expression does not actually require any knowledge of mathematics; all that it requires is knowledge of the meanings of the symbols. To prove that the assertion expressed by the symbols is true does, however, require some knowledge of mathematics.

If $P = 0.5$, the sampling variability of $p$ will be greater than if $P$ has any other value. (This was explained in Sec. 4.4.) If $P$ should be 0.5, samples of 404 would produce values of $p$ between 0.45 and 0.55 (that is, within 0.05 of $P$) just 95 percent of the time, on the average. The farther $P$ is from 0.5, the greater the percentage of the time that $p$ will fall within 0.05 of $P$. Since we do not know $P$, we have allowed for the least favorable case.

plete census of the population.[5] Similarly, in measuring the useful life of the equipment in a telephone plant, the practical choice is not between measurements for a sample of the equipment and equally accurate measurements for all the equipment, but between fairly precise measurements of a sample made carefully by competent engineers, and crude measurements of the whole plant made hastily by less skilled people. Even in laboratory experiments in the sciences, the difficulties of precise measurement are often so great that it is better to reduce the number of items measured in order to take more care with the individual measurements.

A fourth reason for sampling is that a complete survey may be impossible because the population contains infinitely many items. The reactions of clouds to seeding, mentioned in Sec. 4.3.1, is a case in point. Similarly, in studying the effects of a medical treatment, the population ordinarily is all responses to the treatment that will ever occur with patients in a certain condition, an essentially infinite population.

A fifth reason for using samples is that for many data the population is inaccessible, and no more data can be had from it. This is particularly true of time series—historical records giving measurements of some phenomenon at various dates in history. Careful records of the level of Lake Michigan (and of each of the Great Lakes), for example, are available for each month from January, 1860 to the present. For studying the seasonal pattern of the lake's level—that is, the relation between the levels for the various months of the year— or for studying such other characteristics as cycles, trends, and extremes in its level, the records constitute a sample (as of the end of 1956) of 97 years. It may be possible to extend the record prior to 1860 by using less accurate or less systematic observations, but not much reduction in the sampling error can be expected this way except at the expense of introducing errors from the unreliability of the data; and, of course, nothing but watchful waiting can extend the record into the future. Nevertheless, the seasonal variations shown in the 97 years must be regarded as a sample, in that continuation of the same basic forces and processes does not produce an identical

---

5. The reader may wonder why, in view of the advantages of sampling, the entire population of the United States is enumerated completely every ten years. Aside from the over-riding fact that the Constitution requires this, perhaps the most important reason is that information is required for very small groups of the population—such as small towns, individual neighborhoods in cities, etc.—as well as for the country as a whole. Even so, however, about half the questions on the 1950 census were asked only of a sample —for some questions a 20 percent sample, and for some questions a $3\frac{1}{3}$ percent sample (namely, a $16\frac{2}{3}$ percent subsample of the 20 percent sample).

pattern of fluctuation each year. A person facing a problem for which the correct decision depends on the seasonal pattern needs to make allowances for the extent to which the 97 observations reflect sampling variation. (This problem is discussed further in Chap. 18.)

All five of these reasons for sampling fall under the general principle that there comes a point beyond which the increase in information from additional observations is not worth the increase in cost.

## 4.6
## RANDOMNESS IN SAMPLING

### 4.6.1 Meaning of Randomness

So far the point stressed in connection with the sampling demonstration has been *sampling variability*—its pattern and the relation of this pattern to the population from which the samples originate. An equally important matter illustrated by the demonstration is *randomness*. The method of sampling we used is an example of random sampling.

"Random" as used in statistics is a technical word; it has a meaning different from the one given it in popular usage. When a sample is called "random," this describes not the data in the sample, but the process by which the sample was obtained. Thus, randomness is a property not of an individual sample but of the process of sampling, just as in a game of cards a fair hand is not one in which the cards have certain values, but one dealt by a certain process. In fact, what in card games is called a fair hand is precisely what in statistical terminology would be called a random sample of the deck.

A sample of size $n$ is said to be a *random sample* if it was obtained by a process which gave each possible combination of $n$ items in the population the same chance of being the sample actually drawn. Thus, in our demonstration, the thorough shaking of the box before each sample was drawn was intended to give each possible set of 20 beads the same chance as any other set of falling into the 20 holes in the panel. A primitive way to achieve randomness is to assign the numbers 1 to $N$ to the $N$ items in a population, write the numbers on cards or chips, thoroughly mix them, select a set of $n$, and then use as the sample the items corresponding to these $n$ numbers. This way of achieving randomness is mentioned here primarily to clarify the meaning of randomness. Actually, these mechanical manipulations are surprisingly difficult to perform reliably, so in practice use is made of tables of random numbers that have been prepared for this purpose.[6]

## 4.6.2 Reasons for Randomness

Nonstatisticians usually assume that the importance of randomness arises from the "fairness" and lack of bias with which such samples represent the population. This is important, of course, but of more importance is the fact that *the pattern of sampling variability for any population is known if, but only if, the sampling is random*. As we said in Sec. 4.3.5, the basic tool of statistical analysis is knowledge of the patterns of sampling variability that result from various populations. This knowledge can be obtained only through the laws of mathematical probability, and these laws apply only to random samples. Thus, only random samples permit objective generalizations from the sample to the whole population. Two competent statisticians will reach similar conclusions from a given sample if it is known to be random and if they have agreed on methods of analysis; furthermore, they will agree on objective statements about the confidence to be put in their conclusions. They need not argue intuitively and interminably whether 25,000 items are enough or 3 too few to support a given generalization. The statistician thus depends on the fact that the pattern of variability of random samples from any population can be determined through the mathematical laws of probability. He not only recognizes sampling variability, he exploits it.

Perhaps the reason nonstatisticians tend to think of randomness only in terms of fairness is that this is its most obvious role in card games. Even in these games, however, knowledge of "the probabilities," as it is sometimes expressed, or of "the sampling distribution," as a statistician might express it, is typically essential to effective play, even if one is assured his fair share of desirable cards.[7]

To repeat: Randomness is important in statistics primarily because if a sample is random, but not otherwise, the mathematical laws of probability are applicable and make it possible to know the

6. The largest and best of these tables is that of The Rand Corporation, *A Million Random Digits with 100,000 Normal Deviates* (Glencoe, Illinois: Free Press, 1955). A smaller but also excellent table is the *Table of 105,000 Random Decimal Digits* issued free by the Interstate Commerce Commission, Washington (Statement 4914, File No. 261-A-1, Bureau of Transport Economics and Statistics).

7. This point is elucidated in a book by S. W. Erdnase describing methods of cheating at cards. He points out that simply changing the probabilities gives a considerable advantage to the person who knows the true probabilities himself and can profit by others' miscalculations. S. W. Erdnase, *Artifice, Ruse, and Subterfuge at the Card Table: A Treatise on the Science and Art of Manipulating Cards* (Chicago: F. J. Drake and Co., 1905). The cover bears the title *The Expert at the Card Table*.

patterns of sampling variability in terms of which the sample must be interpreted.

### 4.6.3 Randomness vs. Expert Selection

Almost any sampling method will have some pattern of variability, but random samples are the only ones that have a *known* pattern of variability. Consider an expert, whether he is an expert at judging the proportion of red beads in a box or at judging the proportion of Democratic votes in an election. This expert will not, of course, claim to specify the true proportion precisely. His figures are subject to sampling variability. But there is no way of knowing the pattern of variability in his method.[8]

This is not to disparage experts and expert judgments. It is better to rely on expert judgment for most everyday problems, than to make statistical studies of every question that arises. But when a statistical study is made, it should be an independent, objective *statistical* study, which may or may not confirm the expert's judgment. If the method is a mixture of statistics and expertise without the use of randomization, the result will be just one more huff or puff on a windmill which is probably spinning too freely already.

If the sample for a statistical study is selected according to expert judgment it *may* give better results than if it is selected according to statistical principles—provided the expert is so expert that a statistical study was not needed anyway. But such results do not reinforce, they only reiterate, the expert's judgment. It is important to see the contrast between statistical method and expert judgment. When we select data solely by judgment, expert or otherwise, we rely on a man; when we rely on random sampling, we rely on a method. The purpose of collecting facts is to give them full opportunity to support or contradict judgment, thereby adding to the knowledge available.

Example 117   Sampling Castings

A former student has told us of a sampling blunder in his company which stemmed from using a convenient but nonrandom sampling method. The company had found from past experience that about 90 percent of certain

---

8. Before the sampling demonstration described in Sec. 4.3 was actually done, several boxes of red and green beads were shown to one of the observers, and he was asked to estimate the proportion red by eye. Actually all the boxes had the same proportion red as Population I; his estimates were not all the same, but their average was quite close to the correct proportion for Population I. When this same "expert" (if we may take the liberty, for purposes of exposition, of qualifying him too easily as such) was shown a series of boxes all having the same proportion red as Population II, however, his average differed greatly from the correct proportion.

castings it was buying were defective. The defects usually showed up only after some machining had been done. One supplier claimed that he had developed a new method which would virtually eliminate defective castings.

When the first new lot was received it was decided to take a sample of the lot and have the sample items X-rayed before any machining was done. A sample of 20 castings was taken from the top of the box containing the the entire lot and the X-ray inspection did show a great improvement in quality (actually no flaws were detected). On the basis of this the lot was accepted.

The lot was machined and 75 percent of the castings had to be scrapped. Subsequent inquiry showed that by an error of the supplier, the box was filled mostly with castings from the old method, with the new ones only on top.

Had the 20 castings been chosen randomly from the entire box, not just the top layer, there would have been only one chance in a trillion of finding no defective castings if the lot contained 75 percent defectives.

There are, to be sure, circumstances in which nonrandom sampling may be appropriate. (1) Random selection of samples is often more costly than nonrandom selection. This cost argument is not always as valid as it may seem, though, partly because the cost may represent ineptness at random sampling, and partly because random samples may give more valuable results. To put it another way, results of given value may be obtainable with smaller samples if sampling is random; indeed, they may be unobtainable with nonrandom samples of any size. (2) There may be occasions when only very few items can be included in a sample, as when an intensive study of cities is to be made, and even two cities would be too expensive, or when there is only enough of a new drug to treat, say, five cases, and the delay involved in getting more would be prohibitive. In these instances, generalization from the sample to the population will be essentially a matter of judgment anyway, since even a random sample of 1 or 5 will probably not alone justify any but foregone conclusions; so it is best for the expert who will have to make the judgment to select the cases in whatever way he judges will best illuminate the issue in his own mind. (3) Again, the argument that particular nonrandom methods of sampling have led to valid results in a certain kind of problem in the past always deserves serious consideration— though sooner or later such methods usually produce fiascos, as in the case of the *Literary Digest* presidential poll made in 1936 by methods that had proved successful in the previous four elections, or the case of the Gallup, Roper, and other presidential polls made in 1948 by methods that had proved successful in the previous three elections.[9] (4) Another situation where nonrandom sampling is appropriate is where only certain data are accessible, as in studying trends in the frequency of mental disease (Sec. 2.8.2) or the standard of living in 19th-century England (page 12). (5) Finally, random sampling—and even sampling of any kind—may be inappropriate where the object is to locate specific individuals, for example, the particular blood specimens with positive reactions to a test for communicable disease, or the specific income tax returns with errors.

In all these situations except the last, however, an inverted question should be asked: Of what population are the data a random sample? That is, if whatever process produced the data were repeated indefinitely often, what population would it generate? The relation between this sampled population and the target population in which we are interested is then a question for expert consideration. Goldhamer and Marshall, for example, considered such a question in their study of mental disease (Sec. 2.8.2). They would have liked to study the onset of all cases in the United States, but had to study instead hospital first admissions in Massachusetts; but they judged trends shown by these data to be satisfactory representations of trends in the onset of all cases. As a matter of fact, this inverted question—What population was actually sampled?—should be raised even where it has been possible to make a conscientious attempt to obtain a random sample from a clearly specified target population, for the best laid sampling plans (like other plans) can seldom, if ever, be executed perfectly. Subjects refuse to co-operate, animals die, machines break down, experimental materials are interchanged, etc.

Though randomization and expert selection are incompatible, this is by no means true of randomization and expert judgment about sample design. As we shall see in Chap. 15, there are sound statistical methods for getting the best out of expert judgment without sacrificing the advantages of randomization. Suppose, for example, that we are planning a study of employee attitudes toward a firm, and an expert tells us that these attitudes vary with sex, race, union membership, department, and length of service. We could use the expert's judgment by *stratifying* our population—dividing it into a number of smaller populations (called *strata*) that are relatively homogeneous with respect to sex, race, etc.—and then sampling at random from each stratum. Furthermore, expert judgment is, as we emphasized in discussing the study of trends in the frequency of mental disease (Sec. 2.8.2), essential in selecting problems, deciding what to measure, and interpreting the findings of any statistical study.

9. Other factors than the sampling methods contributed to these fiascos.

### 4.6.4 Probability Samples

What we have described as random sampling is sometimes called *simple random sampling,* to distinguish it from various more elaborate sampling procedures, such as the stratified sampling just mentioned. All of these more elaborate sampling procedures are based ultimately on samples that are random in the sense we have described. The essential thing, as we have seen, is that the laws of mathematical probability should govern the pattern of sampling variability—that is, the *sampling distribution*—of such samples. A broader term, *probability sampling,* is used to describe any sampling process in which randomness enters at some stage in such a way that the laws of mathematical probability apply and provide the sampling distribution needed for interpreting a sample.

A probability sample of size $n$ is one for which each set of $n$ items in the population has a known probability of being the set chosen for the sample. More generally, the actual probability of selection for the sample need not be known; knowledge of the relation of this probability to the proportion of such samples in the population is sufficient. In the special case of a simple random sample, the probability of being the set chosen for the sample is the same for each set of $n$ items in the population.

This chapter, confined to basic ideas, is not the place to discuss more elaborate probability sampling methods. Let us, however, briefly recall Example 72B, which dealt with estimating the number of children per family. The fallacy of this example can be viewed as one of analyzing as a simple random sample what is in fact a more complicated type of probability sample.[10] Here the chance of a family being included in the sample is not the same for all family sizes, so we do not have a simple random sample. But since we know the relation between the number of children in a family and its chance of being included in the sample, we do have a probability sample. In fact, the probability of a family's being included is simply proportional to the number of children, and this knowledge makes it possible to estimate the average number of children per family free of the bias discussed in Chap. 3.[11]

10. In Chap. 3, we interpreted the fallacy as one of failing to get a simple random sample when that was intended. The point is that the method of sampling and the method of analysis do not jibe. This can be interpreted either as the wrong analysis for the sampling method or as the wrong sampling method for the analysis; but it should be distinguished from sampling in such a way that the data are totally unanalyzable.

11. In effect, we simply disregard half the twos reported, two-thirds of the threes, three-fourths of the fours, and so on, and average the remaining observations. More precisely, but in terms not explained until Chap. 7, we take a weighted mean of the data, weighting each observation by its own reciprocal.

## 4.7
## LAW OF LARGE NUMBERS

Consider for a moment the 1,000 beads that were drawn from Population I as if they were all one sample from a very large population. We see from Table 105 that this sample shows 548 red beads, or a sample proportion of 0.548. In interpreting this result we must recognize, however, that a second sample of 1,000 would almost surely yield a different result, a third sample still a different one, and so on. Sampling variability could be demonstrated for samples of 1,000 just as it was for samples of 20, and from that point of view the sample of 1,000 reported in Table 105 now plays the same role as the preliminary sample of 20 discussed in Sec. 4.3.2. And if we were to proceed to draw 50 samples of 1,000, all beads drawn could be regarded together as one sample of 50,000, the interpretation of which would have to take account of the pattern of variability in samples of 50,000.

But the laws of probability assure the following. The probability that $p$ will be within a given range of $P$ is greater for samples of 100 than for samples of 20 from the same population, and still greater for samples of 1,000. For example, in samples of 20 from a population in which $P = 0.50$, about half the time $p$ would be farther than 0.05 from $P$. In samples of 100 from the same population, deviations of $p$ from $P$ exceeding 0.05 would occur only a little more than one-fourth of the time. In samples of 1,000, such deviations would occur only about once in 1,000 times, and so on. This illustrates one aspect of what is called "the Law of Large Numbers." The larger the samples, the less will be the variability in the sample proportions. Tosses of pennies illustrate the same thing. If a fair coin is tossed 50 times, the proportion of heads may well be as little as 0.4 or as much as 0.6. But if the coin is tossed 5,000 times, the proportion of heads is unlikely to fall outside the range 0.48 to 0.52.

There is another aspect of this law which must be borne in mind, and which can also be illustrated by coin-tossing. If two men match pennies repeatedly, the Law of Large Numbers does not guarantee, as many people think, that they will break even. The truth is that one of them will go broke. This is obvious if both players start with fortunes of only a penny, fairly plausible if they start with only a dime. It may seem less plausible as the fortunes increase, but nevertheless, the principle is true no matter how large the fortunes—though the time required becomes astronomical as the fortunes become even moderate. As the number of tosses increases, the number of wins for the two contestants may confidently be expected to diverge by larger and larger *amounts*, though by smaller and smaller *proportions*. A divergence of 500 wins would be highly improbable in 5,000 tosses, but could easily happen in 500,000 tosses; an excess of the larger number of at least 50 percent of the smaller number is highly probable (in fact, certain) in 5 tosses, but highly improbable in 5,000.

The arithmetic by which the absolute discrepancy can increase while the percentage discrepancy is decreasing deserves an illustration. Suppose 25 tosses of a fair coin show 15, or 60 percent, heads. This is an excess of $2\frac{1}{2}$, or 10 percent of the number of tosses, above the expected even division. Now suppose 100 tosses show 55, or 55 percent, heads. This is an excess of 5 above the expected number, or twice as many as in the sample of 25, but an excess of 5 percent, or one-half as large a percentage as in the sample of 25. In general, if the departure of the number of heads from expectation increases, but less than in proportion to the number of tosses, the departure of the percentage of heads will decrease.

The naive conception of these principles, commonly spoken of as the "law of averages," is sometimes taken to imply that tosses of a fair coin approach half heads and half tails because after an unusual number of heads it is more likely that tails will turn up. This is not true, since each toss is independent of those before and after. The following anecdote illustrates the same point:

EXAMPLE 122   SEQUENCE OF BOYS

When 18 boys were born consecutively in five days recently at Walther Memorial Hospital it was generally assumed that a run of girl babies would begin. Since the run of 18 boys ended with the birth of a girl last Friday this is how the new arrivals have been recorded.

In the next six births, five were boys. Then four boys in the next six births. Again, four boys in the next five births. From Midnight Tuesday—five more boys and only one girl. Since October 31: 36 boys, 6 girls.[12]

As Tippett has put it, the Law of Large Numbers works by its "swamping" effect rather than by compensation.[13] An unusual result that produces 50 too many heads in 1,000 tosses, for instance, will not be perceptible in the proportion of heads after 99,000 more tosses, unless a similar excess occurs repeatedly.[14] But the basis of the Law

12. *Chicago Daily News*, November 10, 1949.

13. L. H. C. Tippett, *Statistics* (New York: Oxford University Press, 1943), p. 87.

14. If 1,000 tosses show 550 heads, the percentage is 55. Assume that the next 99,000 tosses show 49,500 heads, or exactly 50 percent. Then the 100,000 tosses have shown 50,050 heads, or 50.05 percent. The effect of a large excess in a single thousand is thus "swamped," not compensated for. The other 99,000 tosses, of course, will not show exactly 50 percent heads, but they are as likely to show less as to show more if the coin is really fair. At any rate, whatever discrepancy from 50 percent there is in the whole 100,000 will be due scarcely at all to the first 1,000.

of Large Numbers is that for an improbable event to occur $n$ times is improbable to the $n$th degree.

Thus, while samples vary, averages and proportions vary less in large samples than in small samples from the same population.

## 4.8
## STATISTICAL CONTROL

One reason for constructing Fig. 109 in the way we did is that it resembles charts called *control charts* that are often useful in practical situations. All repetitive processes—no matter how carefully arranged

Number of overdue books per sample of 100

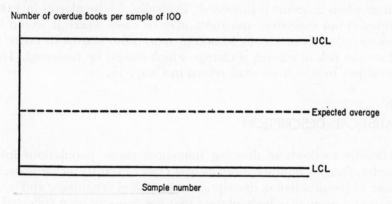

Sample number

FIG. 123. Outlines of a control chart.

—are accompanied by variability that cannot be explained by "assignable" causes. The process is said to be "in control" when the pattern of this variability is like that of independent, random samples from the same population—as in our experiment with the sampling box. Suppose we know from past experience that a process, when in control, leads on the average to a certain proportion, $P$, of items having a certain characteristic. For example, in a certain library at a certain season the average proportion of books not returned by the date due may be known and stable. Then we can plot a control chart resembling Fig. 123, on which are to be plotted the results of samples for different days—say the number of books not returned among a sample of 100 due on a given day. UCL denotes "upper control limit"; LCL denotes "lower control limit."[15] These limits

---

15. Statistical control limits should be carefully distinguished from the terms "specifications" or "tolerance limits" which are commonly used in manufacturing. The latter usually refer to what the process should do; control limits are based entirely on what the process actually has done.

are usually set so that, if the process is in control, only about 3 plotted points in 1,000, on the average, will fall outside the limits. When a point does fall outside the limits, therefore, it is a signal to look for a change in the process for which, presumably, some explanation other than chance can be assigned. In the library example, there might, for instance, have been an increase in thefts or in errors in checking off returned books.

If the process is in control, there will be false alarms from 3 of every 1,000 samples, on the average. If this is too many, the statistical control limits can be set farther apart; the appropriate distance can be determined from the acceptable false alarm rate. If the control limits are set too far apart, the risk of not detecting an important change when it occurs is increased. Typically, 3 false alarms in 1,000 samples is not excessive, and there may be cases where it would be economically advantageous to accept more false alarms in order to reduce the risk of missing a change which should be corrected. This is a subject to which we shall return in Chap. 16.

## 4.9
## STATISTICAL DESCRIPTION

Besides methods of drawing inferences about populations from samples, there is another fundamental class of statistical methods. It relates to the problem of description. The task of organizing and summarizing a particular body of data that has actually been collected is a problem in *statistical description*, in contrast with the task of drawing conclusions from data actually on hand about a larger body of data that have not been completely collected, which is a problem of *statistical inference*. These two problems are by no means completely separate, but it is feasible and useful to discuss them separately.

The value of skillful statistical description is suggested by Winston Churchill in a memorandum written while he was First Lord of the Admiralty in 1939–40:

> Surely the account you give of all these various disconnected Statistical Branches constitutes the case for a central body which should grip together all Admiralty statistics, and present them to me in a form increasingly simplified and graphic.
>
> I want to know at the end of each week everything we have got, all the people we are employing, the progress of all vessels, works of construction, the progress of all munitions affecting us, the state of our merchant tonnage, together with losses, and numbers of every branch of the R. N. and R. M. The whole should be presented in a small book such as was kept for me by Sir Walter Layton when he was my statistical officer at the Ministry of Munitions in 1917

and 1918. Every week I had this book, which showed the past and the weekly progress, and also drew attention to what was lagging. In an hour or two I was able to cover the whole ground, as I knew exactly what to look for and where. How do you propose this want of mine should be met?[16]

We have already given one illustration of the process of statistical description when we prepared Table 106 from Table 105. The data of Table 105 are difficult to assimilate. About all we see is that the number of red beads varies, and usually begins with the digit "1." Table 106, however, presents the data in a way which immediately brings out the pattern in the variation and considerably sharpens our appreciation of the range in which most of the results lie. This is accomplished at the expense of obscuring the sequence in which the results occurred, the one thing that Table 105 does show, but our purpose in Table 106 was to highlight the pattern and shade the irrelevant and distracting features of the data. Fig. 109 is another description of the same data, one readily grasped by the eye, which brings out fairly well the pattern shown in Table 106 while also showing the sequence of occurrence, which would be important if control were in doubt.

Much of the most useful work done by statisticians consists in simply arranging masses of data so that they are comprehensible, or so that they focus attention on patterns and relations that are important, or else in summarizing masses of data by a few significant measures, for instance, an appropriate average. In Example 19A the principal contribution made by the statistician was in transcribing figures from the standard record forms and presenting them on a chart showing the relation between flying hours since overhaul and number of failures. Similarly in Example 19B (Merchant Ship Losses in Relation to Convoy Size), once the data had been properly organized and presented the conclusion was fairly clear.

Successful statistical description, like most successful statistical work, depends greatly on knowledge of the subject matter. Mere manipulation of figures or preparation of standard tables and graphs is seldom fruitful unless guided by a clear conception of the subject matter and of what relations would be worth looking for. To a considerable extent statistical description is an art, rather than a science. As with other arts, however, there are certain basic techniques whose mastery is necessary, though not sufficient, for success. The brevity of our discussion of statistical description here, in this chapter on basic ideas, should not be taken to measure the relative importance

16. Winston Churchill, *The Gathering Storm* (New York: Houghton Mifflin Company, 1948), p. 730.

of descriptive statistics. Rather, it reflects the fact that the field of descriptive statistics is not dominated by a few broad and pervasive principles as is analytical statistics, but is essentially a collection of techniques.

Chaps. 5 to 9 emphasize statistical description, with some related discussion of sampling and inference. That is, they will consider mainly the problems of describing a particular body of data already on hand rather than the problems of how to plan the collection of, and draw inferences from, data. These latter questions, the basic groundwork for which has been laid in this chapter, will be the subject of Chaps. 10 to 19.

## 4.10
## CONCLUSION

*Population* is an abstract concept fundamental to statistics. It refers to the totality of numbers that would result from indefinitely many repetitions of the same process of selecting objects, measuring or classifying them, and recording the results. A population is, thus, a fixed body of numbers, and it is this general body of numbers about which we would like to know. What we actually know is the numbers of a *sample*, a group selected from the population. Because the numbers in the population almost always vary, both inherently and through the variations in measuring and recording, the results of a sample depend on which numbers from the population are included in the sample. In generalizing from the sample to the population, therefore, allowance must be made for the fact that the sample results are partly fortuitous. This allowance is made by considering the pattern of sampling variability, called the *sampling distribution*, of sample results, that is, by considering the various samples that could occur from any particular population and their respective probabilities of occurring. Some populations are then seen not to be likely to produce a sample such as that observed, and others to be likely to produce it. The population from which the sample came is inferred to be one of the latter.

This process depends upon being able to deduce from any assumed population the sampling distribution that would result if samples were drawn from it. This is possible if, but only if, the sampling is *probability sampling*. A probability sample of size $n$ is a sample selected from the population in such a way that there is a known probability of selection for every set of $n$ numbers in the population, or at least that there is a known relation between the probability that this set will constitute the sample and the proportion of all sets of *n* in the population that contain the same numbers. A *random sample* (or better, a *simple random sample*, since frequently "random sample" is used in the sense of "probability sample" here) is one in which each possible set of *n* values in the population has the same probability that it will constitute the sample.

The distribution of samples depends on the nature of the population. For a given sample size, samples will be more variable the more heterogeneous the items in the population. And the *Law of Large Numbers* tells us that for a given population, the variability of sample averages and proportions will be smaller, the larger the sample.

The principal reasons for using samples instead of observing the whole population are associated with the fact that once the sample attains a certain size, additional observations will not reduce the *sampling error*, that is, the allowance for sampling variability in the conclusions, enough to be worth the additional cost.

A direct application of these ideas is the *statistical control chart*. The results of successive samples are plotted on this chart, and as long as the same population is being sampled virtually all of the sample results will lie between two limits, called *statistical control limits*. When a result falls outside the limits, it is evidence that there has been some change in the underlying population, for the sample is not within the probable range for the original population.

In contrast with *statistical inference*, which deals with methods of drawing conclusions or making decisions about populations on the basis of samples, *statistical description* deals with methods of organizing, summarizing, and describing data. Since the purpose of description is to prepare the way for inference, the basic ideas of inference have been outlined in this chapter as essential background for Part II, on description, and Chap. 5, on measurement and observation.

## DO IT YOURSELF

Several types of example are included at the end of this and later chapters. Some are directed primarily at specific points covered in the chapter. Others involve ideas from earlier chapters relevant to a full understanding of the current chapter, or sometimes anticipate points to be raised and developed more fully in subsequent chapters. Thus, the examples provide a continuing review of earlier material and an introduction to new material. Our aim is to help you apply more effectively both your common sense and the things you have learned in earlier chapters, not simply to provide drill on the details covered in the current chapter.

EXAMPLE 128A

In Example 100B the sample is not necessarily too small, although we need more information to know. There is a defect in the *sampling method*, however, which may be very serious. What is it?

EXAMPLE 128B

In example 100C, suppose that a much larger sample had been taken, and the greater decline of cigarette sales in stores with vending machines was statistically well established—that is, more than reasonably ascribable to chance. If you were a restaurant manager thinking of installing a vending machine, what major reason would you find for not being too discouraged by this evidence?

EXAMPLE 128C

Can you think of three more practical problems, besides those mentioned on page 103, which are identical statistically to the bead demonstration?

EXAMPLE 128D

Table 632 is a short table of random digits. In Chap. 10 we will explain its use and the principles underlying it, but you can use it now to get the feel of a sampling experiment without the apparatus described in Sec. 4.3. Here is the first line of the table:

10097  32533  76520  13586  34673  54876  80959  09117  39292  74945

In generating your sampling demonstration, proceed across this line, and succeeding lines, taking 20 digits at a time to comprise each sample. Let the digits 1, 2, 3, 4, whenever encountered, stand for a red bead; the digits 5, 6, 7, 8, 9, 0 for a green bead. In each 20 digits, count the number of "red beads" (that is, digits 1, 2, 3, or 4). For example, there are 8 "red beads" in the first 20, because we find 1, 3, 2, . . ., 3 occurring in the first 20 digits. Similarly there are 6 "red beads" in the second 20, and so on.

Start with the third 20 digits and continue for 50 samples, recording your individual results in a table like Table 105, and a frequency distribution like Table 106. Plot your 50 results in a chart like Fig. 109.

From the visual impression given by your Tables and Chart, would you conclude that 50 samples like yours would be likely to occur in sampling experiments from either Population I or Population II in Sec. 4.3? How do you think you would have felt about this question on the basis of your first sample alone?

You can easily devise variants of this experiment; for example, let the digits 1 and 2 represent red beads, the remaining digits, green beads.

*Do It Yourself*

EXAMPLE 129

Review Example 110. Make a graph like Fig. 109 for (1) the 14 sample results given in the text discussing this example, and (2) the 14 numbers given in our footnote on p. 111. In what ways is the visual impression from the second part of your chart different from that of the first part?

# Chapter 5

# *Observation and Measurement*

## 5.1
### INTRODUCTION

What relation is there between the real world and the mere numbers that the statistician deals with? This is one of the most obvious, yet one of the most frequently overlooked, questions to be raised about any statistical study.

Strictly speaking, statistical analysis deals not with, say, the population of an area but simply with a set of numbers acquired in a certain way which it happens to please someone to designate by demographic terms; it deals not with the value of a certain plant and its equipment but with a set of numbers to which someone has attached words relating to plant and equipment; it deals not with the frequency of red beads in samples of 20 but with a set of numbers, produced by a certain process, which have been described in terms of red and non-red beads. In short, statistical analysis deals with numbers produced by certain operations, and strictly speaking its conclusions relate to the processes producing the numbers. But of course the interest the numbers have arises from their association with—their measurement of—things in the real world. In a sense, the relation between the numbers and the real world is not a problem for the statistician but for the subject-matter specialist—the engineer, sociologist, physicist, epidemiologist, etc. The meaning of the statistician's work is so completely dependent upon the meaning of the numbers with which he works, however, that he cannot wisely draw a sharp line between method and substance.

**130**

*5.1 Introduction*

*aggregation*

These ideas are brought out vividly in the following quotation:

When we speak of "observing" business cycles we use figurative language. For, like other concepts, business cycles can be seen only "in the mind's eye." What we literally observe is not a congeries of economic activities rising and falling in unison, but changes in readings taken from many recording instruments of varying reliability. These readings have to be decomposed for our purposes; then one set of components must be put together in a new fashion. The whole procedure seems far removed from what actually happens in the world where men strive for their livings. Whether its results will be worth having is not assured in advance; that can be determined only by pragmatic tests after the results have been attained.

This predicament is common to all observational sciences that have passed the stage of infancy. An example familiar to everyone is meteorology. The layman observes the weather directly through his senses. He sees blue sky, clouds, snow, and lightning; he hears thunder; he feels wind, temperature and humidity; at times he tastes a fog and smells a breeze; he sees, hears, and feels storms. The meteorologist can make these direct observations as well as a layman; but instead of relying upon his sense impressions he uses a battery of recording instruments—thermographs, barographs, anemometers, wind vanes, psychrometers, hygrographs, precipitation gauges, sunshine recorders, and so on. That is, he transforms much that he can sense, and some things he cannot sense, into numerous sets of symbols stripped of all the vivid qualities of personal experience. It is with these symbols from his own station and with similar symbols sent to him by other observers dotted over continents and oceans that he works. . . .

All of us can observe economic activities as easily and directly as we can observe the weather, for we have merely to watch ourselves and our associates work and spend. What we see in this way has a wealth of meanings no symbols can convey. We know more or less intimately the hopes and anxieties, efforts and fatigues, successes and failures of ourselves and a few associates. But we realize also that what happens to us and our narrow circle is determined largely by what is being done by millions of unidentified strangers. What these unknowns are doing is important to us, but we cannot observe it directly.

A man tending an open-hearth furnace has a close-up view of steel production. But what he sees, hears, smells and feels is only a tiny segment of a vast process. He works at one furnace; he cannot see the hundreds of other furnaces in operation over the country. And smelting is only one stage in a process that includes mining and transporting iron ore, limestone, coal, and alloys; the getting of orders for steel, the erection of plants, and the raising of capital; importing and exporting, hiring and training workers, making and selling goods that give rise to a demand for steel, setting prices, and keeping accounts of outgo and income. No man can watch personally all these activities. Yet those engaged in them and in the activities dependent on the steel industry need an over-all view of what is happening. To get it they, like meteorologists, resort to the use of symbols that bear no semblance to actual processes and that are compiled mainly by other men.

For the intermittent process of making steel in a furnace with its heat and noise, its dim shadows and blinding glares, they substitute a column of figures purporting to show how many tons of steel ingots have been turned out by all the furnaces in a given area during successive days or weeks. That colorless record gives no faintest idea of what the operation looks like or feels like; it does not tell whether the work is hard or easy, well or ill paid, profitable or done at a loss. It suggests continuous operation, which is achieved at no furnace. It hides differences of location and types of product. And it separates the one act of turning out tonnage from all the other activities with which it is interwoven. Many, though not all, of these interrelated changes are likewise recorded in columns of figures; but each record is as devoid of reality and as divorced from its matrix as the record of tons produced.[1]

## 5.2
## THE RELATIONSHIP BETWEEN A NUMBER AND THE REAL WORLD

In considering any statistical analysis, it is always a good idea to stop and think about the relationship that the numbers bear to the subject and questions in which we are interested. Numbers by themselves have no meaning or significance; their significance depends on the circumstances and events that gave rise to them.

Consider a number labeled "number of passenger automobiles produced in the United States during the week ended August 25, 1956." Did someone stand at the end of the assembly lines tallying finished units? If so, was every line included, and was the tallying perfectly accurate? And how were all the separate counts compiled for the total figure? Perhaps each manufacturer reported his own figures, some based on production schedules, some on the number of engines sent to the assembly line, and some on the number of automobiles not only off the assembly line but also approved on a final inspection process. What of cars partially produced during the week; do two half-finished cars count as one, and if so, when is a car half-finished? What of production of parts for shipment abroad unassembled?

Far more complex are the operations resulting in the number labeled "index of consumer prices, October 1956," or that labeled "velocity of light," or that labeled "safe concentration of carbon dioxide," or that labeled "patellar reflex reaction time."

In each case, there is a sequence of operations resulting in a number, and the significance of the number depends on those operations. Thus, in the case of passenger car production, if there is a tally of each unit as it rolls off the line, the resulting figures are only partially a measure of productive activity, for they do not reflect plant maintenance, manufacture of parts, and other productive activities except insofar as these other productive activities are closely correlated with

---

1. Arthur F. Burns and Wesley C. Mitchell, *Measuring Business Cycles* (New York: National Bureau of Economic Research, 1946), pp. 14–16.

## 5.2 Numbers and the Real World

the number of units coming off the line. If a line has just been opened, there may be a large amount of productive activity on incomplete units that is not reflected in the tally at the end of the line; or when the line is about to be closed, there may be tallies recorded for cars on which only a few final steps were taken during the relevant period. On the other hand, if completed units leaving the assembly line is precisely the quantity of interest, tallies at the end of the line will provide excellent measurements. Similarly, the number of people admitted to mental hospitals may or may not be a good measure of the prevalence of mental illness, and the number of crimes reported by the police may or may not be a good measure of the prevalence of crime.

Here are some examples in which numbers did not mean what they seemed to:

EXAMPLE 133A MUSEUM ATTENDANCE

We hear of a museum in a certain Eastern city that was proud of its amazing attendance record. Recently a little stone building was erected nearby. Next year attendance at the museum mysteriously fell off by 100,000. What was the little stone building? A comfort station.[2]

EXAMPLE 133B NITROGLYCERIN

Fuel for rockets was being produced in fairly large batches. The percentage of nitroglycerin in the product was supposed to be of crucial importance to safety and to performance. Therefore, an elaborate chemical analysis was made on each batch. This analysis produced a number for each batch, labeled "percent nitroglycerin," and the batch was accepted or rejected on the basis of the number. Large amounts of material were rejected and reprocessed, at considerable cost. Then a statistical investigation was made of the relationship between the reported percentage of nitroglycerin and actual performance of the fuel. No difference in performance was observed between accepted and rejected batches. This meant that the chemical analysis must have been wrong, or else the theory about the relation between nitroglycerin content and performance. It was discovered that the chemical analyses, despite their impressive appearance, were producing numbers that bore little relation to the actual percentage of nitroglycerin. In fact, the product was less variable than the chemical analysis. The numbers on which decisions were being made bore too rough a relationship to performance.

EXAMPLE 133C LIFE-RAFTS

Rubber life-rafts were tested individually for impact resistance by dropping them from the height of a ship's deck into a pool of water. Only rafts that passed this test were used. Users (or would-be users), however, reported

2. *This Week*, April 17, 1948.

that many of the rafts were bursting on impact with the water. Investigation showed that the test impact was itself weakening good rafts to the point where they failed on a second impact. Thus, the figures collected related to the number of rafts which had survived a particular test in the past, not to the number which would give satisfactory service in real emergencies.

### EXAMPLE 134 PRICE RIGIDITY

An important question of empirical research in economics is the extent of rigidity—or flexibility—of prices in oligopolistic industries—industries in which there are relatively few sellers.

. . . It is not possible to make a direct test for price rigidity, in part, because the prices at which the products of oligopolists sell are not generally known. For the purpose of such a test we need transaction prices; instead, we have quoted prices on a temporal basis and they are deficient in two respects.

The first deficiency is notorious: Nominal price quotations may be stable although the prices at which sales are taking place fluctuate often and widely. The disparity may be due to a failure to take account of quality, "extras," freight, guaranties, discounts, etc.; or the price collector may be deceived merely to strengthen morale within the industry. The various studies of steel prices . . . contain striking examples of this disparity. . . . We cannot infer that all nominally rigid prices are really flexible, but there is also very little evidence that they are really rigid.

The second deficiency is that published prices are on a temporal basis. If nine-tenths of annual sales occur at fluctuating prices within a month (as is true of some types of tobacco), and the remainder at a fixed price during the rest of the year, the nominal price rigidity for eleven months is trivial. . . .[3]

Two aspects of the quality of data are brought out by these examples. One aspect is *precision*, or reproducibility—often called, misleadingly, "reliability." In Example 133B, the numbers produced varied considerably even when there were no differences in the quantity of interest, the percentage of nitroglycerin. In Example 133C, however, the measurements were precise, in that they showed whether the rafts could withstand the test; and similarly in Example 133A, the measurements were precise in that they showed exactly how many people entered the building. The second aspect is *relevance*, often called "validity." In Example 133A, the numbers obtained were not relevant measures of the number viewing the exhibits at the museum, and in Example 133C the numbers were not relevant measures of performance subsequent to the test. In Example 133B, on the other hand, the numbers were perhaps relevant, in the sense of going up on the average when the true nitroglycerin content went up, and down when it went down, but they were not precise.

3. George J. Stigler, "The Kinky Oligopoly Demand Curve and Rigid Prices," *The Journal of Political Economy*, Vol. 55 (1947), p. 69.

### 5.2 Numbers and the Real World

By the Law of Large Numbers, the average of a large number of independent measurements will be more precise and therefore have more reliability than a single measurement. On the other hand, the average will have no more relevance than the individual measurements. In the case of the box of beads described in Chap. 4, the proportion of red beads in a sample of 20 is not a very precise measure of the proportion in the box, but by averaging the results for enough samples we can get as precise a measure as we please. On the other hand, if the sampling is not random, or the beads are counted by a badly color-blind person, the average of the figures given for a large number of samples will be little if any more relevant a measure of the proportion in the box than the figure given for one sample.

There is no safety in numbers. There is something about numbers that lures people into thinking uncritically, as though the number were an intuitively obvious, crystal clear, absolutely true, inherent property of the object. This illusion ought to be dispelled; numbers should be accepted only after a careful examination of their significance. All kinds of ambiguities and complications can prevent numbers from measuring what they are supposed to measure.

In order to evaluate a set of data, it is necessary to know how they were actually obtained. Skepticism is justified when a report fails to provide these details or skims over them with such phrases as "scientific precautions were taken to insure accuracy," "by a depth interviewing technique we were able to get at true motivations, not just rationalizations," etc. Conversely, the inclusion in a report of a detailed account of the methods of measurement is a sign that the writer is at least aware of the difficulties and may therefore have done reasonably well in overcoming them. When statistical evidence is presented in popular sources, such as the daily newspapers, it would be neither appropriate nor possible, of course, to include the technical account of the data. The critical reader will then do well to ask himself the question suggested in Sec. 3.3, "How can they really know?"

While it is possible to become too skeptical, gullibility is far more common. It is particularly easy to be gullible when the purported evidence agrees with what is already believed, or when the facts reported admit of a plausible and ingenious interpretation which is dramatic enough to attract attention, or when the subject is remote from one's own technical knowledge. The Kinsey reports, for example, are widely quoted and discussed by scientists and laymen alike, with little or no critical study of the accuracy of their evidence.

Occasionally, information is collected which might be thought wholly inaccessible. In Sec. 2.8.2 we described a study in which accurate data were obtained on the incidence of the psychoses during the last 100 years. The Federal Reserve Board has for several years been obtaining intimate details of family finances. In these examples, the methods by which the data were collected have been carefully described and checks on precision and relevance have been reported. In the absence of such documentation, however, one might well be skeptical about the possibility of obtaining such hard-to-get data.

## 5.3
## INTERNAL EVIDENCE

Often data contain within themselves evidence about their own quality. Inconsistencies, irregular patterns, and unlikely or impossible values are among the common clues to poor quality, though none is ever sufficient proof of poor quality.

### 5.3.1  Inconsistencies

EXAMPLE 136A  KINSEY ON MALES

A statistician reviewing a widely publicized statistical study reported:

The number of individuals involved in the study is given on p. 10 as 12,214. On p. 5, however, there is an outline map of the United States with the legend "Sources of histories. One dot represents 50 cases." The map contains 427 dots, so presumably represents 21,350 cases. . . .

The . . . column [of Table 41] which is said to distribute 179 males by occupation totals 237 . . . Tables 37 and 41 both include what appears to be the same distribution by age at onset of adolescence, but the frequencies differ. . . .

Table 40, p. 198, shows . . . 11,467 . . . 30 [years] or under. Table 41, p. 208, however, shows the number of cases from adolescence through 30 as 11,985. . . . Tables 104 and 105 . . . show fewer cases 32 years of age and under than are shown as 30 and under in Tables 40, 41, and 44. . . .

The numbers of cases shown in these clinical tables are hard to reconcile with one another, however, for the sum of the numbers shown in various subdivisions sometimes exceeds, and sometimes is exceeded by the number shown for the whole group.

. . . p. 63 . . . suggests that 300 *or less* is the number of items involved for any one person. . . . p. 50 suggests, however, that . . . the maximum history covers 521 items. . . .[4]

EXAMPLE 136B  COMMUNISTS IN DEFENSE PLANTS

In this example, a general impression that the study as a whole may be

---

4. W. Allen Wallis "Statistics of the Kinsey Report," *Journal of the American Statistical Association*, Vol. 44 (1949), pp. 463–484. The paragraphing here does not conform to the original.

### 5.3 Internal Evidence

highly reliable is augmented by the author's careful analysis of the degree of unreliability resulting from inconsistencies.

> Suppose a respondent says he would let an admitted Communist work in a defense plant, but would fire a store clerk whose loyalty is suspected but who swears he has never been a Communist. He *could* mean both of these things. But it is a good bet that one or the other response is wrong. . . . Inconsistencies become especially serious with respect to questions about which opinion is overwhelmingly on one side. For example, only 298 people of the entire national cross-section of 4,933 said they would not fire an admitted Communist in a defense plant. But analysis shows that as many as 100 of the 298 who said they would be lenient in this situation said they would *not* be lenient in *a variety* of situations which most people think are far less dangerous. These 100 people constitute only 2 percent of our entire sample, but they constitute a *third* of all who say they would not fire a worker in a defense plant.[5]

Thus, the analysis of internal inconsistencies suggests that the proportion reported as "no" on a particular proposition was probably at least half as much again as the proportion in the sample who really meant "no."

### 5.3.2 Irregularities

When averages or proportions move smoothly or regularly with respect to some related variable, at least a favorable presumption is established. The following example illustrates this kind of consistency:

EXAMPLE 137 NEONATAL MORTALITY

The following are the death rates in 1951 of infants under one week of age:[6]

| Age (days) | 0–1 | 1–2 | 2–3 | 3–4 | 4–5 | 5–6 | 6–7 |
|---|---|---|---|---|---|---|---|
| Deaths per 1,000 live births: | 9.8 | 3.1 | 2.1 | 1.1 | 0.6 | 0.5 | 0.3 |

If these figures had moved erratically from day to day, instead of changing continuously in the same direction (or perhaps with one reversal of direction), it would have raised doubts about their quality.

The regularity of data by no means proves their quality, of course, for spurious regularity could be introduced in many ways. Similarly, as Example 139 shows, irregularity is not prima facie evidence of poor quality, but only of grounds for special inquiry into its causes, with inadequacy of the data prominently in mind as a possible cause. The following two examples illustrate the fruits of investigating ir-

---

5. Samuel A. Stouffer, *Communism, Conformity, and Civil Liberties: A Cross-section of the Nation Speaks Its Mind* (Garden City, New York: Doubleday and Company, Inc., 1955), pp. 46–47.

6. *Statistical Abstract: 1954*, Table 83, p. 81.

regularities, and the third illustrates the danger of simply assuming that irregularity proves the data are in error.

EXAMPLE 138A  ROUNDING AGES

Ages reported in the Census cluster at numbers divisible by 5, and even more at numbers divisible by 10. There is, of course, nothing in the statistics of births, deaths, and migration to confirm a true clustering at multiples of 5. Undoubtedly it reflects inaccuracies in reporting. The following table shows that clustering has diminished with time, which suggests that the data are improving in quality.

TABLE 138

PERCENT OF POPULATION REPORTING AGES WITH EACH
FINAL DIGIT, SELECTED U. S. CENSUSES, 1880–1950.

| Year | 0 | 1 | 2 | 3 | 4 | 5 | 6 | 7 | 8 | 9 | Population (Million) |
|------|------|-----|------|-----|-----|------|-----|-----|------|------|----------------------|
| 1880 | 16.8 | 6.7 | 9.4  | 8.6 | 8.8 | 13.4 | 9.4 | 8.5 | 10.2 | 8.2  | 50  |
| 1920 | 12.4 | 8.0 | 10.2 | 9.4 | 9.4 | 11.3 | 9.7 | 9.4 | 10.6 | 9.6  | 106 |
| 1950 | 11.2 | 8.9 | 10.2 | 9.7 | 9.7 | 10.6 | 9.8 | 9.7 | 10.2 | 10.1 | 151 |

*Source:* Selected from Table 1 of an article by Ansley J. Coale, "The Population of the United States in 1950 Classified by Age, Sex, and Color—A Revision of Census Data," *Journal of the American Statistical Association*, Vol. 50 (1955), pp. 16–56.

Methods have been developed for correcting for this clustering, which, while not applied to the basic Census tabulations, undoubtedly give a more realistic picture of the ages of the population than do the basic tabulations.

EXAMPLE 138B  LUMINOUS INTENSITY

The same phenomenon is often found in laboratory measurements. For example, here are measurements of the luminous intensity of 12 light bulbs:

| | | | | | |
|------|------|------|------|------|------|
| 17.70, | 17.55, | 17.57, | 17.75, | 17.50, | 18.00 |
| 16.55, | 16.05, | 16.40, | 16.40, | 16.05, | 16.70 |

All but one of these measurements terminates either in 0 or 5. This suggests that measurements were typically made only to the nearest 0.05; thus "17.55" apparently means an intensity between 17.525 and 17.575, not between 17.545 and 17.555. If this tendency to round to 0 or 5 had not been noticed, it might have appeared that the measurements were made to the nearest 0.01. It is advisable to read instruments as accurately as possible, even to the point of having to guess a little about the final digit. The information lost by unnecessary rounding cannot be restored later by statistical analysis, but the inaccuracies in the last digit do tend to average out when a number of readings are made.

*5.3 Internal Evidence*

EXAMPLE 139 MEASLES IN PREGNANCY

... in New South Wales at the Census of 1911 the number of deaf-mutes was curiously and conspicuously high in the age-group 10–14. This, naturally, did not escape the census statisticians at the time, but they were—not unreasonably—inclined to ascribe the excess to a more complete enumeration of the deaf at the school ages.

Ten years later, however, in the census of 1921, the peak had moved on to the age-group 20–24. In other words, it related to the same cohort. The previous explanation would not hold, and the statistician in charge was then moved to write ". . . there is some evidence to suggest that the increase in incidence of deaf-mutism at certain ages synchronizes with the occurrence of epidemic diseases, such as scarlet fever, diphtheria, measles, and whooping cough."

... It is desperately easy to be wise after the event, but in the census returns and in the institutional data there is clear statistical evidence which might, with the aid of an epidemiological survey, have brought to light the phenomenon of the effects of rubella [German measles] in pregnancy many years before the alert clinical mind detected it [early in the second World War, in Australia].[7]

## 5.3.3 Extreme Values

Sometimes data contain extreme values that are patently unreasonable. For example, a person's age recorded as "115" would justify suspicion. An extreme observation is a kind of internal evidence that something went wrong with the data-gathering process, at least insofar as this one observation was concerned. But under many circumstances it would not be known for sure that a given figure is impossible. An investigator who discards data that do not conform to his preconceptions is apt to end up with data tailored to those preconceptions, thus defeating the very purpose of the study. There is no point in using energy and money in collecting data, only to waste them by picking and choosing among them to satisfy preconceptions.

The tendency to discard unusual observations is too prevalent. It is better to abide by a general rule never to discard observations. Like most rules, this one has exceptions. If, while the data are being collected, some accident or unusual event casts doubt upon a particular observation, it may be ignored. But the decision to ignore it should be made before looking at the number, and only on the grounds that the collecting process has failed; and the number should then be ignored whether or not it "looks" reasonable. (An example of good practice in this respect is provided by the vitamin experiment; see Sec. 2.8.3.) The detection of wrong observations casts doubt on the process by which the data have been produced and therefore on

7. A. Bradford Hill, discussion, *Journal of the Royal Statistical Society, Series A (General)*, Vol. 116 (1953), p. 7.

all other observations, even if they appear reasonable. The determination to be objective may be carried to the point of absurdity, but it seldom is; whereas all too often observations are thrown out on essentially subjective grounds.

Even when it is clearly established that certain measurements are wrong and should be excluded, it is important to be aware that discarding these observations, though justifiable, does not solve the problem of interpreting the remaining data. This is because the probability of sour measurements may be greater for certain types of observations than others. For example, a study of corporate assets would encounter the obstacle of inadequate record keeping in some small companies. Even if it were felt that information obtainable from these companies was worthless, exclusion of these companies from the sample would mean that the remaining companies would tend to be larger companies where record keeping was more satisfactory. An estimate of average assets per company would therefore tend to be too high. The statistician should at the least point out this bias and preferably should attempt to allow for it by making a special investigation of the discarded companies, or a sample of them.

### EXAMPLE 140  TRIPLICATE READINGS

Suppose that three chemical analyses are made on a sample of ore, and that two of these analyses agree closely as to the concentration of a particular mineral, while the third measurement differs widely from the other two. A common laboratory procedure is to discard the apparently "wild" measurement. To illustrate the unsoundness of this procedure, we selected at random ten samples of three measurements from a population in which the mean was known to be 2. Here are the results:

| Samples of Three Measurements | | | Mean of Two Closest Measurements | Mean of All Three Measurements |
|---|---|---|---|---|
| 0.724 | 0.782 | 1.547 | 0.753 | 1.018[a] |
| 1.682 | 1.201 | 0.336 | 1.442[a] | 1.073 |
| 0.623 | 0.743 | 2.495 | 0.683 | 1.287[a] |
| 4.334 | 1.663 | 0.045 | 0.854 | 2.014[a] |
| 0.864 | 2.642 | 5.436 | 1.753[a] | 2.981 |
| 2.414 | 1.989 | 2.666 | 2.540 | 2.356[a] |
| 1.506 | 2.364 | 0.763 | 1.134 | 1.544[a] |
| 3.048 | 2.037 | 2.759 | 2.904 | 2.615[a] |
| 2.347 | 4.816 | 1.536 | 1.942[a] | 2.900 |
| 2.637 | 2.563 | 1.893 | 2.600 | 2.364[a] |

[a] Indicates mean closer to population mean of 2.

Five of these samples present a temptation to discard a measurement. Yet in seven of the ten samples, the mean of all three measurements is closer

### 5.3 Internal Evidence

to the true mean of the population than is the mean of the two measurements closest to each other. This is not a sampling aberration of the particular measurements we have shown here; there are sound reasons why this kind of result is typical.[8]

EXAMPLE 141A  PEARL HARBOR

An erroneous rejection of extreme observations profoundly affected the course of history when, shortly after 7 o'clock on the morning of December 7, 1941, the officer in charge of a Hawaiian radar station ignored data solely because they seemed so incredible.

### 5.3.4  Spurious Regularity

The preceding three subsections have stressed irregularities. As we have seen in Chap. 4, there will be a certain amount of irregularity, that due to chance factors, even in the best statistical data. When there is less irregularity than would be expected from chance, there are grounds for suspecting that the regularity is spurious. A common cause of spurious regularity is simple dishonesty: the data have been "cooked." Those experienced in questionnaire studies are often able to detect dishonest—or grossly incompetent—interviewing by the lack of variety in recorded responses.

EXAMPLE 141B  IMAGINARY COIN TOSSES

A class of 65 students was asked to make up the outcomes of 37 imaginary tosses of a fair coin, trying to make them as realistic as possible. The results of the last five "tosses" for each student were tabulated, with the following results:

| Number of Heads: | 0 | 1 | 2 | 3 | 4 | 5 | Total |
|---|---|---|---|---|---|---|---|
| Number of Students: | 0 | 5 | 27 | 27 | 6 | 0 | 65 |

By application of the principles of probability presented in Chap. 10, it can be shown that the average or expected results, if the 65 students had actually tossed fair coins, are:

| Number of Heads: | 0 | 1 | 2 | 3 | 4 | 5 | Total |
|---|---|---|---|---|---|---|---|
| Number of Students: | 2.0 | 10.2 | 20.3 | 20.3 | 10.2 | 2.0 | 65.0 |

---

8. The samples have been formed by adding 2 to each of the first three entries from the first ten lines of the "Table of Gaussian Deviates" in The Rand Corporation, *A Million Random Digits with 100,000 Normal Deviates* (Glencoe, Illinois: Free Press, 1955). We have considered that there will be a temptation to reject the outlying observation whenever it is twice as far from the middle observation as is the remaining observation. On the average, about five-eighths of samples of three from such a population (normal) will show the greater difference more than twice the smaller difference from the middle observation. The precision obtained by averaging a certain number of values of the average-of-the-closest-two can be obtained by averaging only about half as many values of the average-of-all-three.

The students thus produced too many "typical" samples—2 or 3 heads—and too few "atypical" ones— 0, 1, 4, or 5 heads.

## 5.4
## RECORDING DATA

Care should be taken that the basic records show exactly what was observed, including any observations subsequently discarded. A sound practice, common in the natural sciences, is to record data in a way that distinguishes between numbers that were observed and those that represent the units in which these numbers are expressed. Thus, if a gas meter reads in hundreds of cubic feet and shows 2,391 on the dials, this would be reported not as 239,100 but as $2391 \times 10^2$. This indicates that the exact figure was between 239,050 and 239,150, whereas 239,100 might be interpreted as indicating an exact figure between 239,099.5 and 239,100.5. Similarly, if the last dial were judged to be a quarter of the way from 1 to 2, the record would read $2391.25 \times 10^2$, not 239,125.

When data are reported, however, they may be less accurate than the original records, as for example billions of dollars, or millions of people, if more detail would not be useful in the context of the report. In listening to election returns over the radio, for example, it is confusing to hear more than two or at most three figures—for example, 1 million, 8 hundred thousand, or perhaps 1 million, 850 thousand. Full detail, such as 1,853,428, may be so confusing, especially when there are several candidates, when several election areas are reported separately, or when several people in the room are commenting on the results as they are announced, that the hearer does not comprehend even which candidate is leading.

The numbers recorded should correspond to what is actually observed, as distinguished from what is inferred from the observed data or derived from them by computations. A few examples may help to make this point clear.

Example 142  Soap Defects

In recording the number of surface defects on bars of soap, the defects on one side were counted. On the assumption that there would be the same number on both sides, the number of defects was doubled before recording. Whether or not this assumption is correct (and it is at best true only on the average, not for each individual bar), the number of defects actually observed is the number that should be recorded. In this example, the absence of odd numbers in the data would provide internal evidence, similar to that

*5.4 Recording Data*

of Example 138A (Rounding Ages), that the numbers did not correspond to the defects actually present.

EXAMPLE 143A  INSURANCE PREMIUMS

A branch insurance office had not classified its premiums for a certain year in a way that was unexpectedly requested by the home office. The branch office assumed that the division of the total premiums into the various categories would be proportionately the same as in the previous year, and prepared figures on this basis. These data may be worse than none—for example, if the problem involved year-to-year changes in the proportions of premiums in the various categories. The central office would have been better off had it been told that the data were unavailable, and asked whether it wanted to incur the cost of getting them, or would be satisfied with estimates of the kind described; and, at least, the figures submitted should have been accompanied by a clear statement that they were estimates based on the proportions of the previous year.

EXAMPLE 143B  EMPLOYMENT AND PRODUCTION

Employment is often used as a basis for estimating production. The method is a useful one; in many industries, man-hours worked provides a better index than actual finished units (recall the discussion of automobile production in Sec. 5.2). But such an index of production should be accompanied by an explanation of how it was derived. Without this information there would appear to be a remarkably close relation between employment and production in these particular industries. If both the employment index and the production index are based on employment, the close correlation represents a mathematical truism rather than any significant relationship in the industrial world. A further complication in measures labeled "production" which are based on no facts except man-hours is that man-hours must be multiplied by a factor supposed to represent output per man-hour. This factor may be unreliable. Often it contains a more or less arbitrary upward trend—that is, it is increased systematically month by month (three percent per year is currently much in style). Someone comparing the output and employment figures will "discover" an upward trend in output per man-hour (called "productivity") which has simply been put there by the man who produced the output figures.

EXAMPLE 143C  IMPUTED OPINIONS

A friend of the authors was interviewed in a market research study designed to find out what consumers like and do not like about toothpaste. Our friend found himself completely unable to articulate his reasons for preferring the brand of toothpaste he was using. The interviewer suggested that perhaps it was something about the taste or maybe the price, but our friend was sure these were not the reasons. Finally, when the interviewer saw that no concrete reason was forthcoming, she recorded "superior cleansing ability." The interviewer had been instructed that "don't know" was an unacceptable answer. Another market research interviewer reported that respondents were extremely inarticulate and noncommittal, but that he had been able to report specific answers because he knew intuitively what the respondents actually meant. It is entirely appropriate to give the respondent every chance to reply fully, without, of course, actually suggesting answers. If a respondent's reply is recorded (and some record should always be made, even of uninformative responses), it should be as nearly as possible as he gave it and not as "interpreted" to show what he "really" meant.

### EXAMPLE 144A   HEART SIZE

To determine the size of a living person's heart from an X-ray film, a radiologist sometimes measures two diameters, multiplies them together, divides by the average ratio of this product to the projected area of the heart, multiplies by another factor to allow for parallax due to the heart's distances from the film and X-ray source, and expresses the resulting figures as a percentage of the average heart size for individuals of the given height and weight. The only direct observations in all this that have any relation to the particular patient are the two diameters. These, or possibly their product, might as well be expressed directly as a percentage of the average diameters. The intervening steps make the numbers seem more meaningful, and serve a useful purpose when data from different sources are being combined, but no amount of arithmetic will introduce any facts beyond the two diameters.

### EXAMPLE 144B   WATER CHLORINATION

This example, of which the consequences might have been tragic, was given us by a student who had lived for many years in the country involved:

> The water supply of a certain city of about 80,000 was known to be seriously contaminated. An efficient chlorinator was, therefore, always in operation at the intake to the city system. A full-time laboratory technician was stationed in the city to measure the chlorine content of samples of water taken daily at various points throughout the city. On one occasion, however, the chlorination system failed, and various provisions against such a contingency failed also. The failure was not discovered for more than a week. During the whole period that the water supply was not chlorinated, the daily reports of the water tests continued to show approximately the same chlorine content, with only normal variation. It developed that the technician had been submitting made-up data. Fortunately, however, the waterborne-disease rate did not rise—at least the numbers purporting to measure it did not.

## 5.5
## KINDS OF OBSERVATIONS

### 5.5.1   Univariate and Bivariate Observations

The basic facts of any statistical investigation are called *observations*. Sometimes the term *items* or *scores* is used instead. The thing observed is called a *variable*. Thus, an observation is a specific value of a variable. These ideas are best brought out by illustrations:

(1) In a study of the income and spending habits of 1,000 families, the investigator obtains (among other facts) the total income for each family. Income is the variable here, and a specific family's income is a single observation; there are 1,000 such observations in all.

(2) In another study, the investigator obtains the age of the "head" of each of 1,000 families. Age of the head of the family is the variable here, and the specific ages of the heads of the specific 1,000 families are the 1,000 observations.

In each of these samples, only one variable was observed. When the observation gives only one fact (that is, refers to only one variable), it is called a univariate ("one-variable") observation. It would also be possible to obtain both incomes and ages for the same 1,000 families. In this case each observation consists of a pair of numbers and is a bivariate ("two-variable") observation. If we have 1,000 univariate observations giving ages of the head and another 1,000 univariate observations giving incomes of the same families, we cannot obtain the 1,000 bivariate observations from these. Information about how the observations are paired is required. On the other hand, a set of 1,000 bivariate observations is easily converted into two sets of 1,000 univariate observations. The terms *trivariate* and *multivariate* are obvious extensions.

The "variables" represented by observations may be of two kinds: *quantitative* and *qualitative*.

## 5.5.2 Quantitative Variables

Whenever the observations refer to a measurable magnitude, the observations are called "quantitative." Family income and age of heads of families are illustrations of quantitative variables. The numbers describing income or age have two important characteristics: (1) they have operational meanings, and (2) differences, ratios, and other results of ordinary arithmetic operations have definite meanings.

5.5.2.1 *Continuous and Discrete Quantitative Variables.* It is impossible to measure cash income actually received in a given year more accurately than to the nearest cent. A family can have an annual cash income of $7,457.29 or $7,457.30, but nothing in between. Since possible values of the variable, cash income, move in *discrete* steps or jumps—one cent at a time in this example—it is a *discrete* variable, and measurements of income are discrete observations.[9] Similarly, the number of leaves on a tree is a discrete variable.

In measuring age, by contrast, one is not limited in this way. Theoretically, a person's age might be measured to the nearest month, the nearest week, the nearest minute, the nearest millisecond, or any degree of fineness. Hence one may think of age as a *continuous* variable in the sense that between any pair of numbers, however close together, it is possible to have another number. Measurements, however, are limited by the precision of measuring instruments, and records are limited by the number system. It might be possible to measure and record age to the nearest minute, for example, but not much finer. Hence measurements of age actually go in discrete steps or jumps, just as do measurements of income. Regardless of whether the subject of study is described by a discrete variable, or a continuous variable, recorded measurements themselves are discrete.

Often in statistical theory the assumption that variables are continuous is made because mathematical operations are then simpler—though harder for the beginner to understand. This assumption frequently gives results in good enough agreement with those obtained by more complicated methods which allow for the discreteness of actual measurements.

5.5.2.2 *Quantitative Comparisons.* We review here some of the arithmetical principles involved in quantitative comparisons. Joe Louis, when boxing, weighed approximately 220 pounds and Eddie Arcaro, when jockeying, weighs about 110. In what ways can one compare the weights of these two men?

(1) First, Louis is 110 pounds heavier than Arcaro and Arcaro is 110 pounds lighter than Louis. This type of comparison is known as an *absolute comparison*. The "absolute difference" of weights is 110 pounds. It is possible to make absolute comparisons because weight can be expressed in equal units, pounds in this example.

(2) Louis is twice as heavy as Arcaro and Arcaro is one-half as heavy as Louis. This comparison is called a *relative comparison*. Relative comparisons require not only equal units but a definite zero point as well; for example, it is incorrect to describe a temperature of 64° F. as twice as warm as if it were 32°, for the zero point is arbitrary and does not indicate absolute lack of warmth.

---

9. An economic definition of income would call 5 cents received in a two year period an income of $2\frac{1}{2}$ cents per year. When thus defined. as a *rate*, income is not a discrete but a continuous variable.

## 5.5 *Kinds of Observations*

Relative comparisons of this kind are frequently expressed in terms of percentages. Taking Arcaro's weight as the base of comparison, 110 pounds becomes 100 percent. Then Louis's weight of 220 pounds becomes

$$\frac{220}{110} \times 100 \text{ percent} = 200 \text{ percent.}$$

Thus, Louis's weight is 200 percent *of* Arcaro's weight, or, alternatively, 100 percent *greater than* Arcaro's weight. Both of these statements are precisely the equivalent to the earlier statement: "Louis is twice as heavy as Arcaro."

Next, make Louis's weight the base of comparison, by letting 220 pounds correspond with 100 percent. Arcaro's percentage must have the same ratio to 100 as Arcaro's weight has to Louis's weight. That is,

$$\frac{A \text{ percent}}{100 \text{ percent}} = \frac{110 \text{ lbs.}}{220 \text{ lbs.}} \qquad \text{or} \qquad A \text{ percent} = 50 \text{ percent.}$$

Again the result can be expressed in two equivalent ways: Arcaro's weight is 50 percent *of* Louis's weight, or Arcaro's weight is 50 percent *less than* Louis's.

For comparisons involving percentages, five precautions are worth noting:

(1) If one quantity is 0 it is 100 percent smaller than a non-zero quantity, but one positive quantity cannot be *more* than 100 percent smaller than another. Occasionally, such statements are encountered as: "Production of automobiles fell off 120 percent from last year." This is impossible, and it is not clear what may have been meant. Perhaps production last year was 220 percent of some base year's production, and this year returned to the base year level. In other words, the base of the 120 percent may not be last year's production.

Confusion is especially likely to occur in describing changes in percentages. If steel production rises from 80 to 90 percent of capacity, this is a relative rise of 12.5 percent or an absolute rise of 10 percent. Such confusion can be reduced by referring to the absolute rise as "10 percentage points."

It might seem reasonable to describe a change from, say, a profit of 100 in one year to a loss of 100 in the following year as a decrease of 200 percent, since the new level is below the old level by 2 times the old level. But if this definition were applied in reverse, a change from a loss of 100 to a profit of 100 would also have to be described as a decrease of 200 percent, since the new level is also obtained by subtracting 2 times the old level from the old level. It is therefore best not to use percentages to measure changes in quantities whose signs alter.

(2) <u>Percentage changes that are equal in magnitude but opposite in sign do not offset one another</u>. More generally, adding two successive percentage changes does not give the total percentage change. This is illustrated by the following example.

EXAMPLE 148   MOTOR VEHICLE SALES

Factory sales of motor vehicles, which were just over 8 million in 1950, declined in 1951 by nearly 1¼ million, or 15 percent. Had they then increased in 1952 by 15 percent, they would not have returned to 8 million, or 100 percent of the 1950 level, but only to 7.8 million, or 97¾ percent of the 1950 level. After the 15 percent decrease in 1951, it would have required an 18 percent increase for 1952 to regain the 1950 level. The fact is that sales decreased in 1952 by nearly 1¼ million again, but this time the decrease was not 15 but 18 percent, since the base was smaller, 6¾ instead of 8 million. And, though the 1953 figure rose more than 32 percent, this did not offset (or come within 1 percent of offsetting) the 15 and 18 percent falls of 1951 and 1952, but left sales still two-thirds of a million vehicles, or more than 8 percent, below the 1950 level.[10]

(3) <u>Percent comparisons are awkward if one quantity is many times as large as another</u>. The population of the United States is about 11 times as large as the population of Canada. It would be perfectly correct to say that the population of the United States is 1,000 percent larger than that of Canada, but it is hard to grasp the meaning of "1,000 percent" larger, while "11 times as large" is quite easy. When there are large relative differences between quantities being compared, percentages should be avoided.

(4) Instead of saying either that Louis's weight is 200 percent *of* Arcaro's, or 100 percent *larger than* Arcaro's, one might be tempted to state, incorrectly, that Louis's weight is 200 percent more than Arcaro's. A person reading this last statement would be led to believe that Louis is three times as heavy as Arcaro. Often misrepresentations of this kind are deliberate in propagandist writing, and once in a while they creep into disinterested studies because of ignorance or carelessness. In fact, this type of misrepresentation is so common that in the absence of the absolute numbers from which the percentages are computed, there is always doubt as to whether, say, a "200 percent increase" really means a threefold or a twofold increase.

---

10. The sales figures, for the benefit of those who want to calculate more precisely the changes stated in the text, were: 1950—8003 thousand; 1951—6765 thousand; 1952—5539 thousand; 1953—7323 thousand. *Statistical Abstract: 1955*, Table 664, p. 551.

(5) The base of a percentage, when it is not explicitly stated, may not be what you take for granted. Profits expressed as percentages in the real estate business, for example, commonly represent percentages of the purchase price, but in the furniture business they represent percentages of the selling price. Thus, to buy for $1,000 and sell for $1,250 represents a 25 percent profit in real estate parlance but a 20 percent profit in furniture parlance.

### 5.5.3 Qualities or Attributes

Sometimes each observation tells only that a particular quality or *attribute*—such as "red hair," "Democrat," "over 65 years old," etc.—is present or absent. Observations of this kind are still amenable to statistical treatment, simply by counting or enumerating the number of observations in which a given quality is present and the number in which it is absent. The terms *enumeration data* and *attribute data* are sometimes applied to qualitative observations.

The results of individual psychotherapy, for example, might not be describable in terms of quantities, but the percentage of cures achieved, according to some definition of a "cure," is a number that can be interpreted statistically; so is the number of times that a single individual is cured on a particular occasion (necessarily either 0 or 1).

Sometimes qualitative observations can be ranked. The grade "A" can be ranked above the grade "B," "B" above "C," etc., for grades prepared by the same method. Attitudes might be ranked as "very favorable," "favorable," "neutral," "unfavorable," and "very unfavorable." However, there would, except for special purposes, be no meaningful way to rank the categories, "red hair," "blond hair," etc.; drug store, grocery store, clothing store, etc.; or Harvard, Yale, Columbia, Princeton, Cornell, Chicago, Stanford, etc.

Even if qualitative observations are ranked, arithmetic operations on the ranks are meaningless for describing the relations among them. For purposes of calculating "average grades" in high school or college, for example, numbers are sometimes assigned to letter grades: "F" may be $-2$; "D," 0; "C," 1; "B," 2; and "A," 3. Except for the fact that they should either increase or decrease from "F" to "A," these numbers are essentially misleading. It is not true, as the numbers imply, that an "A" represents three times as much of something as a "C." Someone else might prefer to attach the numbers, 0, 1, 2, 3, 4 to these grades, or 1, 2, 4, 8, 16, and there would be no basis for arguing that his scale was less valid than $-2$, 0, 1, 2, 3. This device may be useful as an administrative convention for designating honor students, but it should be understood that it is arbitrary, however precise and objective the arithmetic may look. With a different, equally reasonable, scoring system there would be some differences in the list of honor students.

The same general remarks apply to the field of attitude measurement. It may be useful to rank attitudes and attach numbers to them, but the resulting numbers cannot be interpreted as quantities. In particular, it is meaningless to say that Private K likes the army twice as much as Private P, or that a certain baseball fan prefers the Brooklyn Dodgers four "emotes" more than the New York Giants. While it is conceivable that attitudes might be put into numbers in such a way that the units would be equal, and possibly even so that the zero point would be unique, most attempts at psychological scaling claim no more than to place results in the right order.

## 5.6
## OBTAINING INFORMATION BY COMMUNICATION

Much of the basic statistical data used in the social sciences, in business, and especially in public affairs is obtained by communicating with people either orally or in writing. While methods of obtaining information by communication cannot be covered adequately within the scope of this book, a few general ideas and specific rules of thumb may be pertinent. We shall consider three of the problems which must be faced: interviewing, design of the questionnaire, and summarizing individual responses.

Before going into these matters, however, it is worth while to point out that the process of getting information by communication is usually susceptible to both deliberate and unconscious abuse. The propagandist—and he is everywhere—has the choice of many subtle ways of obtaining answers which seem to prove his point. Even the objective investigator often can be misled by a mishandling of the communication process. It is not at all easy even for the expert to obtain sound measurements by communication, and the novice, who accounts for a high fraction of all questionnaires, is even more prone to error. Moreover, responses to questions, even when there are no faults of the kinds just mentioned, may be irrelevant or seriously misleading for many problems. For example, it has sometimes been concluded that businessmen do not try to maximize profits simply because they say that they do not, or even because they are not familiar with such economic concepts as "marginal cost" and "marginal revenue." Thus, in interpreting the answers to questions, as in interpreting statistical measurements generally, there is need to consider the basic issue of the relationship between numbers and the real world—in this case, the connection between what people say and what they do or what they really think.

## 5.6.1 Interviewing

The person doing the interviewing unavoidably influences the quality of the information collected. Example 70A suggested that white interviewers obtained distorted answers from Negroes on the question of the fairness of the Army toward Negro soldiers. You need only recall the times you have been interviewed to recognize how much influence the interviewer had in determining what you said. Truly skillful interviewers are scarce; not every capable person can become a good interviewer. While such qualifications as honesty and interest in the work are as essential to good interviewing as to good work in other fields, several special qualifications are required, especially two: (1) a genuine understanding and appreciation of the need for objectivity and neutrality on the part of the interviewer and (2) the ability to make the respondent feel at ease so that he will be willing to make a serious attempt to answer the questions carefully. In other words, the interviewer must probe for responses without letting his own opinions intrude and without making the respondent irritated or self-conscious in a way that would lead to inaccuracies. Experience in interviewing, in the absence of proper training and of skill in putting people at ease, does not guarantee good interviewing. General intelligence and good education help, chiefly through making the interviewer aware of the need for objectivity and of the real danger that his own opinions and personality may affect the respondent's answers in subtle ways.

How can the reader of statistical reports based on surveys tell whether or not the interviewing was well done? A few rules of thumb provide some guidance. As in all aspects of statistical work, the greater the detail in which the report describes the methods used, and the fewer the vague claims, such as "highly trained interviewers," the better the work is likely to be. The description of method may tell something about the selection, training, and supervision of interviewers, and this may yield clues to the quality of the interviewing. If two or three days were devoted to training interviewers, if carefully written instructions were prepared, etc., the results are likely to be better than if the interviewers were simply given questionnaires and told to collect interviews. Finally, when one is familiar with a particular field, he is likely to learn the reputation of the organizations who do surveys in that field. The marketing research analyst usually has some idea of the interviewing methods used by firms engaged in marketing research, while the social psychologist learns the strong and weak points of the various centers for attitude and opinion research.

### 5.6.2 Questionnaires

The phrasing and sequence of questions is difficult and important. Questionnaires must always be tried ("pretested") under realistic conditions before being put into use, and often retrials are necessary to test revisions made in the light of the first trials.

(1) *Definition of Objectives.* The information desired should be defined carefully and precisely. If this essential first step is slighted, troubles will multiply as the investigation proceeds: unnecessary or irrelevant data will accumulate, while really vital data are not obtained at all or are lost in a mountain of worksheets and questionnaires. It is necessary to confine the scope of any one questionnaire by establishing priorities for the things that are really needed over those that merely would be "interesting to find out about." Preliminary investigation, including study of previous surveys in the same field, is needed. "Problems" seldom come to the statistician in well labeled packages, while scientists frequently must devote long periods to relatively aimless exploration. But when information-gathering is formally begun, it is essential that objectives be defined clearly. Ideally, the objectives of the study should be formulated explicitly and actually enumerated in the final report.

Only after it has been decided exactly what information is wanted can methods be chosen for getting it. It is necessary to devise a series of operations by which abstract concepts are translated into actual measurements of some kind.

(2) *Ability and Willingness to Answer Questions.* Two basic requirements must be met by any question: the person who is to answer the question must be *able* to furnish the information desired and he must be *willing* to do so. Your neighbor may be willing but unable to divulge certain facts about your personal finances or your family life, while you may be able but unwilling to do so.

Unwillingness may be reflected in a high rate of "refusal" by people approached for interviews. Sometimes, also, people will permit themselves to be interviewed yet answer some questions incorrectly and refuse to answer other questions at all. The reader of a statistical report should try to find out how many people refused to be interviewed and how many did not give answers to individual questions, as evidenced by the number of responses labeled "no answer," "no response," "not ascertained," etc. Failure to report this information is a major fault in a statistical study.

Inability to furnish information is clear-cut when a person is asked about things of which he knows nothing. Special problems arise, however, when the person is asked about something of which he knows a little and can guess more, as when a husband is asked about his wife's expenditures, or when someone is asked about things nearly forgotten, such as a radio program a month ago, or experiences in childhood.

Whenever possible, objective checks should be made to find out how honestly people are replying and how able they are to give the information sought. In a market research study it was discovered that many people, asked what brand of flashlight battery they were using, named the most familiar brand when actual inspection showed they had some other.

There are many devices for overcoming inability and unwillingness. Methods of "aided recall" have been devised by psychologists to aid memory. Numerous questioning methods have been evolved to overcome unwillingness. A classic illustration of the latter is the question "When do you intend to read *Gone with the Wind?*" Actual readers replied that they had already read the book, while the rest were given an opportunity to "save face" by replying that they intended to read it soon.

(3) *Avoiding Ambiguity in Wording.* The wording of the question should be completely unambiguous—though this is sometimes nearly impossible if people of divergent backgrounds are being interviewed. Students, faced with ambiguous examination questions, to which several interpretations are logically possible, rely on their knowledge of the instructor in order to guess which answer he wants. Questionnaires are even more difficult to prepare than examinations, because respondents usually are not as strongly motivated, as well-informed about the field of the questions, or as alert and intelligent as a student taking an examination. But the goal to be sought is a wording that eliminates alternative logical interpretations and suppresses emotional overtones and unusual words which might lead to misinterpretation. It is best to use short sentences, uncomplicated sentence structure, and common, specific words. Finally, it is advantageous to explore important points by several different but related questions, since this not only gives a rounded view of the respondent's opinions

but also allows for such checks of internal consistency as that described in Example 136B (Communists in Defense Plants).

## EXAMPLE 154A  AIR FORCE QUESTION

In the question, "Should the United States have a large air force?" the word "large" is vague. Even if the respondent were asked to choose among several sizes his answer would not have much meaning without some idea of how well he understood the costs and uses of, and the alternatives to, air weapons. To some people one thousand planes might seem a large air force; to others, 143 groups might seem small.

## EXAMPLE 154B  LAW SCHOOL COURSES

Ambiguity that could easily have been avoided by pretesting the question is described in the following quotation:

> Professor Cheatham sent a questionnaire to all the law schools. In one of the questions inquiry was made as to whether the school addressed was offering a course on the professional functions of the lawyer. Approximately two-thirds of the schools answered that they were giving such a course. The remainder said they were not. He felt, however, that these figures were hardly dependable, for, from the answers given, some deemed it a "course" if the school sponsored a few lectures on the formal rules of professional conduct, while others, even if such lectures were given, did not consider that those lectures could be called a "course."[11]

## EXAMPLE 154C  DESIRE FOR REFORMS

A nation-wide Gallup Poll, published August 21, 1943, presented results on the question:

> "After the war, would you like to see many changes or reforms made in the United States, or would you rather have the country remain pretty much the way it was before the war?"

> ... In view of the fact that the particular phrasing of this question seemed likely to render it especially susceptible to heterogeneous interpretations by respondents, it was decided to make a new study of this same question with the objective of ascertaining the variety of contexts in which the question tended to be answered.

> A small scale poll repeating the Gallup question was made in the New York City area between August 30 and September 4, 1943. ... In all, 114 interviews were taken.

> ... The procedure in each interview was as follows: The interviewer introduced himself, explaining that he was ... making a survey among randomly chosen persons in New York. The standard poll question (see above) was then asked. Instead of ending the interview at this point, or going on to another specific question, as in customary poll practice, the interviewer then encouraged the respondent to enlarge upon his answer in an informal conversational way. The interviewer asked no direct questions at this point; he simply urged the

11. Albert J. Harno, *Legal Education in the United States* (San Francisco: Bancroft-Whitney Company, 1953), pp. 156–157.

**5.6 Information by Communication**

respondent to extend and explain his answer. . . . The respondent was then asked two direct questions. If he had voted for "changes or reforms," he was asked:

a) "What sort of changes or reforms would you like to see?"
b) "Are there any changes or reforms that you wouldn't like to see?"

If he had voted for things to "remain the same," he was asked:

a) "What sort of things wouldn't you like to see changed or reformed?"
b) "Are there any changes or reforms you would like to see?"

A full verbatim account of each interview was taken, and the sex, age, income and education of the respondent recorded.

. . . A careful study of the verbatim reports of the free discussion following answers to the direct poll question clearly reveals a wide diversity in the interpretations which people make of the question. On the basis of the spontaneous discussion by the individual and the specific things to which he refers as requiring or not requiring changes or reforms it is possible to ascertain roughly which one of a number of frames of reference seems to be operative for each individual in responding to the poll question. Each of the 114 interviews has been rated by the authors as to frame of reference. Most of the interviews seemed to fall naturally into seven main categories; the remaining interviews were rated as non-ascertainable as to frame of reference. The seven categories . . . are as follows:

1. *Domestic changes or reforms.* This frame of reference would seem to be the one *intended* by the Gallup question. Responses classified under this heading are those referring to various social, economic and political reforms within the United States. . . .

2. *Technological changes.* Respondents classified under this heading are those who seem to be thinking of technological changes only. . . .

3. *Basic political-economic structure of the United States.* Respondents classified here are those who seem to take the question to refer to drastic changes in the United States, such as in the Constitution, in democracy, in "our way of life". . . .

4. *Foreign affairs of the United States.* These respondents seem to interpret the question as referring to changes in our relations with other countries. . . .

5. *Immediate war conditions as the standard of judgment.* Some respondents seem to answer the question in terms of a comparison of the peaceful state of affairs before the war with the current unpleasant war conditions. They do not really express an opinion about *post-war* changes or reforms. . . .

6. *Immediate personal condition as the standard of judgment.* In these interviews the answers are made narrowly in terms of a consideration of whether the individual himself is satisfied, not of whether changes or reforms in general in the country are desirable. . . .

7. *Desirable state of affairs in general as the standard of judgment.* In these cases the question is parried. No real content is embodied in the answer; there is merely an affirmation of the desirability of a good state of affairs. . . .

*Frame of reference non-ascertainable.* . . .

The percentage of people giving each interpretation to the poll question (and hence the relative importance of each frame of reference in determining the final poll tabulation) is given [below]. It appears that roughly one-third of the sample of respondents interprets in frames of reference other than "Domestic changes or reforms," which is, presumably, the *intended* frame of reference for the question.

| Frame of Reference | Percentage of respondents |
|---|---|
| Domestic changes or reforms | 63% |
| Technological changes | 2 |
| Basic political-economic structure of U. S. | 10 |
| Foreign affairs of the U. S. | 3 |
| Immediate war conditions as standard of judgment | 4 |
| Immediate personal condition as standard of judgment | 7 |
| Desirable state of affairs in general as standard of judgment | 4 |
| Non-ascertainable | 7 |
| | 100% |
| | $N = 114$ |

The analysis proceeds to show, among other things, that the desire for "changes or reforms" depended on what kind of changes or reforms the respondent had in mind.[12]

(4) *Neutrality of Wording.* Answers will vary with the wording of questions even though the literal meaning is the same. Consider the following two questions:

"You approve of rent control, don't you?"
"You don't approve of rent control, do you?"

Probably a larger proportion of people would indicate approval of rent control in reply to the first question than to the second, yet literally interpreted the two questions ask the same thing. A more "neutral" wording, such as "Do you approve or disapprove of rent control?" would be preferable. However, no wording is correct in any ultimate sense: the purpose of the study must be considered. If a study purporting to show opposition to rent control had employed the question, "You don't approve of rent control, do you?" its conclusions would be, to say the least, questionable. It is possible, however, that for certain purposes this question might be quite legitimate —for example, to find out how many people are strongly enough in favor of rent control to say so even when asked a leading question suggesting the opposite.

Sometimes surveys use different forms of a question with different respondents. If the results are fairly similar for the different forms, this is often interpreted as showing that opinion is fairly well crystallized, hence not subject to slight influences. Low proportions not answering or answering "don't know" (referred to as NA's and DK's in the jargon of the opinion surveyors) also are sometimes interpreted this way.

12. Richard S. Crutchfield and Donald A. Gordon, "Variations in Respondents' Interpretations of an Opinion-Poll Question," *International Journal of Opinion and Attitude Research*, Vol. 1 (1947), pp. 1–12.

The sequence of questions, as trial lawyers know, may affect responses. If a respondent has just answered that he thinks his landlord would increase his rent if rent controls were removed, he may be more inclined to say that he favors rent control.

(5) *Rule-of-Thumb Checks.* Before studying the results of a survey, the wording and sequence of questions of the questionnaire should be examined. While specialized experience with questionnaires is needed for real expertness, common sense alone is often good enough to detect serious errors.

One good rule-of-thumb is to answer the questions yourself, trying to put yourself in the respondent's place if the questions do not apply to you, to see if you would understand them completely and would be able and willing to answer them. If the questionnaire fails this test, there is reason to question the information obtained from it. If it passes this test, it still may have deficiencies as an instrument for obtaining information from a "cross-section" of the population. College students, for example, would probably have no trouble in understanding the expression "frame of reference," yet it would be meaningless to many people. So if you understand the questions yourself and would be willing and able to answer them, you must then ask if the information requested is of the kind that most people would be willing and able to give.

## 5.6.3 Coding

Questions may be grouped into two categories with respect to the way in which they are to be answered: "fixed-alternative," and "open-ended" or "free-response." In the fixed-alternative question, the respondent is given a limited number of responses, such as "yes," "no," and "don't know." In the free-response question, the respondent answers in his own words and the interviewer records them as nearly verbatim as he can. Sometimes the two methods are combined; a person might be asked if he approved or disapproved of rent control (fixed-alternative), and then asked for his reasons (free-response).

Regardless of the type of question used, there is a problem in classifying the respondent's answer. With the fixed-alternative question, this classification is done jointly by the question-writer, who establishes the categories, and the interviewer, who puts the responses into the categories. With the free-response question the answers actually recorded by the interviewer must be classified later into a relatively small number of groups so that the main findings can be summarized. The technical name for the process of classifying answers to free-response questions is "coding." Some understanding of this process is useful in interpreting statistical findings. A few standard requirements of coding may be mentioned.

(1) Each answer should fit into at least one category. In other words, the categories ought to be exhaustive.

(2) Each answer should fit into only one category. In other words, the categories ought to be mutually exclusive. This condition, while desirable, is not essential and is often not feasible. If, for example, the response to "What kind of things do you worry about most?" is "Paying bills. My husband has been in the hospital and may have to go back again," it might be classified either as "Personal Business or Family Finance" or as "Health of Self and Family" if these are two of the categories.[13] No set of categories is likely to avoid such double classifications for an "open-ended" question of this kind. The drawback to having a single answer classified in more than one category is that the percentages, in the tabulation of number of respondents giving each answer, will exceed 100, thus complicating comparisons among questions which differ in the total number of responses given, and tending to overemphasize the views of respondents who give longer or more comprehensive answers, which are classified under several headings.

(3) If the question itself is ambiguous, such as "Do you think you are better or worse off than before the war?" the separate interpretations of the question (personal health and happiness, national affairs, etc.) should be kept separate. The answers of those who thought the question referred to national affairs, for example, should if possible be tabulated separately from the answers of those who thought it referred to personal affairs. This will bring into the open the fact that the question was ambiguous, and make it possible to salvage something. The answers reveal, at least, the things people think of when "well-being" is mentioned.

(4) The categories should be chosen for their pertinence to the subject being studied. Ordinarily responses would not be classified according to their length, for example—though they are so classified in some psychological tests, for example, the Rorschach test.

(5) The number of categories used must represent a compromise between the need for summarization and the need for knowing the nuances and fine shades of meaning in the individual responses.

13. The question and response are taken from an actual survey made in May, June, and July, 1954. The answer was actually classified as concern over personal business or family economic problems. Samuel A. Stouffer, *Communism, Conformity, and Civil Liberties: A Cross-section of the Nation Speaks Its Mind* (Garden City, New York: Doubleday and Company, Inc., 1955), p. 61.

(6) The process of coding should be "reliable" in the sense that different people would agree pretty well on the category into which each response should be classified.

(7) The number of people who did not answer the questions or whose answers were not recorded by the interviewer should be shown in a "no answer" or "not ascertained" category. Even if this category is not shown in the report of a study, there usually are some responses of this kind. It is often possible to infer from the internal evidence of the findings whether or not there were "no answers." If there is a large number of "no answers," allowance must be made for the fact that people who do not answer are frequently different from those who do. In the 1948 presidential pre-election surveys, for example, a large number of people were undecided, and there is evidence that these people voted more heavily for Truman than did the group of people who had made up their minds.

## 5.7
## CONCLUSION

The meaningfulness of the decisions reached from a statistical analysis depends upon the relation between the numbers that are analyzed and the real world to which the decisions will relate. The numbers that are analyzed arise from a series of operations, sometimes as simple and intuitively meaningful as counting the number of times a stick can be laid along the edge of a table and writing down the number, sometimes so complicated that a vast body of scientific and technological knowledge accumulated over centuries is necessary to understand the relation between the final number and the particular aspect of the world that is of interest. In principle, the statistician, as statistician, need not be an expert on these measuring processes. In practice, his experience is likely to throw light on aspects of the measuring processes that would be overlooked by specialists without statistical training. At any rate, the user of a statistical report must inquire about the basic data before he bases any conclusions on them, and on the whole the statistician who worked with the data is best qualified to do this for him.

Even with controlled conditions, repeated measurements of the same thing typically vary somewhat. The less they vary, the more precise the measurements are said to be. However much individual measurements may vary, averages of groups of them will (by the Law of Large Numbers) vary less; the larger the groups averaged, the less the variation. If, however, the operations producing the numbers are unrelated to the subject about which information is desired, the average of a large number of observations, even though quite precise, will lack relevance. If, for example, we read a large number of clocks thinking they are thermometers, we get an average "temperature" of considerable precision but little, if any, relevance.[14]

Whether numbers are useful thus depends basically on the operations connecting them with the real world, but it is sometimes possible to get clues about their usefulness from the internal evidence within a body of data. Such clues may be provided by the self-consistency or inconsistency of different observations or averages, by the regularity or irregularity of variation in the data when they are classified in certain ways, or by the plausibility and reasonableness of the data.

For proper interpretation of data, they should be recorded exactly as observed, not as inferred from what was observed. They should be expressed so as not to confuse units of measurement with the observations: for example, if one gross (144) is the unit of measurement and 17 are observed, this should be recorded as 17 gross, not as 2,448, or if a length is measured in sixteenths of an inch it should be recorded as, say, $7\frac{4}{16}$ inches, or even 116 sixteenths, but not as $7\frac{1}{4}$ inches. Reports, of course, should show only the amount of detail pertinent, not necessarily the full detail of the original records.

Observations are classified as univariate, bivariate, trivariate, or multivariate, according to the number of variables observed jointly. They are classified as quantitative or qualitative, according as they record measurements (or counts) or simply non-quantitative attributes. Quantitative variables are classified as continuous or discrete, according to whether any value is possible within a certain range, or only particular values; actual recorded measurements are always discrete, even when the variable measured is continuous. Qualitative variables are dealt with statistically by counts of the number of times they occur.

A kind of measurement of general public interest, about which statisticians are especially expected to be informed, is that occurring when information is obtained by asking people questions. Such surveys require expert interviewing; careful attention to the questions, to bring out the information really wanted, to ask about things that people will be willing and able to answer, to avoid ambiguity and to

---

14. The wording "little, if any" is to allow for the fact that (depending on what we are measuring the temperature of) there might be some relation between the time and the temperature, in which case the clock readings would contain some information about the temperature.

avoid influencing answers; and careful coding of answers for statistical summarization.

To comprehend a large body of data, and to communicate them to others, they must be organized, summarized, and presented in compact and intelligible ways that illuminate their salient features. The next four chapters deal with such statistical description.

## DO IT YOURSELF

### Example 161A

Re-study Example 117. It is possible that the quotation is wrong in ascribing the error to the method of sampling alone. Can you think of an alternative explanation consistent with the incident described?

### Example 161B

If you are a baseball fan, try carefully to define "batting performance" so that it could be measured. How would your definition compare with current measurements, such as "batting averages" and "slugging averages"? Do the same thing for pitching and fielding performance.

### Example 161C

More than a dozen athletes are reported to hold jointly the world's indoor record for the 60-yard dash, 6.1 seconds. Does that mean that all these men would run a dead heat if each could duplicate his best performance in the same race?

### Example 161D

Comment on the following interpretation of measurements of height:

At present, 70 percent of the female population of the United States is still 5 feet 5 and under. About 8 percent is 5 feet 8½ and over. This leaves only 22 percent in the size range usually called average.[15]

### Example 161E

The number of freight cars on American railroads is reported to have declined by nearly 20 percent from 1916 to 1951. Does it follow that there was too low a supply of freight cars in service in 1951? Give your reasoning in detail.

### Example 161F

In a study of the effectiveness of a counseling technique in alleviating personality disturbances, a basic measurement of the effectiveness was the number of counseling sessions to which each patient returned voluntarily.

---

15. *Chicago Daily News*, June 9, 1952.

The more sessions, the more effective the technique was assumed to be. What possible strengths and weaknesses of this measure occur to you?

EXAMPLE 162A

Because of alarm about the frequency of accidents in its plants, a large company hired a safety director to institute a special program of safety indoctrination. Almost immediately after the institution of this program, the accident rate decreased sharply. What information would you like to have before concluding that the safety program was a success?

EXAMPLE 162B

Think back to the most recent physical examination you have had. Do you remember any particular measurements which seemed to be inaccurately taken?

EXAMPLE 162C

Comment critically on the following quotation:

In the course of holding post-combat mass interviews with approximately 400 infantry companies in the Central Pacific and European Theaters [in World War II], I did not find one battalion, company, or platoon commander who had made the slightest effort to determine how many of his men had actually engaged the enemy with a weapon. But there were many who, on being asked the preliminary question, made the automatic reply: 'I believe that every man used a weapon at one time or another.' Some added that wherever they had moved and viewed, it had seemed that all hands were taking an active part in the fighting.

Later when the companies were interviewed at a full assembly and the men spoke as witnesses in the presence of the commanders and their junior leaders, we found that on an average not more than 15 per cent of the men had actually fired at the enemy positions, or personnel with rifles, carbines, grenades, bazookas, BARs, or machine guns, during the course of an entire engagement. Even allowing for the dead and wounded, and assuming that in their numbers there would be the same proportion of active firers as among the living, the figure did not rise above 20 to 25 per cent of the total for any action. The best showing that could be made by the most spirited and aggressive companies was that one man in four had made at least some use of his fire power.[16]

EXAMPLE 162D

If you are a college or university student, find out how enrolment at your school is defined and measured. Can you suggest any improvements?

EXAMPLE 162E

The averages of three different methods of testing iron content in crushed blast furnace slag were, for the same sample of slag, 0.03 percent, 0.17 per-

---

16. S. L. A. Marshall, *Men Against Fire: The Problem of Battle Command in Future War* (New York: W. Morrow and Company, 1947), pp. 53–54.

*Do It Yourself*

cent, and 0.095 percent. These differences are much more than would be expected by chance. Does this demonstrate that at least two of the three methods are useless? What further information would you like to have in answering this question? Discuss.

EXAMPLE 163A

A statistician was once shown some experimental data purporting to measure two different performance characteristics of a weapon. Although unfamiliar with the technical details of the experiment or of the measurements, he quickly suggested that (1) only one performance characteristic was actually measured and (2) one of the numbers on the page was inaccurate. Both comments turned out to be correct. The actual data are not available but the following contrived data illustrate the point:

| Weapon | Performance Characteristic (1) | Performance Characteristic (2) |
|--------|--------------------------------|--------------------------------|
| A | 0.00058 | 0.0820 |
| B | 0.00058 | 0.0662 |
| C | 0.00078 | 0.1103 |
| D | 0.00044 | 0.0622 |
| E | 0.00054 | 0.0764 |
| F | 0.00066 | 0.0933 |
| G | 0.00059 | 0.0834 |
| H | 0.00047 | 0.0665 |
| I | 0.00066 | 0.0933 |
| J | 0.00059 | 0.0834 |
| K | 0.00055 | 0.0778 |

What internal evidence in the data led the statistician to make these comments?

EXAMPLE 163B

A famous psychologist is reported to have said, "Intelligence is what intelligence tests measure." What different things could he have meant by this comment? What do you think about each of these?

EXAMPLE 163C

One major source of dissension between the Free World and the Soviet Union has been the failure of the Soviets to return all prisoners of war captured in World War II. There have been numerous disputes over the actual number of prisoners still held in the Soviet Union, the Soviet claim always being much smaller than the Western claim. What possible sources of divergence between claim and fact, aside from deliberate misrepresentation, might there be: (1) in the Western claims and (2) in the Soviet claims?

## Observation and Measurement

### EXAMPLE 164A

According to the *Statistical Abstract: 1955*, p. 71,[17] the two leading causes of death in 1953 were (1) diseases of the heart and (2) malignant neoplasms (that is, cancer). The annual rates per 100,000 were 357.8 and 144.7, respectively. Make and interpret all the percentage comparisons between these two numbers that seem meaningful.

### EXAMPLE 164B

Suppose you are given the job of finding out why employees leave a given company. Draft a questionnaire, and explain just how and when it would be administered.

### EXAMPLE 164C

... Consider Judge Frank's personally conducted opinion poll in a trademark case: "I have asked a dozen American men and women, selected at random, what Touraine means; their invariable reply was 'a part of France'" (*La Touraine Coffee Co., Inc. v. Lorraine Coffee Co., Inc.*, 157 F. 2d 115, 120 [C.C.A. 2, 1946]). "I have tried the same question on members of several of my classes in Trade Regulations and have not yet obtained the reply Judge Frank invariably received" (Kennedy, "Law and the Courts," in *The Polls and Public Opinion* 92, 102, n. 43 [Meier and Saunders (eds.), 1949]).[18]

Comment on possible reasons for the difference in the results obtained by Judge Frank and Mr. Kennedy.

### EXAMPLE 164D

In Example 136B, what reservations would you have about this basic assumption: "He *could* mean both of these things. But it is a good bet that one or the other response is wrong."

### EXAMPLE 164E

The following questionnaire was used in a study of trends in the corporal punishment of children: Age? _____ Male or Female? _____ Were you spanked? _____ How often? _____ Do you spank your children? _____ How often?_____

This questionnaire was mailed to 2,000 persons selected at random from the list of registered voters in a certain city of about 600,000 population.

(1) Criticize the questions, indicating revisions in any that you consider unsatisfactory.

(2) Assuming that necessary revisions in the questions were made, what difficulties would you anticipate in drawing conclusions from the returned questionnaires?

---

17. Original source: National Office of Vital Statistics, annual report, *Vital Statistics of the United States.*

18. Fred M. Kecker, "Admissibility in Courts of Law of Economic Data Based on Samples," *Journal of Business*, Vol. 28 (1955), p. 122.

# PART II
# STATISTICAL DESCRIPTION

# Chapter 6

# *The Art of Organizing Data*

## 6.1
### INTRODUCTION

EXAMPLE 167   TELEVISION AND LIBRARY USE

A sociologist wished to study the effect of television upon the use of library facilities in a community. From the list of card holders at the public library, he selected a sample to whom he mailed a questionnaire. One question asked was whether the respondent owned a television set and, if so, when it was purchased. After the questionnaires had been returned, the sociologist started to organize his information to cast light on the original question. For all respondents who had bought television sets within the preceding two years, January 1, 1950, to December 31, 1951, he compared the rate of borrowing from the public library before and after the purchase of television. Suppose, for example, that a respondent bought a television set on March 1, 1950, and that between January 1 and March 1 of 1950 he had borrowed four books, a rate for those two months of 2 books per month, or 24 per year. For the remaining 22 months from March 1, 1950, to December 31, 1951, this respondent borrowed 33 books, a rate of 1.5 per month or 18 per year. For this particular respondent, then, the rate of use of the library declined after the purchase of his television set, from 24 to 18 books per year.

For all people who had purchased television sets during the period, the sociologist averaged the rates before purchase and compared this with the average rate after purchase. He found a decline of 10 percent, from an average rate of 40 per year to an average rate of 36. (These figures are not exactly those obtained, but they illustrate the principle.)

Next, he computed the average rate of use for the entire two-year period by nonpurchasers of television sets. The average for this group was, say, 41. (The study was made in the early 'fifties, so the number of people in the

**167**

sample who had had television sets throughout the two-year period was too small to be worth analyzing separately.) The sociologist tentatively concluded that his data showed that television had caused a decline in the rate of use of the library. He decided to discuss his results with a statistician before publishing them.

The statician's opinion was that the averages presented were inconclusive and might even be misleading. His two main criticisms were: (1) The average rate for the "control group," those who had not purchased television during 1950 or 1951, may also have declined during the two-year period. For example, while the average was 41 for 1950–1951, it might have been 43 for 1950 and only 39 for 1951. This seemed possible, since there might be a tendency for library use to decline the longer a card is held. (2) The method of calculating "before" and "after" rates for television owners left the possibility that a seasonal effect might distort the findings. This possibility may be illustrated by an extreme case. Suppose that all television purchasers had obtained their sets on March 1, 1950, and suppose that people tend to use the library more in winter than at other seasons. Then the "before" rate of use would be based solely on two winter months, while the "after" rate would include both winter and summer months. Thus, even if acquisition of television had no effect on library use, there would be an apparent decline in use after purchase, solely because of the seasonal pattern.

The statistician suggested the following procedure: For each television purchaser, compare the rate of library use during the twelve months prior to purchase with the rate during the twelve months after purchase. (This could be done without returning to the respondents, since the library-use data were available in the library's records.) For each purchaser, choose a nonowner at random from those matching the purchaser in such characteristics as age, neighborhood, etc. For the nonowner, calculate separate rates of library use for exactly the same two time periods as for the purchaser with whom he was paired. Then compare the averages of owners and nonowners.

The sociologist naturally felt somewhat foolish, for the statistician's suggestions seemed to represent only common sense rather than technical knowledge; but they represent a kind of common sense that becomes highly developed in good statisticians through varied experience in analyzing data. Failure to use such common sense is not at all uncommon, as many of the statistical misuses in Chap. 3 attest. In the library study, the sociologist was unusually fortunate in being able to reorganize the data he had obtained, though he regretted the extra work and delay. Often, people are not so fortunate: no reorganization of data will bring out facts that were not collected in the first place. Even in the library study, there were several ways in which more useful information could have been obtained by advance planning that could not be obtained later by reorganizing the data.

*6.1 Introduction*

For example, only about 30 percent of the recipients of questionnaires actually returned them, and these people may have been atypical in many ways. With more careful advance planning, follow-up questionnaires might have been sent out to the 70 percent who did not respond to the initial mailing, or other provisions might have been made to study the representativeness of the respondents. In any event, it is always preferable to plan in advance and in detail how the data to be collected will eventually be organized. The plans may later be revised; in fact, many new ideas for analysis are bound to arise when the data are actually available. But wasted effort and erroneous conclusions are much less likely when the organization of data is well planned in advance. In the example we have discussed, the statistician had, in a sense, an unfair advantage, in that he knew of the efforts of the sociologist and had an opportunity to look at actual data; but the two suggestions we have emphasized are such as almost any good statistician would have made in advance.

Intelligence, imagination, and experience are the main ingredients in the ability to organize data in such a way as to answer the questions for which the data were collected. The organization of data is, therefore, often one of the hardest things in an investigation. To understand the difficulties of organizing data, consider the following problems:

(1) A large private university experiences a decline in enrolment. Is this decline due to widespread conditions that affect all institutions of higher learning, or can it be attributed to conditions peculiar to that particular university? The appropriate action would probably be quite different in the two cases.

(2) How would one discover the effect of age on the probability that an airline pilot will have an accident? The answer may seem relatively obvious. But suppose that the frequency of accidents diminishes with age, at least in a certain range of ages. The question then arises whether this decline is due to age per se or to the fact that older pilots, on the average, have more flying experience. If experience has something to do with the decline, how can this be allowed for in assessing the effect of age?

(3) A large national advertiser, now spending 20 million dollars yearly for advertising, wants to know whether he is spending too much or too little.

(4) How would different business forecasting services be evaluated? Different weather forecasting services?

(5) Is it true that a person's attitudes toward foreigners tend to be more favorable the greater his contacts with them?

(6) Is psychoanalysis more effective than nondirective counseling in alleviating a particular neurotic condition?

It is helpful, in casting about for a plan of procedure, to set up statistical tables, fill them in with imaginary data, and then ask how well this particular table would answer the question. By this kind of trial and error, successful plans for collecting and organizing data may be evolved. Hence the collector of statistics needs to know something about the planning of statistical tables and, more broadly, the tools of statistical description and inference. Likewise, the user or reader of statistics needs to know something about these tools.

To put the problem of statistical description in sharper focus, suppose that a questionnaire study has just been completed and a large pile of questionnaires is waiting to be digested. It might be possible for a person to read through some or all of the questionnaires and acquire a subjective impression of the information contained in them. When the study is a large one, even this is not easy. Besides, the subjective impression is likely to be inaccurate, since the questionnaires that make the most impression are apt to be those which conform with anticipations or prejudices, or which involve especially vivid answers. Subjective impressions, moreover, are hard to communicate accurately to others, and even when communicated may not carry conviction. Hence the need for statistical tools to reduce the mountain of data to a brief, objective, and comprehensible form.

It is helpful to think of each questionnaire as a multivariate observation consisting of many facts about a particular individual. Some of these, such as his income, age, expenditures on newspapers and magazines, number of arrests for speeding, or number of children, are quantitative, while others, such as his religious affiliation, attitude toward farm price supports, or country of birth, are qualitative. Two devices that can be used to describe and summarize the information contained in a collection of observations are frequency distributions and descriptive measures. Frequency distributions are examined in detail in this chapter, and descriptive measures are examined in Chaps. 7 and 8. These three chapters are important not only because statistical description is important in itself, but also because the ideas involved are basic to statistical inference or analytical statistics.

One brief caution is in order as you read these chapters, and indeed the rest of the book. We cannot avoid spending a good deal of time on specific techniques for handling data. This necessary emphasis on techniques can easily give the impression—contrary to the one we have been trying to convey—that statistical applications involve mechanically looking up the right recipe for the problem at hand. In-experienced users of statistics, for example, will often ask questions like, "Should I compute the arithmetic mean or the median of my data?" The prior question, as we have tried to emphasize, is, "What am I trying to find out?" Only after that question is clearly answered can specific techniques be chosen. Thus, the answer may be that *neither* the arithmetic mean *nor* the median is appropriate, but that something else—for example, the entire frequency distribution—is required. Particular statistical methods are simply tools, the skilled use of which requires insight into the subject matter and a clear conception of the problem.

## 6.2
## UNIVARIATE FREQUENCY DISTRIBUTIONS: QUANTITATIVE OBSERVATIONS

### 6.2.1 Organizing Data

We will describe a simple classroom investigation that brings out some important principles of organizing quantitative data. Each student—all were men—was given the following instructions: "Record your weight, to the nearest pound, on the card provided." Sixty-four cards were received. They were divided haphazardly (that is, with makeshift attempts at randomization) into two packs. The following 32 numbers were obtained from one pack, in the sequence listed:

| | | | | | | | |
|---|---|---|---|---|---|---|---|
| 198 | 165 | 189 | 155 | 148 | 155 | 203 | 151 |
| 178 | 142 | 160 | 174 | 152 | 183 | 180 | 197 |
| 158 | 164 | 142 | 170 | 175 | 239 | 175 | 180 |
| 200 | 190 | 185 | 186 | 175 | 180 | 191 | 189 |

What idea is conveyed about the variable "weight" by looking at this set of numbers? It is clear that the average is somewhere between 142 and 239 pounds, which are the largest and smallest weights reported. But with the data in this form, it is difficult to judge where within this range the average comes, or how the individual weights are distributed about the average. Even the largest and smallest would be difficult to find if the sample were a large one, say, 320 or 3,200 observations instead of 32.

Before proceeding with the description of these data, we may recall and illustrate some of the ideas of Chap. 5. We saw there that it is well to consider carefully the relationship that recorded numbers bear to the matter under investigation. The numbers in the preceding list were intended to represent "weights" to the nearest pound of the individuals in the class. Do they? We can think of many reasons why they might not. Nothing was said to indicate whether we were speaking of weight dressed or undressed. Many people may not have weighed recently enough to know their weights to the nearest pound, especially since weight varies during a day. The numbers are undoubtedly associated with weight, but in part they may represent misunderstandings, whims, lack of knowledge, ambiguity in the question, mistakes, and other distortions.

As we saw in Chap. 5, there often is evidence within the data themselves about the significance of the numbers. In the weight data, 14 of the 32 numbers end in 0 or 5. We would expect on the average only 20 percent, that is, 6.4 per 32, to end in these two digits. This suggests that there may have been a tendency to round weights to the nearest five pounds rather than one pound. But is this apparent tendency "real," or is it only a chance peculiarity of the particular 32 observations we happened to observe? Principles for attacking this problem were explained in Chap. 4. This one result that we have actually observed should be interpreted against the pattern of sampling variability that would result if many more samples were observed. Suppose we use a sampling box like that described in Chap. 4, but with one-fifth of the beads red, and perform a sampling experiment like that of Chap. 4, drawing, however, not 50 but perhaps 50,000 samples, each of 32 beads. Would samples containing 14 or even more red beads appear fairly often? If so, the fact that the one sample we have actually observed has shown 14 would not signify that the average of the population exceeds 6.4, for it would be within the range of sampling variation to be expected if the population average were 6.4. Or would samples containing 14 or more red beads be rare? If so, we would conclude that the true population average must be more than 6.4, our reasoning being that since 14 red beads actually appeared, this contradicts the assumption according to which fewer would, in all probability, appear. It is not necessary to perform the sampling demonstration, for the theory of probability makes it possible to calculate (using procedures like those described in Chaps. 10 and 11) that, on the average, only one sample in 500 will contain as many as 14 or more red beads if samples of 32 are drawn at random from a population in which 20 percent are red. Thus, we conclude quite confidently that the tendency to round the weights to the nearest five pounds is real, not just a fortuitous characteristic of our sample.

We now return to the main thread of the analysis. We want to get some picture or pattern from the data as a whole. This usually

*6.2 Quantitative Observations*

requires rearranging the numbers. One way to get a clearer picture
from the data is to array them in order of size. This has been done
in Table 173. We see at once that the range is 97, the difference be-

TABLE 173

ARRAY OF 32 WEIGHTS

| Rank | Weight (pounds) | Rank | Weight (pounds) |
|---|---|---|---|
| 1 | 142 | 17 | 178 |
| 2 | 142 | 18 | 180 |
| 3 | 148 | 19 | 180 |
| 4 | 151 | 20 | 180 |
| 5 | 152 | 21 | 183 |
| 6 | 155 | 22 | 185 |
| 7 | 155 | 23 | 186 |
| 8 | 158 | 24 | 189 |
| 9 | 160 | 25 | 189 |
| 10 | 164 | 26 | 190 |
| 11 | 165 | 27 | 191 |
| 12 | 170 | 28 | 197 |
| 13 | 174 | 29 | 198 |
| 14 | 175 | 30 | 200 |
| 15 | 175 | 31 | 203 |
| 16 | 175 | 32 | 239 |

tween 239 and 142; that half of the weights are 175 pounds or less
and half are 178 or more, and that more weights fall in the 170's
and 180's than in other ten-pound intervals formed by the first two digits.

Nevertheless, some things can be seen only with difficulty in the
array—for example, the tendency of the numbers at the center to
bunch together, and of those at the ends to scatter apart. If there
were 3,200 observations, or even 320, the array would take too much
space and be hard to comprehend.

Table 174 illustrates a useful device for avoiding these limitations
of the array without losing its advantages. This table is an example
of a *frequency distribution* (see also Table 106), and is essentially a
condensed version of the array, giving not the individual observations, but the frequencies of observations within small ranges or intervals. The upper limit of an interval is identical with the lower
limit of the next interval. This causes no confusion in this example,
because weights were reported to the nearest pound. We will indicate in a moment why we chose to center the intervals at 142.5,
152.5, etc., instead of at 140, 150, etc., or 145, 150, etc.

The main problem in preparing a frequency distribution is that
of selecting the intervals. The following considerations bear on the
problem.

*The Art of Organizing Data*

TABLE 174

FREQUENCY DISTRIBUTION OF 32 WEIGHTS

(Equal Intervals)

| Weight | Number of Persons |
|---|---|
| Under 137.5 | 0 |
| 137.5–147.5 | 2 |
| 147.5–157.5 | 5 |
| 157.5–167.5 | 4 |
| 167.5–177.5 | 5 |
| 177.5–187.5 | 7 |
| 187.5–197.5 | 5 |
| 197.5–207.5 | 3 |
| 207.5–217.5 | 0 |
| 217.5–227.5 | 0 |
| 227.5–237.5 | 0 |
| 237.5–247.5 | 1 |
| 247.5 and over | 0 |
| Total | 32 |

Suppose that a man thought he weighed exactly 148.5 pounds, and that he was asked to report his weight to the nearest whole pound. What weight should he report? In the first place, if he knew his weight *exactly*, it certainly would not be 148.5; if it were measured accurately enough it would turn out to be at least, say, 148.-5000000001, or else at most 148.4999999999. Then he would know whether to report 148 or 149. This illustrates the idea of a *continuous variable* mentioned in Sec. 5.5.2.1. With a continuous variable, an infinite number of results are possible, out to any number of decimals, and accurate enough measurement would in principle prevent the number from coinciding exactly with the boundary of an interval.

In the practical world, however, we can neither attain nor utilize such accuracy. Measurements must be of limited accuracy, and this means that only certain numbers can possibly arise. As we saw in Sec. 5.5.2.1, a variable that can take on only certain numbers is called a *discrete variable*. Some penny weighing machines print a whole number, and some print a continuous scale, but to make it slightly more interesting for our purposes, consider one that prints weights to tenths of a pound. The man discussed in the last paragraph gets a slip of paper from the scale saying that he weighs 148.5 pounds, and he has no means of getting a more accurate weighing. He is asked his weight to the nearest pound. There are at least three things the statistician might advise him to do: (1) report half a man at 148 and half at 149; (2) toss a coin, on the grounds that he is so close to the boundary that it makes no difference which interval his weight is put into, provided there is no systematic tendency to put such boundary values predominantly into the higher or predominantly into the lower interval; or (3) round to the nearest even number. The first alternative is probably the most useful, but the idea of half a man is so diverting that it is usually avoided. The second alternative has the disadvantage that it cannot be verified if someone else checks the work, while the third does not have these disadvantages and is generally employed. The reader will be able to think of other alternatives, but the long history of the subject shows that none of them will be preferable to all three of those mentioned.[1]

Now that we have seen how quantitative observations can be assigned to *some* interval, even when they appear to lie exactly on the boundary, let us see how to decide *which* intervals to use. One question is the *width* of the intervals or, to put it differently, the *number* of intervals to be used. The number of intervals is decided by balancing two extremes. On the one hand, all the observations might be included in a single class, extending from, say, 135 to 250. This would overdo summarization. We would not even know the range of the observations. On the other extreme, each distinguishable observation could be put into a class by itself. This brings us back essentially to the array, which does not summarize sufficiently. A good rule of thumb for compromising between these two extremes is to form from 5 to 15 classes. Like all rules of thumb, this suggestion is only for guidance, and is not a substitute for common sense. Some statisticians prefer a working rule that the interval should be about one-half of the standard deviation of the observations. (The standard deviation is a measure of dispersion or variability that will be discussed in Chap. 8.) Whatever working rule is followed, however, it should be remembered that nothing is lost except the work of further summarization if in the early stages of a study too small an interval is used, but valuable information may be lost, perhaps irretrievably, if too large an interval is used. A small class interval often provides internal checks on the care with which the original measurements were taken. The preponderance of weights ending in 0 or 5 would suggest (if we did not know it already) that the data were obtained by questioning rather than by careful, standardized measurements.

---

1. Rounding to the nearest *odd* number might seem just as good a practice as rounding to the nearest even number. It is not as good, however, simply because uniformity of practice is highly desirable, to avoid misunderstandings or innumerable explanations, and rounding to the nearest even number is the accepted convention. Should special circumstances warrant a departure from this rule, a clear explanation should be given.

This fact is no longer evident with the broader class intervals of Table 174.

Once the approximate width of the intervals has been determined, it should be changed to some convenient number like 1, 5, 10, 25, etc., and ordinarily not a number like 7, 13, 27, etc. In the weights, 10 pounds is a convenient interval length, whereas 11 pounds would be awkward, though both lead to a convenient number of intervals, 9 to 11.

After the width and number of the class intervals are fixed, there is the problem of the location of their centers. Since computations are likely to be made treating the data as if they were all concentrated at the center of the intervals (see Chaps. 7 and 8), the intervals should be chosen to make this assumption as nearly true as possible. For example, since there is a tendency to report weights as numbers ending in 0 or 5, placing numbers like 140, 145, 150, etc., at the centers of the intervals will result in the least distortion when it is assumed that all the observations in an interval fall at the center. Thus, the boundaries of the intervals should be at numbers like $142\frac{1}{2}$, $147\frac{1}{2}$, $152\frac{1}{2}$, etc., not 140, 145, 150, etc. Another advantage, since weights ending in 0 or 5 were reported only some, not all, of the time, is that observations fall in the same group whether they were originally reported to the nearest pound (say 152) or to the nearest five pounds (say 150).

EXAMPLE 176    CLUB DUES

A private club classifies its members according to their salaries in order to adjust the dues accordingly. The following class intervals are used:

Under $3,000
$3,000 and under   5,000
5,000 and under   7,000
7,000 and under 10,000
10,000 and over.

This is a poor set of intervals because of the tendency of salaries to bunch near the "round" thousands. One man whose contract calls for $7,000, found that his checks really add up to $6,999.96 because he is paid $583.33 per month, and never receives the extra one-third of a cent owed him each month. Which class should he be in?

In Table 174 the last five lines together include only one man. It is tempting just to put "207.5 and over" and omit the last four lines. This is called an "open-ended" interval and is exasperating.[2]

---

2. The open-ended intervals of Example 176 are not objectionable, but those intervals are not for the purpose of constructing a frequency distribution.

**6.2 Quantitative Observations**

If it is not known, even roughly, how much the heaviest man weighed, there is no way to make certain calculations, such as the average (to be discussed in Chaps. 7 and 8). Table 177 gets around the difficulty by combining two intervals into one (197.5–217.5) and the next three into one (217.5–247.5). Then it assures us that none exceeded 247.5 or fell short of 137.5. Since the purpose of a frequency distribution is to reveal the pattern in the data, this smoothing-out technique really emphasizes the message in the table, although there is some danger of confusion in interpretation when unequal intervals are used, and some inconvenience in further analysis.

TABLE 177

FREQUENCY DISTRIBUTION OF 32 WEIGHTS

(Unequal Intervals)

| Weight | Number of Persons |
|---|---|
| Under 137.5 | 0 |
| 137.5–147.5 | 2 |
| 147.5–157.5 | 5 |
| 157.5–167.5 | 4 |
| 167.5–177.5 | 5 |
| 177.5–187.5 | 7 |
| 187.5–197.5 | 5 |
| 197.5–217.5 | 3 |
| 217.5–247.5 | 1 |
| 247.5 and over | 0 |
| Total | 32 |

**6.2.2 Graphs**

Fig. 178 gives a graphical picture of the array of weights. We have first drawn a line and scaled it for weight. Then a dot has been made on this line to indicate each observation. Where the marks are fairly dense is where most of the weights fall. Such a line chart becomes impractical when there are many marks coinciding, or nearly coinciding, but a way to get around this is to pile them up above the line—to use a second dimension, in other words. To do this we divide the line into segments, and above each segment draw a bar whose height represents the number of observations within that segment, as in Fig. 178 in which the segments correspond to the intervals of Table 174. The heights of the bars thus vary with the density of points along the line. (The idea that *density* is the fundamental quantity presented should be kept in mind, for we will revert to it later.) The area of each bar corresponds to the number of cases in the segment on which it stands.

A chart like Fig. 178 is called a *histogram*; the divisions of the bars into units are not usually shown. Each rectangle within a bar

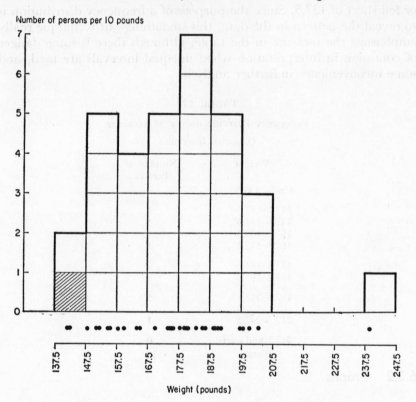

FIG. 178. Histogram and array of 32 weights. (Equal intervals.)
Source: Tables 173 and 174.

may be thought of as representing a single observation. For example, the shaded rectangle just above the interval bounded by 137.5 and 147.5 represents a measurement of weight between 137.5 and 147.5. Another rectangle is piled on top, making the area of the first column that of two rectangles and indicating that there are two measurements in the first interval. Similarly, there are five measurements between 147.5 and 157.5, four between 157.5 and 167.5, and so on. Since there are 32 measurements in all, there are also 32 rectangles. If we call the area of each rectangle one unit, the area of each bar is equal to the frequency in the interval on which it stands, and the total area is 32. The important point, however, is that *the areas above the various intervals are proportional to the frequencies in those intervals.*

The big gap on the right-hand side of Fig. 178 probably obscures the real pattern in the data through mere chance effects. This can be avoided by making the histogram from Table 177, the table with unequal intervals. But to do this correctly we must remember that

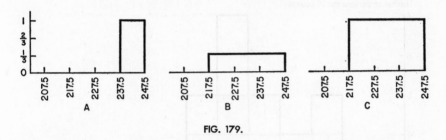

FIG. 179.

each observation is to be represented by a rectangle of the same *area*, in order that the areas of the bars will be proportional to the frequencies in the intervals. With these unequal class intervals, the single observation which appeared before in the interval from 237.5 to 247.5 will now fall in the interval 217.5 to 247.5. The length of the interval at the base of the rectangle is now 30 (247.5 minus 217.5) rather than 10 (247.5 minus 237.5). Hence the height must be only one-third as great, as in Figs. 179B and 180. Similarly, in Fig. 180 we see that the original bar of height 3 units for the interval 197.5–207.5 has been changed to a bar of height $1\frac{1}{2}$ spread out over the interval 197.5–217.5. If we had represented the frequency not as in Fig. 179B, but as in Fig. 179C, it would have been quite misleading. It would have appeared that there were three measurements rather than just one between 217.5 and 247.5.

This leads us to define a little more precisely the quantity measured along the vertical axis of a histogram. "Frequency" is a loose description of this quantity. More precisely, it represents the *density* of the observations, that is, it represents the number of cases *per unit* of the horizontal axis. In our illustration, it is the number of cases per ten pounds, ten pounds being a unit in which the intervals can conveniently be expressed. One case in an interval 30 pounds wide is 1 case per 30 pounds or $\frac{1}{3}$ case per 10 pounds. If we want the vertical scale in Figs. 178 and 180 to show number of cases per pound, instead of per ten pounds, we need only divide the numbers on the vertical scale by 10; thus, the density in the interval 217.5 to 247.5 may

*The Art of Organizing Data*

also be expressed as $\frac{1}{30}$ case per pound. Similarly, the three cases in the interval 197.5 to 217.5 pounds may be expressed as a density of 3 per 20 pounds, 1.5 per 10 pounds, 0.15 per pound, etc. These are simply different ways of saying the same thing, as are "144 pounds per square foot" and "one pound per square inch," or "360 miles per hour" and "528 feet per second."

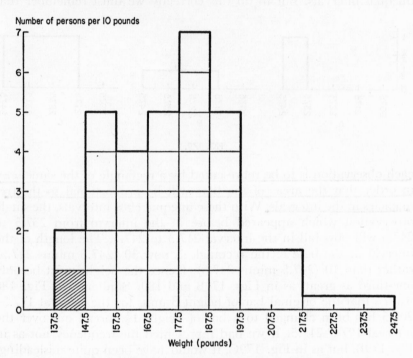

Number of persons per 10 pounds

FIG. 180. Histogram showing 32 weights. (Unequal intervals.)
Source: Table 177.

Sometimes, usually ill-advisedly, instead of a histogram a *frequency polygon* is used. This is a series of lines connecting the points which in a histogram would be the midpoints of the bar tops. Fig. 181A shows a frequency polygon superimposed on the histogram of Fig. 180. The frequency polygon is not a good graphic representation of the basic data, for areas are not proportional to frequencies. Notice particularly how the frequency in the highest, or *modal*, class is under-represented. The one advantage of frequency polygons is that when several sets of data are to be shown on the same graph it is a little clearer to superimpose frequency polygons than to superimpose histograms, especially if the class boundaries coincide.

### 6.2 Quantitative Observations

There is another reason for mentioning the frequency polygon here: it suggests the use of a smoothed curve as an idealized representation. Thus, a quick impression of the distribution of weights is

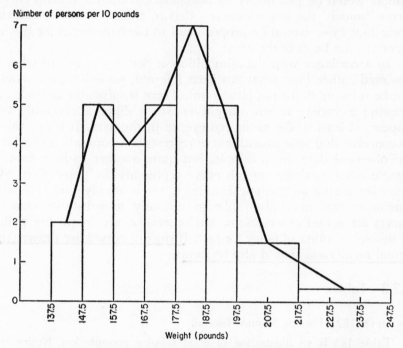

FIG. 181A. Histogram and frequency polygon showing 32 weights. (Unequal intervals.)
Source: Table 177.

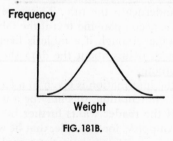

FIG. 181B.

conveyed by the smooth curve in Fig. 181B. Fig. 181B represents the kind of population usually supposed to underlie sample data such as those shown in Figs. 178, 180, and 181A. If an extremely large sample were taken, so that the bars could be made very narrow and still contain substantial numbers of observations, and if the vertical or "frequency" scale were reduced so as to keep the area of the histogram for this extremely large sample the same as the area for the observed sample, then the histogram of the extremely large sample would be practically indistinguishable from a smooth curve. Areas "under" the smooth curve—that is, areas between it and the horizontal axis—would be proportional to the frequencies for the intervals at the bases of the areas.

In accordance with the admonition of Sec. 5.4 to record what is observed rather than what has been inferred, smooth curves should not be substituted for the histograms. There is surprising latitude for drawing a variety of smooth curves which differ appreciably but appear, at least to the untutored eye, to fit the data. It is one thing to conceive that *some* smooth curve represents a population of which the observed data are a sample, but quite another to infer from a sample what *particular* smooth curve represents the population; the latter is a matter for analytical statistics, and is usually quite difficult. Engineers are especially prone gratuitously to substitute smooth curves for actual observations, but others too are frequently guilty of this substitution of fancy for fact. If smooth curves are drawn, the actual histograms should also be shown.

### 6.2.3 Tables

EXAMPLE 182 URBAN FAMILY INCOME

Table 183 is an illustration of good tabular presentation. Notice the following points:

(1) The title is concise, yet clear and informative. Greater detail is given in the headnote just underneath the title. This type of information would probably be given in the accompanying text if the table were included in an article or book on income—though if a table is likely to be referred to as source material, the main points about the data should be summarized in such a headnote or footnote.

(2) The source of the information is given in a footnote at the bottom of the table. Like other documentation, this may be a nuisance to provide and is too often omitted. If the reader wants further information, however, he may be lost without it. Suppose, for example, that he wants to know precisely how total money income is defined, how the sampling was done, or how reliable the figures are. He would refer to the source and expect to find these questions answered—and he would probably find there answers to important questions he had not thought to raise. To appreciate the importance of these apparent refinements, it is necessary only to consider using the data of Table 183 in a study of changes in the inequality of income distribution, or in a comparison of the distribution of income in various countries.

## 6.2 Quantitative Observations

TABLE 183[3]

MONEY INCOME OF URBAN FAMILIES, 1952—
PERCENT DISTRIBUTION OF FAMILIES BY INCOME LEVEL

Data based on sample. See source for evaluation of sampling reliability. Includes data for families in quasi-households (hotels, large rooming houses, etc.) as well as households. "Family" refers to a group of two or more related persons residing in the same household. For definition of urban areas, see p. 13, footnote 4.

| Total Money Income | Families |
|---|---|
| Number (thousands) | 26,786 |
| Percent | 100.0 |
| Under $500 | 2.4 |
| $   500–$   999 | 2.8 |
| $ 1,000–$ 1,499 | 4.1 |
| $ 1,500–$ 1,999 | 4.4 |
| $ 2,000–$ 2,499 | 5.8 |
| $ 2,500–$ 2,999 | 7.1 |
| $ 3,000–$ 3,499 | 9.7 |
| $ 3,500–$ 3,999 | 9.3 |
| $ 4,000–$ 4,499 | 8.7 |
| $ 4,500–$ 4,999 | 7.8 |
| $ 5,000–$ 5,999 | 13.7 |
| $ 6,000–$ 6,999 | 8.5 |
| $ 7,000–$ 9,999 | 10.8 |
| $10,000–$14,999 | 3.4 |
| $15,000 and over | 1.5 |
| Median Income (For definition, see Chap. 8) | $4,249 |

Source: Department of Commerce, Bureau of the Census, Current Population Reports, Series P-60, No. 15.

(3) The body of the table is clear and easy to read. It is impractical to give precise and comprehensive directions for making a clear table. The common faults in tabular exposition are analogous to those in literary exposition: incompleteness, too much detail, obscurity, ambiguity, poor sequences, etc. The main requirements are accuracy, simplicity, and orderly appearance. A help in achieving a satisfactory format is to follow a good published table: the Statistical Abstract is a good source not only of varied and excellent data, but of varied and excellent table formats.

Table 183 illustrates also a difficult problem in forming class intervals. Solution of this problem resulted in both unequal classes and open-ended classes at both extremes. The unequal classes were necessary in order to show

3. This table has been adapted from Statistical Abstract: 1955, Table 357, p. 300.

the detail at the middle range of the income scale where most of the families are, without creating an unwieldy number of classes at the upper end. The open-ended classes resulted from limitations of the data. People may be willing and able to indicate simply that their incomes are above $15,000, but unwilling or unable to specify them more exactly. Similar considerations apply to people with negative incomes. Furthermore, sampling errors would be considerable if the top or bottom classes were subdivided. Had the data been available, class limits going by successive powers of ten, or else a supplementary detailed table for incomes over $15,000, would have been helpful.

### 6.2.4   Relative or Percentage Distributions

The frequency distribution of Table 183 differs in another important respect from the distributions of weights shown in Tables 174 and 177. Instead of absolute frequencies, it shows relative frequencies, for example, that 2.4 families per 100 had incomes less than $500. Sometimes the terms *absolute frequency distribution* and *relative* or *percentage frequency distribution* are used to distinguish these two cases.

When percentage distributions are shown alone, as in Table 183, it is important that the total number of observations be shown somewhere. Thus, Table 183 shows that there were 26,786 thousand urban families in 1952. From this, anyone who needs them can compute absolute frequencies.

Histograms, frequency polygons, and smooth curves can be drawn for relative as well as absolute distributions. The area of the histogram or smooth curve is equal to 100 percent, or to 1.00 if decimal fractions are used instead of percents. The density interpretation of the vertical or frequency axis is then percent (or fraction) of total frequency per unit of the horizontal axis. The vertical axis for a histogram of Table 183, for example, would represent the number of groups of 268 thousand (that is, 26,786 thousand divided by 100) per 500 dollars of income.

Relative frequency distributions and their histograms, polygons, or smoothed curves facilitate comparisons of different sets of data when the number of observations varies from one set to another. For example, there would be considerable difficulty in comparing the curves in Fig. 185A, representing absolute frequency distributions. Certain comparisons are, of course, easy, namely those involving absolute numbers; for example, between points *a* and *b* there are approximately equal numbers of observations from the two distributions. It is not so readily apparent, however, that while nearly all the observations in distribution *B* fall between *a* and *b*, only a small

### 6.2 Quantitative Observations

fraction of the observations in distribution *A* fall within this interval. But by using relative frequency curves, this second kind of comparison can be brought out; that is, it is possible to compare the *shapes* of the curves without having to allow for their great difference in size. Fig. 185B illustrates this; the *areas* under both curves are now the same, and comparisons can be made readily. In the first place, distribution *B* is located at larger values of the variable than *A*. Second, the variation in distribution *A* is greater than in *B*. Finally, only a small proportion of the observations in *A* are larger than any of the

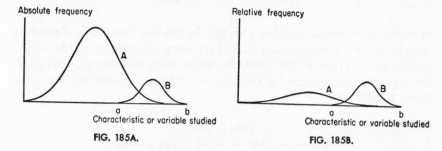

FIG. 185A.    FIG. 185B.

observations in *B*. Comparisons of this kind are extremely useful, and it is desirable to acquire some facility in making them. One way to do this is to sketch other situations and draw verbal conclusions. Such comparisons illustrate the usefulness of statistical concepts even when precise numerical data are not available. For example, the situation depicted in Figs. 185A and 185B makes it easy to understand why there may be as many Cadillacs in low-income neighborhoods as in exclusive suburbs: Cadillacs are much rarer in low-income than in high-income families, but high-income are much rarer than low-income families.

Sometimes, particularly in investigations based on samples, percentage distributions are shown without any mention of the absolute number of observations on which the distribution is based. Sometimes the omission is due to carelessness; sometimes it is due to the desire to divert attention from the smallness of a sample. In this case, a little detective work will sometimes reveal the sample size, as in the case of Table 186A. First, find the smallest percentage in the table (3.8 in this case) and the smallest difference between two percentages (3.8 in this case), and take whichever is smaller (3.8 here). Tentatively consider that this smallest percentage represents one case. Then 100 divided by the smallest percentage ($100 \div 3.8 = 26.3$ here) and rounded to an integer (26) may be conjectured to be the total number

TABLE 186A

FICTITIOUS DATA ON TIME SPENT WATCHING
TELEVISION DURING ONE WEEK

| Time (in hours) | Percent of Respondents |
|---|---|
| Less than 1 | 23.1 |
| 1 but less than 5 | 15.4 |
| 5 but less than 10 | 30.8 |
| 10 but less than 15 | 19.2 |
| 15 but less than 20 | 7.7 |
| 20 or more | 3.8 |
| Total | 100.0 |

of cases. The actual number can not be smaller than this, though it may be larger. Conjecturing that there were 26 cases in Table 186A, we reconstruct it as Table 186B, obtaining each frequency by dividing the reported percentage by 100/26, or 3.846, and rounding the quotients to integers.

TABLE 186B

RECONSTRUCTION OF ABSOLUTE FREQUENCIES
UNDERLYING TABLE 186A

| Percent | Number |
|---|---|
| 23.1 | 6 |
| 15.4 | 4 |
| 30.8 | 8 |
| 19.2 | 5 |
| 7.7 | 2 |
| 3.8 | 1 |
| Total | 26 |

As a check, we may recompute the percentages from these reconstructed frequencies; in this case they agree to the number of decimals reported in the original table.

It is possible, of course, that 3.8 percent actually represents two, three or some other number of observations. But this would imply that all the frequencies were multiples of two, or of three, or of some other number, which seems unlikely unless the number of classes is very small. For example, it would be an extraordinary coincidence if the actual frequencies were 60, 40, 80, 50, 20 and 10, and the total, 260.

If the smallest percentage, or difference between two percentages, divided into the reported percentages yields quotients which are not all integers but are all multiples of $\frac{1}{2}$ (except for discrepancies attributable to rounding) it would be conjectured that the smallest percentage represents two observations. Then 200 divided by the smallest percentage would be conjectured to be the total number of observations. Similarly, if the reported percentages divided by the smallest percentage yields quotients which are not all integers but are all multiples of $\frac{1}{3}$ (except for discrepancies attributable to rounding) it would be conjectured that the smallest percentage represents three observations, and that 300 divided by this smallest percentage is the total number of observations. It is impractical to determine very large sample sizes by this procedure unless the percentages are reported to several decimals, but even so it will be possible to determine that the sample is not smaller than a certain size.

The method can obviously be extended, although it works best when the smallest percentage is actually computed from a small number.[4] Sometimes, in the report of an investigation, several tables will be given, all clearly based on the same observations, and one should then look through these tables to find the smallest percentage (other than zero), or the smallest difference between percentages (not necessarily in the same table), and divide this percentage into the reported percentages.

EXAMPLE 187  MOVIE RATINGS

The magazine *Consumer Reports* formerly published each month a series of movie ratings made by readers. The following are taken from the August, 1949, issue, p. 383.

> CU [Consumers Union] presents these ratings with the aid of some 2000 subscribers. Each participant, as soon as he sees a picture, notifies CU by special

---

4. We can even work with the smallest difference between differences in the percentages, or the smallest difference between a percentage and a difference between percentages. More generally, any arithmetic operation on the percentages that consists only of multiplying them by whole numbers (positive, negative, or zero) and then adding them (subtracting is included as multiplying by $-1$ and adding) will result in a percentage that corresponds with a whole number of observations. Inaccuracies in the percentages, however, are magnified in this process, so unless the original percentages are given to a considerable number of decimals the results of elaborate calculations will be unreliable. For an analysis of a case where neglect of this "propagation of error" led to unjustifiable conclusions, see Russell T. Nichols, "Soviet Production Estimates," *Journal of Political Economy*, Vol. 57 (1949), pp. 249–250, which shows that certain deductions that had been made about Soviet postwar industrial production were unreliable. The deductions had been based on published percentages of changes from period to period. For example, period 1 might be 18 percent higher than period 0; period 2, 22 percent higher than period 1, etc. An ingenious analysis of relationships among such percentages seemed to show that production had declined very sharply in the spring of 1946, 1947, and 1948. This, in turn, suggested a variety of interesting speculations about reasons for such changes. Nichols' article, however, shows that the conclusion about declines in the second quarter could easily be due entirely to the propagation of errors resulting from the fact that the original figures were rounded off, so a figure reported as 18 percent might actually be as low as 17.5 percent or as high as 18.5 percent.

card whether he considers it to be "Excellent (E)," "Good (G)," "Fair (F)," or "Poor (P)." The tabulation shows the percentage of replies in each category. . . .

| PICTURE | E | G | F | P |
|---|---|---|---|---|
| ADVENTURE IN BALTIMORE | 0 | 71 | 29 | 0 |
| ALIAS NICK BEAL | 6 | 27 | 47 | 20 |
| BAD BOY | 11 | 67 | 22 | 0 |
| BADMEN OF TOMBSTONE | 0 | 25 | 50 | 25 |
| BARKLEYS OF BROADWAY | 26 | 58 | 16 | 0 |
| BEAUTIFUL BLONDE FROM BASHFUL BEND | 0 | 17 | 60 | 17 |
| THE BRIBE | 0 | 50 | 36 | 14 |
| BRIDE OF VENGEANCE | 11 | 22 | 56 | 11 |

It is relatively simple to make inferences by inspection about the sample sizes for some of the films. For "Badmen of Tombstone," for example, the sample size is probably 4, though it could be 8 or any larger multiple of 4; for "Bride of Vengeance" it is probably 9. Others are less obvious. For example, "Barkleys of Broadway" shows percentages of 26, 58, 16, 0. The smallest difference is 10, that is, $26 - 16$. This suggests a tentative sample size of 10, but this is obviously inconsistent with the percentages 26, 58, and 16. Since there has been rounding, $26 - 16$ could mean as much as $26.5 - 15.5 = 11$ or as little as $25.5 - 16.5 = 9$. Hence we may try 9 and 11 as sample sizes, but a little calculation shows that neither of these will fit the observed percentages. The next step is to double these minimum sample numbers and try 18, 19, 20, 21, 22. It turns out that 19 fits perfectly, as shown below:

| Rating | Number | Percent | Rounded Percent | CU Percent |
|---|---|---|---|---|
| E | 5 | 26.3 | 26 | 26 |
| G | 11 | 57.9 | 58 | 58 |
| F | 3 | 15.8 | 16 | 16 |
| P | 0 | 0.0 | 0 | 0 |
| Total | 19 | 100.0% | 100% | 100% |

Expressing data like these as percentages without showing the sample size can be very misleading; certainly the reference to "the aid of some 2000 subscribers" in this case is misleading.

Frequencies can be handled by a device called the *cumulative distribution*, as well as by the frequency distribution. We illustrate this in Table 189, which shows two cumulative distributions, one cumulated downward (that is, showing for each income level the sum of the frequencies for that income level and all lower levels) and one upward. Ordinarily, only one of the cumulated distributions is shown, since the corresponding percentage of the other is simply 100 minus the figure shown.

For many purposes, a cumulative distribution is more useful than a noncumulative one. For example, we can see more readily

*6.2 Quantitative Observations*

from either column of Table 189 than from Table 183 that about
a quarter (26.6 percent) of the families had incomes below $3,000
and about a quarter (24.2 percent) had incomes above $6,000, or
that the percent having incomes between $5,000 and $10,000 was 33
(95.1 minus 62.1, or 37.9 minus 4.9).

TABLE 189

MONEY INCOME OF URBAN FAMILIES, 1952—
CUMULATIVE PERCENT DISTRIBUTION OF FAMILIES
BY INCOME LEVEL

| Total Money Income | Percent of Families Receiving | |
|---|---|---|
| | Less than this much | This much or more |
| $ 500 | 2.4 | 97.6 |
| 1,000 | 5.2 | 94.8 |
| 1,500 | 9.3 | 90.7 |
| 2,000 | 13.7 | 86.3 |
| 2,500 | 19.5 | 80.5 |
| 3,000 | 26.6 | 73.4 |
| 3,500 | 36.3 | 63.7 |
| 4,000 | 45.6 | 54.4 |
| 4,500 | 54.3 | 45.7 |
| 5,000 | 62.1 | 37.9 |
| 6,000 | 75.8 | 24.2 |
| 7,000 | 84.3 | 15.7 |
| 10,000 | 95.1 | 4.9 |
| 15,000 | 98.5 | 1.5 |

*Source:* Computed from Table 183.

## 6.2.5   An Example of Interpretation

EXAMPLE 189   GOLDBRICKING

Table 190 is based on the experience of one man during nine months in a
machine shop. He was paid on a piecework basis, but calculated his earnings
on an hourly basis. If the piece rate amounted to less than 85 cents per hour,
workers in this shop received 85 cents per hour anyway. If the piece rate
amounted to more than 85 cents, the workers received the piece rate.

The most striking feature of the table, other than the astonishingly large
range, is the concentration of observations at two ranges: 35–54 and 125–134
cents per hour. While there are other possible interpretations, it seems likely
that the lower concentration arises from jobs in which it is difficult to earn
85 cents per hour at piece rates. The worker, therefore, tends to slacken off,
since he will get 85 cents anyway. The higher concentration in the range

*The Art of Organizing Data*

TABLE 190

PRODUCTION PIECEWORK HOURS WORKED, BY TEN-CENT EARNING INTERVALS

| Earnings per Hour (in cents) | Hours Worked | | Percent | |
|---|---|---|---|---|
| Unknown | 103.9 | | 7.7 | |
| 5–14 | 3.0 | | .2 | |
| 15–24 | 51.0 | | 3.8 | |
| 25–34 | 49.8 | | 3.7 | |
| 35–44 | 150.1 | | 11.1 | |
| 45–54 | 144.5 | | 10.7 | |
| 55–64 | 57.7 | | 4.3 | |
| 65–74 | 63.8 | | 4.7 | |
| 75–84 | 57.7 | | 4.3 | |
| Total under 85 cents | | 681.5 | | 50.4 |
| 85–94 | 51.2 | | 3.8 | |
| 95–104 | 19.5 | | 1.5 | |
| 105–114 | 17.9 | | 1.3 | |
| 115–124 | 83.0 | | 6.1 | |
| 125–134 | 496.3 | | 36.7 | |
| 165–174 | 1.5 | | 0.1 | |
| Total 85 cents or more | | 669.4 | | 49.6 |
| Total | | 1,350.9 | | 100.0 |

*Source:* Donald Roy, "Quota Restriction and Goldbricking in a Machine Shop," *American Journal of Sociology,* Vol. 57 (1952), p. 428. All "unknown" hourly earnings were below 85 cents.

125–134, with practically no observations greater than 134, also suggests an interpretation: whenever the worker finds he can earn more than 134 cents an hour, he works only hard enough to attain a rate between 125 and 134 cents because of fear that the standards department would revise the piece rates downward if it found, say, that workers could frequently make 180 cents an hour, or even 140 cents an hour.

The example illustrates how the organization of data into a simple table may bring out patterns that would otherwise be overlooked, and how these patterns may suggest interpretations. The interpretations, of course, require further investigation before they can be regarded as proved—recall the discussion of Sec. 1.2. Actually, the interpretations suggested here are the correct ones, as the article from which the table was taken makes clear in fascinating detail. A simple organization of statistical data which must have been readily accessible to the management of the shop would have shown that the incentive system was not working the way it was intended to work, and that changes in policy, such as revision or abandonment of the piece-rate system, might well have been considered. It might or might not have been easy to find the same information by direct observation. Workers might not have behaved at all typically when members of the standards department were on the floor of the shop, and they might not have been willing to tell an outsider, even though every worker understood exactly what was going on. The 1.5 hours during which the author earned a rate of 165–174 represents his very early experience, before he was told of the "system" by his fellow workers.

Table 190 also shows another interesting fact: all the "unknown" hours were less than 85 cents. This illustrates the always-present possibility that the "not-ascertained" observations may be systematically different from the others, and that they certainly cannot be either ignored or assumed to be typical.

Still another point illustrated by Table 190, one that is sometimes confusing to readers of statistical tables, is that the individual percentages do not add up precisely to the subtotals. For example, the first nine percentages add up to 50.5 instead of the 50.4 shown. This is due to rounding the individual figures to the nearest tenth of a percent. A similar rounding discrepancy accounts for the difference in Table 186A between the smallest percentage reported, 3.8, and the difference between 23.1 and 19.2, equal to 3.9, even though both represent one observation. Thus 1/26 rounded to the nearest one tenth of a percent is 3.8 percent; but 2/26 is not 7.6 percent but 7.7 percent.

## 6.3
## UNIVARIATE FREQUENCY DISTRIBUTIONS: QUALITATIVE OBSERVATIONS

### 6.3.1 Tables

EXAMPLE 191   LAND USE

Table 192 needs little comment. The variable is qualitative, the classification given to each small unit of land. Obviously, the problem of classification is not easy, but we shall not try to discuss the establishment of classes for qualitative distributions, since this almost always involves detailed knowledge of the particular subject matter. In Sec. 5.6.3 we did discuss some similar complications arising in questionnaire studies.

### 6.3.2 Graphs

In showing frequency distributions of qualitative variables, a device called a *bar chart* is often used. The bars in a bar chart may be either vertical or horizontal. Fig. 192 presents the data of Table 192 in a bar chart.

Notice that here the frequency scale has no other interpretation than number, relative or absolute; it does not represent density. Similarly, the thicknesses and areas of the bars have no meaning, nor do the locations of the bars along the vertical axis. In short,

*The Art of Organizing Data*

TABLE 192[5]

LAND UTILIZATION IN THE UNITED STATES, 1950
(In Millions of Acres)

| Land Use | Number | Percent |
|---|---|---|
| Total | 1,904 | 100.0 |
| In farms | 1,159 | 60.9 |
| Pasture | 485 | 25.5 |
| Cropland harvested | 345 | 18.1 |
| Forests and cut-over waste | 220 | 11.6 |
| Crop failure and cropland lying idle or fallow | 64 | 3.4 |
| Farmsteads, lanes, and waste | 45 | 2.4 |
| Not in farms | 745 | 39.1 |
| Forest land capable of producing timber of commercial quality and quantity | 311 | 16.3 |
| Pasture, including arid woodland (piñon, juniper, chaparral) | 290 | 15.2 |
| Roads, railroads, cities, parks, ungrazed desert, and other waste land not in farms | 144 | 7.6 |

*Source:* Department of Commerce, Bureau of Census and Department of Agriculture, Bureau of Agricultural Economics—co-operative report, *Graphic Summary of Land Utilization in the United States,* and records.

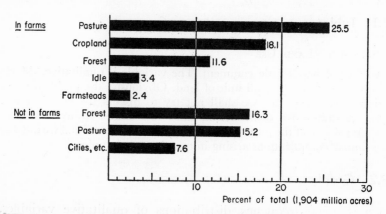

FIG. 192. Land utilization in the United States, 1950.
Source: Table 192.

though bar charts and histograms are often confused, they have no relation except the fortuitous one that both involve bars—as do musical scores and saloons.

5. Again we have modified and rearranged a table from the *Statistical Abstract: 1955.* This one is from Table 767, p. 628.

### 6.3 Qualitative Observations

A minor variant of the bar chart is the pictogram. Here the bars are replaced by rows of small schematic pictures depicting the characteristic represented by the bar, each picture standing for a given amount of the characteristic. A pictogram of Table 192, for example, might use small pictures of pastures, forests, etc., each picture representing, say, 20 million acres if an absolute scale is used, or representing one percent if a percentage scale is used, as in Fig. 192.

The choice between the bar chart and the pictogram is a matter of cost and audience. The pictogram is more costly, but is becoming increasingly common in presentations to the general public and even in technical reports for administrators. In a report intended for statisticians or technical people, on the other hand, the bar chart is probably preferable.

Another useful device for presenting qualitative data is the pie-chart. Here a circle is divided into pie-shaped pieces whose areas and circumferences are proportional to the quantities to be represented.

An objectionable variation of the pictogram or pie-chart is to use a series of pictures, each proportional in height to the quantity it represents. Figures of different sizes are difficult to interpret. The viewer does not know whether the height, width, area, or apparent volume of the object pictured is the significant measure. Consider, for example, the following two circles:

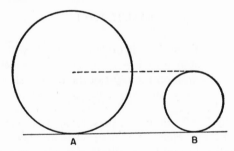

If it is intended to show that quantity $A$ is twice quantity $B$, the reader will be misled. While the height of circle $A$ is twice that of $B$, its area is four times as great; and the eye tends to make comparisons in terms of *area* rather than height. Lines or bars rather than circles or solid figures should be used if graphic comparisons are to be made. Circles are useful in the pie-chart because the components of a total are illustrated as the slices of a single pie, and it makes no difference whether the slices are judged by their circumferences, areas, or central angles—though even here, a pie shown in perspective can confuse.

## 6.4
## BIVARIATE FREQUENCY DISTRIBUTIONS

### 6.4.1 Bivariate vs. Two Univariate Distributions

EXAMPLE 194    EDUCATION OF ADULT POPULATION

TABLE 194[6]

SCHOOLING OF PERSONS 25 YEARS OLD AND OVER, BY SEX, 1950
Marginal Totals Only
(Numbers in thousands)

| Sex | Schooling Completed | | | | Total[a] |
| --- | --- | --- | --- | --- | --- |
| | Less than grade school (under 8 years) | Grade school (8–11 years) | High school (12–15 years) | College (16 or more years) | |
| Male | | | | | 41,286 |
| Female | | | | | 43,784 |
| Total[a] | 23,357 | 32,507 | 23,922 | 5,285 | 85,070[b] |

[a] Totals are rounded directly from exact figures, so may differ by 1 from the sum of the entries.

[b] Omits 2,413 thousand persons 25 years old and over for whom years of schooling was not reported.

Source: *U. S. Census of Population: 1950*, Vol. II, Part 1.

The reason for not filling in this table completely will be apparent in a few minutes. For the moment, concentrate on the numbers actually given. First, in the "total" column on the right of the table is shown the number of persons 25 and over by sex—actually, only those persons 25 and over who reported years of schooling, but we shall ignore this qualification hereafter. That is, each of the 85,070 thousand persons 25 and over is classified by a qualitative variable, sex. Hence this column is another example of a univariate frequency distribution for qualitative observations.

In the "total" row at the bottom of the table, the 85,070 thousand persons 25 and over are classified by a quantitative variable, the number of years of schooling completed: did not complete grade school (that is, did not complete the 8th grade), completed grade school but not high school, completed high school but not college, and completed college and perhaps more. This is also a univariate frequency distribution, this time of a quantitative variable. These two univariate distributions are called *marginal distributions* since they occupy the "margins" of the table.

---

6. Adapted from *Statistical Abstract: 1955*, Table 129, p. 112.

### 6.4 Bivariate Frequency Distributions

Now let us inquire about the eight spaces or "cells" in the body of the table which have been left blank. See if you can devise some way of filling in these cells solely on the basis of the information given in the two univariate frequency distributions. This is worth puzzling over a bit, before going on to the next paragraph.

An observation would be classified in the upper left-hand cell if and only if it would be classified both as "male" and "did not complete grade school." It is possible, on one extreme, that all of the 23,357 thousand with less than grade school educations might be male. At the other extreme, none of them might be male. Similarly, each other cell entry, as far as we can tell from the table as given, might be as large as its column total or as small as 0.[7]

The table might be filled in by making some assumptions. For example, it might seem reasonable (does it to you?) to assume that the number having each amount of schooling should be divided between the males and females in the ratio 41,286 to 43,784 that applies to the totals. On this assumption we get the following figures for the cell entries:

|        |        |        |       |
|--------|--------|--------|-------|
| 11,336 | 15,776 | 11,610 | 2,565 |
| 12,021 | 16,731 | 12,312 | 2,720 |

Figures computed this way are sometimes called "expected frequencies," indicating that they are what would be expected if each row (or column) had the same relative distribution as the total row (or column). Note that we get exactly the same figures if we assume that the proportion of males with any given amount of schooling is the same as the proportion of the total with that amount of schooling. For example, the over-all proportion completing college is $5,285/85,070 = 6.213$ percent, and this proportion of 41,286 is 2,565. Such figures are no better than the assumption, and there is no way, with the information so far given, to tell whether the assumption is good or bad. You may find it interesting to make a guess, on the basis of your general knowledge or impressions, about the accuracy of these figures.

The only way to get the correct answer is to "cross-classify" each observation according to these two variables. By cross-classification is meant the procedure of classifying each observation by both variables, here sex and schooling. In effect, it is necessary to look at each observation and decide which row *and* column, and hence which cell, it should be assigned to. It is not enough, as we have explained, to know only the totals of the rows and columns. When this is done for the 85,070 thousand persons 25 and over, the eight cells have the numbers shown in Table 196 (from the same source).

----

7. In many tables, incidentally, some of the cell entries can be seen to be necessarily greater than 0. To illustrate this, suppose that the total for the first column in Table 194 were 42,000. Then the lower left cell entry could not be less than 714. If it were less than 714, the upper left cell entry would have to be bigger than 41,286 to make the first column total 42,000; but if the upper left cell entry exceeds 41,286, the total for the first row cannot be right. If the largest total of one set, say rows, lies between the largest and smallest totals of the other set, columns, there will be some cells that cannot have entries as small as 0.

*The Art of Organizing Data*

TABLE 196

CELL ENTRIES FOR TABLE 194

| Sex | Less than Grade School | Grade School | High School | College | Total |
|---|---|---|---|---|---|
| Male | 12,047 | 15,798 | 10,414 | 3,027 | 41,286 |
| Female | 11,309 | 16,708 | 13,509 | 2,258 | 43,784 |
| Total | 23,357 | 32,507 | 23,922 | 5,285 | 85,070 |

The eight cell frequencies constitute the *joint distribution:* each observation has been cross-classified, that is, classified *jointly* by sex and schooling. From this information it is easy to reconstruct the total column and total row by simple addition.

The joint distribution tells about the association (relationship, correlation) between the two variables, sex and schooling. For example, we can compute from it that only 5.2 percent of women 25 and over had completed college, but 7.3 percent of men had done so. Again, although men constitute only 48.5 percent of people 25 and over, they constitute 57.3 percent of college graduates 25 and over.

One fact should be particularly emphasized from this discussion. To study association, it is necessary to cross-classify the data and tabulate the joint frequency distribution. It is impossible to tell anything about the *association* between two variables from the two univariate distributions. Even this cross-classification leaves much to be desired in comparing the schooling of males and females, however. There is, for example, a difference in the age distributions of the sexes, and the figures for men may reflect more strongly than do those for women recent practices with respect to education. Thus, a three-way cross-classification, by sex, by age, and by schooling completed might be advantageous.[8] This kind of thing has been discussed in Sec. 3.6, and will be again in Chaps. 7 and 9.

### 6.4.2 Absolute and Relative Joint Frequency Distributions

So far we have treated only the absolute joint distribution—that is, the distribution which gives absolute numbers of observations. There are *three* different kinds of *relative* joint frequency distribution. These different kinds should be understood thoroughly in order to read tables intelligently. All three are shown in Table 197.

From any of the three cases in Table 197, it is possible to find the absolute frequencies, at least approximately, since the numerical equivalent of 100 percent is given. For example, in Case I the "100

8. Table 194 was, in fact, obtained by condensing such a trivariate table. See *Statistical Abstract: 1955*, Table 129, p. 112.

## 6.4 Bivariate Frequency Distributions

TABLE 197

THREE METHODS OF EXPRESSING THE DATA OF TABLE 196
AS RELATIVE FREQUENCIES

CASE I
As Percentages of Column Totals

| Sex | Schooling Completed | | | | Total |
|---|---|---|---|---|---|
| | Less than grade school | Grade school | High school | College | |
| Male | 51.6 | 48.6 | 43.5 | 57.3 | 48.5 |
| Female | 48.4 | 51.4 | 56.5 | 42.7 | 51.5 |
| Total (thousands) | 100.0 (23,357) | 100.0 (32,507) | 100.0 (23,922) | 100.0 (5,285) | 100.0 (85,070) |

CASE II
As Percentages of Row Totals

| Sex | Schooling Completed | | | | Total (thousands) |
|---|---|---|---|---|---|
| | Less than grade school | Grade school | High school | College | |
| Male | 29.2 | 38.3 | 25.2 | 7.3 | 100.0 (41,286) |
| Female | 25.8 | 38.2 | 30.9 | 5.2 | 100.0 (43,784) |
| Total | 27.5 | 38.2 | 28.1 | 6.2 | 100.0 (85,070) |

CASE III
As Percentages of Table Total

| Sex | Schooling Completed | | | | Total |
|---|---|---|---|---|---|
| | Less than grade school | Grade school | High school | College | |
| Male | 14.2 | 18.6 | 12.2 | 3.6 | 48.5 |
| Female | 13.3 | 19.6 | 15.9 | 2.7 | 51.5 |
| Total | 27.5 | 38.2 | 28.1 | 6.2 | 100.0 (85,070) |

*Source:* Computed from Table 196.

percent" at the bottom of the first column represents 23,357 thousand persons 25 and over. From this information, simple arithmetic gives the (approximate) absolute numbers pertaining to each of the two cells in that column. Thus, 51.6 percent of 23,357 is 12,052. This agrees closely with the figure 12,047 shown in Table 196. Since the percentage is given only to one decimal place, the figure may be in error by as much as 0.05, or nearly 12 thousand people.

Percentages are used to facilitate relative comparisons. There are three such comparisons of possible interest in the example given here, each giving rise to one of the three tables shown. In Case I it is seen immediately that 51.6 percent of all persons 25 and over with less than a grade school education are males, and so on. This table facilitates the comparison of the sex compositions of various educational levels.

In Case II the comparison is between the sexes with respect to education. For example, 29.2 percent of males but only 25.8 percent of females had not completed grade school. In the present example, this comparison is probably of more interest than the first. The comparison to be emphasized determines which components of the table are to add up to 100 percent—that is, whether the 100 percent's are to appear in the right-hand column or the bottom row.

In Case III the whole joint frequency distribution with 85,070 thousand observations has been in effect reduced to a relative distribution. This form is no more valuable in facilitating a study of association (that is, a comparison between variables) than the original numbers. What it does show is the importance of each of the eight cells in relation to the total number of observations.

As a digression, we may note that it is true that most (meaning more than half) of those with less than a grade school education were males, but it is not true that most of the males had less than a grade school education. Confusion between these kinds of statement is common.

EXAMPLE 198   CLERGYMEN

In a recent public controversy it was charged that the largest single group of sympathizers with a certain position were members of the clergy. This was apparently interpreted by many as a charge that the particular position was the one most commonly held by clergymen. The patent absurdity of this resulted in discrediting the charge and discharging the discreditor.

One or even both variables of a bivariate table might be cumulated. This would be particularly relevant to comparing the educational levels attained by each sex. Table 199, which shows Case II

*6.4 Bivariate Frequency Distributions*

of Table 197 cumulated upward, shows quite clearly, for example, that women generally have completed more schooling than men, except at the college level.

TABLE 199

SCHOOLING COMPLETED BY PERSONS 25 YEARS OLD AND OVER, BY SEX, 1950
(Cumulated Percents)

| Sex | Schooling Completed | | | Number of Persons (thousands) |
|---|---|---|---|---|
| | Grade school or more | High School or more | College or more | |
| Male | 70.8 | 32.5 | 7.3 | 41,286 |
| Female | 74.3 | 36.1 | 5.2 | 43,784 |
| Total | 72.5 | 34.3 | 6.2 | 85,070 |

*Source:* Computed from Table 197, Case II.

### 6.4.3 Extension to Multivariate Frequency Distributions

The general principles of interpreting bivariate distributions apply also to frequency distributions with three or more variables. There is considerably more detail, and as the number of variables increases the number of relationships to consider grows rapidly. Thus, in a trivariate distribution we may want to consider not only whether, say, schooling completed is related to sex and whether it is related to age, but also whether there is *interaction* between the effects of sex and age, that is, whether the relation of schooling to sex differs from one age level to another (or, what comes to the same thing, whether the relation of schooling to age differs between the sexes). It is extremely difficult to grasp all the relations of this kind among four variables, and it requires a good deal of experience to handle even three with assurance; but the underlying approach is the same.

### 6.4.4 Graphical Representation of the Bivariate Frequency Distribution

(1) Fig. 200 shows one way to represent a bivariate distribution by a bar diagram. Bars are drawn to represent each class of one variable, and each bar is divided into sections representing the classes of the other variable. The bar lengths can be the same, as they are

*The Art of Organizing Data*

in Fig. 200, to portray the percentage distributions—in this case the data of Table 197, Case II—or they can vary in proportion to the actual numbers of cases represented by each bar. Either variable can be represented by the bars, and the other by the segments. This type

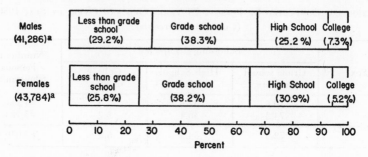

FIG. 200. Schooling completed by persons 25 years old and over, by sex, 1950.
*Source:* Table 197, Case II.

   a. In thousands.

of bar diagram can be used whether one, both, or neither variable is qualitative. Its main disadvantage is the difficulty of comparing lengths of inside bars, for example, grade school, especially when there are many classes.

(2) When one variable is quantitative and one qualitative, a good way to represent a bivariate distribution is shown in Fig. 201. The quantitative variable here is represented on the horizontal axis, and either numbers of observations or percents are represented on the vertical axis; in this instance we have chosen cumulative percents. A separate line is drawn for each class of the qualitative variable. One interesting thing brought out clearly by Fig. 201 is that the number dropping out per year is about the same during the last three years of grade school and of high school, but only about half as much during the last three years of college. This is shown by the fact that the lines from 5 to 8 and from 9 to 12 have the same slope, but the lines from 13 to 16 are only about half as steep. The sharp drops come between finishing grade school and completing even a year of high school, and between finishing high school and completing a year of college. Furthermore, the number dropping out is about the same for both sexes during the last years of grade school, during high school, and during college, and about the same for both sexes between stages except between high school and college, where it is appreciably greater for females.

### 6.4 Bivariate Frequency Distributions

(3) When both variables are quantitative, the graphical representation most commonly used is the *scatter diagram* or *correlation dia-*

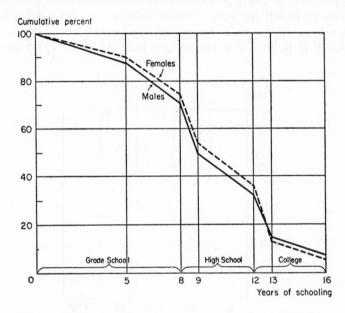

Cumulative percent

Females

Males

Grade School    High School    College

Years of schooling

**FIG. 201.** Schooling completed by persons 25 years old and over, by sex, 1950.

(Percent completing various numbers of years or more.)

Source: *Statistical Abstract: 1955,* Table 129, p. 112. More stages of schooling were used for this chart than for Table 199.

*gram.* We have already seen a special kind of scatter diagram in Fig. 109 which shows the relationship between the number of red beads and the sequence in which the sample was drawn. Another example is given by Fig. 202.

### EXAMPLE 201  INTELLIGENCE TESTS

Fig. 202 shows the relationship between intelligence quotients (IQ's) taken at two times for the same ten children. Each dot represents a single student's score on the two tests. These scores are read as follows, taking the circled dot as an example: By dropping a perpendicular line to the horizontal axis, we see that this child scored 143 on Test 1. By carrying a perpendicular over to the vertical axis, we see that his score was 142 on Test 2. By examining the entire scatter diagram in Fig. 202 we see that (1) there is a tendency for a child to perform consistently on the two tests, and (2) there is only a little variability in performance on the two tests.

*The Art of Organizing Data*

   Another way of charting two quantitative variables is as follows: Class intervals are formed on one variable, and, at the middle of each class is plotted the average value of the second variable for those observations in which the first variable falls within the class. Suppose, for example, that for each weight shown in Table 173 we knew the corresponding height. For the weight class 147.5 to 157.5 there are

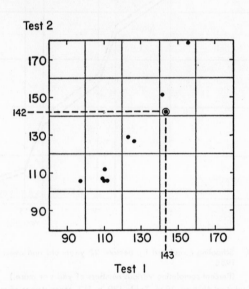

FIG. 202.  IQ's of 10 children on two tests, using the Stanford-Binet test.

Source: Albert K. Kurtz, "Different IQ's for the Same Individual Associated with Different Intelligence Tests," *Science*, Vol. 119 (1954), p. 611.

five observations. We would average the five corresponding heights and plot this average above 152.5 on the weight axis—or perhaps, as a refinement, above 152.2, which is the mean of the five weights in the class. Alternatively, we could form height classes, and plot the corresponding average weights; this will *not* give the same picture, however, as is brought out in Sec. 17.4.1. Often the plotted points are connected by lines, simply to help the eye locate and follow the points, as in Fig. 201; and of course the lines *between* the points should not be confused with observed data, which are represented only by the points.

## 6.5

## THE USE OF FREQUENCY DISTRIBUTIONS IN THE SUMMARIZATION OF INVESTIGATIONS

Sec. 6.1 ended by posing a problem of summarizing the information contained in a group of multivariate observations obtained in a questionnaire study. Suppose that the questionnaire included 50 questions asked of each person in a sample. The problem is how to prepare a set of tables and charts to describe adequately the information contained in these observations.

There are 50 univariate distributions which could be prepared, one for each question in the questionnaire. Probably all of these should be prepared for every study. Most of them will be by-products of bivariate or multivariate tabulations, and some of them may be misleading unless relationships to other variables are taken into account.

To study the relationships or associations between answers to different questions—that is, the relationships among the variables under study—it is necessary to form bivariate or multivariate distributions. Each of these distributions tells about one relation or association. But there are a great many possible associations or interrelations among the 50 variables being studied, and to each of these associations corresponds one multivariate frequency distribution. There are 1,225 possible bivariate frequency distributions, 19,600 trivariate distributions, 230,300 four-variate distributions, etc.—all told, there are 1,125,899,906,842,573 (more concisely, about $10^{15}$) joint distributions.[9] So even with the aid of the frequency distribution as a tool for summarization, it is possible to explore only a small proportion of the possible associations which might be examined.

To avoid being swamped by tabulations, the investigator must have an idea in advance of what he is looking for. Fortunately, investigators usually are interested in only a relatively small number of possible interrelationships among the variables under study. Usually, they are content with the simpler ones—bivariate and trivariate relationships—and only a few of these. The ones actually selected are determined by the objectives of the study. For example, in a study of the relationship of certain variables to the incomes of lawyers, it was decided that age and years of professional experience should be taken into account. A trivariate frequency distribution showing

---

9. This number is $2^{50} - 1$, which is the total number of tables possible, minus 50 for the univariate tables.

the interrelationship among age, experience, and income was therefore planned. (This example is discussed in Chap. 9.)

Hence we return to one of our major themes, the importance of careful advance planning. In too many studies, data are collected with only the vaguest of plans and the method of summarization is not considered until the data are actually collected. As a result, needless data are collected, important or even essential data are omitted, much energy is wasted attempting to analyze the data, and needlessly inconclusive results are obtained. It is essential to draw up, in advance of the study, blank forms of the tables that will be prepared from the data, and to consider the interpretations that would be put on various possible sets of data in these tables.

Even with the most carefully planned studies, however, additional tabulations will often be suggested by examination of the data, for the insights gained from analyzing actual data should suggest explorations that were not anticipated in advance. These explorations ought not to be neglected, but special caution is required in interpreting them. We shall return to this point in Sec. 12.11.

It is clear, then, that typically only a part of the information collected in any investigation is actually used. Even less information is actually presented in the final report, since compression is desirable if it does not involve suppression of relevant information. However, a study should always include a statistical appendix, either attached or available on request, which contains all the basic data that have been tabulated. To analyze a study fully, it is necessary to have the original data.

Often data may be useful for some purpose that the original investigators did not have in mind. Researchers seldom realize how often they may obtain exactly the information they need by reanalysis of someone else's basic data. A striking example is the Goldhamer and Marshall study on mental disease described in Sec. 2.8.2. These investigators were able to penetrate a century into the past because hospital records in Massachusetts had originally been collected carefully and had been published. The people who initially made the hospital records, in at least some important instances, were competent statisticians—one of them, Dr. Edward Jarvis, was president of the American Statistical Association from 1852 to 1882—and they had in mind the idea that some day someone would build further on the foundations they laid. They therefore not only did their work carefully and thoroughly, they reported it carefully and thoroughly to posterity.

## 6.6
## CONCLUSION

The data from any statistical study should usually be reduced to compact, comprehensible, and communicable form. The proper form depends on the nature of the data available, of the problems under investigation, and of the analyses to be made. Skeleton tables and charts should be drawn up in advance of a study, for data that will not fit into meaningful tables and charts will not be interpretable, and data that are not interpretable are not useful.

Sometimes data may be organized for a single variable at a time. For quantitative variables, this involves selecting appropriate class intervals and distributing the observations among them. For qualitative variables, it involves careful definitions and distribution of the observations among the different categories. The resulting distributions may be tabulated as absolute or relative (percentage) frequencies, either directly or cumulated upward or downward. Quantitative variables may be charted by histograms, qualitative variables by bar diagrams.

Often the influence of a variable may be concealed or exaggerated, or even reversed, if other variables are not taken into account simultaneously. This may be done through tables and charts involving two or more variables. Such tables present several columns, each column showing for one class of one variable the distribution according to the second variable. The data tabulated may be absolute frequencies; or they may be proportions of the respective columns, of the respective rows, or of the total table. They may also be cumulated. Charts may show bars for one variable, each bar broken into segments according to the second variable; they may show histograms or cumulative distributions according to a quantitative variable, one for each category of the other variable; or, when both variables are quantitative, they may be scatter diagrams or correlation diagrams —that is, ordinary plots on rectangular co-ordinates.

The charts and tables of frequency distributions represent the first step in analyzing the data. Usually, they too require further summarization, for compactness, comparability, or interpretability. (In one of our illustrative distributions, Table 183, we were unable to resist inserting such a summary measure, the median.) The next three chapters deal with methods of describing by a single number or two the salient features of frequency distributions. First, in Chap. 7,

*The Art of Organizing Data*

we consider numbers to characterize the location of a univariate distribution, in the sense that the median of Table 183 may be said to characterize the location of the series of frequencies relative to the income scale. In Chap. 8 we consider methods of characterizing the dispersion of a distribution. In Chap. 9 we consider methods of characterizing the association between variables in joint distributions.

## DO IT YOURSELF

### EXAMPLE 206A

(1) For the following data, form a frequency distribution and draw a histogram.

TABLE 206

ROCKWELL HARDNESS TEST

100 SAMPLES OF STEEL COIL LISTED IN ORDER OF OCCURRENCE

(Read left to right, and down)

| | | | | | | | | | |
|---|---|---|---|---|---|---|---|---|---|
| 58 | 49 | 58 | 57 | 50 | 60 | 64 | 65 | 64 | 59 |
| 65 | 65 | 45 | 54 | 52 | 59 | 65 | 57 | 63 | 54 |
| 65 | 60 | 61 | 47 | 60 | 52 | 63 | 61 | 54 | 63 |
| 62 | 56 | 56 | 65 | 56 | 64 | 65 | 55 | 59 | 65 |
| 64 | 49 | 65 | 50 | 65 | 61 | 64 | 61 | 59 | 63 |
| | | | | | | | | | |
| 58 | 57 | 65 | 60 | 55 | 64 | 65 | 59 | 62 | 65 |
| 64 | 54 | 56 | 58 | 40 | 85 | 53 | 61 | 56 | 65 |
| 58 | 58 | 55 | 52 | 65 | 60 | 65 | 63 | 64 | 63 |
| 60 | 61 | 61 | 65 | 56 | 62 | 65 | 54 | 64 | 63 |
| 57 | 64 | 62 | 58 | 60 | 52 | 53 | 62 | 56 | 65 |

(2) Do you find any internal evidence that makes you suspicious of the accuracy of the original laboratory measurements? If you do, explain the grounds for your suspicion.

(3) In what other way, besides the one you used for part (1), could these data be organized to bring out further important aspects of the data? Carry this out, and tell exactly what you have gained by this additional work. [Hint: Review Sec. 4.8.]

### EXAMPLE 206B

Comment on the following chart and its interpretation.

In Figure [207] are presented on doubly logarithmic co-ordinates the *x*-number of different pages that contained the same *y*-number of pictures per product. Except for the bottom 4 or 5 points, which indicate an avoidance of pages with very few pictures, the distribution is rectilinear.[10]

10. George Kingsley Zipf, "Quantitative Analysis of Sears, Roebuck and Company's Catalogue," *Journal of Marketing*, Vol. 15 (1950), p. 10.

*Do It Yourself*

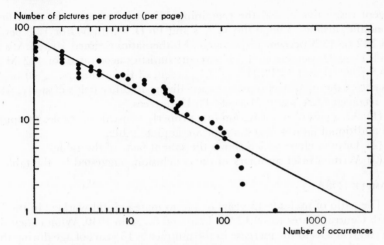

Number of pictures per product (per page)

Number of occurrences

FIG. 207. Distribution of pictures (per product per page). (Sears, Roebuck Catalog, Spring-Summer Edition, 1949.)

EXAMPLE 207A

In a study for a certain major league baseball team, a research firm selected a random sample of fans at each of three late-season games. Among other questions, each fan was asked where he lived and how many games he had attended during the year. Here are some of the results: There were 405 people in the sample, 225 of whom lived within 50 miles of the city. Of those who lived within 50 miles, 63 were seeing their first game on the day of interview; 42 had seen one other game; 37, 2 other games; 21, 3; 16, 4; 15, 5; 12, 6; 14, 7; 0, 8; 2, 9; 1, 10; and 2, 13. The corresponding figures for the other group were 98, 0; 75, 1; 2, 2; 1, 3; 3, 5; 1, 9.

(1) Present the above data in a well-organized, well-labelled, neat table.

(2) If the team's total annual attendance was 1,000,000, how much of this total would you estimate to consist of people living within 50 miles?

(3) State concisely any qualifications about your answer to (2).

EXAMPLE 207B

An article about membership of the American Statistical Association[11] shows that in 1945, 2,563 members of the A.S.A. reported that they had received degrees from colleges or universities. The highest degree received by 861 or 33.6 percent of the members reporting was a B.A.; 910 or 35.5 percent held an M.A., and the remaining 792, or 30.9 percent, held a Ph.D. degree; of the B.A.'s 266, or 10.4 percent, had specialized in economics, while 160 and 92 had majored in mathematics or statistics, 6.2 percent and 3.6

---

11. Abner Hurwitz and Floyd C. Mann, "Membership of the American Statistical Association—An Analysis," *Journal of the American Statistical Association*, Vol. 41 (1946), pp. 155–170.

percent respectively, and the remaining 343 members were in other fields. Economics also led among the M.A.'s and Ph.D.'s with 307 or 12.0 percent and 342 or 13.3 percent respectively. Mathematics claimed 100 M.A.'s and 79 Ph.D.'s, 3.9 percent and 3.1 percent; statistics accounted for 122 M.A.'s or 4.8 percent and 55 Ph.D.'s or 2.1 percent of the members. 1,040 or 40.6 percent of the members reported specializing in other fields of study, where 381 received M.A.'s and 316 held Ph.D. degrees.

(1) Arrange this information in a properly constructed table, computing any additional figures necessary for a complete table.

(2) Draw a chart which shows the salient facts of the table.

(3) Write a brief summary of the conclusions suggested by the table.

EXAMPLE 208A

The ratio of males 9–13 years of age to male gainful workers in the East North Central States was 0.129 in 1930 and 0.156 in 1940. What conclusions can you reach as to the increase in the number 9–13 years of age during those ten years?

EXAMPLE 208B

Which of the following class limits seems best for a frequency table classifying the sales of a 5 and 10 cent store?

| (1) | (2) | (3) | (4) | (5) |
|-----|-----|-----|-----|-----|
| 0–5 | 1–6 | 0.5– 5.5 | 2.5– 7.5 | 4–9 |
| 5–10 | 6–11 | 5.5–10.5 | 7.5–12.5 | 9–14 |
| 10–15 | 11–16 | 10.5–15.5 | 12.5–17.5 | 14–19 |
| etc. | etc. | etc. | etc. | etc. |

Why?

EXAMPLE 208C

A distributor of major home appliances (refrigerators, ranges, etc.) found that some customers were refusing to accept delivery of appliances, even though they had previously agreed not only to purchase the appliance, but to accept delivery on the day the appliance actually arrived. The company considered instituting a policy of telephoning each customer two days in advance of delivery to find out whether the original delivery date was still acceptable. In order to find out whether this policy was worth instituting, they designed the following experiment.

During the period of the experiment, every other customer was to be telephoned two days in advance of scheduled delivery; the remaining customers were not to be telephoned.

The information which follows relates only to customers who had originally scheduled more than 10 days between agreement to purchase and the date of delivery. There were 78 customers in the group to be telephoned. Of these, 57 were actually reached, 10 did not answer and 11 did not have telephones. In this group there were 13 refusals at the time of phone call and

*Do It Yourself*

only 7 refusals at the time of delivery, or 20 refusals in all. Of these 20 refusals, 14 were classified as "hold until notified," 5 were outright cancellations, and 1 was not at home.

In the non-telephoned group, there were 70 customers. There were 11 refusals, all of them, of course, at the time of delivery since no one was telephoned. Seven of these refusals were "hold until notified," 1 was an outright cancellation, and 3 were not at home.

(1) Decide what part of the information given above is needed for a decision. Present this concisely in good tabular form.

(2) State the decision, if any, that is warranted.

(3) Give concise reasons for your answer to (2). Include any qualifications you think necessary. If you think you need more information, state concisely what you would want to know and how it might be obtained.

EXAMPLE 209A

Select one of the six problems mentioned on pages 169 and 170, and try to outline the kind of data that might be useful and how it would be organized.

EXAMPLE 209B

What is the minimum sample size for the films cited but not analyzed in Example 187?

EXAMPLE 209C

Write a summary of the salient conclusion of the following table. What would you like to know besides the information given in the table? What criticisms, if any, of the table itself do you have?

TABLE 209

DISTRIBUTION OF RESPONDENTS BY THEIR PRIVATE OPINIONS ON RUSSIA
AND THE DIRECTION OF THEIR ESTIMATE OF THE GROUP OPINION
IN RELATION TO THEIR OWN PRIVATE OPINIONS

| Individual's Private Opinion | Direction of Estimate of Group Opinion | | Total |
|---|---|---|---|
| | More Pro-Russian | More Anti-Russian | |
| Pro-Russian | 2 | 10 | 12 |
| Anti-Russian | 10 | 2 | 12 |
| Total | 12 | 12 | 24 |

*Source:* Raymond L. Gorden, "Interaction between Attitude and the Definition of the Situation in the Expression of Opinion," in Dorwin Cartwright and Alvin Zander (eds.), *Group Dynamics* (Evanston, Ill.: Row, Peterson and Company, 1953), p. 167.

EXAMPLE 209D

Write a summary of the salient conclusions of the following table. Which percentage distribution(s) of the three types mentioned in Sec. 6.4.2 would

*The Art of Organizing Data*

be most helpful? What more would you like to know besides the information given in the table?

TABLE 210A

COMPARISON OF THE WAY HUSBANDS AND WIVES RATED
THE HAPPINESS OF THEIR MARRIAGE

| Wife's Rating | Husband's Rating | | | | | Total |
|---|---|---|---|---|---|---|
| | Very Unhappy | Unhappy | Average | Happy | Very Happy | |
| Very happy | 1 | 0 | 3 | 24 | 112 | 140 |
| Happy | 0 | 0 | 12 | 38 | 12 | 62 |
| Average | 0 | 3 | 14 | 7 | 6 | 30 |
| Unhappy | 1 | 11 | 2 | 0 | 0 | 14 |
| Very unhappy | 5 | 1 | 0 | 0 | 0 | 6 |
| Total | 7 | 15 | 31 | 69 | 130 | 252 |

*Source:* Ernest W. Burgess and Leonard S. Cottrell, Jr., "The Prediction of Adjustment in Marriage," in Paul F. Lazarsfeld and Morris Rosenberg (eds.), *The Language of Social Research* (Glencoe, Ill.: Free Press, 1955), p. 271.

EXAMPLE 210A

In an attempt to find out the effect of telecasts on attendance at football games, an investigator interviewed a number of people in Philadelphia in 1949. He classified his respondents into TV owners and nonowners. TV owners were further classified into groups according to length of ownership. For each of these groups, he ascertained the percentage of people who attended at least one football game in 1949. His chief results are shown below.

TABLE 210B

TELEVISION OWNERSHIP AND FOOTBALL ATTENDANCE

| TV Status | Percentage Attending At Least One Game in 1949 |
|---|---|
| TV owners | |
| 1–3 months | 24 |
| 4–11 months | 41 |
| 1–2 years | 45 |
| over 2 years | 54 |
| Non-TV owners | 46 |

He concluded that TV's initial effect was bad for attendance, but that this effect tended to wear off as people got used to their sets: TV had only a novelty effect.

Comment critically but constructively.

EXAMPLE 210B

Make a histogram of the data of Table 183, excluding the open-ended intervals.

# Chapter 7

# *Averages*

## 7.1
### DESCRIPTIVE STATISTICS

Chap. 6 dealt with the frequency distribution as a device for organizing and summarizing data. This chapter and the next deal with specific characteristics of frequency distributions.

A single number describing some feature of a set of data is called a "descriptive statistic." It is thus an even more compact description than the frequency distribution. Indeed, the frequency distribution is essentially a set of descriptive statistics, for each number showing

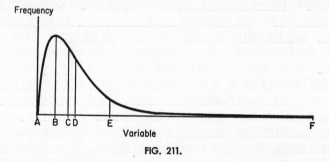

Frequency

A  B  C D    E                                                              F

Variable

FIG. 211.

how many observations are in a class is a descriptive statistic. A descriptive statistic focuses attention more sharply than a frequency distribution on the feature of the data which it measures, but ignores more of the complexity of the data.

The descriptive statistics considered in this chapter are those called "measures of location," or, more commonly, "averages." We are often interested in knowing where a distribution is located, or centered, on the scale of conceivable values of the variable being studied, more or less independently of its other characteristics, such as its dispersion or shape.

As examples of measures of location, consider Fig. 211. This is to be considered as a histogram, but from so large a sample and with such narrow class intervals, that it is essentially a smooth curve. The location of this distribution could be described by any of the following:

(1) Its *minimum, A*.

(2) Its *mode, B*, the value of the variable at which the greatest concentration of observations occurs.

(3) Its *median, C*, the value of the variable which exceeds half of the observations and is exceeded by half.

(4) Its *mean, D*, the measure that the layman ordinarily refers to by the term "average."

(5) Its *ninetieth centile, E*, the value of the variable which exceeds 90 percent of the observations and is exceeded by 10 percent.

(6) Its *maximum, F*.

Innumerable other measures could be constructed; in fact, (3) and (5) are two examples of a type which alone could produce innumerable measures—for example, the 17th, 83rd, or 99th centiles.

If the curve of Fig. 211 could be thought of as the population, having a rigid and known shape and size, any measure of location would be as useful as another, since from any one the others could be deduced. In selecting a measure of location, only one criterion would concern us, the meaning or interpretation to be placed on the measure, and its suitability for the problem at hand. This would be the only question within the framework of descriptive statistics.

When, as in most practical statistical work, the curve of Fig. 211 has to be thought of as a schematization of a histogram based on a sample, and the exact shape of the population is not known, it is not possible to deduce one measure from another. Furthermore, we must take into account the sampling variability of various measures, and also the extent to which conclusions drawn from them would be affected by any errors in assumptions we may make about the shape of the population distribution. It thus becomes necessary to distinguish two questions: (1) What measures will best describe the features of the population pertinent to a specific problem? (2) What computations from the sample will best enable us to estimate the measures in which we are interested? Consideration of the second question belongs in Part III, on Statistical Inference. In the present chapter, our attention will be focused primarily on the meanings of various measures of location.

*7.2 The Mode*

We shall discuss three measures of location, the mode, the median, and the mean, then consider certain important special kinds of mean, and finally discuss the problem of choosing and interpreting the proper measure.

## 7.2
## THE MODE

Roughly speaking, the mode is the value that occurs most often. More accurately, it is the value of the variable at which the concentration of the observations is densest.

Consider Tables 106 and 107B. These show that the modal number red was 12 (with a frequency of 10) for the samples of beads from Population I, and 3 (with a frequency of 15) for the samples from Population II. Again, Table 192 shows that the modal type of land use was pasture.

The maximum frequency may be attained by two (or more) classes, in which case both (or all) are modes and the distribution is called "bimodal" (or "multimodal"). With a discrete quantitative variable, any value is called a "mode" (more properly, a local mode)  if its frequency is not exceeded by either of the adjoining values. Thus, the distribution of Table 106 is bimodal, a secondary mode occurring at 10. It is, in fact, trimodal, for there is a minor mode at 18. In practice we would ordinarily ignore such secondary modes as these, but on grounds of statistical inference—that they are probably not present in the population—rather than of statistical description. The introduction of considerations of inference should lead us to ask also how accurately even the major mode of the sample describes the population.

If classes are combined, the modal class may not be the one which includes the value that was the mode in the uncombined data. In Table 106, for example, if 6 and 7 red beads are combined, 8 and 9, etc., the modal class is 10–11, with a frequency of 17 in contrast to a frequency of 15 for the class 12–13. Similarly, in Table 192, if all land not in farms were combined into a single class, this would be the modal class. On the other hand, if pasture in farms were divided into three categories, say ploughable land used for pasture (3.6 percent), forest and woodland used for pasture (7.1 percent), and other pasture (14.8 percent), the modal use would not be pasture at all but cropland.

EXAMPLE 214  PRESCRIPTIONS AND SODAS

An executive of a large chain of drug stores, when asked whether soda fountains do not represent the largest volume of his company's sales, replied that prescriptions do. Surprised interrogation brought out the fact that the second largest volume is in "carry-out ice cream," and the third, fourth, and fifth largest volumes are in other divisions of the soda fountain. The sales of any two of the soda fountain departments exceeded those of the prescription department.

With continuous quantitative variables there would be no mode in a sample unless the observations were grouped. As we saw in Sec. 5.5.2.1 and Sec. 6.2.1, sufficiently accurate measurement would reveal a difference between any two observations, so no frequencies would exceed 1. Limitations on the accuracy of measurement result in grouping, however, as does organizing the data into a frequency distribution, and this produces one or more modal classes. For example, the modal income class of urban families in 1952 is shown by Table 183 to be $3,000–$3,499, with 9.7 percent of the families. (Recall the discussion in Sec. 6.2.2 and avoid the trap of taking the 13.7 percent in the class $5,000–$5,999 as indicating the mode; maximum *density* is the criterion, and this class, being of double width, represents only 6.85 percent per $500.) Here, too, when classes are combined the mode may be shifted. Thus, forming $1,000 classes starting at $500 results in a mode in the $3,500–$4,500 class.

Sometimes rules are given for use with grouped data to find a single number to be called the mode. The idea underlying these rules can be illustrated by Table 183. There are many more cases in the class just above the modal class, $3,000–$3,499, than in the class just below it, so it seems likely that the upper half of the modal class contains more observations than the lower half, so that $3,250–$3,499 may be a more precise number for the mode. Proceeding on such assumptions, we can converge to a single number. Any result more precise than a $500 class will, however, be a matter of assumption rather than a description of the data. Furthermore, there are few if any purposes for which a single-valued number for the mode is appreciably more useful than an interval.

The mode indicates the winner in an election involving several candidates or issues (barring special requirements, such as a majority of all votes cast). The value which is most common may, however, be very uncommon, and may be scarcely more common than several other values. If a set of 1,000 fair coins is tossed at random indefinitely many times, for example, the modal number of heads will be

500; but 500 will occur only 2.52 percent of the time, and 499 will occur 2.50 percent of the time, as will 501. <u>The mode is much less frequently useful than either the median or the mean</u>. <u>We have taken it up first because it is the easiest of the three measures of location to find from a frequency distribution, and because it is meaningful for both quantitative and qualitative data.</u>

EXAMPLE 215   PLATAEAN ESCAPE FROM PELOPONNESIAN CIRCUMVALLATION

One of the most common and useful applications of the mode is illustrated by the following example of its employment for military purposes 24 centuries ago:

> The same winter [428 B.C.] the Plataeans, who were still being besieged by the Peloponnesians and Boeotians, distressed by the failure of their provisions, and seeing no hope of relief from Athens, nor any other means of safety, formed a scheme with the Athenians besieged with them for escaping, if possible, by forcing their way over the enemy's walls [of circumvallation]. . . . Ladders were made to match the height of the enemy's wall, which they measured by the layers of bricks, the side turned towards them not being thoroughly whitewashed. These were counted by many persons at once; and though some might miss the right calculation, most would hit upon it, particularly as they counted over and over again, and were no great way from the wall, but could see it easily enough for their purpose. The length required for the ladders was thus obtained, being calculated from the breadth of the brick.[1]

The everyday, contemporary version of this application of the mode is in checking calculations. If a calculation is repeated several times, the value accepted is that which occurs most often, not the median, mean, or any other figure. Even in these cases, a majority, or some more overwhelming preponderance, rather than merely a mode, is usually required for a satisfactory decision.

# 7.3
# THE MEDIAN

The median is often said to be the "middle" observation, or the value above which half of the observations lie and below which half lie. Both statements convey correct impressions, but both need to be made more precise.

The median number of red beads in Table 106 is 11. We find this by counting the observations from either end of the distribution to the 25th and 26th observations in order of size. These are among

---

1. Thucydides, *The Peloponnesian War*, III, 20, Crawley translation (New York: Modern Library, 1951), pp. 155–156.

the eight observations which showed 11 red beads. Similarly, in Table 107B the median number of red beads is 3. Had the 25th and 26th observations had different values, any value between them could have been called the median; it is customary in such cases to take the number halfway between them. In Table 173, for example, the 16th and 17th of the 32 weights are 175 and 178 pounds, so the median would be taken as 176.5.

Notice that in the two bead examples, the medians we have given are exactly equal to the middle observation—or rather to the middle observations, for among an even total number there are two middle observations. In these examples it is not true, however, that as many observations lie above the median as below. In Table 106, there are 22 observations below 11 and 20 above; and in Table 107B, there are 18 below 3, and 17 above. In the weight example, on the other hand, exactly half the observations are below and half are above the median, but the median is not the middle observation— it is not any of the observations.

A more precise definition of the median is this: The median of a set of observations is any number that neither exceeds nor is exceeded by more than half of the observations. When there is an odd number of observations, this leads to the value of the middle observation. When there is an even number of observations, it leads to any of the numbers at or between the two middle observations.

When observations are grouped in a frequency distribution and the variable is continuous, all that can be done accurately (except under unusual circumstances, such as in Table 174, where one of the class boundaries happens to divide the distribution into two equal parts) is to find the class in which the median lies. From Table 183, or more easily from Table 189, we see that the middle income is in the group $4,000–$4,499. A single number for the median is often desirable, and may be obtained as follows: Note that below $4,000 lie 45.6 percent of the observations. To reach 50 percent another 4.4 percent would be required, but 8.7 percent are included in the next $500 interval; so to the lower limit of this interval, $4,000, we add 4.4/8.7 (50.6 percent) of $500, getting a median of $4,253. This results in a reasonable guess, but only a reasonable guess, at the true median. The median shown in Table 183, presumably computed from the ungrouped data, is $4,249.

The median is a good measure when what is wanted is a figure for the "typical" individual. It is meaningless for completely non-quantitative data, but meaningful so long as the data can be ranked, as, for example, the grades A, B, C, D, F. It is not affected by the particular values of the largest or smallest observations, so it can be computed even when a frequency distribution involves open-ended intervals, like the "under $500" and "$15,000 and over" intervals in Table 183.

On the other hand, the median suffers the disadvantage that there is no way to tell from the medians of separate groups what the median of the combined group would be. In 1952, the median income of urban white families was $4,484 and of urban nonwhite families $2,631; but without combining the distributions and recalculating there would be no way to compute the median for all urban families.[2] This disadvantage is sometimes important, for often medians but not the full distributions are published.

An interesting and sometimes useful property of the median is that the sum of the absolute deviations of the observations from their median is less than from any other number. An *absolute deviation* of one number from another is the difference without regard to its algebraic sign.

Suppose there are 5 observations, 1, 2, 3, 4, 6. Then the median is 3. The absolute deviations of the observations from the median are obtained as follows:

| Observation | Algebraic Deviation from Median | Absolute Deviation from Median |
|---|---|---|
| 1 | $1 - 3 = -2$ | 2 |
| 2 | $2 - 3 = -1$ | 1 |
| 3 | $3 - 3 = 0$ | 0 |
| 4 | $4 - 3 = 1$ | 1 |
| 6 | $6 - 3 = 3$ | 3 |

The sum of the absolute deviations is 7. To see that the sum of the absolute deviations from any other number would be larger than 7, consider some number between 2 and 3, say 2.3. The absolute deviation of each of the two observations smaller than 3 is now 0.7 less than before. But this is precisely offset by the fact that the absolute deviations of the two observations greater than 3 are both increased by 0.7. And the absolute deviation of the remaining observation, 3, is now 0.7 instead of 0. The net result is that the sum of the absolute deviations is increased by 0.7. For any substitute median between 2 and 4, the sum of the absolute deviations is increased by the difference between the substitute median and the actual median. If the substitute were less than 2 or more than 4 (the observations adjacent

---

2. Note that the difference between $4,484 and $2,631 exaggerates the color difference in income because the two groups are distributed differently by other factors affecting income, such as region of the country and size of city. See Sec. 3.6 and Chap. 9.

*Averages*

to the median) the increase in the sum of the absolute deviations would be even larger than the difference between the substitute and 3. If this reasoning is not clear, perhaps it will be clarified by the discussion in the following example, which is a simplified version of one practical application of this property.

EXAMPLE 218   OPTIMUM LOCATION

Suppose that an enterprise has six stations located on a road as follows:

```
         A       B           C  D  E     F
    |____|_____|_____|__|__|_____|
    0    3       6          11 12 13    15
```

Each of these six stations must send a messenger frequently to a central station for supplies. Where should the central station be located to minimize the sum of the distances from the six stations to the central station? Obviously it should not be west or east of all the stations. But just where should it be?

To solve this, measure the distance of each station from any convenient point on the same road (if the zero point is within the range of the stations, distances will have to be counted plus or minus, according to their direction from zero), and take the median distance as the place for the central station.

| Station | Location | Distance | |
|---------|----------|----------|--------|
|         |          | From C   | From D |
| A       | 3        | 8        | 9      |
| B       | 6        | 5        | 6      |
| C       | 11       | 0        | 1      |
| D       | 12       | 1        | 0      |
| E       | 13       | 2        | 1      |
| F       | 15       | 4        | 3      |
| Total   | 60       | 20       | 20     |

The median distance is between 11 and 12 miles, so the total amount of travel will be minimized (at 20 miles each way) by putting the central station at or between stations C and D. This leeway to locate anywhere within a mile can be used to choose an economical location, or to meet other requirements. The sum of the absolute deviations from C, from D, or from any spot between C and D is 20 miles. If the station is west of C or east of D, this sum will exceed 20 miles. To see this by a different argument than the one just presented, suppose that the central station is at A. Then moving it from A to B would decrease by 3 miles the distance from five stations, and increase by 3 miles the distance from one station, a net reduction of 12 miles. Moving it from B to C decreases by 5 miles the distance from four stations, and increases by 5 miles the distance from two stations, another net reduction, this time of 10 miles. Moving it from C to D makes no difference, because there are now three stations on each side of the interval. For every mile that it is moved west of D, there are more stations suffering an increase in distance than there are gaining in distance, so there will be a net increase in the sum of the distances.

Sometimes the median is erroneously thought to be the number midway between the smallest and largest observations. For the weight data this is 190.5, halfway between 142 and 239. The correct (though, as we have seen, somewhat imprecise) statement, by contrast, is that the median is the *observation* midway between the lowest and highest observations in the *array* of the observations.

## 7.4
## THE MEAN

### 7.4.1 The Arithmetic Mean

The mean is the sum of the observations in a sample divided by the number of observations in the sample.

Let us introduce some notation which will condense such statements. To make the transition from words to symbols easy, consider the data on weights given in Chap. 6. For convenience we repeat here the weights of the 32 men:

| | | | | | | | |
|---|---|---|---|---|---|---|---|
| 198 | 165 | 175 | 190 | 239 | 142 | 170 | 151 |
| 178 | 148 | 155 | 189 | 180 | 185 | 186 | 197 |
| 158 | 152 | 142 | 160 | 203 | 155 | 175 | 180 |
| 200 | 175 | 164 | 183 | 180 | 174 | 191 | 189 |

The total of these observations is simply

$$198 + 178 + 158 + \cdots + 180 + 189 = 5,629.$$

Hence the 32 men weighed a total of 5,629 pounds, or nearly three tons—a statistic that might be useful if the group wanted to charter an airplane.

Now let a general symbol, say $x$, stand for the variable being studied, in this example, weight. Let $x_1$, that is, $x$ subscript 1, read "$x$ sub 1", represent whatever particular number one observation in the sample happens to be; let $x_2$ represent a second observation, and so on to $x_{32}$. To speak more generally, we would say that $x_i$ represents the $i$th observation in the sample, where $i$ takes the values 1, 2, 3, . . . , 32. To generalize this further, let $n$ represent the number of observations in the sample. Then the observations are represented by $x_i$ where $i = 1, 2, 3, \ldots, n$.

The sum of the observations in a sample of 32 may be written

$$x_1 + x_2 + x_3 + \cdots + x_{32},$$

or the sum of the observations in a sample of $n$

$$x_1 + x_2 + x_3 + \cdots + x_n.$$

Usually these two sums would be written more compactly as

$$\sum_{i=1}^{32} x_i \quad \text{and} \quad \sum_{i=1}^{n} x_i.$$

The sign "$\sum$" is the Greek capital sigma, the counterpart of $S$, the initial letter of "sum." It is called a summation sign, and it means "add, or sum, what follows." Below the summation sign we find "$i = 1$." This means that the quantity following the summation sign will also contain an "index of summation," $i$, and that this is to be replaced by every integer in turn from $i = 1$ to whatever we find above the summation sign—32 in the first case and $n$ in the second.

Thus $\sum_{i=1}^{32} x_i$ can be evaluated as follows: Write $x_1$, $x_2$, $x_3$, etc., each time replacing $i$ by one of the integers from 1 to 32. Then add up the observations numbered from 1 to 32. For the weights,

$$\begin{aligned} \sum_{i=1}^{32} x_i &= x_1 + x_2 + \cdots + x_{32} \\ &= 198 + 178 + \cdots + 189 \\ &= 5629. \end{aligned}$$

To complete our rewriting of the first sentence of this section, let us introduce the symbol $\bar{x}$, read "$x$ bar", to represent the mean value of a sample of observations on the variable $x$. This notation, incidentally, is so uniform in statistics that it is common simply to take the meaning of $\bar{x}$ for granted (or some other letter with a bar over it, if the variable has been represented by another letter) without definition.

The first sentence of this section may now be rewritten:

$$\bar{x} = \frac{\sum_{i=1}^{n} x_i}{n}.$$

For a while you will find it necessary, or at least helpful, to translate back into everyday language whenever you encounter the $\sum$ notation—which will be often throughout the remainder of the book. Thus, the equation above would be translated: "The mean of a sample is equal to the sum of the $n$ observations divided by $n$." If

## 7.4 The Mean

you make such a translation conscientiously every time you encounter a $\Sigma$, you will soon acquire facility at it and be able to use the $\Sigma$ notation correctly yourself. Nothing more is involved than a different way of saying or writing simple things that have long been familiar to you. If you do not master the $\Sigma$ notation, you will have difficulty following the remainder of this book, and will probably complain that it is all Greek to you.

Try yourself out right now. Can you show that for the weight data

$$\sum_{i=1}^{12} x_i = 2{,}010,$$

and

$$\sum_{i=13}^{32} x_i = 3{,}619?$$

You will not encounter any $\Sigma$ notation more complicated than that in this book, though you will encounter some equally simple $\Sigma$ notation that leads to more complicated arithmetic; for example, in Chap. 8 you will encounter, for these same data,

$$\sum_{i=1}^{32} x_i^2 = 1{,}003{,}483.$$

Since in elementary statistics summations almost invariably include all the observations in the sample, it is more common to omit than to include the range of summation and the index of summation (our subscript $i$), writing, for example,

$$\bar{x} = \frac{\Sigma x}{n}.$$

For many people, the greatest obstacle to learning statistics or to reading literature which makes use of statistics is an allergy to algebra. Our detailed treatment of the symbolic definition of the arithmetic mean is useful in suggesting a cure for this ailment. When confronted by any unfamiliar formula, write down in words the meaning of each component symbol. Then try to state in words the meaning of the whole formula. Finally, make up a very simple numerical example to see if you can carry through the operations implied by the formula. For example, you might substitute the numbers 1, 2, and 6, in the formula for the mean, and actually write out

$$\bar{x} = \frac{\Sigma x}{n} = \frac{1 + 2 + 6}{3} = 3.$$

Numerical examples often clarify otherwise formidable algebraic expressions. There is some danger that simple arithmetical examples will have misleading special features, but surprisingly often they contain all the essential features of general mathematical arguments.

From the weight data we have, then,

$$\bar{x} = \frac{\sum x}{n} = \frac{5,629}{32} = 175.91.$$

If each man had weighed 175.91 pounds, the total weight of the class would have remained 5,629 pounds; or, conversely, if the 5,629 pounds were divided equally among the members of the class, each would have 175.91 pounds. That the arithmetic mean has this property can be seen by a simple algebraic operation. We begin with the formula for the arithmetic mean, $\bar{x} = \sum x/n$, and multiply both sides of the equation by $n$: $n\bar{x} = \sum x$. This equation says that the mean multiplied by the number of observations gives the total. This simple algebraic property is perhaps the most important feature of the arithmetic mean, and helps explain the use of the mean instead of other averages like the median and mode, which have no analogous algebraic property.

EXAMPLE 222   PARACHUTES

A simple practical application of this property of the mean is described in the following quotation:

> Among the many patriotic industrialists called to Washington [in 1940] was Robert T. Stevens, one of the country's leading textile manufacturers. . . . Trying to consider every phase of war activity that might involve the use of textiles he thought of parachutes . . . he learned that the average requirement was four parachutes per war plane—figuring the heavy bombers (eleven men) plus the pursuit planes (one man) plus the essential reserves. He consulted the procurement officers in the Army and Navy and was told that they estimated they would need a total of 9,000 parachutes for the coming year, 1940–41—6,500 for the Army Air Corps, 2,500 for the Navy. Stevens did some multiplication of his own and told the officers that he figured they would need 200,000 parachutes instead of the 9,000 for which they were asking. They asked him how he had arrived at this fantastic figure. He replied, "The President has asked for 50,000 war planes. I just multiplied that by four."
>
> So the number of parachutes on the production program was boosted from 9,000 to 200,000. . . .[3]

A property which follows directly from the fact that $n\bar{x} = \sum x$ is that the sum of the deviations of the observations from the mean, taking due account of algebraic signs, is always 0. Algebraically,

---

3. Robert E. Sherwood, *Roosevelt and Hopkins* (revised ed.; New York: Harper and Brothers, 1950), pp. 161–162.

## 7.4 The Mean

$$(x_1 - \bar{x}) + (x_2 - \bar{x}) + \cdots + (x_n - \bar{x})$$
$$= x_1 + x_2 + \cdots + x_n - \bar{x} - \bar{x} - \bar{x} \cdots (n \text{ times})$$
$$= \sum x - n\bar{x} = \sum x - \sum x = 0.$$

By rewriting the first two of these lines in $\sum$ notation you will learn something more about using the $\sum$ notation, namely, that the sum of variables, each reduced by a constant, is equal to the sum of the variables, minus the constant multiplied by the number of terms in the sum; that is,

$$\sum(x - C) = \sum x - nC$$

where $C$ stands for any constant. You might try this out on the simple numerical example previously mentioned, with $n = 3$, $x_1 = 1$, $x_2 = 2$, $x_3 = 6$. Let $C$ equal, for example, 2.

Another property of the arithmetic mean that makes it desirable for certain statistical purposes is that the sum of the squared deviations from it is smaller than the sum of the squared deviations from any other number. The deviation of the $i$th observation is $x_i - \bar{x}$; hence the sum of the squared deviations is $\sum(x_i - \bar{x})^2$. Then $\sum(x_i - \bar{x})^2$ is less than it would be if $\bar{x}$ were replaced by any other value. Consider the three observations 1, 2, and 6. Their mean is 3, and the sum of the squared deviations from 3 is 14, as you should verify. The sum of the squared deviations from various other trial values is as follows:

Trial value: 1 2 2.5 2.9 3.1 3.5 4
Sum of squared deviations: 26 17 14.75 14.03 14.03 14.75 17

### 7.4.2 Weighted Means

We have seen that the mean when multiplied by the number of cases gives the total. This property is also the basis for *weighted means*. Suppose we are told that the mean of 12 observations is 167.50 pounds and that the mean of another 20 is 180.95 pounds. To find the mean of all 32, we *weight* these two means by their sample sizes; that is, we compute as if we had 12 observations each equal to 167.50 and 20 each equal to 180.95, thus:

$$\frac{167.50 + 167.50 + \cdots (12 \text{ times}) \cdots + 180.95 + 180.95 + \cdots (20 \text{ times})}{32}$$
$$= \frac{12 \times 167.50 + 20 \times 180.95}{32} = \frac{2010.00 + 3619.00}{32}$$
$$= \frac{5629.00}{32} = 175.91.$$

To put the calculations of the preceding paragraph into more general form, let $n_1$ and $n_2$ represent the sizes of the two samples, $\bar{x}_1$ and $\bar{x}_2$ their means, and $\bar{x}$ the mean of all $n$ observations, where $n = n_1 + n_2$. Then

$$\bar{x} = \frac{n_1\bar{x}_1 + n_2\bar{x}_2}{n_1 + n_2}.$$

Note that

$$n_1\bar{x}_1 = \sum_{i=1}^{n_1} x_i,$$

$$n_2\bar{x}_2 = \sum_{i=n_1+1}^{n} x_i,$$

and

$$\sum_{i=1}^{n_1} x_i + \sum_{i=n_1+1}^{n} x_i = \sum_{i=1}^{n} x_i.$$

Hence the numerator of the formula for $\bar{x}$ just above is algebraically identical with the numerator of the formula we gave earlier for $\bar{x}$; and the denominators are identical, too.

If there were several samples, say $k$, of sizes $n_1, n_2, \ldots, n_k$ and means $\bar{x}_1, \bar{x}_2, \ldots, \bar{x}_k$, the general mean would be

$$\bar{x} = \frac{\sum n_i\bar{x}_i}{\sum n_i}.$$

This may be written in a way that is more appealing intuitively if we let $w_i$ (choosing $w$ from "weight") be the proportion of all the observations that are in the $i$th group; that is,

$$w_i = \frac{n_i}{\sum n_i}.$$

Then

$$\sum w_i = 1$$

and

$$\bar{x} = \sum w_i\bar{x}_i.$$

If 10 percent of the observations are in the first group, 30 percent in the second group, and 60 percent in the third group, we get the mean by adding together 10 percent of the mean of the first group, 30 percent of the mean of the second group, and 60 percent of the mean of the third group.

**7.4 The Mean**

To illustrate, consider three samples, as follows.

| Sample | Observations | $n$ | $\bar{x}$ |
|--------|--------------|-----|-----------|
| 1 | 1 | 1 | $\bar{x}_1 = 1$ |
| 2 | 2, 6, 4 | 3 | $\bar{x}_2 = 4$ |
| 3 | 1, 0, 5, 1, 1, 4 | 6 | $\bar{x}_3 = 2$ |
| Sum | | 10 | |

$$w_1 = \tfrac{1}{10} \qquad w_2 = \tfrac{3}{10} \qquad w_3 = \tfrac{6}{10}$$
$$\bar{x} = w_1\bar{x}_1 + w_2\bar{x}_2 + w_3\bar{x}_3 = 0.1 \times 1 + 0.3 \times 4 + 0.6 \times 2 = 2.5$$

The fact that adding a single observation to a sample will not change the mean much can be looked at in terms of weighted means. Let the mean of $n - 1$ observations be $\bar{x}_1$ and let the $n$th observation be $x_n$. Then $\bar{x}$, the mean of all $n$ observations, can be regarded as a weighted mean of $\bar{x}_1$ and $x_n$, the weights being $(n - 1)/n$ and $1/n$, respectively. That is,

$$\bar{x} = w_1\bar{x}_1 + w_2 x_n$$

$$= \left(\frac{n - 1}{n}\right)\bar{x}_1 + \frac{1}{n}x_n = \frac{n}{n}\bar{x}_1 - \frac{1}{n}\bar{x}_1 + \frac{1}{n}x_n$$

$$= \bar{x}_1 + \frac{1}{n}(x_n - \bar{x}_1).$$

So, as we stated in discussing Example 110 (Motives), an $n$th observation changes the previous mean by one $n$th of its difference from the previous mean. For example, the mean of the three observations of Sample 2 above is 4; adding a fourth observation equal to 6 would increase the mean by $(6 - 4)/4 = 0.5$, to 4.5.

The weights used in combining means are not necessarily those used for computing the separate means. Suppose, for example, that the mean incomes of dentists, of lawyers, and of physicians have been measured in samples, and that the numbers in these professions in the country are known from a census or other source. To get a mean for the three professions combined, we proceed as though each dentist in the country had the income shown by the sample, and thus obtain a figure for the aggregate income of dentists. Adding to this similar aggregates for lawyers and physicians, we obtain a figure for the aggregate income of the whole group. Dividing this aggregate by the number of individuals in the whole group gives the weighted mean we want.

### 7.4.3 Proportions as Means

Proportions, though ordinarily thought of as fractions or percentages, can be thought of as special cases of the arithmetic mean. They can be obtained by the same formula, $\bar{x} = \sum x/n$, and the properties of the mean applied to them.

In the sampling demonstrations of Sec. 4.3, suppose we had defined a variable $x$ as the number of red beads in any one of the positions of the panel. Then $x_i$ would be the number of red beads in the $i$th position. Now there will be one and only one bead in each position; if it is red, the value of $x$ is 1, and if it is not red, the value of $x$ is 0. Then $\sum x$ is simply the number of red beads among the whole 20 positions, or in general among the whole $n$. $\sum x$ is what we called $X$ in Sec. 4.3.1; and $\bar{x} = \sum x/n$ is what we called $p$, the sample proportion.

Example 226   Change of Residence

Each year approximately 20 percent of the people in the United States change residence.[4] This can be interpreted as meaning that an average of 0.2 is obtained when to each person is attached a number showing the number of "movers" he represents, that is, 1 if he has moved, however often, and 0 if he is living in the same dwelling as a year ago.

### 7.4.4 Other Means

What we call the arithmetic mean, or simply the mean, is usually just called the "average" in everyday language. Properly speaking, however, any number representing a "typical" value is an average, so modes and medians, as well as means, are averages. There are, furthermore, many different kinds of mean. The arithmetic mean, while only one of a large class, is usually intended if the term "mean" is used alone. Three other means that are used from time to time are the root mean square, the geometric mean, and the harmonic mean.

7.4.4.1   *Root Mean Square.*   If each observation is squared, the arithmetic mean of the squares is computed, and the square root of this mean taken, the result is a mean known as the *root mean square*. The root mean square is larger than the arithmetic mean of the same observations. The standard deviation, which we shall discuss in Chap. 8, is essentially a root mean square.

7.4.4.2   *Geometric Mean.*   If the observations are all multiplied together and the $n$th root is taken, the result is known as the geometric

---

4. *Statistical Abstract: 1955*, Tables 38 and 39, pp. 43–44.

### 7.4 The Mean

mean. For actual computations the arithmetic mean of the logarithms is computed, and the anti-logarithm of this mean is taken. The geometric mean is less than the arithmetic mean of the same observations. It is meaningful only for sets of observations which are all positive. It has the property that taking its nth power (that is, multiplying together n numbers all equal to the geometric mean) gives the same result as multiplying together all of the original observations. Consider the figures on factory sales of motor vehicles referred to in Example 148. Sales, actual and as a percent of the previous year, were:

| Year | 1950 | 1951 | 1952 | 1953 |
|---|---|---|---|---|
| Sales | | | | |
| Thousands | 8,003 | 6,765 | 5,539 | 7,323 |
| Percent of previous year | | 84.53 | 81.88 | 132.20 |

The arithmetic mean of the three sales rates is 99.54 percent, suggesting (erroneously) an average rate of decline of 0.46 percent per year. Had sales declined uniformly by 0.46 percent per year from 8,003 thousand in 1950, they would have been 7,893 instead of 7,323 thousand in 1953; so clearly the actual average rate of decline was larger than 0.46 percent. The geometric mean of the percentages, however, is useful here. This can be computed either as the anti-logarithm of the arithmetic mean of the logarithms, or as the cube root of the product of percentages, and is 97.08, indicating an average decline of 2.92 percent. A uniform decline of 2.92 percent each year would have meant sales of 7,322 thousand in 1953, the discrepancy from 7,323 being due to calculating the geometric mean only to two decimals. The principal use of the geometric mean is in averaging a sequence of ratios.

7.4.4.3 *Harmonic Mean.* If the reciprocal of each observation (the number of times the observation goes into 1) is taken, the arithmetic mean of the reciprocals is computed, and the reciprocal of this mean is taken, the result is known as the harmonic mean. The harmonic mean is less than the geometric mean of the same observations. It is useful when the observations are expressed inversely to what is required in the average, for example, when the average hours per mile is required but the data show miles per hour.

EXAMPLE 227   UP THE HILL AND DOWN AGAIN

Nearly everyone has erroneously used the arithmetic mean in some variant of the following puzzle, the correct solution of which requires the harmonic mean: A motorcyclist wants to ride to the top of a hill and back at an over-all average speed of 60 miles per hour. If his mean speed going up is 30 m.p.h., what must be his mean speed coming down? The answer is *not* 90 m.p.h. No speed for the second half of a trip can be great enough to bring the over-all average up to double the average speed for the first half of the distance. If the total distance is *d* miles, an over-all mean speed of 60 m.p.h. requires that the trip be completed in *d* minutes; but traveling half the distance at 30 m.p.h. has already used up the full *d* minutes. The problem should be analyzed through the harmonic mean. If the harmonic mean is to be 60, its reciprocal must be 1/60. For this, the sum of the reciprocals of the two speeds must be 1/30. Since the reciprocal of the first speed is 1/30, the reciprocal of the second speed must be 0. But no finite speed has a reciprocal as small as 0.

The principle involved in Example 227 will be examined in some detail, to make the nature and uses of the harmonic mean clear. We shall examine it in the context of a trip from Chicago to Denver and return, going the 1,000 miles west by train at 50 m.p.h. and returning by plane at 200 m.p.h. What mean speed would be appropriate for finding the average time for the 1,000-mile trip? The arithmetic mean speed is 125 m.p.h., suggesting that the average time for the 1,000-mile trip is 8 hours, and the total time for the round trip 16 hours. But the trip out at 50 miles per hour takes 20 hours and the return at 200 miles per hour takes 5 hours, which gives us a mean of $12\frac{1}{2}$ hours instead of 8, and a total for the round trip of 25 instead of 16.

What we neglected is the fact that an average refers to some class of units, which must be appropriate to the use to which the average is to be put. In this case, the units can be either units of distance or units of time. We can think of recording a speed for each mile (or kilometer, furlong, or other unit of distance) and averaging these. Or we can think of recording a speed for each hour (or minute, second, or other unit of time) and averaging these.

Suppose that we record the speed for each of the 2,000 miles traveled. There are 1,000 50's and 1,000 200's, and the arithmetic mean of these 2,000 numbers is 125. This is the mean speed per mile traveled.

But suppose that we record the speed for each of the 25 hours traveled. Then there are 20 50's and 5 200's, and the mean of these 25 numbers is 80. This is the mean speed per hour traveled.

Which do we want for computing the time required for the 1,000-mile trip? Fundamentally, we intend to utilize the relation $n\bar{x} = \sum x$, but this time $\sum x$ and $\bar{x}$ are known, and we will find $n$ as $\sum x/\bar{x}$. Since the $n$ we want is in hours, the $n$ used in computing $\bar{x}$ must also be in hours; that is, $\bar{x}$ should be the mean speed per hour.

### 7.4 The Mean

Thus, if the total distance is to be 2,000 miles, as here, and will build up at a mean rate of 80 miles per hour, the number of hours required will be 25. For any distance of which half is to be traveled at 50 m.p.h. and half at 200 m.p.h., the mean of 80 will be appropriate for calculating the time required.

We obtained our mean of 80 as a weighted arithmetic mean of the two speeds,

$$\bar{s} = \left(\frac{t_1}{t_1 + t_2}\right)s_1 + \left(\frac{t_2}{t_1 + t_2}\right)s_2 = w_1 s_1 + w_2 s_2$$

where $\bar{s}$ is the mean speed, $t_1$ and $t_2$ are the times traveled at each speed, and $s_1$ and $s_2$ are the two speeds, and $w_1$ and $w_2$ represent the proportions of the total time spent traveling at each speed. We could, however, have rearranged the formula as[5]

$$\bar{s} = \frac{2}{\dfrac{1}{s_1} + \dfrac{1}{s_2}}$$

and this is simply the harmonic mean of $s_1$ and $s_2$.

As another illustration of the harmonic mean, suppose that you buy a dollar's worth of oranges at 10 cents each and a dollar's worth at 5 cents each. It is easy to make the mistake of saying that the mean price was $7\frac{1}{2}$ cents. This is the mean price per dollar spent ("dollar-averaging," as in Example 94C), but the mean price per orange bought is the harmonic mean, $6\frac{2}{3}$ cents. This could also be obtained as the weighted arithmetic mean of the two prices, with weights proportional to the numbers of oranges bought at the two prices.

### EXAMPLE 229   TRANSPORT CAPACITY

During World War I, in order to get an estimate of the transport capacity of U. S. ships, the round trip distance to Europe was divided by the arithmetic mean speed of the ships to be used. This figure, taken in conjunction with turn-around time, allowances for delays, etc., gave an estimate of shipping capacity that proved unattainable. Once again, the harmonic

---

5. If you want to verify the algebra of this rearrangement, start by letting $d$ represent the distance that is traveled at each speed and replacing $t_1$ and $t_2$ by $d/s_1$ and $d/s_2$, respectively. This gives, since $d = s_1 t_1 = s_2 t_2$,

$$\bar{s} = \frac{d}{\dfrac{d}{s_1} + \dfrac{d}{s_2}} + \frac{d}{\dfrac{d}{s_1} + \dfrac{d}{s_2}} \cdot$$

On dividing numerator and denominator of each term by $d$, and noting that both terms are alike, the formula for the harmonic mean appears as in the text.

mean should have been used. The arithmetic mean number of hours per mile could have been calculated and multiplied by the number of miles to get the mean number of hours per trip. This is algebraically equivalent to dividing the number of miles by the harmonic mean speed, since the harmonic mean speed is the reciprocal of the arithmetic mean number of hours per mile.

To avoid pitfalls, it is necessary to remember that an average is necessarily taken over some definite type of unit. Example 72B brought this out, too. In that example, one might have been concerned either with children averaged over families or children averaged over children; the purpose of the investigation determines which. So long as one thinks of "average *per such-and-such*," he is likely to avoid these errors.

### 7.4.5 Computing Arithmetic Means

It is easy enough to add up 32 numbers, but for a sample of 320 or 3,200, it would become tedious even with an adding machine. Time may be saved by computing the mean from the grouped data of a frequency distribution. This is done by working as though the mean of the observations in each class has the same value as the midpoint of the class. If this were really true, the computation would lead to the same result as the computation from ungrouped data. The assumption will ordinarily not be true, of course, but fortunately the errors tend to cancel. Unless the classes are too wide, or badly chosen, the approximation will be satisfactory. (The proper choice of classes was discussed in Sec. 6.2.1.)

Table 231 shows all the steps involved in computing the mean from a frequency distribution. Note that the first two columns (Weight and Number of Persons) constitute the frequency distribution of the sample of 32 weights, shown previously in Table 174. Columns 3 and 4 ($x$ and $fx$) are included to show the computation of the mean. To obtain the total weight of the sample, the mid-value, $x$, of each class is multiplied by the number $f$ in that class, giving the total weight for the class. Then the total weights for the various classes are summed to obtain the total weight of the sample.

The total weight of the sample computed this way is 21 pounds more than the total we obtained in Sec. 7.4.1 by adding the 32 observations individually. Dividing the total by 32 gives a mean of 176.56, instead of 175.91 obtained from the individual observations.[6]

6. Although the mean *could* be in error by half the width of the classes (assuming they are all of equal width), or 5 pounds in this case, in fact it is most unlikely (about 3 chances

### 7.4 The Mean

This calculation can be thought of as a weighted mean, such as was discussed in Sec. 7.4.2. Here the individual means are approximated by the mid-values of the classes, and the weights are the frequencies.

TABLE 231

CALCULATION OF MEAN

| (1)<br>Weight | (2)<br>Number<br>of<br>Persons<br>$f$ | (3)<br>Mid-value<br>of<br>Class<br>$x$ | (4)<br>Total Weight<br>of Class<br>$fx$ |
|---|---|---|---|
| 137.5–147.5 | 2 | 142.5 | 285.0 |
| 147.5–157.5 | 5 | 152.5 | 762.5 |
| 157.5–167.5 | 4 | 162.5 | 650.0 |
| 167.5–177.5 | 5 | 172.5 | 862.5 |
| 177.5–187.5 | 7 | 182.5 | 1277.5 |
| 187.5–197.5 | 5 | 192.5 | 962.5 |
| 197.5–207.5 | 3 | 202.5 | 607.5 |
| 207.5–217.5 | 0 | 212.5 | 0 |
| 217.5–227.5 | 0 | 222.5 | 0 |
| 227.5–237.5 | 0 | 232.5 | 0 |
| 237.5–247.5 | 1 | 242.5 | 242.5 |
| Total | 32<br>$n$ | | 5650.0<br>$\sum fx$ |

$$\bar{x} = \frac{\sum fx}{n} = \frac{5650.0}{32} = 176.56$$

*Source:* Table 174.

We can simplify the calculation still further by noticing the following about Table 231: If each weight is reduced by some fixed amount, the mean will be reduced by the same amount. This will simplify computing the mean, and we can then add back the required constant. In Table 231, 182.5 is a convenient number to subtract from each weight. The mid-values, $x$, then become $-40$, $-30$, $-20$, $-10$, 0, $+10$, $+20$, $+30$, $+40$, $+50$, and $+60$. These are the values whose mean we must find, using the frequencies in the $f$ column; and we will add 182.5 to this mean. But these values can themselves be simplified by dividing by an appropriate constant; the resulting mean will then have to be multiplied by the same constant before we add back the 182.5. The class interval, in this

in 1,000) to be in error by as much as $\sqrt{3/n}$ times this maximum. In this case, $\sqrt{3/n}$ is 0.306, so even if we had not known the exact value we could have been rather confident of being within 1.53 pounds of it.

*Averages*

case 10, is generally an appropriate constant by which to divide. Thus, we define the *coded* mid-value, *m*, as the mid-value of a given class minus some constant, divided by the class interval; in this case,

$$m = \frac{x - 182.5}{10}.$$

For example, the mid-value *x* of the first class, 137.5 to 147.5, is 142.5. Hence the coded mid-value is

$$m = \frac{142.5 - 182.5}{10} = -4.$$

Thus we obtain the numbers $-4, -3, -2, \ldots +5, +6$ to be averaged by the frequencies in the *f* column. This has been done in Table 232. The result, $-19/32$, then has to be multiplied by 10 in

TABLE 232

SHORTCUT CALCULATION OF MEAN

| Weights | Number of Persons $f$ | Coded Mid-value $m$ | Total Coded Weight $fm$ |
|---|---|---|---|
| 137.5–147.5 | 2 | −4 | −8 |
| 147.5–157.5 | 5 | −3 | −15 |
| 157.5–167.5 | 4 | −2 | −8 |
| 167.5–177.5 | 5 | −1 | −5 |
| 177.5–187.5 | 7 | 0 | 0 |
| 187.5–197.5 | 5 | +1 | 5 |
| 197.5–207.5 | 3 | +2 | 6 |
| 207.5–217.5 | 0 | +3 | 0 |
| 217.5–227.5 | 0 | +4 | 0 |
| 227.5–237.5 | 0 | +5 | 0 |
| 237.5–247.5 | 1 | +6 | 6 |
| Total | 32 $n$ | | −19 $\sum fm$ |

$$\bar{x} = 182.5 + \frac{10\sum fm}{32} = 182.5 + \frac{10(-19)}{32} = 182.5 - 5.94 = 176.56$$

*Source:* Table 231.

order to compensate for the earlier division by 10 in establishing the coded mid-values. This gives $-190/32$. Then we have to add 182.5 to this result in order to compensate for the fact that we had originally subtracted 182.5 from the uncoded mid-values. Thus we obtain

**7.4 The Mean**

$182.5 - (190/32) = 176.56$, exactly as in the computation illustrated in Table 231.

Table 233 shows the shortcut method of calculation when the class intervals are not all equal. It is based on the same data as the calculations just shown, except that now several intervals have been consolidated into single classes, for example, 197.5 to 207.5 and 207.5 to 217.5 into 197.5 to 217.5, with mid-value 207.5. To obtain the coded mid-value, we subtract (for variety) 172.5 (instead of 182.5, as before), and divide this difference by 10; that is,

$$m = \frac{x - 172.5}{10}.$$

For the class 197.5 to 217.5, with mid-value 207.5,

$$m = \frac{207.5 - 172.5}{10} = 3.5.$$

Similarly, $+6$ is found to be the coded mid-value of the interval 217.5 to 247.5.

TABLE 233

SHORTCUT CALCULATION OF MEAN

(Unequal Intervals)

| Weights | Number of Persons $f$ | Coded Mid-value $m$ | Total Coded Weight $fm$ |
|---|---|---|---|
| 137.5–147.5 | 2 | −3 | −6 |
| 147.5–157.5 | 5 | −2 | −10 |
| 157.5–167.5 | 4 | −1 | −4 |
| 167.5–177.5 | 5 | 0 | 0 |
| 177.5–187.5 | 7 | +1 | 7 |
| 187.5–197.5 | 5 | +2 | 10 |
| 197.5–217.5 | 3 | +3.5 | 10.5 |
| 217.5–247.5 | 1 | +6 | 6 |
| Total | 32 $= n$ | | 13.5 $\sum fm$ |

$$\bar{x} = 172.5 + \frac{10(13.5)}{32} = 172.5 + 4.22 = 176.72$$

$$m = \frac{x - 172.5}{10}$$

*Source:* Table 231.

These methods do not work—nor does any method—if any class is open-ended. For example, Table 183 shows 1.5 percent of urban families with incomes of $15,000 or more in 1952. It would not be possible to compute a mean income for all families from such a table unless one were willing to guess the mean income of the families with incomes of $15,000 or more. Since this class is open-ended, it is no longer possible to use a mid-value as a guess. When problems like this are encountered in practice, and it is not possible to compute a mean directly from the original data, it is necessary to use some other measure of central tendency—such as the mode or median. A minimum possible value for the mean could be found by assuming that all those in the top class have exactly $15,000.

7.4.5.1 *A Digression on Computations.* We have discussed only a few of the ways to cut down the work needed to calculate averages. It is appropriate at this point to dispel the common misconception that statistical calculations are necessarily tedious. This misconception stems from the image many people have of all arithmetical operations being performed with pencil and paper, and as clumsily as they do occasional bits of arithmetic themselves. This misconception leads to the erroneous belief that the ability to do arithmetic quickly and accurately is the prime requirement for a statistician (or for that matter for an accountant or a mathematician). While such ability may be useful, most statistical computations are—or should be—done on machines. For most investigations, desk calculating machines suffice, and anyone seriously interested in applying statistics in his own work should learn how to do the operations of addition, subtraction, division, multiplication, cumulative multiplication, and perhaps square roots with these machines.

There are also machines that make other parts of statistical routine relatively easy, for example, the various sorting, counting, and tabulating machines. For large-scale calculations, a wide variety of electronic calculators are coming into use. No one should undertake extended statistical work without first finding out about the technological devices that can materially lessen the cost in time and money, nor should anyone decide that a projected statistical analysis is too difficult without exploring the availability of machines.

Skill in the use of machines is also important in extensive calculations. It pays to devise systematic, efficient procedures. Many people make statistical calculations without being aware of the shortcuts that would make the calculations quicker and, often, more accurate.

Certain repetitive operations can be eliminated by reference to mathematical tables. One book of tables particularly useful even to those who have ready access to machines, and indispensable to those who do not, is *Barlow's Tables of Squares, Cubes, Square Roots, Cube Roots, and Reciprocals*,[7] originally prepared in England during the Napoleonic Wars. For every integer from 1 to 12,500 this shows the square, square root, square root of ten times the number, cube, cube root, and reciprocal; and various other useful information is also included. The coverage is extended to all numbers by appropriate use of decimal points.

There are many practical aspects of computations that are hard for nontechnicians to appreciate. For example, the best way to check a calculation is for a second person to do it from scratch, independently. If you just look over what you have done, you can easily make the same error twice, or even ten times. Unless the checks are independent, even the modal result may be wrong. Of course, many calculations are not checked at all, sometimes out of self-confidence, sometimes laziness, and perhaps occasionally despair. All skilled computers make mistakes, and they know they do. The more skilled the computer, the more likely he is to insist upon independent checks. He is also likely to be energetic about preventing errors in the first place, by setting up worksheets neatly and systematically, planning the routine so as to minimize the need for copying numbers (copying errors are particularly prevalent), and so on.

The reading of statistical studies almost never demands extensive computations with elaborate equipment. Another kind of calculation —a "mental approximation"—is much more useful. It is unfortunately true that many producers of statistical studies are careless about calculations, and that even with great care, as we have seen, mistakes sometimes occur. Hence, for serious reading one ought to form the habit of checking calculations roughly, when this is possible, to see if gross errors have been made. For example, if 14.1 is shown as the square root of 20, one can immediately see that since even 10 times 14.1 exceeds 20, 14.1 times it will also do so. The discovery of gross errors, in turn, arouses misgivings about all the calculations in the study.

In this book, we do not always present the "best" computing methods. This is especially true of Part III. Computing methods involve technical ideas and apparatus whose explanation would detract from, or become intertwined with, the statistical ideas. Furthermore, it may be easier to make an occasional calculation "the hard way" than to remember, or restudy, the "best" way each time, and we

---

7. New York: Chemical Publishing Company, 1944.

have in mind in this book the reader who is only occasionally a computer.

## 7.5
## WHICH AVERAGE?

The choice of a particular average is usually determined by the purpose of the investigation. Within the framework of descriptive statistics, the main requirement is to know what each average means and then select the one that fulfills the purpose at hand. In Example 222, the arithmetic mean of the number of parachutes per plane, 4, was the only average appropriate for obtaining the total requirements for parachutes. To describe the "typical" number of parachutes per plane, however, the two modes, 1 (fighters) and 11 (bombers), would be appropriate. In predicting an individual observation on the basis of a frequency distribution, the mode of the appropriate distribution has the highest probability of being a correct prediction, "right on the nose." The median has the smallest mean absolute error for a number of predictions. The mean gives the smallest discrepancy between the algebraic sum of the predictions and the actual sum, if a number of predictions are made.

The mean length of life is used by insurance companies for calculating rates for simple whole-life policies; the median is often used by census or public health officials for descriptive purposes. A businessman who wanted to estimate the total cost of a wage increase would use the mean wages before and after the change; someone who wanted to convey an idea of the standard of living of a group of workers by a single statistic might use the median income.

Besides meaningfulness, there are other considerations, usually of secondary importance, in choosing an average.

(1) In certain commonly encountered applications, the mean is subject to less sampling variability than the mode and the median. This reason, however, is in the realm of statistical inference, and we shall explore the sampling variability of statistics in Chap. 11.

(2) Given only the original observations, the median is sometimes easiest to calculate, especially if a calculating machine is not available. Sometimes when there is no strong advantage for the mean, this advantage is enough to indicate the use of the median. If calculating equipment is available, however, it may be easier to compute the mean than to array a large number of observations.

(3) Once a frequency distribution has been formed, the mode and the median are more quickly calculated than the mean. More-

over, when some classes are open-ended the mean cannot be calculated from the frequency distribution.

(4) <u>The median is not a good measure when there are very few possible values for the observations</u>, as with number of children or size of family.

(5) <u>The mode and median are relatively little affected by</u> "extreme" observations. In an income study, for example, one individual making $50,000 per year would have a substantial effect on the mean of a group of 100 people. If the other 99 had a mean income of $4,000, the $50,000 man would raise the mean of the group of 100 to $4,460, while the median, and probably the mode, would be the same as if his income had been $5,000 or even $4,001. For many purposes, of course, the fact that the mean is affected by extreme observations in proportion to their frequency and "extremeness" is one of its chief advantages.

The first and fifth of these reasons may seem contradictory, but they are not. The fifth reason refers to desirable properties of a descriptive measure of a population. The first reason refers to the merits of certain sample statistics as a basis for generalizations about the population. Thus, sometimes when the population median is of interest, the sample mean provides the best basis for estimating it, and vice versa. The first consideration properly belongs in Part III, on statistical inference, rather than in this part, on statistical description.

## 7.6
## INTERPRETATION OF AVERAGES

We now consider a few common problems in the interpretation of averages. Later, in Chap. 9, we shall discuss the interpretation of comparisons between the arithmetic means of different groups. The problem of comparisons is deferred because many of the principles apply not only to the comparison of averages, but also to the comparison of other characteristics, such as dispersion or association.

### 7.6.1 Allowance for Dispersion

The first requirement in the interpretation of an average is to remember that any average gives only a partial picture of the information contained in a set of data. In particular, no inkling is given of the amount of dispersion about the average. Several examples of unfortunate disregard of dispersion were given in Sec. 3.8, and might well be reviewed now.

## 7.6.2  A Mean Has Both a Numerator and a Denominator

EXAMPLE 238  ARIZONA DEATH RATES

In 1937 the number of deaths per thousand in Arizona was 16.8; in 1943 the death rate had declined to 8.4. A major explanation for the reduction was the large number of military posts established there. These added appreciably to the small population of the state, which enters the denominator of its death rate, but added virtually no deaths, because of the ages and physical selection of military personnel. Thus the rate fell because of the changed composition of the population.

This example illustrates an important point in interpreting any mean (or proportion or ratio); it must be remembered that there is *both* a numerator and a denominator to be considered. A common error is to attribute all changes in ratios to changes in the numerators. Thus, in the preceding example, many people might have jumped to the conclusion that Arizona had become a more healthful place to live. Many ratios or means are used in business and administration as measures of performance. For example, there is the ratio of sales to inventory (turnover), the ratio of current assets to current liabilities, and the ratio of operating costs to sales. It is often required that such ratios be higher (or lower) than a given number. While these ratios have some usefulness, they must be handled with caution. Let us take one simple example. A retail store manager decides that his ratio between operating costs and sales, say 0.70, is too high and should be reduced. He reduces it to 0.65, yet finds that his operating profit to be applied against overhead cost is *less*, not more. Here are some hypothetical numbers that suggest how this could have happened:

|                      | Before   | After  |
|----------------------|----------|--------|
| Sales                | $10,000  | 8,500  |
| Operating Costs      | 7,000    | 5,525  |
| Operating Profit     | $3,000   | 2,975  |
| Ratio, Costs to Sales| 0.70     | 0.65   |

The reduction of costs (the numerator of the ratio of costs to sales) may itself have had an effect on sales (the denominator). For instance, the store manager might have reduced his sales staff and cut down his inventory, thereby losing sales because of inadequate service or failure to carry desired items in stock.

Many of the most important ideas for the user of statistics are as simple as this point about the numerators and denominators of ratios, yet a surprising number of practical mistakes are due to confusion about such ideas.

### 7.6.3 Distinction between "Change of the Average" and "Average of the Changes"

In comparing sets of incomes for two different years, it is possible for the median income to increase and the median change in income to be negative. This is demonstrated by Table 239.

TABLE 239

HYPOTHETICAL INCOME DATA FOR TWO YEARS

| Family | First Year | Second Year | Change |
|--------|-----------|-------------|--------|
| A | 5 | 2 | −3 |
| B | 10 | 8 | −2 |
| C | 15 | 18 | +3 |
| D | 20 | 19 | −1 |
| E | 25 | 20 | −5 |
| Sum | 75 | 67 | −8 |
| Mean | 15 | 13.4 | −1.6 |
| Median | 15 | 18 | −2 |

The median income has increased by 3 units, yet the median change is a decrease of 2 units. This is an example of the sort of seeming paradox that may end in such confusion as: "Statistics show that the average income has increased"; "No, statistics prove that the average change in income is downward!" Of course this does not show that statistics can be arranged to prove anything, but only that statistical data cannot arrange to protect themselves against stupid, careless, or deceptive interpretations. Notice that here both statements are true; "common sense" is wrong in supposing that they are contradictory. The quantity that the first statement says has increased is a different quantity from the one that the second statement says has decreased. For some purposes one figure would be more meaningful than the other; and for almost any purpose it would be desirable to know it if the two quantities were moving in opposite directions. And if only one of them is available, it is important to realize that it does not guarantee anything about the direction of change in the other.

This example provides another good reason for using the designation "median" or "mean" instead of the ambiguous term "average." The reason is that the change of the arithmetic mean is always equal to the arithmetic mean of the changes. Note in Table 239 that the mean of the first year (15) plus the mean change (−1.6) equals the mean of the second year (13.4). The median lacks this useful property.

## 7.7
## CONCLUSION

Often it is desirable to describe a frequency distribution by some one of its characteristics. Usually the most significant characteristic is its location: How large or small are the observations? The principal measures of location, usually called averages, are the mean and the median, but the mode is also useful for some purposes.

There are many different means. Usually the first desideratum in choosing among them is to single out a characteristic corresponding to, and giving a definite interpretation of, the vague idea of the location of the distribution. The root mean square, used in measuring dispersion, exceeds the arithmetic mean (which is the mean referred to if no modifier is used) of the same data. The geometric mean, useful for measuring mean rate of change, is less than the arithmetic mean. The harmonic mean, useful when the observations are expressed inversely to what is required for a certain problem—as when mean hours per mile is required but the data show miles per hour, or when mean dollars per unit is required but the data show units per dollar—is in turn less than the geometric mean. Other means, not discussed here, can be as large as the largest observation or as small as the smallest observation.

One of the useful properties of the arithmetic mean is that when multiplied by the appropriate number of units, it shows the corresponding total of the variable studied. This property makes it possible to combine the means of several groups, through a weighted mean, to find the mean of the combined groups. Another property is that the sum of the deviations of the observations from the mean, taking due account of algebraic signs, is zero; that is, the positive and negative deviations are equal in magnitude. Still another property is that the sum of the squares of the deviations is less for the mean than for any other number. In general, the mean has many useful arithmetic properties not possessed by the median or mode; these are conveniently expressed through $\Sigma$, or summation, notation.

Proportions can be regarded as means by the device of letting a variable be 1 when a characteristic is present, 0 when it is absent. Then summing the observations is equivalent to counting those with the characteristic.

Whereas the mean could be described as the typical value of the variable, in the sense that the sum of the deviations above it and below it are equal, the median could be described as the value of the variable for a typical individual, in the sense that as many individuals are above as below it. The median can be computed in some cases where the mean cannot, for example when the frequency distribution contains open-ended classes; and, unlike the mean, it is meaningful when the observations are ranked in order of magnitude but not measured in fixed units. A useful property of the median is that the sum of the deviations from it, all taken as positive, is less than from any other value.

The mode is useful for indicating the most common result. One of its applications is in selecting the correct value after a series of repetitions of the same calculations. The mode can be computed for any data for which the mean and median can be computed, and in addition for qualitative data.

Problems related to the comparison of averages are discussed in Chap. 9, and problems of drawing inferences about population means from sample means in Chaps. 11, 12, 13, and 14. In the next Chapter, 8, however, we leave measures of location and turn to measures for describing the next most significant general feature of a frequency distribution, its variability or dispersion.

## DO IT YOURSELF

### EXAMPLE 241A

(1) What are the medians of Table 106 and 107B, respectively?

(2) Combine the two tables into a single frequency distribution, and find the new median.

(3) How is this median related to the medians of the two original distributions?

(4) Is this relationship just a coincidence, or does it illustrate a general property of the median?

[Hint: Replace Table 107B by one in which all the observations are 7's, and then compute the medians of the separate and combined samples.]

### EXAMPLE 241B

Show that for the weight data of Table 173

$$\sum_{i=1}^{16} x_{2i} = 2{,}850$$

and

$$\sum_{i=0}^{15} x_{2i+1} = 2{,}779.$$

*Averages*

EXAMPLE 242A

Analyze Example 222 in the same way you were asked to analyze the examples at the end of Chap. 3.

EXAMPLE 242B

Prove the following relations algebraically if you can, and in any case demonstrate them in several numerical examples:

$$(1) \qquad \frac{1}{n}\sum(x - c_1) = \bar{x} - c_1$$

$$(2) \qquad \frac{1}{n}\sum\left(\frac{x - c_1}{c_2}\right) = \frac{1}{c_2}(\bar{x} - c_1),$$

where $c_1$ and $c_2$ are two constants.

How are these two relations used in Sec. 7.4.5? How might they be used in facilitating computations based on individual observations rather than grouped data?

EXAMPLE 242C

Verify the computations given at the end of Sec. 7.4.1, and plot them on a graph.

EXAMPLE 242D

Make up a simple numerical example to illustrate the point in Sec. 7.4.3.

EXAMPLE 242E

You take a trip which entails traveling 900 miles by train at an average speed of 60 miles per hour, 3,000 miles by boat at an average of 25 m.p.h., 400 by plane at 350 m.p.h., and finally 15 miles by taxi at 25 m.p.h. What is your average speed for the entire distance (4,315 miles)?

EXAMPLE 242F

What is misleading about the claim for dollar averaging in the second paragraph of Example 94C? [Hint: work out a numerical illustration involving just two time periods.]

EXAMPLE 242G

Compute the mean, median, and mode for the Rockwell hardness data of Table 206, using both the original data and at least two different groupings of the data into frequency distributions.

EXAMPLE 242H

In Sec. 7.5 we give this example: If 99 people have a mean income of $4,000, the addition of a man with an income of $50,000 will raise the mean to $4,460. Verify this statement.

*Do It Yourself*

EXAMPLE 243A

The following data represent travel expenses (other than transportation) for 7 trips made during November by a salesman for a small firm:

| Trip | Days | Expense | Expense per day |
|------|------|---------|-----------------|
| 1 | 0.5 | $13.50 | $27.00 |
| 2 | 2.0 | 12.00 | 6.00 |
| 3 | 3.5 | 17.50 | 5.00 |
| 4 | 1.0 | 9.00 | 9.00 |
| 5 | 9.0 | 27.00 | 3.00 |
| 6 | 0.5 | 9.00 | 18.00 |
| 7 | 8.5 | 17.00 | 2.00 |
| Total | 25.0 | $105.00 | $70.00 |

An auditor criticized these expenses as excessive, asserting that the average expense per day is $10 ($70.00 divided by 7). The salesman replied that the average is only $4.20 ($105 divided by 25) and that in any event the median is the appropriate measure and is only $3.00. The auditor rejoined that the arithmetic mean is the appropriate measure, but that the median is $6.00.

(1) Explain the proper interpretation of each of the four averages mentioned.

(2) Which average seems to you appropriate?

EXAMPLE 243B

Comment on the following statement: "The arithmetic mean has the disadvantages of being affected by extreme values and may, therefore, give a biased figure."

EXAMPLE 243C

In Example 222, suppose every fighter plane required one parachute and every bomber required 11. What proportion of the planes were fighters and what proportion were bombers?

# Chapter 8

# *Variability*

## 8.1
### VARIABILITY AND ITS IMPORTANCE

No average by itself tells anything about dispersion. Even if several of the usual averages are computed for a distribution, ordinarily they reveal little about dispersion. There is, for example, no way to tell solely on the basis of the information given below which of the two distributions has the greater variability:

|  | Distribution *A* | Distribution *B* |
|---|---|---|
| Mean | 12 | 12 |
| Median | 12 | 9 |
| Mode | 12 | 6 |

But whatever average value we use, there is typically some dispersion or scatter about it. In fact, that is why we compute an average: to see through the variability, so to speak, to the general location. We will ordinarily be nearly, or even fully, as much interested in measuring the scatter about the average, in order to know how representative the average is of the individual cases.

The Tombigbee River example (Example 80B) illustrated the need for knowledge of variability as well as general level. The fact that the river averaged only one foot in depth did not exclude the possibility that a wader might have to swim. Again, it is proverbial that even if the average strength of the links of a chain is high, the variability is also important to its strength.

In engineering problems, measures of variation are often especially important. The amount of actual variability in dimensions of supposedly identical parts is critical in determining whether or not the components of a mass-produced item really are interchangeable. The variability in length of life of light bulbs may be even more important than the average if the bulbs are used in an inaccessible location and can be replaced only at regular intervals.

A classical problem in the social sciences requiring the measurement of variability is the measurement of "inequality" of the distribution of income, or personal property, corporate assets, etc., and, in principle, less quantifiable things, such as "power" of one kind or another.

## 8.2
## THE RANGE

The range is the difference between the highest and the lowest value in a group of observations. In the sample of 32 weights shown in Table 173, the smallest weight was 142, the largest 239. The difference between these numbers, 97, is the range, as we have seen (Sec. 6.2.1).

The range is easy to calculate and is certainly the most natural and commonly encountered way of describing variability. It is often used as a measure of variability in engineering and medical reports. Nevertheless, the range is difficult to interpret because *its interpretation depends upon the number of observations from which it is computed*, and is difficult and complicated even then. This weakness is particularly dangerous in comparing the variability of different groups of observations.

EXAMPLE 245   AGE AND SEXUAL ACTIVITY

The following comparison is included in a study of the sexual activity of men:[1]

| Age Group | Number of Cases | Next to Highest Rate of Sexual Activity (frequency per week) |
|---|---|---|
| Adolescence to 15 | 3,012 | 29.0 |
| 61–65 | 58 | 4.0 |

The "next to highest rate" is the next to highest observation in the group; since the lowest rate is undoubtedly zero, or close to it, this measure is much like the range. It is concluded from these data (and figures for intermediate ages) that the maximum (or rather, next-to-maximum) rate of sexual activity declines markedly with age. This may or may not be true, but these data do not prove it and the decline, if any, is almost certainly less than these figures suggest. If two groups of the same age were studied, one a group of 3,012

1. Alfred C. Kinsey, Wardell B. Pomeroy, and Clyde E. Martin, *Sexual Behavior in the Human Male* (Philadelphia and London: W. B. Saunders Company, 1948), p. 234.

*Variability*

and the other of 58, it is almost certain that the group of 3,012 would show a considerably higher maximum or next-to-maximum, than the group of 58.

We once used 64 members of a class to illustrate the point on which the preceding example foundered (the same 64 from whom the 32 weights of Table 173 were taken). Each recorded his weight, to the nearest whole pound, on a card. Four groups of eight cards each were selected from the 64 shuffled cards on which the weights were recorded. The range in the first group; the first and second combined; the first, second, and third combined; in all four groups combined; and in the entire group of 64 is given in Table 246.

TABLE 246

HIGHEST AND LOWEST WEIGHTS AND THE RANGE

Samples of 8, 16, 32, and 64

| Number of Weights | Lowest Weight | Highest Weight | Range |
|---|---|---|---|
| 8 | 148 | 183 | 35 |
| 16 | 148 | 189 | 41 |
| 24 | 148 | 200 | 52 |
| 32 | 148 | 203 | 55 |
| 64 | 142 | 239 | 97 |

The larger the number of weights, the greater the range tends to be. It is clear that increasing the number of people in the sample cannot possibly decrease the range already observed. On the other hand, it is likely that as the group increases someone will turn up with a weight exceeding the previous maximum or falling short of the previous minimum—and each time this happens the range will increase. So, on the average, the larger the sample, the larger the range. It is impossible therefore to interpret the range properly without allowing for the number of observations involved. The proper allowance, however, depends very much on the shape of the population, something that cannot be determined reliably except from very large samples.

Though many current uses of the range are unsatisfactory, there is one common use that is valid, namely in statistical quality control. In statistical quality control, the same sample size is used repeatedly, so that comparisons between ranges are not distorted by differences in sample size. Moreover, since the sample size is usually very small— for example, four—the range tells nearly as much about variability as would the magnitudes of all the observations.

The range is also useful for quick, approximate checks on the accuracy of calculations of other measures of variability. Even though it has no fixed relation to them, certain orders of magnitude can be expected and there are certain limits on the relations among them. We shall give an example of this in Sec. 8.5.2.

## 8.3
## THE MEAN (OR AVERAGE) DEVIATION

Since the range is of limited usefulness, let us consider other possible measures of variation. Suppose that we have a sample of six observations, 0, 2, 3, 4, 4, 5. The mean of these observations is

$$\frac{0 + 2 + 3 + 4 + 4 + 5}{6} = 3.$$

Now we obtain the deviation of each observation from the mean of 3. For the first observation, for example, this gives a deviation of $0 - 3 = -3$. Similarly, the other deviations are $-1$, $0$, $+1$, $+1$, and $+2$.

It may seem that a good way to measure dispersion would be to take the mean of these six deviations. But this gives

$$\frac{-3 - 1 + 0 + 1 + 1 + 2}{6} = 0.$$

The result would be zero for any group of observations because, as we saw in Sec. 7.4.1, the algebraic sum of the deviations of a group of observations from their own mean is always zero. To avoid this, we take the absolute values of the differences, that is, disregard the signs, and get the arithmetic mean of these absolute values. This is called the *average deviation* or (better) the *mean deviation*. In our example, we get

$$\frac{3 + 1 + 0 + 1 + 1 + 2}{6} = 1.33$$

for the mean deviation.

For practice in using symbols, we give the symbolic definition of the mean deviation. First take any observation, say the "$i$th," $x_i$. Then subtract the mean from this observation, $x_i - \bar{x}$. Take the absolute value of this deviation, $|x_i - \bar{x}|$, where the vertical lines stand for "absolute value." Now add up all $n$ deviations:

$$\sum_{i=1}^{n} |x_i - \bar{x}|.$$

*Variability*

Finally, divide by $n$:

$$\text{Mean deviation} = \frac{\sum\limits_{i=1}^{n} |x_i - \bar{x}|}{n}.$$

We might omit the index and range of summation, as is customary when all observations are included, and write the mean deviation as

$$\frac{\sum |x - \bar{x}|}{n}.$$

The mean deviation is interpreted as are arithmetic means generally, since it is the arithmetic mean of the absolute deviations. This ease of interpretation is its chief advantage. It has two important disadvantages. From the mean deviations of several groups of observations, it is not possible to find the mean deviation of the combined group. Among the averages discussed in Chap. 6 only the means are free of this limitation; among the measures of dispersion, only the standard deviation is free of it. A second drawback of the mean deviation is that it is not often useful for statistical inference.

### 8.3.1 Calculation of the Mean Deviation

The calculation of the mean deviation does not require that the mean be subtracted separately from each observation. First calculate the arithmetic mean, and then divide the observations into two groups, those that are above and those that are below the mean. (Those that are equal to the mean are omitted from both groups.) Let $A$ be the sum of the observations above, and $B$ the sum of those below, $\bar{x}$; and let $a$ be the number of observations above, and $b$ the number below, $\bar{x}$. Then the mean deviation is given by

$$\frac{A - B - (a - b)\bar{x}}{n}.$$

In our example, where the observations are 0, 2, 3, 4, 4, and 5, we find $A = 13$, $B = 2$, $a = 3$, $b = 2$, $\bar{x} = 3$, and $n = 6$, so we have

$$\frac{13 - 2 - (3 - 2)3}{6} = 1.33,$$

just as before.

## 8.4
## POSITIONAL MEASURES OF DISPERSION

At the beginning of Chap. 7 we mentioned that there is a large class of measures of location like the median or the ninetieth centile, measures defined by a certain position on the scale of cumulative frequency. Any two or more such positional measures, or *quantiles*, give some idea of variability; enough of them, in fact, reproduce the frequency distribution. Thus, forty percent of the observations lie between the median and the ninetieth centile.

Some quantiles are used frequently enough to have special names, for example, quartiles, deciles, and centiles (often called percentiles). These are numbered from smallest to largest. The first quartile, for example, is that value of the variable at or below which one-fourth of the observations lie. The second quartile is the same as the median. The third quartile is that value at or below which three-fourths of the observations lie. A similar ordering applies to deciles, which are tenths, and to centiles, which are hundredths; thus, the 99th centile is a value exceeded by only one percent of the observations.

As with the median (see Sec. 7.3), a certain amount of hairsplitting is necessary for precision. The 65th centile, for example, may fall between two observations. For the first column of the weight data of Table 173, there are 16 observations, and 65 percent of 16 is 10.4. The 10th and 11th weights are 164 and 165. Strictly speaking, there is no 65th centile for this distribution, for any weight from 164 to 165 is a 62.5 centile. An approximate figure for the 65th centile might be desirable, however, for comparison with other distributions. Such a figure can be obtained by considering the "10.4th" observation to be 0.4 of the way between the 10th and the 11th, that is, 164.4. With a discrete variable, on the other hand, there may be too many observations at a given centile. For the bead data of Table 106, for example, the 65th centile falls at the "32.5th" observation. This is 12; so is every other centile from the 60th to the 80th, for the observation occurred ten times.

Note also the difference between the first quarter, say, and the first quartile. The first quarter is a group of observations or a range of the variable, while the first quartile is the boundary point between the first and second quarters. An observation can be within the highest tenth but not within the highest decile—it is at or above the highest decile. Unfortunately, the inaccurate expression—"in the $x$th centile"—is common.

*Variability*

Measures of variation are obtained from quantiles by selecting two quantiles symmetrical about the median and taking the difference between them, or sometimes half the difference. One of the best of these measures, which we may call the *decile range*, is based on the first and ninth deciles.

From the cumulative distribution of urban family income shown in Table 189, we find that 9.3 percent of the observations are less than $1,500. The additional 0.7 percent needed to equal 10 percent represents 0.7/4.4 of the cases in the next class, whose length is $500, so we estimate the first decile as

$$\$1,500 + \frac{0.7}{4.4} \times \$500 = \$1,580.$$

Similarly, the ninth decile is estimated as

$$\$7,000 + \frac{5.7}{10.8} \times \$3,000 = \$8,583.$$

Then

$$D = \$8,583 - \$1,580 = \$7,003$$

is the decile range. Thus, a span of $7,003 includes the central 80 percent of incomes—all but the upper and lower tenths.

A more common measure is the *semi-interquartile range*, or *quartile deviation*, which is half the difference between the first and third quartiles. The difference between the first and third quartiles is ordinarily subject to wider sampling variability than the difference between the first and ninth deciles, and is no more descriptive or interpretable. Half of an interquantile difference has no special significance unless the distribution is symmetrical, but it invites misinterpretation. Had we shown $3,502 as the "semi-interdecile range," for example, it might have suggested the *false* interpretations (1) that 80 percent of the observations are within $3,502 of the median ($4,249), (2) that 10 percent are further than $3,502 below the median (that is, below $747, where actually there are only 3.8 percent if we assume that half of those from $500 to $999 are below $747—which is undoubtedly an overestimate), and (3) that 10 percent are further than $3,502 above the median (that is, above $7,751, where actually there are about 13 percent if we assume that three-fourths of those from $7,000 to $9,999 are above $7,750—which, however, is undoubtedly an overestimate). If we know the decile range, the appropriate measure of location to which to anchor it is not the median but either the first or ninth decile.

The advantage of the decile range, or other quantile range, is that it is easy to interpret. The disadvantages are similar to those of the mean deviation: it is not amenable to algebraic treatment, hence to combination for several groups, and it is, on the whole, less useful in statistical inference than is the standard deviation.

# 8.5
# THE STANDARD DEVIATION

## 8.5.1 Nature and Interpretation of the Standard Deviation

The arithmetic mean of the deviations is always zero, but the other means (see Sec. 7.4.4), in general, are not. The geometric and harmonic means of the deviations are not meaningful since not all the deviations can be positive. The root mean square of the deviations, however, is positive, and this is called the *standard deviation.*

The standard deviation can be obtained in four steps: (1) take the deviation of each observation from the mean, (2) square each deviation, (3) obtain the mean of the squared deviations, and (4) extract the square root of this mean. Instead of obtaining exactly the mean in the third step, we divide by $n - 1$ rather than $n$, for reasons that will be indicated later.

Consider the six observations with a mean of 3 that were presented in Sec. 8.3. With their deviations and squared deviations they are:

|  |  |  |  |  |  |  | Total |
|---|---|---|---|---|---|---|---|
| Observation: | 0 | 2 | 3 | 4 | 4 | 5 | 18 |
| Deviation: | −3 | −1 | 0 | +1 | +1 | +2 | 0 |
| Squared deviation: | 9 | 1 | 0 | 1 | 1 | 4 | 16 |

Then the mean squared deviation, dividing by 5 instead of by 6, is 3.2. The square root of this, 1.79 (see Table 626), is the standard deviation.

Now let us use symbols to express this more generally. Consider any observation, say the $i$th, $x_i$, and take its deviation from the mean, $x_i - \bar{x}$. Then square this deviation, $(x_i - \bar{x})^2$. Add the squared deviations for all $n$ observations and divide by $n - 1$:

$$s^2 = \frac{\sum_{i=1}^{n} (x_i - \bar{x})^2}{n - 1}$$

where $s$ is the standard deviation. (Its square, $s^2$, is called the *variance*; the variance is of little importance as a descriptive statistic but is important in statistical inference and in combining samples.) The standard deviation, $s$, is then

$$s = \sqrt{\frac{\sum\limits_{i=1}^{n} (x_i - \bar{x})^2}{n - 1}}.$$

The standard deviation has the least intuitive appeal or obvious interpretation of any statistical measure so important and so widely used. The reasons for its usefulness relate both to its practical merits for describing variability and to its role in statistical inference, that is, drawing conclusions about populations from samples.

As a descriptive statistic, the standard deviation is essentially a mean, as we have seen. It is the simplest type of mean of the algebraic deviations that gives positive numbers that increase when variability increases and are in the same units as the observations. It is especially convenient for computing variability of composites. Two kinds of composite are important, and it is applicable to both. To illustrate, consider the bead data of Tables 106 and 107B. First, we could combine the distributions into a single distribution of 100 observations. The standard deviation of this combined distribution can be computed directly from the standard deviations and means of the individual distributions. Second, suppose our demonstration had been set up to obtain 50 samples of 40 beads each, 20 from Population I and 20 from Population II. Such a distribution can, in fact, be formed by adding the numbers of red beads in corresponding samples from the two populations, obtaining the sequence 15, 16, 10, . . ., 12, and 17 (see Fig. 109). The standard deviation of these observations can be computed from the standard deviations of the two distributions.

The analytical properties of the standard deviation center around its role in the so-called normal distribution, the shape of which is shown in Figs. 361. To base decisions on a statistic from a sample it is necessary, as Chap. 4 emphasized, to allow for sampling variability. Commonly the pattern of variability in a statistic, if computed from indefinitely many samples of the same kind, is well enough described by a normal distribution, which in turn is completely described by its mean and variance. Thus, the variance, or squared standard deviation, plays a major part in statistical inference. The reason for dividing by $n - 1$ instead of $n$ is that then the variance plays its role in statistical inference with greater ease and facility, while the

*8.5 The Standard Deviation*

standard deviation plays its role in statistical description as effectively with $n - 1$ as with $n$.

Finally, many distributions resemble the normal distribution. To the extent that they do, the standard deviation is a good gauge of the proportions of the observations falling within given distances from the mean. For a normal distribution, about two-thirds of the observations differ from the mean by less than the standard deviation, about 95 percent by less than twice the standard deviation, and practically all by less than three times the standard deviation.

Be sure to use $s$, not $\sigma$ as is sometimes done, to denote the standard deviation computed from a sample. The symbol $\sigma$ (Greek lower-case "sigma") is, in standard practice, used to represent the standard deviation of a population only. Recall the emphasis in Sec. 4.2 on the importance of the distinction between samples and populations.

## 8.5.2 Computation of the Standard Deviation

While the formula given for the standard deviation is the one that makes it easiest to understand what the standard deviation is, it is *not* a good formula for computing. For computing, it is better to make use of the fact that

$$\sum(x - \bar{x})^2 = \sum x^2 - \frac{(\sum x)^2}{n} .$$

The two quantities to compute from the observations are, therefore, $\sum x$, which is already available if the mean has been computed, and $\sum x^2$. The squares of the six observations used in Sec. 8.5.1, for which $\sum x = 18$, are 0, 4, 9, 16, 16, and 25, so $\sum x^2 = 70$. Hence,

$$\sum(x - \bar{x})^2 = 70 - \frac{18^2}{6} = 70 - 54 = 16,$$

the same result we found earlier.

Not only is this method usually easier, it also is more accurate. When the mean has to be rounded, each deviation is slightly in error due to rounding, the squares are slightly in error too, and the results of the calculation may be less accurate than when the "shortcut" formula is used.

The same shortcuts we applied for computing means in Sec. 7.4.5 can be used also for the standard deviation. Reducing (or increasing) each observation by the same amount does not affect the standard deviation at all—the new set of numbers has the same variability but is simply located at a different place along the scale.

*Variability*

Multiplying (or dividing) each observation by the same positive constant multiplies (or divides) the standard deviation by the same constant; this is equivalent to changing the units in which the observations are measured.

Ordinarily the standard deviation is computed at the same time as the mean, and with the same coding. Tables 232 and 233 are repeated here as Tables 254 and 255, with the extra calculations necessary to find the standard deviation shown in boldface. In the first table, equal class intervals are used; in the second, the intervals are unequal.

TABLE 254

SHORTCUT CALCULATION OF MEAN AND STANDARD DEVIATION
(Equal Class Intervals)

| Weights | Number of Persons $f$ | Coded[a] Mid-value $m$ | $fm$ | $fm^2$ |
|---|---|---|---|---|
| 137.5–147.5 | 2 | −4 | −8 | **32** |
| 147.5–157.5 | 5 | −3 | −15 | **45** |
| 157.5–167.5 | 4 | −2 | −8 | **16** |
| 167.5–177.5 | 5 | −1 | −5 | **5** |
| 177.5–187.5 | 7 | 0 | 0 | **0** |
| 187.5–197.5 | 5 | +1 | 5 | **5** |
| 197.5–207.5 | 3 | +2 | 6 | **12** |
| 207.5–217.5 | 0 | +3 | 0 | **0** |
| 217.5–227.5 | 0 | +4 | 0 | **0** |
| 227.5–237.5 | 0 | +5 | 0 | **0** |
| 237.5–247.5 | 1 | +6 | 6 | **36** |
| Total | 32 $n$ | | −19 $\sum fm$ | **151** $\sum fm^2$ |

*Source:* Table 232.

[a] The coded mid-value, $m$, is the mid-value of the class minus a constant (usually the midpoint of some central class), divided by a constant (usually the class interval); in this case $m = \dfrac{x - 182.5}{10}$, since the class interval is 10.

$$\bar{x} = 182.5 + \frac{10 \sum fm}{32} = 182.5 + \frac{10(-19)}{32} = 182.5 - 5.94 = 176.56,$$

$$s = (\text{Class interval}) \sqrt{\frac{\sum fm^2 - \frac{(\sum fm)^2}{n}}{n - 1}} = 10 \sqrt{\frac{151 - \frac{(-19)^2}{32}}{31}}$$

$$= 10 \sqrt{\frac{151 - 11.2812}{31}} = 10 \sqrt{\frac{139.7188}{31}} = 10 \sqrt{4.5071} = 10(2.123)$$

$$= 21.23.$$

## 8.5 The Standard Deviation

TABLE 255

SHORTCUT CALCULATION OF MEAN AND STANDARD DEVIATION
(Unequal Class Intervals)

| Weights | Number of Persons $f$ | Coded[a] Mid-value $m$ | $fm$ | $fm^2$ |
|---|---|---|---|---|
| 137.5–147.5 | 2 | −3 | −6 | 18 |
| 147.5–157.5 | 5 | −2 | −10 | 20 |
| 157.5–167.5 | 4 | −1 | −4 | 4 |
| 167.5–177.5 | 5 | 0 | 0 | 0 |
| 177.5–187.5 | 7 | +1 | 7 | 7 |
| 187.5–197.5 | 5 | +2 | 10 | 20 |
| 197.5–217.5 | 3 | +3.5 | 10.5 | 36.75 |
| 217.5–247.5 | 1 | +6 | 6 | 36 |
| Total | 32 | | 13.5 | 141.75 |

*Source:* Table 233.

[a] Here $m = \dfrac{x - 172.5}{10}$,

$$\bar{x} = 172.5 + \frac{10(13.5)}{32} = 172.5 + 4.22 = 176.72,$$

$$s = 10 \sqrt{\frac{141.75 - \dfrac{(13.5)^2}{32}}{31}} = 10 \sqrt{\frac{141.75 - 5.6953}{31}}$$

$$= 10 \sqrt{\frac{136.0547}{31}} = 10\sqrt{4.3889} = 10(2.095)$$

$$= 20.95.$$

The standard deviation is thus approximated as 21.23 or 20.95, according to which grouping is used.

These two results may be compared with the result which would have been obtained had we started with the original ungrouped data. This is 20.72, as is shown in the calculation below (in which no coding was used):

$$s = \sqrt{\frac{\sum x^2 - \dfrac{(\sum x)^2}{n}}{n - 1}} = \sqrt{\frac{1,003,483 - \dfrac{(5629)^2}{32}}{31}}$$

$$= \sqrt{\frac{1,003,483 - 990,176}{31}} = \sqrt{429.26}$$

$$= 20.72.$$

In general, the standard deviation tends to be slightly larger computed from grouped than from ungrouped data. All three of these results agree to the nearest pound, 21, which is sufficient for descriptive purposes. Class intervals not larger than one-fourth the standard deviation (five pounds in this instance) will give accurate enough results for analytical purposes, and the errors introduced by intervals as wide as one-half the standard deviation (as here) are ordinarily acceptable.

Here we may illustrate the use of the range as a quick check on gross error in the computations. As a rough guide especially useful for the statistical reader we suggest questioning the accuracy of the computations if the range is less than the standard deviation or more than seven times the standard deviation. In our example, the range is 97 and the standard deviation is 21, so there is no reason to suspect an error. Similarly, the mean is comparable in magnitude to the median, 177.5 by Tables 254 and 255. This is, of course, far from a proof of correctness; it simply affords partial protection against gross errors. A person responsible for calculations should, however, have all work done twice as nearly independently as possible.

### 8.5.3 The Relative Standard Deviation

Sometimes when all of the observations are by nature positive, it is desirable to measure variability relative to the mean rather than in absolute units. A common way to do this is to divide the standard deviation by the mean, that is, $s/\bar{x}$. For example, for the weight data we found a mean of 176 and a standard deviation of 21. Hence,

$$\frac{s}{\bar{x}} = \frac{21}{176} = 0.12.$$

That is, the standard deviation is 12 percent of the mean. This statistic is called the *relative standard deviation*, or, more commonly though less descriptively, the *coefficient of variation*. The coefficient of variation is usually expressed as a percent rather than as a decimal; thus, in the example above, one might speak of the coefficient of variation as 12 rather than 0.12.

This measure is especially useful in planning experiments or surveys, for specifying the accuracy desired in the results. The absolute sampling error of the results is expressed as a special kind of standard deviation which we shall discuss later. The absolute sampling error divided by the mean to be estimated is then the coefficient of variation. For example, in a survey of the incidence of unemployment, it might be desirable that the relative standard deviation be 10 percent or less.

## 8.6
## THE LORENZ CURVE

EXAMPLE 257   INCOME BEFORE AND AFTER TAXES

Fig. 257 is a Lorenz curve showing the degree of inequality in the distribution of income in 1953 before and after federal income taxes.

If we look, for example, at the point on the horizontal scale marked "40" we find a reading on the vertical scale of "15" for the "before taxes" curve. This means that the 40 percent of spending units who had the lowest incomes before taxes received 15 percent of the aggregate money income before taxes of all spending units. After taxes, however, the lowest 40 percent received 17 percent of the aggregate income after taxes. The 40 percent of spending units with the lowest income after taxes are not necessarily identically the same spending units who constitute the 40 percent with the lowest income before taxes.

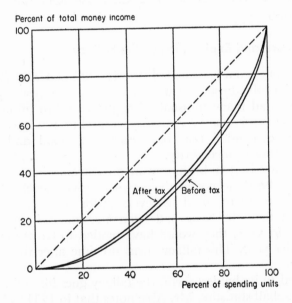

FIG. 257. Lorenz curves of distribution of money income by spending units, before and after federal income tax, 1953.[a]

*Source: Statistical Abstract: 1955,* Table 353, p. 298. Based on data from the 1954 Survey of Consumer Finances, conducted by the Federal Reserve System in co-operation with the Survey Research Center of the University of Michigan.

[a] A spending unit is defined as all persons living in the same dwelling, and belonging to the same family, who pool their income to meet their major expenses.

*Variability*

It is customary to draw a diagonal line of "complete equality" such as the dashed line of Fig. 257. This would represent a situation in which each spending unit received the same income. What would be a meaningful line of perfect equality is a difficult question. Suppose, for example, that every spending unit received the same income at the same age—its life-pattern of earnings was the same—but that this income was low in the early years of the unit, rose to a high in middle-life, declined slightly until retirement, then declined sharply after retirement. Many would consider this complete equality. If so, the diagonal line should be replaced by one reflecting such inequality as is due only to differences in stage of career. Where such a line would fall is a subject of current research by income statisticians.

Comparison of relative standard deviations also provides a good indication of differences in the inequality of distributions.

## 8.7
## "REGRESSION" FALLACY IN THE COMPARISON OF VARIABILITIES

### 8.7.1 Nature and Explanation of the Fallacy

Some years ago a book called *The Triumph of Mediocrity in Business* reported that the businesses which had made unusually high and those which had made unusually low profits in an initial year both tended to be nearer the average in profitability at a later year. From this it was concluded that both extremes—good and bad—were coming together at mediocrity.

The fallacy is that businesses at an extreme in any given year are likely to be those not at their own long-run levels. As a matter of fact, the argument could be reversed; for if the businesses at the extremes in the later year had been chosen, and then compared with an *earlier* year, they would have tended to be nearer the mean in the earlier year. This fallacy turns up time after time in all sorts of contexts.

Reduced to its baldest form, the fallacy goes like this. In looking through baseball statistics, Mr. Able notes that in 1954 Mays (Giants) led the major leagues with a batting average of 0.345, while Williams (Giants) came in last with an average of 0.222, but that in 1955 Mays' average dropped to 0.319 while Williams' rose to 0.251. From this Mr. Able concludes that major league batters are becoming mediocre: good batters are getting worse, the bad ones better— just as he suspected.

At the same time, Mr. Baker has a suspicion that differences in the caliber of major league batters are becoming greater. He observes that Kaline (Tigers) led the major leagues in 1955 with an average of 0.340, O'Connell (Braves) trailed with an average of 0.225. Checking further, he finds that Kaline averaged 0.276 in 1954 and O'Connell averaged 0.279. His conclusion: good batters are getting better, poor ones worse—just as he suspected.

Mr. Able and Mr. Baker have both committed the regression fallacy. Put this way, the fallacy may seem too transparent to be credible, but then many common statistical fallacies look absurd in their simpler manifestations. If Able and Baker had started with an average of the ten (or fifty) top and the ten (or fifty) bottom batters their procedure would appear more reasonable, yet they would still have been committing the fallacy and they would have reached the same contradictory conclusions.

We shall explain the fallacy still further with a hypothetical example. Suppose that the 1951 and 1956 percentage profits of 46 companies were as in Table 259.

TABLE 259

DISTRIBUTION OF 46 COMPANIES BY PERCENTAGE
PROFITS IN 1951 AND 1956

(Hypothetical)

| 1956 Percentage | 1951 Percentage | | | | | Number | Mean Profit in 1951 | Mean Profit in 1956 |
|---|---|---|---|---|---|---|---|---|
| | 0–5 | 5–10 | 10–15 | 15–20 | 20–25 | | | |
| 0–5 | 2 | 2 | | | | 4 | 5.0 | 2.5 |
| 5–10 | 1 | 4 | 6 | | | 11 | 9.8 | 7.5 |
| 10–15 | | 2 | 12 | 2 | | 16 | 12.5 | 12.5 |
| 15–20 | | | 6 | 4 | 1 | 11 | 15.2 | 17.5 |
| 20–25 | | | | 2 | 2 | 4 | 20.0 | 22.5 |
| Number | 3 | 8 | 24 | 8 | 3 | 46 | | |
| Mean Profit in 1956 | 4.2 | 7.5 | 12.5 | 17.5 | 20.8 | | 12.5 | |
| Mean Profit in 1951 | 2.5 | 7.5 | 12.5 | 17.5 | 22.5 | | | 12.5 |

From Table 259, Mr. Able would prepare Table 260A. Able shows that those making below-average profits in 1951 have, on the average, done better in 1956, and those making above-average profits in 1951 have, on the average, done worse. There has been a leveling-out process, he firmly believes, eliminating both extremes in favor of mediocrity. From the same Table 259, Mr. Baker might

*Variability*

### TABLE 260A
#### FIRMS GROUPED ACCORDING
#### TO 1951 PROFITS

| Number of Firms | Mean Profit (percent) | |
|---|---|---|
| | 1951 | 1956 |
| 3 | 2.5 | 4.2 |
| 8 | 7.5 | 7.5 |
| 24 | 12.5 | 12.5 |
| 8 | 17.5 | 17.5 |
| 3 | 22.5 | 20.8 |
| 46 | 12.5 | 12.5 |

prepare Table 260B. Baker shows that those making below-average profits in 1951 have, on the average, done worse in 1956, and those making above-average profits in 1951 have, on the average, done better. Thus he concludes that there is increasing inequality: "Statistics prove it," he says.

### TABLE 260B
#### FIRMS GROUPED ACCORDING
#### TO 1956 PROFITS

| Number of Firms | Mean Profit (percent) | |
|---|---|---|
| | 1951 | 1956 |
| 4 | 5.0 | 2.5 |
| 11 | 9.8 | 7.5 |
| 16 | 12.5 | 12.5 |
| 11 | 15.2 | 17.5 |
| 4 | 20.0 | 22.5 |
| 46 | 12.5 | 12.5 |

Mr. Charley says this shows that you can prove anything by statistics. Mr. Dog says figures don't lie but liars figure. An air of general confusion, mystery, and suspicion is generated. Able and Baker may come to blows, with Charley and Dog egging them on.

The crux of the situation is that the bulk of the firms tend to be generally around the average—12 of the 46 were in the middle (10–15) group both years, and 16 more firms were in it one year and in an adjacent (5–10 or 15–20) group the next year. In any given year, a small fraction of those firms which generally tend to be around average will be off at an extreme. Of those few firms which are generally at an extreme, most will in any one year be at the extreme. So when we select the firms which are at an extreme in a particular year, we have firms of two kinds: those which are generally at the extreme and can be expected to stay there, and those which are generally not at that extreme and can be expected to move back toward the center of the distribution. These latter move the average of the group toward the average of the whole distribution. But their places at the extreme are taken by other firms affected by temporary fluctuations:

> ... while the concerns at the margins of the group, if they remain in business, often go toward the center, those in the center of the group also go toward the margins. Some go up and some down; the average of the originally center group may, therefore, display little change, since positive and negative deviations cancel in averaging; while for an extreme group, the only possible motion is toward the center.[2]

### EXAMPLE 261A   HEIGHTS OF FATHERS AND SONS

The regression phenomenon was first noted by Sir Francis Galton in connection with heights. The situation may be schematized this way: Tall people represent three groups: (1) typical individuals from tall stock; (2) unusually tall individuals from medium stock; (3) unusually short individuals from extremely tall stock. (Extremely unusual individuals from short stock can be omitted from this schematization; taking account of them would only strengthen the conclusion.) Group 2 is more numerous than group 3, simply because medium stock is more common than extremely tall stock. Thus, if we select a group of tall people, and measure the heights of their relatives, we find the relatives shorter, on the average, than the original group. The relatives are, to be sure, taller than the general average of the population, because of the effects of groups (1) and (3).

Galton gave this phenomenon the name "regression." The phenomenon is real enough; the fallacy is in interpreting it to indicate a decrease in the variability of the whole population.

### EXAMPLE 261B   CONSUMPTION FUNCTION

Still another example of the regression fallacy arises in studying the relationship between family income and family expenditures, the so-called "consumption function." The difficulty is that the average consumption expenditures of families with incomes less than, say, $1,500 in any particular year tend to be inflated by the fact that some families whose incomes are normally larger happen (perhaps because of illness or retirement) to have low incomes in the particular year, yet (by borrowing or drawing on savings) maintain expenditures more like their normal pattern than like the families for whom $1,500 is a normal income. Similarly, the average expenditures of those with incomes over $15,000 are pulled down by some who are in the

2. Harold Hotelling, review of Horace Secrist, "The Triumph of Mediocrity in Business," *Journal of the American Statistical Association*, Vol. 28 (1933), pp. 463–465. Also see Secrist's reply and Hotelling's rejoinder in the same journal, Vol. 29 (1934), pp. 196–199,

group only by virtue of "windfalls" in the particular year, so do not greatly increase outlays above the level appropriate to their normal income. Hence, one would get the impression that people tend to spend a smaller fraction of their income the larger that income is, even if in fact everyone tended to spend the same fraction of his "normal" income. What the basic relation really is is a matter currently under study by economic statisticians.

### EXAMPLE 262A  MIDTERMS AND FINALS

Teachers—except, of course, statistics teachers—sometimes commit the regression fallacy in comparing grades on a final examination with those on a midterm examination. They find that their competent teaching has succeeded, on the average, in improving the performance of those who had seemed at midterm to be in precarious condition. This accomplishment naturally brings the teacher keen satisfaction, which is only partially dampened by the fact that the best students at midterm have done somewhat less well on the final—an "obvious" indication of slackening off by these students due to overconfidence.

### EXAMPLE 262B  SALES OR POLITICAL CAMPAIGNS

The regression fallacy has led some people to conclude that the effect of a sales campaign or a political campaign is to make those previously hostile less hostile, but to make those previously friendly less friendly. People are rated on an attitude scale, say from 1 (hostile) to 7 (enthusiastic), then rated again after a campaign. Some of those initially enthusiastic are now less so. It is fallacious to draw conclusions from these facts alone; if there is variability, a similar result must appear. For the whole group, variability may well be as large after as before.

### EXAMPLE 262C  LEADERS OF SUCCESSIVE GENERATIONS

The following quotation illustrates a correct avoidance of the regression fallacy in a context where many have fallen into it. The quotation serves also to illustrate the use of statistical reasoning without numerical data, in an analysis which would not ordinarily be thought of as involving statistical principles.

> Do you look to the leading families to go on leading you? Do you look to the ranks of men already established in authority to contribute sons to lead the next generation? They may, sometimes they do, but you can't count on them; and what you are constantly depending on is the rise out of the ranks of unknown men, the emergence of somebody from some place of which you had thought the least, of some man unanointed from on high, to do the thing that the generation calls for. Who would have looked to see Lincoln save a nation?[3]

3. From a speech by Woodrow Wilson in Madison, Wisconsin, October 25, 1911, quoted in Saul K. Padover, *Wilson's Ideals* (Washington: American Council on Public Affairs, 1942), pp. 18–19.

*8.8 Conclusion*

We may put our conclusion about the regression fallacy in the form of a general principle: Take any set of data, arrange them in groups according to some characteristic, and then for each group compute the average of some second characteristic. Then the variability of the second characteristic will usually appear to be less than that of the first characteristic. For example, the ten percent of individuals with the greatest incomes will have a lower proportion of the total wealth than they do of the total income. Mathematically, this "principle" need not hold invariably, but the conditions necessary for it to hold are very commonly fulfilled in practice.

A valid way to compare the variabilities of two populations is to take a random sample from each, examine the frequency distributions, and calculate the standard deviations, or some other measures of variability. If there is a real tendency toward mediocrity, the second sample will show less dispersion.[4]

## 8.8
## CONCLUSION

The dispersion of a distribution is sometimes described by such measures as the range, which is the difference between the largest and smallest observations, or the decile range, which is the difference between the ninth and first deciles (that is, the points below which lie nine-tenths and one-tenth, respectively, of the observations). More commonly, dispersion is described by various averages of the deviations of the individual observations from the mean, especially their root mean square (that is, the square root of the mean of their squares), which is called the standard deviation. In computing this mean, one less than the number of cases (that is, $n-1$) is used as the divisor, rather than the number of cases ($n$), because this makes the measure more useful for statistical inference, and just as useful for statistical description. The principal advantages of the standard deviation are that when samples are combined in various ways, the standard deviations of the combined data can be calculated from the means and standard deviations of the original data; and that it plays a central role in the normal distribution, which in turn is useful not only for approximating the distributions of many kinds of data,

---

4. If two independent samples are not available, but only a single sample measured twice, comparison of variabilities is possible but more complicated, since it depends on the correlation between the sums and differences of corresponding observations. See J. F. Kenney and E. S. Keeping, *Mathematics of Statistics, Part II* (2d ed.; New York: D. Van Nostrand Company, 1951).

but especially for approximating the sampling distributions of many of the measures computed from samples.

The degree of inequality in a distribution is sometimes described by a Lorenz curve. For this, the observations in the sample are arranged in order, and a chart is prepared showing the proportion of the sum of the observations which is accounted for by various proportions of the observations, cumulated either from the smallest or from the largest observation.

In comparing the variability of the same, or corresponding, individuals at different times, an illusion of decreasing variability is often caused by what is known as the regression fallacy. If the observations are fluctuating from time to time, a particular class interval of a frequency distribution formed at any given time will contain some individuals who have "strayed" there from higher class intervals, and some who have strayed there from lower class intervals. If the individuals in the class interval are looked at later, the strays tend to have returned to their home classes. The average of the individuals who were in the class at the original time is higher at the later time if more of the strays had come from, so have now returned to, higher classes than had come from, hence returned to, lower classes. Correspondingly, the average at the later time will be lower if more of the strays in the class interval at the original time had come from below than had come from above. The larger number of strays will have come, generally, from the direction in which the larger number of individuals lies, that is, from the direction of the mode of the distribution. Thus, the effect of concentrating on the individuals in a given class at a given time and computing the average of these individuals at a later time, is to find that their average has moved toward the mode. This "regression" is a real phenomenon if the individuals are fluctuating in time; the fallacy lies in interpreting it as showing anything about the dispersion of the entire distribution at the two times.

Location and dispersion are the two most important properties of univariate variables. For bivariate or multivariate variables, however, not only the locations and dispersions of the individual variables are important, but also the association between the variables. Measures for describing association or correlation are, therefore, our next subject, and will complete our discussion of statistical description.

## DO IT YOURSELF

EXAMPLE 265A

For the Rockwell hardness data of Table 206, compute (*a*) from the original data and (*b*) from a frequency distribution, the following measures of variability: decile range, semi-interquartile range, mean deviation, variance, standard deviation, coefficient of variation.

EXAMPLE 265B

The meteorologists of a certain commercial airline claim an average error in their forecasts of flight time of −10 minutes, by which they mean that the aircraft arrive on the average 10 minutes earlier than forecast. Can the forecasting performance be evaluated on the basis of this information? Discuss.

EXAMPLE 265C

What descriptive devices would be most useful in characterizing the following distribution? Demonstrate.

TABLE 265

BIRTH WEIGHT DISTRIBUTION, SINGLE BIRTHS:
JANUARY–MARCH, 1950

| Birth Weight | Percent |
|---|---|
| 2 lbs. 3 oz. or less | 0.4 |
| 2 lbs. 4 oz. to 3 lbs. 4 oz. | 0.5 |
| 3 lbs. 5 oz. to 4 lbs. 6 oz. | 1.1 |
| 4 lbs. 7 oz. to 5 lbs. 8 oz. | 4.4 |
| 5 lbs. 9 oz. to 6 lbs. 9 oz. | 17.9 |
| 6 lbs. 10 oz. to 7 lbs. 11 oz. | 38.2 |
| 7 lbs. 12 oz. to 8 lbs. 13 oz. | 27.6 |
| 8 lbs. 14 oz. to 9 lbs. 14 oz. | 7.8 |
| 9 lbs. 15 oz. or more | 2.1 |
| Total | 100.0 |
| (Number) | (820,618) |

*Source: Statistical Abstract: 1954*, Table 69, p. 71.

EXAMPLE 265D

What descriptive devices would be most useful in characterizing Table 183? Demonstrate.

EXAMPLE 265E

What descriptive devices would be most useful in characterizing the distribution of Table 266? Demonstrate.

TABLE 266

ESTIMATED POPULATION OF CONTINENTAL UNITED STATES,
BY AGE: JULY 1, 1954

| Age | Number (Thousands) | Age | Number (Thousands) |
|---|---|---|---|
| Under 1 year | 3,531 | 40 to 44 years | 11,091 |
| 1 and 2 years | 7,193 | 45 to 49 years | 9,884 |
| 3 and 4 years | 7,083 | 50 to 54 years | 8,674 |
| 5 to 9 years | 16,347 | 55 to 59 years | 7,743 |
| 10 to 14 years | 12,886 | 60 to 64 years | 6,575 |
| 15 to 19 years | 11,055 | 65 to 69 years | 5,259 |
| 20 to 24 years | 10,899 | 70 to 74 years | 3,973 |
| 25 to 29 years | 11,900 | 75 years and over | 4,482 |
| 30 to 34 years | 12,343 | All Ages | 162,414 |
| 35 to 39 years | 11,495 | | |

*Source: Statistical Abstract: 1955*, Table 23, p. 32.

EXAMPLE 266A

Verify all the computations in Table 259.

EXAMPLE 266B

There is a widespread belief in major league baseball that the second year is an unlucky year for new players who have successfully finished their first year in the major leagues. Why do you think this belief exists? What data would you want if you were investigating whether the second year really is less successful?

EXAMPLE 266C

What kind of measure of variability of rainfall would be most relevant for engineers planning a storm sewer system for a city?

EXAMPLE 266D

Prove the following relation algebraically if you can, and in any case demonstrate it by a numerical example.

$$\sum (x - \bar{x})^2 = \sum x^2 - \frac{(\sum x)^2}{n}.$$

EXAMPLE 266E

Prove the following statements algebraically if you can, and in any case demonstrate each by several numerical examples.

(1) Reducing (or increasing) each observation by the same amount does not affect the standard deviation.

(2) Multiplying (or dividing) each observation by the same positive constant multiplies (or divides) the standard deviation by the same constant.

*Do It Yourself*

EXAMPLE 267A

Suppose the mean height of men is 68 inches and the standard deviation is 3 inches, and that the corresponding numbers for women are 64 inches and 2.5 inches. If men and women are equally numerous, what is the mean and standard deviation for men and women? [Hint: The relation given in Example 266D helps.]

EXAMPLE 267B

Does the regression phenomenon arise in Example 209D? Explain.

# Chapter 9

# *Association*

## 9.1
### ASSOCIATION AND CAUSE AND EFFECT

If changes in one variable are accompanied systematically by changes in another, the variables are described as *associated* or *correlated*. We have already given an illustration in Table 196, which shows an association between sex and schooling. It is association in this sense that we are interested in describing.

Association may or may not indicate a cause-and-effect relation. Whether it does is often a matter of interpreting "cause and effect." There is probably no direct physiological cause for males' receiving less primary and secondary education, but more college education, than females. The explanation of the association surely runs in social terms. But social differentiations between the sexes are themselves at least partly related to physiological differences or to social adaptations to physiological differences. Whether sex should be called a cause of the difference in schooling is therefore a question of whether this kind of linkage should be called causal.

Suppose—contrary to fact—that when males and females are classified as urban and rural it should turn out (i) that within urban areas or within rural areas males and females get the same schooling, (ii) that they get more in urban areas than in rural areas, and (iii) that a larger proportion of females than of males live in urban areas. Then the apparent sex difference would be interpreted by most people as "really" an urban-rural difference. But there would still remain the question why sex is related to the urbanism characteristic and whether, granted that it is, sex should be regarded as "causally" linked to schooling.

If we do interpret the sex-schooling relationship as causal, there is no question as to which causes which. For the data of Table 269,

**268**

however, which shows a definite association between income and schooling, the direction of the causation is by no means clear. Low income causes some families to remove their children from school and high income enables other families to keep their children in school, so income may be said to have a causal effect on schooling. Just as clearly, education enables some families to earn high incomes, and lack of it restricts others to low incomes, so schooling may be said to have a causal effect on income.

EXAMPLE 269   INCOME AND SCHOOLING

TABLE 269

MEDIAN INCOME IN 1949 BY YEARS OF SCHOOL COMPLETED,
MALES 25 YEARS AND OVER

Based on a $3\frac{1}{3}$ percent sample of 1950 census returns

| Total[a] | Years of Schooling Completed | | | | | | | |
|---|---|---|---|---|---|---|---|---|
| | None | Elementary | | | High school | | College | |
| | | 1 to 4 | 5 to 7 | 8 | 1 to 3 | 4 | 1 to 3 | 4 or more |
| $2699 | 1108 | 1365 | 2035 | 2533 | 2917 | 3285 | 3522 | 4407 |

*Source: Statistical Abstract: 1955*, Table 126, p. 110. Original source: *U. S. Census of Population: 1950*, Vol. IV, Part 5B.

[a] Includes a small number for whom years of school completed was not reported.

In any case, a real association between variables points the way to investigations to account for it. Each of the possible chains of connection that we have mentioned in discussing Tables 196 and 269 points to further inquiries which will help illuminate the relationships, however the terms "cause" and "effect" are used.

We shall first discuss association as revealed in tables like 196 and 269 or more complicated ones. Tables involve sets of numbers, but the methods and principles are essentially the same as in comparing a pair of numbers, for example median years of schooling for males and females, or median income for high school graduates and for college graduates. Then we shall discuss tables that show simply frequency distributions, rather than means or other descriptive measures of distributions. We shall then discuss a technique, known as "standardized averages," for adjusting for differences in the composition of two groups in respects other than those of direct interest. Finally, we shall present an account of a classroom discussion of a specific problem.

**Association**

We defer until Chap. 17 a group of methods—regression and correlation—which are useful for certain problems when both variables are quantitative. The descriptive aspects of these methods might well be included in this chapter, but this would entail excessive duplication between this chapter and Chap. 17. We therefore postpone them to Chap. 17, where we can discuss both description and analysis.

## 9.2
## HOW TO READ A TABLE

Information can be packed into a table like sardines into a can, and if you cannot read a table, it is as if you had a can of sardines but no key. Ordinary reading ability is no more effective in reading a table than an ordinary can opener in opening a can of sardines, and if you go at it with a hammer and chisel you are likely to mutilate the contents.

We will try to extract information from Table 270 about the association of illiteracy with age, color, and sex. We urge that before you read further you study Table 270 and jot down your own conclusions in the sequence in which you reach them.

EXAMPLE 270   ILLITERACY

TABLE 270

ILLITERACY RATES, BY AGE, COLOR, AND SEX, 1952

Based on a sample of about 25,000. Persons unable both to read and to write in any language were classified as illiterate, except that literacy was assumed for all who had completed 6 or more years of school. Only the civilian, noninstitutional population 14 years of age and over is included.

Percent Illiterate

| Age (years) | White | | | Nonwhite | | | Both Colors | | |
|---|---|---|---|---|---|---|---|---|---|
| | Male | Female | Both | Male | Female | Both | Male | Female | Both |
| 14 to 24 | 1.2 | 0.5 | 0.8 | 7.2 | 1.4 | 3.9 | 1.8 | 0.6 | 1.2 |
| 25 to 34 | 0.8 | 0.6 | 0.7 | 9.7 | 3.8 | 6.4 | 1.6 | 0.9 | 1.2 |
| 35 to 44 | 1.2 | 0.5 | 0.8 | 7.5 | 5.9 | 6.6 | 1.7 | 1.0 | 1.3 |
| 45 to 54 | 2.2 | 1.4 | 1.8 | 12.8 | 10.4 | 11.5 | 3.2 | 2.3 | 2.7 |
| 55 to 64 | 3.6 | 3.4 | 3.5 | 19.4 | 16.9 | 18.1 | 4.7 | 4.4 | 4.5 |
| 65 and over | 5.6 | 4.4 | 5.0 | 35.8 | 31.2 | 33.3 | 7.6 | 6.2 | 6.9 |
| 14 and over | 2.1 | 1.5 | 1.8 | 12.7 | 8.2 | 10.2 | 3.0 | 2.1 | 2.5 |

*Source: Statistical Abstract: 1955*, Table 132, p. 115. Original source: Bureau of the Census, *Current Population Reports*, Series P-20, No. 45.

## 9.2 How to Read a Table

You will not extract any information from the table if you continue to divert your gaze from it in embarrassed bewilderment. Don't stare at it blankly, either—focus your eyes and pick out some detail that is meaningful, then another, then compare them, then look for similar comparisons, and soon you'll know what the table says.

There are at least two good reasons for learning to read tables. The first is that once the reading of tables is mastered (and this does not take long), the reader's time is greatly economized by reversing the usual procedure, that is, by studying the tables carefully and then just skimming the text to see if there is anything there that is not evident in the tables, or not in them at all. This not only saves time but often results in a better understanding: a verbal description of any but the simplest statistical relationship is usually hard to follow, and besides, authors sometimes misrepresent or overlook important facts in their own tables. A second reason for learning to read tables is that users of research can better describe the data needed to answer their administrative or scientific problems if they can specify the types of tables needed, and this requires an understanding of tables. Research workers, in turn, can plan investigations more effectively if they visualize in advance the statistical tables needed to answer the general questions that motivate the research.

Consider, then, Table 270. By following a systematic procedure it is possible to grasp quickly the information presented. Here are the main steps:

(1) *Read the title carefully.* One of the most common mistakes in reading tables is to try to gather from a hit or miss perusal of the body of the table what the table is really about. A good title tells precisely what the table contains. In this case, the title shows that the table tells about illiteracy, in relation to age, color, and sex, in 1952, and that the data are presented as rates—percent illiterate.

(2) *Read the headnote or other explanation carefully.* In the headnote to Table 270 we get a more precise indication of the basis for classifying people as illiterate. We see, in fact, that the rates are slightly too low because it was taken for granted that any person who had completed six or more years of school was literate; but it is reasonable to suppose that the error from this source is negligible. We note also that the mentally deficient, criminals, and others in institutions have been excluded, as have the armed forces, so that the data relate to people in everyday civilian life. Finally, we note that the data are based on a sample, so we make a mental note not to attach too much importance to any single figure, or difference between figures, without first looking up the sampling error.

Information of the kind given in the headnote of Table 270 is often not attached directly to the table, but must be sought elsewhere in the text. Those who prepare reports that include statistical tables should, but frequently do not, keep in mind not only the reader who reads straight through the report without putting it down, but also the user making a quick search for a specific piece of information.

(3) *Notice the source.* Is the original source likely to be reliable? In this case, the answer is definitely "yes," for the Bureau of the Census is one of the most competent statistical agencies in the world. The secondary source, the *Statistical Abstract*, is a model of its kind. But *you* are getting the data from a tertiary source, this book. What about its reliability? Unless you have checked some of our previous data against their sources, you really do not know about that, and even if you did it would be a mistake to put complete reliance on the data without verifying them.[1] Of course *we* assure you of our reliability; but we would not trust your infallibility, or even our own, no matter who gave *us* assurances.

(4) *Look at the footnotes.* Maybe some of them affect the data you will study. Sometimes a footnote applies to every figure in a row, column, or section, but not every figure to which it applies has a footnote symbol. This is the case with Table 269, in a sense. The footnote indicates that some individuals are included in the total but not in any years-of-school class, which implies that some, and probably all, of the classes lack a few observations. The footnote, incidentally, would have been better if "a small number" had been specified more precisely, preferably as a percent of the total number.

(5) *Find out what units are used.* Reading thousands as millions or as units is not uncommon. Long tons can be confused with short tons or metric tons, meters with yards, degrees with radians (as in Example 82A), U. S. with Imperial gallons, nautical with statute miles, rates per 1,000 with rates per 100,000, "4-inch boards" with boards 4 inches wide,[2] fluid ounces with ounces avoirdupois, and so on. In Table 270 illiteracy is expressed in percent—incidence per 100—and age in years.

The foregoing steps are, in a sense, all for preliminary orientation before settling down to our real purpose—as a dog turns around two

---

1. Please let us know of the inaccuracies you find, here or elsewhere in this book.

2. A "4-inch board" is $3\frac{3}{4}$ inches wide, the 4 inches referring to the width of the rough lumber.

A useful compilation of units in common use is *World Weights and Measures: Handbook for Statisticians*, prepared by the Statistical Office of the United Nations in collaboration with the Food and Agriculture Organization of the United Nations (provisional ed.; New York: United Nations, 1955).

or three times before settling down for a nap. They do not take long, and ought to be habitual, but if you omit them you may suffer a rude awakening later—or never awaken at all.

(6) *Look at the over-all average.* The illiteracy rate for all ages, both colors, and both sexes—the whole population, in other words— is shown in the lower right hand corner of Table 270 as 2.5 percent, or one person in 40. This may surprise you, for probably not one in 400 and perhaps not even one in 4,000 of your acquaintances 14 years of age or older is illiterate. On a matter like this, for a country of 165 million people and three million square miles, neither one's own impressions nor the consensus of one's friends' impressions is valid.

(7) *See what variability there is.* It is quickly evident that there are percentages less than 1 and more than 30 in the table. There is, therefore, extraordinary variation in illiteracy among the 24 basic groups into which the population has been divided (two sexes, two colors, six age classes).

(8) *See how the average is associated with each of the main criteria of classification.*

(a) *Age.* Looking in the section for "both colors" and down the column for "both" sexes, we see that the illiteracy rate is essentially constant at about $1\frac{1}{4}$ percent from ages 14 to 44, but then rises sharply through the remainder of the age classes to a rate in the highest age class 5.7 percentage points larger than, and $5\frac{3}{4}$ times as large as, the rate in the lowest age class. (Avoid phrases such as "illiteracy increases with age," which suggest that given individuals change as they age.)

At this point, some competent table-readers, especially if they were particularly interested in the association between age and illiteracy, would pursue this path further. We shall, however, complete our survey of the gross associations with the three variables, then take up each in detail. Probably neither route has any general advantage over the other.

(b) *Sex.* In the "both colors" section, comparison of the entries at the bottoms of the "male" and "female" columns, which apply to all ages, shows that the illiteracy rate for males (3.0 percent) is over 40 percent larger than that for females (2.1 percent). In view of our finding about age, we make a mental note to consider the possibility that this is merely the association with age showing up again in the guise of a sex difference, through the medium of a difference in the age distributions of the sexes. Correspondingly, we make a note to check on the possibility that the apparent association with age is due to differences in the sex ratio at different ages. More generally, we recognize that the associations with age and sex may be *confounded*, that is, mixed together in what looks like an association with age and an association with sex.

The idea of confounding is important enough for a digression. Suppose illiteracy rates by sex and age were:

| Age | Male | Female | Both Sexes |
|-----|------|--------|------------|
| Young | 1.0 | 1.0 | 1.0 |
| Old | 10.0 | 10.0 | 10.0 |

These hypothetical illiteracy rates are identical for young males and young females. They are also identical for old males and old females. But they differ greatly between the young and the old. In other words, there is a strong relation between age and illiteracy, but none at all between sex and illiteracy. Now suppose that the frequencies are as shown below:

| Age | Male | Female |
|-----|------|--------|
| Young | 100 | 300 |
| Old | 200 | 100 |

The over-all illiteracy rate for males would be (see Sec. 7.4.2)

$$\tfrac{1}{3} \times 1.0 + \tfrac{2}{3} \times 10.0 = 7.0;$$

for females it would be

$$\tfrac{3}{4} \times 1.0 + \tfrac{1}{4} \times 10.0 = 3.25.$$

Males show a higher over-all illiteracy rate, simply because relatively more of the males are old and the illiteracy rate is higher for the old of either sex. In such a case, the age and sex effects are said to be *confounded*. That is, what is really an age effect appears in the totals as a sex effect, because the age effect has had a different influence on the two sexes due to their different age distributions.

It is usual in statistics to refer to an association with, say, age, as an "age effect," or as the "effect of age," without intending the cause-and-effect implication that this term tends to carry in ordinary usage. All that is meant in statistics is association, and we will use the term "effect" that way.

(c) *Color.* To see the effect of color, we compare the entries at the bottoms of the "both" sexes columns in the "white" and "nonwhite" sections, and find the nonwhite rate (10.2 percent) to be $5\tfrac{2}{3}$ times the white rate (1.8 percent). Again, however, we resolve to investigate possible confounding of all three effects.

### 9.2 How to Read a Table

The main effects, then, seem to be that *illiteracy rates are higher for older people, for males, and for nonwhites.*

(9) *Examine the consistency of the over-all effects and the interactions among them.*

(a) *Age.* The increase of illiteracy with age holds separately for whites and nonwhites. Some difference in detail does appear. For one thing, the nonwhite rate is not constant from ages 14 to 34, but is noticeably lower from 14 to 24. More conspicuous, the increase from the lowest to the highest age class is much larger for nonwhites than for whites: the differences are 29.4 percent and 4.2 percent, and the ratios[3] 8.5 and 6.2. Thus, it appears that age has a greater effect on illiteracy for nonwhites than for whites. For the two sexes, on the other hand, age has about the same effect, as measured by the absolute change (5.8 percent for males and 5.6 percent for females) from the lowest to the highest age class; since females have a lower rate, this makes the ratio higher for females (10.3) than for males (4.2).

A still more careful study of the table would test whether these conclusions hold if we compare, say, the next-to-lowest age class with the next-to-highest (the conclusions are the same), thus guarding against aberrations in individual rates.

Before we italicize these conclusions derived from comparing the separate section totals, let us see whether they hold within sections, that is for each sex of a color, or for each color of a sex. Here, for the first time, we use the real core of the table, the rates for the 24 basic cells. Heretofore we have used only data combined by age, by sex, or by color, or by two of these, or (in step 6) by all three.

First, compare the males of the two colors. Then compare the females. Both comparisons confirm the conclusion that *the increases in illiteracy associated with increases in age are greater for nonwhites than for whites* and that *they are about the same for males as for females.* These statements are equivalent to saying that *the excess of nonwhite over white illiteracy rates is greater in the older age classes* and that *the difference between the sexes is not systematically related to age.*

(b) *Sex.* Similar detailed study leads to the conclusion that *the excess of the male over the female rate is higher for nonwhites than for whites.* Put the other way around, this says that *the difference between the colors is larger for males than for females.*

---

3. Ratios are not very satisfactory for describing changes in percentages unless the percentages remain small, because of the fixed upper bound of 100. The nonwhite rate of 33.3 percent at 65 years and over, for example, could not be multiplied by 8.5 again. Furthermore, the ratios depend on which percentage is used, that for occurrences or that for nonoccurrences. The literacy rates corresponding with the illiteracy rates mentioned in the text, while they have the same numerical differences as the illiteracy rates, have the ratios 1.44 and 1.04.

(c) *Color.* Our conclusions about the interaction between color and sex and between color and age have already been recorded in discussing age and sex.

(10) *Finally, look for things you weren't looking for—aberrations, anomalies, or irregularities.* The most interesting irregularity that we have noticed in Table 270 is in the age class 25–34. For white males this is below—in fact, one-third below—the rates for the preceding and following age classes. For the nonwhite males, however, the rate is above that of the adjacent age classes by about one-third. (The white females also show a higher rate in this age class than in the adjacent ones, but only by 0.1, which might be almost all due to rounding the figures to the nearest tenth of a percent, and in any case is less than the necessary allowance for sampling error.) In attempting to form a plausible conjecture to explain this peculiarity, we first note that the period when this age class was at ages 6 to 8, and therefore learning to read and write, was 1924 to 1935. This suggests nothing to us, though it might to an expert on the subject matter. As a second stab, we note that during the period of World War II, 1942–45, this age class was 15 to 27 years old. It is, therefore, the group that provided the bulk of the armed forces. This lead seems worth investigating. Did the armed forces teach many illiterates to read and write? If so, did this affect white males more than nonwhite? Even so, why would the rate for nonwhite males be increased? Could it be that mortality among whites was higher for illiterates than for literates, but for nonwhites the reverse? We should be surprised if any of these is the explanation, but investigating them would probably lead us to the explanation. A possible explanation, of course, is that the aberration is due to sampling error, or even clerical or printing error, and that the search for substantive explanations would be in vain. But such anomalies are often worth pursuing; this is one of the secrets of serendipity, from which the most fruitful findings of research often result. We would certainly pursue these questions if we were investigating illiteracy instead of explaining how to read a table.

In summary, then, here is what can be read from Table 270, and in considerably less time than it has taken us to tell about it:

Illiteracy in 1952 among the civilian, noninstitutional population 14 years of age and older—

(i) Averaged 2.5 percent.

(ii) Varied greatly with age, color, and sex.

(iii) Was higher at the higher ages, for nonwhites, and for males, with

(a) the age differences larger for nonwhites—that is the color differences larger at the higher ages;

(b) the sex difference larger for nonwhites—that is, the color differences larger for males;

(c) no interaction between age and sex.

(iv) Was, in the 25–34 year age class, anomalously lower for white males, but higher for colored males, than in the age classes just above and just below.

EXAMPLE 277  BRAINS AND BEAUTY AT BERKELEY

TABLE 277

MEAN GRADES OF COLLEGE WOMEN, BY APPEARANCE AND YEAR IN COLLEGE

Data on 643 women students of the University of California who had completed two or more years of college, classified by beauty of face. Grades averaged by scoring $A$ as 3, $B$ as 2, $C$ as 1, $D$ as 0, $E$ or $F$ as $-1$. Frequencies on which averages are based are shown in Table 280.

| Year | Homely | Plain | Good Looking | Beautiful | All Appearances |
|------|--------|-------|--------------|-----------|-----------------|
| Junior | 1.58 | 1.45 | 1.34 | 1.16 | 1.37 |
| Senior | 1.56 | 1.52 | 1.45 | 1.57 | 1.50 |
| Graduate | 1.67 | 1.70 | 1.70 | 1.53 | 1.68 |
| All Years | 1.62 | 1.56 | 1.44 | 1.42 | 1.51 |

Source: S. J. Holmes and C. E. Hatch, "Personal Appearance as Related to Scholastic Records and Marriage Selection in College Women," *Human Biology*, Vol. 10 (1938), pp. 65–76. The means shown here have been recomputed from the original data, loaned by the authors, and in a few instances differ by one unit in the last decimal place from those given in the source.

Repeating the same steps as in reading Table 270, we find at stage 8 that grades are higher in later years in college and with poorer appearance (which, to repeat earlier warnings, does not necessarily mean that given coeds get better grades as they progress in college or regress in appearance). At stage 9, however, we find it necessary to introduce such strong qualifications to the appearance effect as almost to withdraw the finding. All we can say is that for juniors grades decrease with better appearance, but for seniors and graduate students there is no systematic relation. The main effect of appearance is partly a manifestation of the year-in-college effect, in conjunction with different distributions by appearance for the three college classes.

The mean for the plain, for example, is

$$\tfrac{68}{250} \times 1.45 + \tfrac{100}{250} \times 1.52 + \tfrac{82}{250} \times 1.70 = 1.56$$

and for the good looking it is

$$\tfrac{108}{236} \times 1.34 + \tfrac{84}{236} \times 1.45 + \tfrac{44}{236} \times 1.70 = 1.45$$

where the weights are from Table 280 and the method of calculation is from Sec. 7.4.2.[4] The difference between these two means is partly due to the fact that the juniors, who have the lowest grades in both appearance groups, constitute 46 percent of the good looking and only 27 percent of the plain. Similarly, the graduates, who have the highest scores in both appearance groups, constitute 33 percent of the plain but only 19 percent of the good looking. Thus, the difference between these two appearance groups is partly due to the fact that the class effect operates differently in one than the other. The difference between the averages for the plain and good looking is not wholly due to the class effect, however, for among the plain the average for each class is as high as or higher than the average for the same class among the good looking.

Since the appearance effect is not present for the seniors or graduates, we conclude that its presence for all classes combined reflects partly the effect for the juniors and partly confounding of the class effect—that is, heavier representation in some appearance groups than in others of those classes which receive low grades. It would be possible for the appearance effect to work in one direction in all three classes, but in the opposite direction for all classes combined. For the data of Table 277 this is only barely possible, since no set of weights will result in a mean outside the range of the individual means. For the beautiful mean to exceed the homely mean, for example, virtually all of the beautiful and all of the homely would have to be seniors.

In interpreting data of this kind it is necessary to keep in mind selective factors that have determined whether individuals are available for such a sample. This is discussed in Sec. 9.4. The possibilities in connection with Table 277 are varied.

EXAMPLE 278   CAR PURCHASE PLANS

In reading Table 279 there are two things to be kept in mind besides things of the kinds already mentioned. (i) The income classes are so broad that even within a class average income may, and undoubtedly does, increase substantially with education (see Table 269); hence the education effect is not completely separated from the income effect. (ii) Of those with, say, under $2,000 income, some were in this income class in 1948 only because of temporarily low incomes.

---

4. The discrepancy between the result just obtained, 1.45, and the corresponding number in Table 277, 1.44, is due to the fact that the row and column averages in Table 277 were computed from rates more accurate than those shown in the body of the table.

*9.3 Association in Frequency Data*

TABLE 279

PERCENT OF CONSUMER UNITS PLANNING IN 1948 TO PURCHASE A NEW CAR IN 1949
AND PERCENT PURCHASING ONE, BY EDUCATION AND 1948 INCOME

For definition of consumer unit, see footnote to Fig. 257. Data are for nonfarm units and are from a sample survey covering about 2,500 farm and nonfarm units, conducted by the Federal Reserve System in cooperation with the University of Michigan Survey Research Center.

| Education of Head of Unit | 1948 Income | | | | | |
| --- | --- | --- | --- | --- | --- | --- |
| | Under $2,000 | | $2,000 to $4,999 | | $5,000 and Over | |
| | Planned | Purchased | Planned | Purchased | Planned | Purchased |
| Grammar School | 1 | 1 | 4 | 5 | 20 | 16 |
| High School | 2 | 4 | 7 | 8 | 21 | 24 |
| College | 4 | 8 | 15 | 11 | 26 | 27 |

*Source:* Irving Schweiger, "The Contribution of Consumer Anticipations in Forecasting Consumer Demand," in National Bureau of Economic Research, *Short-Term Economic Forecasting*, Studies in Income and Wealth, Vol. 17 (Princeton: Princeton University Press, 1955), p. 461.

Such consumers probably base purchase plans, and also purchases, to a considerable degree on expected or "normal" income. Hence if "normal" income could have been used in place of actual income in 1948, the variation among income classes (and, correspondingly, among education groups within broad income classes) would have been more than is shown in the table. This is another manifestation of the principle underlying the "regression" phenomenon (see Sec. 8.7).

# 9.3
# ASSOCIATION IN FREQUENCY DATA

Table 280 shows a cross-classification by appearance and year in college of the 643 women students whose average grades are shown in Table 277. We have already seen that there is association in these frequencies, in the sense that the percentage distribution by class varies with appearance, or the percentage distribution by appearance varies with class. Sometimes it is useful to compute a single descriptive statistic to measure the amount of association. This would, for example, facilitate comparing the degree of association shown in Table 277 with that for similar data on men, on University of California women at another time, or on women at other universities.

Association

TABLE 280

COLLEGE WOMEN, BY APPEARANCE AND YEAR IN COLLEGE

See headnote to Table 277.

| Year | Homely | Plain | Good Looking | Beautiful | All Appearances |
|------|--------|-------|--------------|-----------|-----------------|
| Junior | 17 | 68 | 108 | 25 | 218 |
| Senior | 27 | 100 | 84 | 33 | 244 |
| Graduate | 39 | 82 | 44 | 16 | 181 |
| All Years | 83 | 250 | 236 | 74 | 643 |

*Source:* Same as Table 277.

The proper measurement of association, as of other properties of statistical data, depends on the use to which the measure is to be put. We shall present three measures appropriate to different uses.

### 9.3.1 One Variable Prior

Suppose one of the variables is to be regarded as prior to the other, either in the sense that it is cause and the other effect, or in the sense that it will be used as a basis for estimating the other. We would regard year as the prior variable if, say, we had data showing year but had no way of ascertaining appearance except insofar as it can be inferred from year. The purpose of a measure of association, under these circumstances, is to indicate how well appearance can be inferred from year in college.

As a measure of association in such cases we use a number indicating how much better we can guess appearance if we take account of year than if we do not take account of year. More specifically, suppose one of the 643 women is selected at random and her appearance guessed as best we can from the data of Table 280 without knowing her year. Our best guess would be the modal appearance, plain. On the average we make a certain proportion of errors in classifying the women this way. In fact, on the average we would be wrong 61.1 percent of the time, which is the proportion of the women in Table 280 who are not plain: $1 - \frac{250}{643} = \frac{393}{643} = 0.611$.

How many of these errors can we eliminate if we take account of year? On the average, 33.9 percent (218/643) of the time we will be told that the woman is a junior. We will then guess "good looking," the modal class for juniors. We will be wrong for 50.5 percent of the juniors, that is 17.1 percent (33.9 times 0.505) of all cases. Seniors constitute 37.9 percent of the women and we err, by guessing them all as plain, with 59.0 percent of them, equal to 22.4 percent (37.9 times 0.590) of all cases. Likewise, errors made for graduates constitute 15.4 percent of all our guesses (28.1 times 0.547). Altogether, when we take account of year we err in our guess about appearance 54.9 percent of the time (17.1 + 22.4 + 15.4), on the average. Thus, the proportion of errors eliminated by taking account of year is

$$g_{a.y} = \frac{61.1 - 54.9}{61.1} = 0.10,$$

where the subscript $a.y$ indicates "appearance, given year," indicating that we are measuring the improvement in our ability to predict appearance that results from taking year into account. We shall call $g_{a.y}$ the *index of association of a with y*.

The interpretation of this index is, of course, direct: ten percent of the errors about appearance are eliminated if we take year into account. If there were perfect association, $g_{a.y}$ would be 1; if there were none, it would be 0.

In general, to compute this measure of association when the classification by rows is prior—that is, used to estimate the classification by columns—proceed as follows:

(1) Add the maximum frequencies from the individual rows (here 108 + 100 + 82 = 290).

(2) Subtract from this sum the maximum frequency in the total row (here 250, leaving 40).

(3) Find the number of cases in the total row, excluding the maximum frequency (here 643 − 250 = 393).

(4) Divide the result of step (2) by that of step (3) (here 40 ÷ 393 = 0.10).

This is the measure sought, say $g_{c.r}$ where $c$ represents the variable by which the columns are classified and $r$ the variable by which the rows are classified, and $r$ is the prior variable. In line with our previous terminology we shall call $g_{c.r}$ the *index of association of column with row*.

If, in Table 280, we regard appearance as prior and measure its ability to improve our guesses about class, the same formula could be made to serve by turning the table sideways, calling rows columns and columns rows. From Table 280 we find

$$g_{r.c} = \frac{39 + 100 + 108 + 33 - 244}{643 - 244} = \frac{36}{399} = 0.09.$$

Again, the association is small.

### 9.3.2 Neither Variable Prior

If neither variable is prior, we may compute a measure which is in principle like the previous two but proceeds as though half the time we will know the row classification and have to guess the column classification, and half the time know the column and have to guess the row. Of the errors that we would make if we knew neither row nor column, but half the time guessed one and half the time the other, the proportion that will be eliminated by knowing row or column, is simply the sum of the numerators of $g_{c.r}$ and $g_{r.c}$, divided by the sum of their denominators. We shall call $g$ the *index of mutual association*.

For the data of Table 280 we find

$$g = \frac{(108+100+82)+(39+100+108+33)-250-244}{(2 \times 643)-250-244} = \frac{76}{792} = 0.10.$$

Had Table 280 been presented as relative frequencies totaling 100 percent for the entire table (as in Table 197, Case III), the relative frequencies would have been treated just as we have treated absolute frequencies.

### 9.3.3 Measures Based on Predictability of Order

A limitation of the measures presented so far is that they treat a prediction as successful only if it scores a bull's-eye, and do not differentiate between near misses, such as predicting beautiful when good looking is correct, and far misses, such as predicting beautiful when homely is correct. The advantage offsetting this is that the $g$ measures can be used when one or both variables are qualitative, as well as when one or both are quantitative.

When both variables are quantitative, at least in the sense of being ranked, a good measure of association is the *index of order association*, denoted by $h$, which tells how much more likely it is that two individuals chosen at random from those in the table will have both variables in the same order, than that they will have them in opposite orders. If we select at random two of the 643 women of Table 280, how much more likely is it that the one with the later year in school will have the better appearance than that she will have the worse appearance? We ignore pairs in which both have the same year or same appearance.

#### 9.3 Association in Frequency Data

The answer to this question is −0.27. That is, the proportion of pairs in which the student of the later year has the better appearance is less by 0.27 than the proportion of pairs in which the student of the later year has the worse appearance. To compute this, we proceed in the following steps:

(1) Multiply each cell frequency in the table by the sum of all the frequencies that are *both* below *and* to the right of it. For example, the frequency 17 is multiplied by $100 + 84 + 33 + 82 + 44 + 16 = 359$, and the 84 is multiplied by 16. None of the frequencies in the last row or last column is used as the first factor of a product, since there are no figures *both* below *and* to the right of them by which to multiply them.

(2) Add these products. Call the sum $S$ (for "same"). For Table 280, $S = 34,609$.

(3) Multiply each cell frequency by the sum of all the frequencies that are *both* below *and* to the left of it. For example, the 84 is multiplied by $39 + 82 = 121$. None of the frequencies in the last row or first column is used as the first factor of a product.

(4) Add the products computed in step (3). Call the sum $D$ (for "different"). For Table 280, $D = 60,181$.

(5) Compute

$$h = \frac{S - D}{S + D}.$$

For our example,

$$h = \frac{34,609 - 60,181}{34,609 + 60,181} = \frac{-25,572}{94,790} = -0.27.$$

(6) As a check, instead of repeating the same calculations,

(a) Add the squares of all the row sums and column sums (but not the table sum) and subtract the squares of all the cell entries. Call this $T$ (for "tied"). From Table 280, $T = 223,869$.

(b) If the calculations are correct,

$$2(S + D) + T = n^2$$

where $n$ is the total frequency for the table. For our example,

$$2(94,790) + 223,869 = 413,449,$$

$$(643)^2 = 413,449,$$

which gives us confidence in our calculations.

Values of $h$ range from $-1$, which occurs when increases in order for one variable are always associated with decreases for the other (ignoring cases where one variable or the other is the same in both observations) to $+1$, which occurs when differences in order are the same for both variables. Also, $h$ is 0 when pairs chosen at random are exactly as likely to have both variables in the same order as to have them in reverse order.

While $h$ measures directly the *difference* between the proportion of pairs in which the two variables are in the same order and the proportion in which they are in reverse order, some prefer to interpret it in terms of a *ratio*. In our example, for instance, pairs in which the two variables are in the same order occur about $\frac{4}{7}$ times as often as pairs in which the two variables are in the opposite order. To compute this, we divide $1 + h$ by $1 - h$ or, alternatively, divide $S$ by $D$.

It must be kept in mind that the bulk of the pairs may have the same value for one or both variables; the proportion of such "ties" can be computed by dividing $T$ by $n^2$; in our case, 54.1 percent of the pairs would be ties.

### 9.3.4 Which Measure to Use?

Of the measures of association presented here, probably $h$ is most often the best if the variables are quantitative. When one variable is qualitative, $g$ is probably best. In cases where it is known that one variable will be used to predict the other, however, $g_{r.c}$ or $g_{c.r}$ should be used. Additional measures, suitable when both variables are quantitative, are presented in Chap. 17.

### 9.4 INTERPRETING ASSOCIATION

In Sec. 9.1 we discussed the difficulties of interpreting association as causation. In this section we are concerned with interpreting the nature of an association itself, whether it is genuine or spurious. We can illustrate the problem by a simple example.

EXAMPLE 284    EFFECT OF TRAVEL ON POLIO

*Travel Held Raising Polio Fatality Rate.*—A study indicating that transportation over long distances greatly increases the death rate among victims of infantile paralysis in the acute stage is described in the new issue of the *Journal of the American Medical Association.*

The results were reported by Dr. M. Bernard Brahdy of Mount Vernon, New York, and Dr. Selig H. Katz of New York, both of whom are associated with the Willard Parker Hospital in Manhattan.

The report is based on a study of the records of 493 polio victims admitted to the hospital. Of these, 380 were local patients who travelled an average of seven miles to the hospital. The other 113 were transported an average of eighty-five miles. The study covered the polio epidemic in the summer and fall of 1949.

There were twenty deaths among the 380 local patients, a fatality rate of 5.2 percent. In contrast, there were eighteen deaths among the 113 transported patients, a death rate of 16 percent.

Whereas only one-fifth of the deaths among the local group occurred within twenty-four hours after admission to the hospital, fully one-half of the fatalities among the transported group occurred in the same period. Average duration of illness before admission was about the same in both groups—three and a fraction days.

"It seems," the report said, "that the greater mortality in the transported group, occurring shortly after admission to the hospital, is a manifestation of the effect of long transportation during the acute stage of illness."[5]

Obviously another variable, seriousness of the initial attack, ought to be taken into account. It seems plausible that the local polio victims are more likely to be hospitalized at Willard Parker Hospital than are victims of polio who live farther away; the latter are likely to be brought to Willard Parker Hospital (which, until it closed, was a famous center for treating contagious diseases) only if their condition is unusually serious. Hence the local victims may tend to have milder cases on the average than the patients brought from a distance, and therefore to have a lower fatality rate, independent of any effect of transportation. What is needed to meet this criticism is data that can be organized into a table of the following form:

TABLE 285

DEATH RATES FROM POLIO

| Severity of Attack | Local Patients | Distant Patients | Total |
|---|---|---|---|
| Mild | $a$ | $b$ | |
| Severe | $c$ | $d$ | |
| Total (Number of cases) | | | |

Then $a$ would be compared with $b$, and $c$ with $d$, to get a more valid measure of the effect of travel. Actually, it would not be easy to organize the data in this form unless the original diagnosis were available (or perhaps the diagnosis just before the patient was transported to the hospital), and unless it was possible to make valid assessments of the seriousness of the attack by the initial diagnosis.

5. *New York Times*, June 30, 1951, p. 31. Both percentages cited in the article are slightly in error if the absolute numbers given there are correct.

Association

Even if the tabulation suggested above were made, there would still be reservations about the nature of the association. Perhaps even for patients whose objective condition appears identical, the physician recognizes the seriousness of some cases by intuitive skill acquired by experience. Or perhaps the basic health of the distant patients before the attack was poorer on the average than the health of those who live nearby. We could easily continue in this way, pointing out other variables that might render non-comparable two groups that at first seemed comparable. If these other variables can be determined, the process of further and further subdivision of the original data can continue. With each successive subdivision according to another variable, another element of non-comparability is removed or reduced. But no matter how far it is possible to go with subdivisions, it is never safe to assume that all the disturbing variables that affect the comparison have been provided for. The following strikingly illustrates the fact that similarity in respect to some characteristics does not guarantee similarity in respect to others:

In 1926 or 1927 two Italian statisticians, Gini and Galvani . . . had to deal with the data of a general census. The data were worked out, a new census was approaching, and the room had to be cleared for the new data. The old data were to be destroyed, but the statistical office wanted to keep a representative sample so as to have material for future studies, as yet unanticipated. Gini and Galvani were responsible for the method of obtaining a sample which would represent the situation in the whole of Italy. What they did is a good example of how not to sample human populations.

The two authors carefully considered the problem . . . and decided to apply the method of purposive selection. The whole of Italy was divided into 214 administrative districts called *circondari* and out of these 29 *circondari* were selected to form the sample. Some of the *circondari* are large districts with more than a million inhabitants. . . .

Various averages for each *circondario* had been calculated previously. Gini and Galvani selected 12 characters of the *circondari* to serve as controls and subdivided these into essential and secondary controls. They tried to select the 29 *circondari* so that the means of the essential controls calculated from the sample would be practically identical with those for the whole population. They also tried to reach a reasonable agreement between the population and the sample means of the secondary controls. If you will look at the figures, you will find that the agreement of the mean of each control in the sample with the mean of the same control in the population is very good.

From the paper by Gini and Galvani, it is uncertain whether or not the old Italian census data were destroyed and the sample was left for future reference. However, the two authors decided to check the goodness of the sample by comparing its various characteristics with those known for the whole population of Italy. . . . Gini and Galvani found that the distributions of various characteristics of the individuals, the correlations, and, in fact, all statistics other than the average values of the controls showed a violent contrast between the sample and the whole population. . . .[6]

We will develop this fundamental point further by another example.

EXAMPLE 287  SMOKING AND CANCER

In 1954, widespread publicity was given to the preliminary report of an extensive statistical study by the American Cancer Society of the relation of smoking, especially cigarette smoking, to lung cancer.[7] The study had been made by classifying 187,766 men 50 to 70 years of age according to their smoking habits, and then 20 months later determining which men had died and for what causes. The lung-cancer death rate was about nine times as high for men smoking one pack or more of cigarettes per day as for non-smokers. A number of earlier but smaller and less publicized studies in several countries showed similar results. The American Cancer Society is still continuing to follow the original group.

A medical statistician reviewing these results noted that the percent of smokers in the sample was less than that in the comparable portion of the U. S. population, that the death rates from all causes for the sample were about 30 percent below the corresponding U. S. rates (even the heavy smokers in the sample showed a lower cancer death rate than the corresponding U. S. rate), and that for the sample, death rates not only from cancer but from all causes were higher for cigarette smokers than for nonsmokers.[8] All these facts indicated to him the presence of substantial selectivity in the sample, reminded him of previous incidents in medical statistics that had proved fiascos, and led him to suggest that a simple selective process like the following *could* have produced a spurious association.

Suppose, first, that the population consists of two groups: Group I, constituting 3 percent of the population, which is on the verge of death and has a death rate during the period of the study of 99 percent; and Group II, constituting 97 percent of the population, which has a death rate of 0.03 percent. Assume that 80 percent of each group are smokers, and that the death rates are the same for smokers and nonsmokers within each group. Thus, the death rates would be (in percents):

6. Jerzy Neyman, *Lectures and Conferences on Mathematical Statistics and Probability* (2d ed.; Washington; Graduate School of the Department of Agriculture, 1952), pp. 105–106.

7. E. Cuyler Hammond and Daniel Horn, "The Relationship between Human Smoking Habits and Death Rates," *Journal of the American Medical Association*, Vol. 155 (1954), pp. 1316–1328.

8. Joseph Berkson, "The Statistical Study of Association between Smoking and Lung Cancer," *Proceedings of the Staff Meetings of the Mayo Clinic*, Vol. 30 (1955), p. 319.

|  | Group I | Group II |
|---|---|---|
| Nonsmokers | 99 | 0.03 |
| Smokers | 99 | 0.03 |

The relative frequencies in the population are:

|  | Group I | Group II |
|---|---|---|
| Nonsmokers | 0.006 | 0.194 |
| Smokers | 0.024 | 0.776 |

Using again the formula for weighted means (Sec. 7.4.2), we obtain the following death rates:

Nonsmokers:

$$\frac{0.006 \times 99 + 0.194 \times 0.03}{0.006 + 0.194} = 0.03 \times 99 + 0.97 \times 0.03$$
$$= 3.00 \text{ percent.}$$

Smokers:

$$\frac{0.024 \times 99 + 0.776 \times 0.03}{0.024 + 0.776} = 0.03 \times 99 + 0.97 \times 0.03$$
$$= 3.00 \text{ percent.}$$

Now introduce the assumption that the proportion of smokers cooperating in the study differs between Groups I and II. Suppose that half the individuals in Group I are included in the sample, the omissions being those whose poor health is obvious or prevents their being interviewed, whether they smoke or not. For Group II, assume that 99 percent of nonsmokers and 65 percent of smokers cooperate. Such differential selectivity would result in a death rate in the sample that is 50 percent higher for smokers than for nonsmokers, being 2.33 percent for smokers and 1.55 percent for nonsmokers. This last step is contained in the bottom half of Table 289, which summarizes the entire analysis.

The crucial assumption, which accounts for the results, is that there is differential selectivity between smokers and nonsmokers, and that *the differential is greater for the healthy than the unhealthy.* In such a study many of the very unhealthy are simply not available for interview, and are therefore eliminated on grounds of ill-health, regardless of smoking habits; there is no harm in that alone. The healthy differ in their availability and cooperativeness and it is not at all implausible that more smokers than nonsmokers will refuse to cooperate in a study of the ill-effects of smoking; again, there is no harm in that alone. It is the combined effect that is damaging. In any event. as Berkson says.

### 9.4 Interpreting Association

... The fact that the exact mechanism of such selective association is not readily visualized is not an adequate reason for considering the suggestion of its possible existence to be—as it has been characterized—"far fetched." ... Nor is it conclusive that the considerable number of statistical studies that have been published all agree in showing an association between smoking and cancer of the lung. On the contrary, ... if correlation is produced by some elements of the statistical procedure itself, it is almost inevitable that the correlation will appear whenever the statistical procedure is used.[9]

TABLE 289

HYPOTHETICAL ILLUSTRATION OF POSSIBLE EFFECTS OF SELECTIVITY
IN PRODUCING SPURIOUS ASSOCIATION

| Group | Smoker | Number in Group | | | Number of Deaths | | | |
|---|---|---|---|---|---|---|---|---|
| | | Group I | Group II | Both groups | Group I | Group II | Both groups | Rate (percent) |
| Popula- | No | 600 | 19,400 | 20,000 | 594 | 6 | 600 | 3.0 |
| tion | Yes | 2,400 | 77,600 | 80,000 | 2,376 | 24 | 2,400 | 3.0 |
| | Total | 3,000 | 97,000 | 100,000 | 2,970 | 30 | 3,000 | 3.0 |
| | No | 300 | 19,206 | 19,506 | 297 | 6 | 303 | 1.55 |
| Sample | Yes | 1,200 | 50,440 | 51,640 | 1,188 | 16 | 1,204 | 2.33 |
| | Total | 1,500 | 69,646 | 71,146 | 1,485 | 22 | 1,507 | 2.12 |

*Source:* Berkson, *op. cit.*, Appendix Table 1.

In bringing up this example, it is not our intention to express or imply an opinion as to whether the association between lung cancer and smoking is spurious. Rather, our intention is to point out the issues involved, for they apply to virtually all associations based on analyses of data from experience rather than experiment.

The example illustrates also the danger of assuming that substantial nonresponse or non-cooperation will have no serious effect on statistical studies. If you try out various assumptions about percentages of differential selectivity on the basic data of Table 289, you will find that the observed association between smoking and lung cancer will be considerably different for different assumed percentages. The rates of selectivity in this example reflect both the original sampling process and the success in obtaining cooperation from those designated by the sampling process. The problem of failure to get information from those designated by the sampling process is called the problem of nonresponse or non-cooperation. The problem of nonresponse may be a serious one even if the original sampling process is completely sound. For example, in the Salk polio vaccine tests of

9. Berkson, *op. cit.*, p. 332. Footnotes omitted.

1954 there is some evidence that children whose parents withheld permission to participate in the trials may have been less susceptible to polio than children who did participate.[10]

Unfortunately, a frequent tendency in practice is to ride rough-shod over the nonresponse problem and pretend either that it does not exist or at least that it could not possibly be serious. The most fruitful way of attacking the problem, once it is recognized, is to spend relatively more resources in getting information on the potential non-respondents by repeated calls, special interviewers, etc. This means that the total number of interviews will be smaller than otherwise, but that the data obtained will be better.

## 9.5
## STANDARDIZED AVERAGES

A method of allowing, or "adjusting," for differences in the composition of groups which are to be compared is to compute what the means would be if the groups had the same standard composition.

In the method of standardized means, the means for the subgroups are combined in the group mean not on the basis of different weights for each group but on the basis of standardized weights. As an illustration, let us turn to Table 277 and for each appearance group combine the three averages of individual years on the basis of equal weights. We have for the beautiful (see Sec. 7.4.2 on computing weighted means)

$$\tfrac{1}{3} \times 1.16 + \tfrac{1}{3} \times 1.57 + \tfrac{1}{3} \times 1.53 = 1.42.$$

Proceeding in the same way for the other appearance groups we find the results shown in the "weights equal" column of Table 291A. This table also shows the results of using as the standard weights the frequencies in the last column of Table 280. With these weights, the standardized mean for the beautiful, for example, is

$$0.339 \times 1.16 + 0.379 \times 1.57 + 0.281 \times 1.53 = 1.42.$$

In this example, unlike some others, there is little difference between the two sets of standardized means, or between the standardized and the unadjusted means. All indicate that after adjusting for the effect of year in college, mean grades are consistently lower when appearance is better. Such standardized means have an advantage

---

10. K. A. Brownlee, "Statistics of the 1954 Polio Vaccine Trials," *Journal of the American Statistical Association*, Vol. 50 (1955), pp. 1005–1013.

of compactness over Table 277, but they do not reveal such facts as that the association is primarily among the juniors.

TABLE 291A

STANDARDIZED MEAN GRADES OF COLLEGE WOMEN,
BY APPEARANCE

| Appearance | Weights | |
|---|---|---|
| | Equal | From all appearances combined |
| Homely | 1.60 | 1.60 |
| Plain | 1.56 | 1.55 |
| Good Looking | 1.50 | 1.48 |
| Beautiful | 1.42 | 1.42 |
| All Appearances[a] | 1.53 | 1.52 |

*Source:* Computed from data of Table 277. Weights based on all appearances combined taken from Table 280.

[a] Appearance groups weighted by frequencies for all years in Table 280.

Another example will illustrate the potential effect of the weights used for standardizing.

EXAMPLE 291   DEFECTIVE OUTPUT OF TWO PLANTS

The data of Table 291B are hypothetical but might represent the number of defective items produced by two plants making the same product.

TABLE 291B

PERCENT OF DEFECTIVE ITEMS PRODUCED
AT TWO PLANTS, BY LOT SIZE
(Hypothetical)

| Plant | Lot Size (units) | Lots Produced (number) | Items Produced (units) | Defectives | |
|---|---|---|---|---|---|
| | | | | Units | Percent |
| A | 100 | 27 | 2,700 | 81 | 3 |
| | 300 | 25 | 7,500 | 375 | 5 |
| | 500 | 6 | 3,000 | 240 | 8 |
| | 700 | 4 | 2,800 | 280 | 10 |
| | 900 | 1 | 900 | 135 | 15 |
| | Total | 63 | 16,900 | 1,111 | 6.6 |
| B | 100 | 1 | 100 | 15 | 15 |
| | 300 | 3 | 900 | 99 | 11 |
| | 500 | 7 | 3,500 | 315 | 9 |
| | 700 | 15 | 10,500 | 420 | 4 |
| | 900 | 20 | 18,000 | 360 | 2 |
| | Total | 46 | 33,000 | 1,209 | 3.7 |

*Association*

While plant *A* has a higher over-all percent defective, 6.6 as compared with 3.7 for plant *B*, examination of the table shows that the rates vary greatly with the size of lot, and that production in the two plants has been distributed very differently by size of lot. To adjust for this, we might take plant *A*'s distribution of items by lot sizes as standard weights. Another possibility is to take plant *B*'s distribution as standard. The two sets of standardized means are shown in Table 292.

TABLE 292

STANDARDIZED MEAN PERCENTAGES DEFECTIVE, PLANTS *A* AND *B*

| Standardized Mean | Plant *A* | Plant *B* |
|---|---|---|
| *A*'s Weights | 6.6 | 9.6 |
| *B*'s Weights | 12.4 | 3.7 |

*Source:* Computed from Table 291B.

Thus, which plant has the lower rate of defectives depends on which weights are used for standardizing. The facts are such that no single mean can describe them adequately, namely that plant *A* does better with small lots and plant *B* does better with large lots.

While the data of this example are contrived to produce an exaggerated effect, in practical cases the magnitude of the difference may depend very much on the weights used, even if its direction is not actually reversed. If, for example, the average price of consumer goods is computed for the United States and for Switzerland, weighting various commodities according to the quantities bought in the United States, the price level may appear higher in Switzerland; yet it may appear lower there if the weights correspond with Swiss consumption. In fact, one of the most important uses of standardized means in public affairs is in just such problems as comparing average prices or production, or average changes in prices or production, at different times or places, and the choice of the standardizing weights is critical to the results.

## 9.6
## AN EXTENDED EXAMPLE

Some of the problems and pitfalls of interpretation that we have already mentioned may be highlighted by the following summary of a classroom discussion of a statistical investigation of the incomes of Chicago lawyers.[11] We shall present this discussion essentially as it

11. Leonard R. Kent, *Economic Status of the Legal Profession in Chicago* (unpublished doctoral dissertation, University of Chicago, 1950). For a summary, see Kent, "Economic

*9.6 An Extended Example*

was reported by an observer who was present for the purpose. The paragraphs marked "Comment" are summaries of remarks by students; the rest summarizes what the instructor said.

## 9.6.1 Lawyers' Income and Military Service

The following table shows the differences in the average incomes of veterans of World War II and nonveterans in a sample of 812 Chicago lawyers:

EXAMPLE 293   LAWYERS' INCOME

TABLE 293

INCOME FROM LEGAL PRACTICE, BY VETERAN STATUS,
CHICAGO LAWYERS, 1947

| Classification | Mean Income | Median Income |
|---|---|---|
| Veterans | $ 7,208 | $5,684 |
| Nonveterans | 10,307 | 8,091 |

*Source:* Kent, unpublished material.

The mean income is the average of the lawyers' earnings obtained by totaling their incomes and dividing by the number in the group. It is not the income of an "average" lawyer. The mean is pulled upward by relatively few very high incomes. If you are interested in the "typical" lawyer's income, the median is a better measure. Half of the incomes exceed the median, half fall short of it. The median is not affected by how high the upper half of incomes are, and is therefore lower than the mean. Neither measure is the "best" or the "right" one in any general or absolute sense. Their usefulness simply depends upon the kind of information wanted or the kind of question asked. What does Table 293 show?

*Comment:* Obviously, it shows that the nonveterans had higher incomes than the veterans.

But can we conclude from this fact that having been in the services was a detrimental factor with respect to 1947 income? Could the data be used to argue that veterans should receive a bonus to compensate them for impairment of earning capacity, and to estimate the proper amount of such a bonus?

*Comment:* The two groups should be classified according to age and experience, which affect income very much.

Status of the Legal Profession in Chicago," *Illinois Law Review*, Vol. 45 (1950), pp. 311–332. Some of the data reported here do not appear in the final dissertation, but are taken from preliminary drafts.

This is an important suggestion. The two groups may differ in regard to several factors that affect income, so some or all the difference between them may be due to factors other than the one under consideration. Since it is reasonable to think that income is related to experience (other data in the same study show this clearly) and that veterans are on the whole younger than nonveterans, this might account for the difference in income between veterans and nonveterans. To isolate the effects of age and experience, a two-way table similar to Table 294 was made up. Within each cell of this table were

TABLE 294

INCOME OF CHICAGO LAWYERS, 1947

| Experience Group (years) | Age Groups (years) | | | |
|---|---|---|---|---|
| | Under 30 | 30–34 | 35–39 | . . . . |
| 0.0 to 1.5 | | | | |
| 1.5 to 2.5 | | | | |
| 2.5 to 4.5 | | Veterans Nonveterans | | |
| 4.5 to 6.5 | | | | |
| . . . | | | | |

recorded two averages, one for veterans and one for nonveterans. Now a comparison of veterans and nonveterans within one of the cells of this table eliminates the effects that might be associated with age and experience.

*Comment:* Shouldn't we take into account the fact that the veterans' years of experience were not all consecutive, but were interrupted? For a certain amount of experience, it may make a difference whether it is interrupted.

If veterans earn less than nonveterans of the same age and experience, this may be the reason. But we would not want to eliminate this cause, because the disconnected nature of a veteran's experience really is due to his service. If being in the service had an effect, that may be the way it worked.

This kind of a comparison gives a slight, but consistent, edge in favor of the nonveterans. The nonveterans had slightly higher incomes in every cell, suggesting that differences in age and experience will not explain all of the difference shown in Table 293. Thus, a small superiority for nonveterans exists after age and experience are taken into consideration. Can this difference be accepted as the effect of military service?

*Comment:* Those age groups are pretty broad. You say veterans are younger than nonveterans, so couldn't the 30 to 35 year old veterans average enough younger than the 30 to 35 year nonveterans to make the difference? The groups ought to be broken down finer.

*Comment:* Perhaps the two groups (veterans and nonveterans) should be broken down further, say into (a) independent practitioners, and (b) employees or members of firms. The latter may have stepped back into their prewar jobs while the independents may have had to start from scratch.

Both good points and the data are available for such classifications. Don't forget, though, that with too much cross-classification there will not be enough cases for comparing averages. Income has wide variability even when everything in sight is held constant, so differences between averages of small samples are not reliable.

*Comment:* You said the differences came out all in the same direction. Even if one alone isn't reliable, wouldn't 15 or 20 practically all in the same direction be convincing?

Yes—if they were all in the same direction. A sound point.

Suppose the suggested subdivisions were made, and a slight but consistent superiority for nonveterans still persisted.

*Comment:* There may have been more incentive for the lawyers with low incomes to go into the service. Furthermore, lawyers with large families were more likely to be deferred, and it is possible that these same lawyers have higher incomes than the rest. In other words, it seems likely that the veterans may have had lower incomes before the war, and the difference observed now may be from lower income-earning ability, rather than the effects of being in the service.

*Comment:* It seems that a valid comparison cannot be made without a before-and-after study. Why weren't the 1940 incomes studied as well as the 1947 incomes?

Actually it would be best to have a comparison of the two groups before as well as after. The study was not conceived until after the war. And probably most people could not report accurately on their income of eight or ten years ago. Furthermore, other complications would turn up. Many lawyers have moved in or out of the Chicago area during this long period, many have died, retired, or entered the profession. This kind of situation is very common to research. It is impossible to foresee the questions that are going to come up ten years from now, and collect the data now. In this case, the assumption that veterans and nonveterans were comparable before the war is implicit. No conclusion is any better than this assumption. Special supplementary investigations could be made to try to judge how nearly correct it is.

*Comment:* It might be possible to check up on this assumption a little by seeing what the dependency status of the two groups was before the war, and what the relation is now between income and number of dependents for lawyers in the same age group. Prewar dependency status would probably be reported more accurately than prewar income.

From this discussion it is probably becoming obvious that the observed difference in the averages can be "explained" in many ways. For every explanation we have thought of thus far, a dozen others could be thought up if enough people tried long enough. The difficulties stem from the fact that we are not sure the thing being studied—income differential between veterans and nonveterans—is independent of the agencies or forces that divided the lawyers into the two groups. To reach a conclusion we have to investigate these possibilities carefully. Such difficulties do not usually prevent a conclusion, though they prevent certainty, and they make care, expertness, and objectivity important.

Perhaps this point will be made clearer by imagining a simple experiment with this class. Suppose that we wish to find out whether this class could learn just as much statistics by reading the book and doing the exercises, but skipping the class sessions completely. We might divide the class into two groups for an experiment. How should we form the groups?

*Comments* (numerous): At random. Probability sample.

It is suggested that we divide the class into two groups at random. This will work, although a few refinements (which we can ignore for our present purposes) might improve the experiment. One group goes to class, one does not. At the end all take the same examination. Suppose the stay-at-homes do better on the average. Then it is objected that the best students were included in the stay-at-home group. Other objections are added to this one, etc. What can we say to these objections? The crucial, built-in safeguard to which we turn is the fact that *the class was divided objectively at random.* In other words, we can assert that the basis for dividing the class into two groups is *known* to have been unrelated to ability, previous knowledge, amount of free time, etc., except by chance—and in comparing the two averages we would, of course, require a bigger-than-chance difference before we concluded that the stay-at-homes had done better. Remember that many of the comments about the veteran-nonveteran income comparison stem from the fact that we can't be sure the process by which lawyers were divided was independent of inherent earning capacity. One of the lessons we should learn from our hypothetical class experiment is that a random procedure for dividing a class into two groups permits definite conclusions from the final comparisons. This is another lesson in the importance of randomness in the methods of statistics.

Suppose we try to measure the effect of the lectures by comparing the performance on a final test of those members of the class who have been absent or drowsy most frequently with the performance of those who have been present and alert most frequently. Then, if the absentees do better, it may be because they already knew so much statistics that they found the class sessions too elementary. And if the absentees do worse, it may be because they found the subject so uninteresting or so incomprehensible that they couldn't bear it, or because they were under heavy outside pressure which not only prevented attending class but prevented doing the reading too. These things could be investigated, and no doubt a conclusion reached, but it would be a long job, and the conclusion would be shaky, in comparison with one obtained by the use of a random division.

Let's digress a moment to consider refinements in our hypothetical study. We might divide the whole group into five subgroups according to how well they can be expected to do in statistics. This assumes we have some basis, for example college grades, age, intelligence, score on a statistics aptitude test, earlier performance, amount of time for study, etc., for dividing the class into prospective excellent, good, medium, poor, and incompetent statistics students. Then we would divide each of the five groups at random into two subgroups, one subgroup to stay away from class and the other to attend. This is an example of stratified sampling, showing how expert judgment could validly be used in conjunction with random sampling. (See Sec. 4.6.3.) The advantage to this refinement is that the amount of chance difference between the averages for the two groups would be reduced, hence a smaller difference between them could be regarded as significant of a true population difference. In other words, our test is more "sensitive" or more "powerful"—capable of detecting smaller differences.

Coming back to the lawyers, we conclude that by and large the nonveteran lawyers in the Chicago area are making considerably

Association

more money than the veterans, but that this difference is almost entirely explained by differences in the compositions of the two groups, and is not due to being veterans or not.

### 9.6.2 Lawyers' Income and Education

Table 298A gives some more data from the same study.

TABLE 298A

INCOME FROM LEGAL PRACTICE, BY EDUCATION,
CHICAGO LAWYERS, 1947

| Education | Mean Income | Median Income |
|---|---|---|
| High School Graduates | $12,095 | $8,955 |
| College Graduates | 11,373 | 6,938 |

*Source:* Kent, *op. cit.,* pp. 94–95.

If you wanted to be a lawyer, and were trying to decide whether to get your training in college or by an office apprenticeship—with income-producing ability as the sole criterion—would you find the results of Table 298A convincing for an office apprenticeship? Probably not. Let's look at still another of the tables:

TABLE 298B

INCOME FROM LEGAL PRACTICE,
BY YEAR OF ADMISSION TO THE BAR AND EDUCATION,
CHICAGO LAWYERS, 1947

| Year of Admission to the Bar | High School, No College | | College Graduates | |
|---|---|---|---|---|
| | Mean income | No. of cases | Mean income | No. of cases |
| Prior to 1910 | $13,559 | 17 | $22,132 | 19 |
| 1910–1914 | 19,188 | 16 | 21,705 | 22 |
| 1915–1919 | 10,577 | 26 | 19,053 | 19 |
| 1920–1924 | 17,100 | 10 | 16,095 | 21 |
| 1925–1929 | 8,206 | 34 | 13,066 | 76 |
| 1930–1934 | 5,500 | 1 | 12,111 | 81 |
| 1935–1939 | —— | 0 | 9,050 | 107 |
| 1940 and after | 2,000 | 1 | 4,696 | 97 |

*Source:* Kent, *op. cit.,* p. 94.

Table 298B shows that every experience category except one has a higher income for college graduates. Yet the high school graduates as a whole have the higher mean income. The point is, of course, that in seeming to compare education groups we are also (to some extent) comparing age and experience effects; for the high school graduates, on the whole, have been practicing much longer than the college graduates. Over 60 percent of the college graduates, but less than two percent of the high school graduates, started practice after 1929.

### 9.6.3 Lawyers' Income and Law School

Again, suppose that you intend to be a lawyer, and wish to choose a school on the basis of future income prospects. Would the results of Table 299 convince you that you should go to one of the two top schools in this table?

TABLE 299

INCOME FROM LEGAL PRACTICE, BY LAW SCHOOL ATTENDED, CHICAGO LAWYERS, 1947

| Law School Attended | Mean Income | No. of Cases | Average Experience |
|---|---|---|---|
| University of Michigan | $18,523 | 22 | 21.1 |
| Harvard University | 18,294 | 46 | 18.0 |
| University of Chicago | 11,306 | 116 | 18.1 |
| Northwestern University | 11,247 | 88 | 16.2 |
| Chicago-Kent | 10,130 | 129 | 20.1 |
| Chicago College of Law | 9,512 | 20 | 28.3 |

*Source:* Kent, *op. cit.*, pp. 122–123.

*Comment:* These data would not help in selecting between Michigan and Harvard; the difference can undoubtedly be explained by chance.

First you would want to investigate whether any hidden factors needed to be allowed for. But even if you were satisfied on this count, it would still be important to realize that the data do not show whether the law school attended bears a cause, or an effect, relationship to earning capacity. For in "earning capacity" we must include "connections" of all kinds. It may be that students with good connections are more likely to attend certain schools.

*Comment:* Students from Chicago who go to Harvard, or any other expensive college, or one a long way off, are probably already better off financially. Some of those Chicago schools are night schools where the students are not as well off in the first place as the ones who go to Northwestern and Chicago, much less the ones who go out of town. If the same study were made of Ann Arbor or Detroit lawyers, Michigan graduates might rank below Chicago and Northwestern.

### 9.6.4 The Achilles Heel of This and Similar Studies

We should say a little bit about the method of collecting these data. There are about nine thousand practicing lawyers in the Chicago area. A reasonably complete list of these lawyers was obtained from the Sullivan and the Martindale-Hubbell directories, supplemented by Bar Association lists. A carefully designed and pretested questionnaire was sent to every fourth lawyer on the alphabetical list. Of 2,444 questionnaires mailed out, only 812 were returned— about the usual rate for studies like this. The questionnaires were returned anonymously, but a postcard was included with each questionnaire and respondents were asked to report their names on the postcards.

If the 2,444 lawyers selected were a representative sample of all the lawyers qualified to be included in the sample, it is still likely that some bias was introduced by the circumstances that selected the 812 respondents from the 2,444 queried. The 1,632 who did not respond could totally reverse the picture if the probability of responding is somehow related to income. We could conjecture about this forever, but there really isn't much hope of coming to a conclusion without some empirical evidence on the matter.

The postcards provide some evidence in this way. Knowing the names of the responding group, it is possible to go back into the law directories and see whether the respondents were representative with respect to age, experience, education, and possibly a few other categories. The respondents were found to be fairly representative with regard to these attributes, and this fortified the assumption that the 812 are representative with respect to income. But ultimately you are just lifting yourself with your own bootstraps this way. The important thing is selection by income, and you don't find this out if you didn't get a response.

Another check was provided by the fact that the Department of Commerce independently made a small study of Chicago lawyers' income at about the same time. The results of the two studies agree reasonably well. Either both studies were getting close to the truth, or—and this is at least as likely—both were affected by the same biases.

The questions raised in this discussion reveal the sort of problems that abound in practical statistical work. On the other hand, the discussion has also indicated that many of these problems can be surmounted by intelligent methods of collecting and interpreting data.

### 9.7 Conclusion

Statistics can yield a lot of useful information when the difficulties are appreciated; it may yield misleading information if the difficulties are witlessly or willfully disregarded.

## 9.7
## CONCLUSION

Two variables are said to be associated if changes in one are, on the average, accompanied by changes in the other. The pattern of change in one variable is often termed the "effect" of the second, but this refers simply to the effect of classifying the data according to the second variable, not to effect in a causal sense.

The first requirement for studying the association between two or more variables is the ability to read a table perceptively. This can be learned easily by adopting systematic procedures. The first steps are orienting ones, finding out from the title, headnote, footnotes, source notes, and other explanatory matter, as well as from the row and column headings, the nature of the information tabulated. A second group of steps involves starting with simple, over-all features of the table: the general level of the entries, their variability, and the association indicated by the summary, or marginal, rows and columns. Next, the core of the table is examined to see if the effects suggested by the marginal rows and columns are true within separate rows and columns, and if so whether to the same extent. As a final step, the table is examined for any unusual relations or cell entries that may be suggestive of ideas or questions other than those with which the table was approached.

The data in tables may be either frequencies—counts of the number of observations in each cell—or summary measures, such as averages, for the values of a quantitative variable that have been recorded for each observation in the cell. For frequency data we can get a descriptive measure of the association by determining how much more accurately the observations in the table could be classified on the basis of both sets of marginal frequencies than on the basis of one set alone. Several indexes of this general type are particularly useful. For quantitative variables, either ranked or measured, an index is recommended which depends on how frequently the order, or rank, of a pair of observations with respect to one variable can be determined by knowing their order with respect to another variable. For qualitative data, the indexes depend on the frequency with which the correct cell of the table can be predicted for an observation on the basis of one variable only, as compared with the frequency with which the correct cell could be predicted if neither variable were known for the observation to be classified.

A device for measuring more sharply the effect of a variable is the standardized mean. The purpose of a standardized mean is to adjust for the effects of other variables, so that groups can be compared with regard to a variable of special interest without the effects of this special variable being confounded with the effects of other variables. Means are standardized by using a standard set of weights to combine the means of subgroups into a single mean. The extent, and even the direction, of an effect can vary from one set of standard weights to another. One of the most important applications of standardized weights is in the construction of index numbers of such things as prices and production.

In interpreting association two questions are foremost: Is it real or spurious? Does it signify a cause and effect relation? The second question is one to be dealt with primarily by experts in the subject matter of the table, the statistician's role being primarily to point out the kinds of possibilities besides simple cause and effect.

The first question involves subtle and delicate issues, hinging on the possibility of selectivity in the data. Selectivity may be of two kinds. In the first place, if the data arise from experience rather than from experiment—that is, the individuals fall into one class or another (smoker, nonsmoker; veteran, nonveteran) according to forces neither controlled by the researchers nor fully understood by them—there may be differences between the groups that induce a spurious association. Only if the data arise by experiment—that is, if the researcher assigns individuals by probability methods to the different categories whose effects he is studying—can there be complete confidence in the absence of spurious association. This kind of selectivity thus occurs in the population itself. The second kind arises in the process of obtaining sample observations from the population. Observations of one kind may be more difficult to make than those of another kind, and the degree of differential selectivity may be unknown. This type of selectivity is especially important in studies of human populations.

This completes our survey of statistical description. We turn now to statistical analysis or inference. First the material of Chap. 4 should be reviewed. There the fundamental role of probability considerations in basing decisions on samples is clearly brought out, and probability will be our first subject (Chap. 10). With this tool it is possible to construct the sampling distributions needed to interpret sample results, and Chap. 11 does this. Chaps. 12 and 13 apply these materials to testing the conformity of samples to hypotheses, and

*Do It Yourself*

Chap. 14 to estimating the characteristics of populations from samples. Chap. 15 deals with the planning of studies based on samples in order to make inferences about populations.

## DO IT YOURSELF

EXAMPLE 303A

For Table 279, (1) summarize the Table's findings and (2) indicate what further information would be helpful in its interpretation.

EXAMPLE 303B

Same requirement as Example 303A, but for the following table:

TABLE 303A

CIVILIAN LABOR FORCE AND UNEMPLOYMENT,
BY SEX AND COLOR, 1950

| | Labor Force (thousands) | | Unemployed (thousands) | | Unemployed per 1,000 | | |
| | Male | Female | Male | Female | Male | Female | Both |
|---|---|---|---|---|---|---|---|
| White | 38,607 | 14,387 | 1,769 | 587 | 46 | 41 | 44 |
| Nonwhite | 3,992 | 2,086 | 310 | 166 | 78 | 80 | 78 |
| Total | 42,599 | 16,473 | 2,079 | 753 | 49 | 46 | 48 |

*Source: Statistical Abstract: 1954*, Table 221, p. 197. Original source: *U. S. Census of Population: 1950*, Vol. II.

EXAMPLE 303C

Same as Example 303A, but for the following table:

TABLE 303B

DEATH RATES FROM TUBERCULOSIS,
RICHMOND AND NEW YORK, BY COLOR, 1910

| Color | Population | | Tuberculosis Deaths | | Tuberculosis Death Rate per 100,000 | | |
| | New York | Richmond | New York | Rich-mond | New York | Rich-mond | Total |
|---|---|---|---|---|---|---|---|
| White | 4,675,174 | 80,895 | 8,365 | 131 | 179 | 162 | 179 |
| Colored | 91,709 | 46,733 | 513 | 155 | 559 | 332 | 483 |
| Total | 4,766,883 | 127,628 | 8,878 | 286 | 186 | 224 | |

*Source:* Modified from Morris R. Cohen and Ernest Nagel, *An Introduction to Logic and Scientific Method* (New York: Harcourt, Brace and Company, 1934), p. 449.

Association

EXAMPLE 304A

Same as Example 303A, but for the following table:

TABLE 304

PERCENTAGE DISTRIBUTION OF REPOSSESSED AND NOT
REPOSSESSED NEW CARS, BY AMOUNT OF DOWN PAYMENT
IN PERCENT OF CASH SELLING PRICE[a]

| Experience | Ratio of Down Payment to Cash Selling Price | | | | | | |
|---|---|---|---|---|---|---|---|
| | Less than 30% | 30–34% | 35–39% | 40–44% | 45–49% | 50–59% | 60% and over |
| Not Repossessed | 4.3 | 17.5 | 13.7 | 11.3 | 9.4 | 19.9 | 23.9 |
| Repossessed | 16.8 | 45.2 | 19.4 | 8.3 | 5.2 | 3.6 | 1.5 |

*Source:* David Durand, *Risk Elements in Consumer Instalment Financing* (New York: National Bureau of Economic Research, 1941), p. 61.

[a] Data based on 4 samples from 3 automobile finance companies.

EXAMPLE 304B

For Table 209, compute: (a) the index of association of column with row; (b) the index of association of row with column; (c) the index of mutual association.

EXAMPLE 304C

Same requirements as Example 304B, for the table required by Example 207A.

EXAMPLE 304D

Same requirements as Example 304B, for the table required by Example 207B.

EXAMPLE 304E

"The . . . table . . . [which] follows . . . refers to . . . white Protestant married couples living in Indianapolis. . . . The table is condensed from a more detailed cross-classification given in [P. K. Whelpton and Clyde V. Kiser, *Social and Psychological Factors Affecting Fertility*, Vol. 2: *The Intensive Study; Purpose, Scope, Methods, and Partial Results* (New York: Milbank Memorial Fund, 1950), pp. 286, 389, and 402]. . . .[12]

(1) What does the table say?
(2) What further information would be helpful in interpretation?
(3) Compute the index of order association, $h$ (Sec. 9.3.3.).
(4) Indicate how $h$ is to be interpreted.

12. Leo A. Goodman and William H. Kruskal, "Measures of Association for Cross Classifications," *Journal of the American Statistical Association*, Vol. 49 (1954), p. 752.

*Do It Yourself*

TABLE 305

Cross-Classification Between Educational Level of Wife
and Fertility-Planning Status of Couple

Numbers in Body of Table are Frequencies

| Highest level of formal education of wife | Fertility-planning status of couple | | | | Row totals |
|---|---|---|---|---|---|
| | A<br>Most effective planning | B | C | D<br>Least effective | |
| One Year College or More | 102 | 35 | 68 | 34 | 239 |
| 3 or 4 Years High School | 191 | 80 | 215 | 122 | 608 |
| Less than 3 Years High School | 110 | 90 | 168 | 223 | 591 |
| Column Totals | 403 | 205 | 451 | 379 | 1,438 |

Example 305A

For the marriage adjustment data, Example 209D, compute the index of order association and state in words how it is to be interpreted.

Example 305B

Compute standardized death rates for each of the two cities of Example 303C, using the following weights: (a) equal; (b) New York population; (c) Richmond population; (d) combined population of the two cities.

Why do your answers appear to suggest contradictory interpretations?

Example 305C

The study by the American Cancer Society mentioned in Example 287 still continues. Suppose Berkson's explanation initially accounts for the entire observed difference in rates between smokers and nonsmokers. Would you expect the difference to narrow, widen, or stay the same as the study continues? Explain.

Example 305D

Why could the method of standardized averages not be applied in Example 287?

Example 305E

For one of the principal index numbers published in the *Survey of Current Business*, find out as much as you can about its purpose and method of construction. Write a short paper summarizing what you have learned.

TABLE 30.8

Cross-Classification by Race, Formal Education, Joint Fertility-Planning Status, and Fertility Planning Status, for Certain "Numbers in both of Table Are Frequencies"

| Race Total | Fertility-planning status of couple | | | | Median level of formal education of wife |
|---|---|---|---|---|---|
| | D Least effective | C | B | A Most effective planning | |
| 250 | 34 | 68 | 54 | 103 | One Year College or More |
| 608 | 132 | 214 | 60 | 181 | 5 or Years High School. Less than 2 or 3 Years High School. |
| 491 | 223 | 168 | 60 | 110 | |
| 1,345 | 389 | 401 | 203 | 302 | Column Total |

EXAMPLE 305A

For the marriage adjustment data, Example 30.D, compute the index of order association and state in words how it is to be interpreted.

EXAMPLE 305B

Compute standardized death rates for each of the two cities of Example 305C, using the following weights: (a) equal; (b) New York population; (c) Richmond population; (d) combined population of the two cities.
Why do your answers appear to suggest contradictory interpretations.

EXAMPLE 305C

The study by the American Cancer Society mentioned in Example 283 will continue. Suppose Pearson's explanation initially led you to expect the entire observed difference in rates between smokers and nonsmokers. Would you expect the difference to narrow, widen, or stay the same as the study continues. Explain.

EXAMPLE 305D

Why could the method of standardized averages not be applied in Example 305C.

EXAMPLE 305E

For one of the numerical index numbers published in the Survey of Current Business, find out as much as you can about its purpose and method of construction. Write a short paper summarizing what you have learned.

# PART III
# STATISTICAL INFERENCE

# Chapter 10

# Randomness
# and Probability

## 10.1
### STATISTICAL INFERENCE

Chap. 4 introduced the concepts of *sample* and *population* as the fundamental concepts in statistics and outlined the ideas that would pervade the greater part of the book. Chaps. 5 to 9 considered the problems of organizing and summarizing data. Now we resume the development begun in Chap. 4, which should be reviewed at this point.

Almost any group of observations, however they are presented or compressed, should be considered a *sample* from some larger group of observations called a *universe* or *population*. In most applications of statistics the sample is important only in that it may cast some light on a population. What is exactly true for the sample is typically only approximately true for the population and the degree of "approximation" may be excellent or may be worthless.

In principle, the distinction between description and inference is simple and hardly needs to be belabored. In practice, however, it is easy to lose sight of the distinction. Too often, in practical affairs and in science, the evidence available from a sample is so engrossing that the need for drawing inferences about a population is forgotten, or worse, inferences are drawn without regard to the possibility of discrepancy between the state of the sample and the true state of the population. A commonplace example will illustrate this danger.

**309**

EXAMPLE 310  EASTERN FOOTBALL

ORANGE BOWL TO FORGET EAST. Miami, Fla. (UP)—K. D. (Buck) Freeman, head of the Orange Bowl schedule committee, said Saturday that it will be "hard to sell me on an Eastern team" again.

Alabama swamped Syracuse 61–6 Thursday in the most lopsided victory in bowl history.

Freeman admitted his disappointment and said that "I guess we'll have to forget about the East as a section from which to choose one of our principals.

"I know if I am on future Orange Bowl schedule committees I'll be hard to sell on an Eastern team."[1]

Such an example would be merely amusing and trivial, were it not that the basic fallacy involved occurs in matters even more important than football, and much oftener than most of us realize. For example, scientists sometimes use elaborate statistical techniques with names like "multiple correlation" or "factor analysis," and having gone through the elaborate computations involved, feel that they have paid proper tribute to the demands of statistics. But the process of finding the degree of multiple correlation in a particular sample, however laborious and seemingly esoteric, results only in the description of one aspect of that particular sample. The coefficient of multiple correlation, taken by itself, is only a descriptive statistic. The problem of inference still remains after it has been computed. That is, the evidence from the sample must be interpreted in terms of the multiple correlation prevailing in the population from which the sample was drawn. If the problem of inference is ignored, all the calculations may serve only to encourage false security in the sample findings.

We turn, then, to the study of the process by which decisions may be based on the data of samples.

## 10.1.1  Uncertainty

In making nontrivial inferences about populations on the basis of information contained in samples, it must be recognized that the inferences may be wrong. For example, on the basis of the evidence contained in a sample, it might be inferred that in a certain month between 3 million and 4 million members of the labor force were unemployed. If the true figure for the entire population were actually 2.6 million, then the inference would be wrong. Of course, if the true unemployment were actually known to be 2.6 million, there would have been no need for the sample in the first place. But exact knowledge of the population is always missing in statistical investigations

---

1. *Chicago Sun Times*, January 4, 1953.

at the time the investigations are made, except in laboratory investigations aimed at testing the sampling methods themselves instead of finding out things about the population. Uncertainty is inevitable in inferences from samples, and statistical inference is oriented toward the objective of evaluating and, through proper planning, controlling the degree of uncertainty surrounding inferences.

Uncertainty goes against the grain of many; it seems intolerable, an evidence of lack of rigor, and of weakness of character. During the war, for example, generals and admirals found it virtually impossible to face the fact that there should be any uncertainty at all as to whether a given batch of, say, shells was up to specified standards of performance and safety of handling. Yet in the existing state of technology there was no feasible method of determining the quality of shells without firing them, so uncertainty could have been dispelled only by firing the entire batch of shells![2]

A little reflection, however, should indicate that uncertainty is inevitable whenever an inference is drawn that goes beyond the evidence actually at hand. Rigor in drawing inferences is not achieved by completely eliminating uncertainty. It is achieved by taking explicit account of uncertainty. Statistical inference embraces tools for reducing uncertainty and for dealing wisely with what remains.

We hope that these introductory remarks about uncertainty, though they are at once banal and vague, will take on both substance and usefulness by the time you finish Part III.

## 10.1.2 Populations

Let us now review and amplify some terminology introduced in Chap. 4.

A *population* (sometimes called a universe) consists of all conceivable observations relevant to some particular question. In a census study of the incomes of families, for example, there might be one observation for each family in the United States. If it is assumed that each of these observations measures without error the income of a family, then no additional observation conceivably would throw

---

2. At least one general faced the uncertainties of ammunition performance squarely and without flinching, and mastered them. See Leslie E. Simon, "Sampling and Sorting Ammunition for the Attack on Normandy," *Conference Papers, First Annual Convention American Society for Quality Control and Second Midwest Quality Control Conference* (Chicago: John S. Swift Company, Inc., 1947), pp. 275–281. Simon had an unfair advantage over other generals, though, in that he is a good statistician, in fact the author of a leading manual on engineering statistics. Besides he was not a general at the time, but became one only later.

further light on family incomes. This collection of observations, therefore, would be an example of a population.

A population is defined in terms of observations rather than in terms of people or objects. The heights of United States citizens constitute a population. The attitudes of United States citizens toward the Taft-Hartley Act, defined, say, by responses to a single specific question, constitute another population. In these two populations the people are the same, but the observations are different. The fact that the statistical problems of drawing and analyzing a sample would be different for the two populations is the chief reason for distinguishing between people or objects and observations made on them. In ordinary speech, however, it is not always necessary to make the distinction explicitly. When there is no question of misunderstanding, therefore, we shall "personalize" or "objectify" populations by speaking of them as if they were composed of people or objects. We may refer, for example, to the population of "United States citizens" with the tacit understanding that we really refer to observations of an unspecified kind made on United States citizens. Such verbal shortcuts are justified only when there is no danger of confusion.

The preceding paragraph reads as if the number of observations in a population is always finite, although possibly some very large number. Such populations, called *finite populations*, are often the object of study. But as we saw in Sec. 4.3.1, the term population can also be interpreted in a different sense, in which the number of observations is *infinite*. If a laboratory technician were to repeat a weighing many times, he would obtain a somewhat different reading each time, assuming that he tried to read his balance as accurately as possible. Any number of weighings the technician actually makes, whether one, three, or 20, is only a sample. But a sample of what? Of all measurements which might be taken if the process of measurement were continued indefinitely under the same basic conditions. The number of such measurements would be infinite. An infinite population is, then, simply a conceptual device. No one could ever actually enumerate all the observations, as could be done (at least in principle) with a finite population.

In Sec. 4.3.1 we presented a few examples of finite and infinite populations. Two more follow.

EXAMPLE 312  BEAD POPULATION

The sampling demonstration of Sec. 4.3 used a box containing more than a thousand beads. These beads, or more precisely, the *colors* of these beads, comprised a finite population. The numbers of red beads in the 50 successive random samples of 20 beads each, which were described in Sec. 4.3.3, comprise a sample of 50 from an infinite population. The infinite population comprises all repetitions of the sampling process by which sets of 20 beads were obtained, if the sampling process were continued indefinitely. (Recall that the 20 beads were replaced after each sample so that the process might in principle have been continued indefinitely.)

EXAMPLE 313A   FAMILY INCOME

We have already seen that the population in an income study of United States families is finite though very large (about 45 million—the exact figure depends on the date and the definition of a family) so long as one assumes each observation to be made without error. But we can imagine that if a given family were asked its income on two separate occasions (and could somehow forget completely about the first response on the second occasion), the two answers might not be the same. Hence, for each family we can think of an infinite population generated by the indefinite repetition of this question under the conditions just described. Thus, the 45 million families give rise to a finite number of infinite populations, one infinite population for each family.

EXAMPLE 313B   MEDICAL EXPERIMENT

Suppose a medical experiment is performed on 100 patients, divided at random into a control group of 50 and an experimental group of 50. The outcome of the experiment is expressed, say, as the difference in mean recovery time for the two groups. The particular experiment comes out with a specific number for this mean difference, say, 9.7 days. Again, we must raise the question, Of what population is this number a sample? There are many ways we could define this population. We might (probably unrealistically) assume that there was no measurement or experimental error, that is, that the response of any patient is measured without error and that no matter how many times a treatment were given to any one patient (if it could be repeated under fixed conditions), his response would be identical. Then we could think of the population as all possible patients in a certain geographical area at a certain time, for example, all people in the United States on January 1, 1957, who are ill from a given disease. This population would be finite. On the other hand, we might think only of the 100 patients and define the population as the mean differences which would result from all possible repetitions of the experiment; there would be as many possible repetitions as there are different combinations of 50 patients to be selected from 100, that is, more than $10^{29}$. Still another definition of the population would combine these two and take the population to consist of the results of an indefinite number of repetitions of the processes of (1) selecting a sample of 100 people from, say, all persons in the United States ill with the given disease and (2) performing the experiment on these 100 people.

These examples should make clear that the concept of "population" is a relative one which will depend on the purposes of the investigation and, also, on the resources available to the investigator. For example, a medical experimenter may not be able to draw a random sample from the population of all persons ill with a given disease, but he may have 100 patients ill with this disease at a medical clinic. He may have to confine his study to the population consisting of all repetitions of the experiment on the 100 patients, assuming fixed basic conditions. He may be able to make a statistical inference from the results of his experiment about this latter population, even though he could not draw inferences about the population of all persons ill with the disease. Of course, what he learns from the 100 people at his clinic might help him make a better guess about the broader population, but he could not draw an inference on purely statistical grounds beyond the population he actually studied. Precisely the same point was involved in the Goldhamer-Marshall study of mental disease; see Sec. 2.8.2. Still another example would be an investigator who studied the population of family incomes in Boston. Statistics would help him make an inference about incomes in Boston if he had drawn his sample from this population. Statistics per se would not help him in drawing an inference about family incomes elsewhere in the United States.

A useful distinction, as we have seen in Sec. 4.6.3, is that between a *target* population (for example, all patients, or all families) and a *sampled* population (for example, clinic patients, Boston families). Statistics enable us to make inferences about the sampled population. Expert knowledge and sound judgment of the subject matter is required to bridge the gap between the sampled and target populations. Often the most important differences between the sampled and target populations arise from selectivity, such as was discussed in Sec. 9.4.

### 10.1.3  Samples

We have already used the word "sample" frequently. A sample is any part of a population. A sample from a given population therefore consists of any portion of the totality of observations comprising the population. The term "sample," strictly interpreted, need not imply anything about the manner in which the observations are selected. As we saw in Sec. 4.6, however, and as we shall see repeatedly as we go on, *randomness* in the selection process is essential in drawing *statistical* inferences about populations.

From one viewpoint, any sample is a random sample from some population, namely the population that would be generated by indefinitely many independent repetitions of the process that produced the sample.

### 10.1.4 Parameters and Statistics

A mean, proportion, median, standard deviation, index of association, or other summary measure of the observations comprising a population is called a *parameter*. The analogous summary measure for a sample is a *statistic*. Hence the term "descriptive statistic" will be replaced by two more specialized words, "parameter" and "statistic." It is essential to keep clearly in mind at all times whether a sample or a population is under discussion. It is, therefore, essential to use different symbols to represent parameters and statistics. We have already mentioned in Sec. 8.5.1 the custom of using $\sigma$ to represent a population standard deviation and $s$ to represent a sample standard deviation. One system of notation is to use Greek letters for parameters and the corresponding Roman letters for statistics. We, however, generally use capital letters for parameters and lower case letters for statistics, as in Sec. 4.3.1 we used $P$ for the population proportion and $p$ for the sample proportion. Similarly, we use $M$ for a population mean; but in deference to a well-nigh universal statistical custom we shall denote the sample mean of a variable by a bar over the symbol for the variable—$\bar{x}$, $\bar{y}$, $\bar{r}$, etc. Also in deference to custom, we use $\sigma$ and $s$ for population and sample standard deviations. Thus, we use these pairs of symbols:

|  | Mean | Proportion | Standard Deviation |
|---|---|---|---|
| Population | $M$ | $P$ | $\sigma$ |
| Sample | $\bar{x}$ | $p$ | $s$ |

Parameters, being characteristics of populations, are fixed, though usually unknown in statistical problems. Statistics, on the other hand, being characteristics of samples, are subject to sampling fluctuations, but are usually known. In Chap. 4, for example, the proportion of red beads in the box, the parameter, was unknown but remained the same. The proportion of red beads in a sample, the statistic, was known, but varied from sample to sample.

## 10.2
## PROBABILITY

Probability has an everyday meaning conveyed by sentences such as, "His chance of winning is pretty small," "It's pretty likely that we'll have rain before tomorrow," or "You're probably right." In each of those examples the idea of uncertainty is acknowledged and appraised.

The use of the word "probability" in statistics, however, is somewhat different from, and more precise than, its nontechnical or popular use. The probability of a particular outcome of a certain event is equal to the *relative frequency* of this outcome among all events of the same kind. Since about 97.5 percent of the infants born alive in the United States reach the age of one year, the probability that a particular infant will survive to the age of one year in the United States is 0.975. If a certain coin tossed in a certain way comes up heads in 40 percent of all tosses, the probability of a head on a particular toss is 0.40. Relative frequency can be as small as 0 if a particular outcome does not occur at all and as high as 1 if the outcome always occurs. Probability, correspondingly, varies from 0 to 1. It is customary to speak of an outcome with a probability of 0 as *impossible* (since it does not happen at all, though it may be logically possible or conceivable). An outcome with a probability of 1 is spoken of as *sure* (since it always happens).

The phrase "a certain event" needs emphasis. To illustrate the need, consider the following misuse of the definition, which results from ignoring this phrase: Since about half of the adults in the United States are men (actually, the percentage in the 1950 census was 49.1 for persons 21 years old and older), the probability is about one-half that a person named Beryl Sprinkel is a man. This is wrong: Beryl Sprinkel is or is not a man, and the probability is certainly not one-half. "A certain event" refers to certain given conditions—to a population, in fact; "a particular outcome" refers to a stated characteristic which some members of the class, or population, possess; and probability applies to a member of the population to be selected at random. If we inquire about the probability that Beryl Sprinkel is a man, we have narrowed the population to one member, who either has or lacks the characteristic. If we know nothing about Beryl Sprinkel but have to address a letter "Dear Mr." or "Dear Miss," the appropriate population is all persons named Beryl, or perhaps all persons with names like Beryl, such as Burl, and we want to know the probability (or at least to which side of 0.5 the probability lies) that a person named Beryl is a man. If we know that Beryl Sprinkel is the economist for a large bank, the relevant population is economists for large banks whose first names are Beryl, or perhaps economists for large banks whose first names do not indicate their sex. But none of this changes the fact that Beryl Sprinkel either is or is not a man.[3]

Similarly with the babies. The probability of a baby's surviving its first year is 0.975 only if the baby is to be selected at random from the whole number of babies born in a certain period. If we narrow our attention to a single baby, the probability is either 0 or 1—either it will live until its first birthday or it won't—though until it dies or reaches its first birthday there is no way to be certain which. Any judgments we have to make will have to be based on the probability for a class of which we judge the specific baby to be a member.

In these examples we have laid the foundation—or at least built the forms into which to pour the foundation—for a point that will assume practical importance in Chaps. 12 to 14: the things about which decisions are made (generally parameters or relations among them) are fixed and not subject to probability calculations. The decisions we make, while fixed once made, are subject to probability measures that describe the processes by which they were made.

Our definition of probability really has meaning only in connection with random selection of individuals from a population. The idea of randomness therefore needs detailed consideration.

## 10.3
## RANDOMNESS

In Chap. 4, our first presentation of the idea of randomness was in terms of thorough shuffling or stirring of the beads in the sampling box prior to the selection of a sample. Then we defined random selection in Sec. 4.6.1 as a process which assigns to all sets of $n$ members of a population the same chance of constituting the sample. This conveys the notion, but probability has crept into the definition under the guise of chance, and probability was defined in terms of randomness.

We are trapped in a kind of circularity common to all linkages between abstract mathematical concepts and the real world. If we attempt, for example, a logical definition of that class of phenomena in the real world to which the propositions of plane geometry apply, we quickly find ourselves defining the class in terms of the applicability of plane geometry. In geometry, a straight line has the property of being the shortest distance between two points. When is a real line such that the properties of a straight line apply to it? When experi-

---

3. In fact, Beryl Sprinkel is a man. He often gets letters to "Dear Miss Sprinkel."

ence shows that it has those properties to an approximation good enough for some practical purpose. We wind up expressing annoyance with "philosophers" and confidence in "what works."

Similarly, to define randomness is to define the class of phenomena to which the propositions of mathematical probability apply. We come down to the fact of experience, that in repeated selection by a given process, the relative frequencies of various events converge to fixed numbers. If these numbers are the relative frequencies of the target population, the sampling is random. Conversely, the numbers to which the empirical relative frequencies do converge describe the sampled population, from which the process is producing random observations.

FIG. 318. A "ten-sided die" or random digit generator.

To gain an intuitive idea of the way randomness is achieved in practical statistics, imagine a ten-sided "die," on each side of which appears one of the digits 0 to 9. Such dice have in fact been constructed,[4] and one is shown in Fig. 318. Now suppose this die is tossed into the air spinning rapidly about its longitudinal axis, and caught with the thumb pressing on one of the numbered sides. The number on that side is recorded and the die tossed again. If the die is spinning rapidly, and is tossed high enough to make several dozen complete revolutions, minute differences in speed of rotation, height of the toss (which largely determines duration of rotation), and timing of the catch result in completely unpredictable outcomes. An actual series of 50 tosses with such a die produced the following sequence:

22831   52643   28455   55940   36680   56646   79990   60016   69540   47286

The digits produced in this manner are random digits; the process by which they are produced is a random process. The population is one in which each digit has probability 0.1.

Randomness inheres in the process itself, not in any particular sequence produced by it. Ultimately, however, the only way to judge whether a process is random is by the properties of the sequences it produces. We will discuss, therefore, some of the properties of random sequences; but first we consider one additional concept important in probability, independence.

4. By H. C. Hamaker of the Philips Research Laboratories, Eindhoven, Holland. This is not literally a die, nor is it ten-sided. It is a right cylinder whose cross section is a ten-edged regular polygon.

*TO BE FIXED ELEMENT OF ATTRIBUTE*

## 10.4
## INDEPENDENCE

The digits produced by our ten-sided die are not only random but *independent*. Events are independent if the probability that one of them will have a certain outcome is the same no matter what the outcome of the others. In recording ages of people, for example, the probability that a person selected at random from the United States population of 1950 would be from 20 to 24 years old is 0.076—about 1 in 13. If, however, we know that his father's age was 30, the probability drops to 0; so these events are not independent. Similarly, Table 280 showed that the probability that one of the 643 college women selected at random would be beautiful was 0.14 if she were a senior and 0.09 if she were a graduate student; so the events "appearance" and "class" are not independent, but *dependent*.

In selecting 20 beads for a sample in the sampling demonstration of Sec. 4.3, the probability that the bead in any one position in the panel would be red was slightly less if the bead in some second position was red than if it was not, for if the bead in the second position was red, it left one less red bead in the part of the population available for filling the first position. In this example, in fact, the probability that the bead in the first position will be red depends on the colors in all the other 19 positions.

Note that in all these examples the sampling is from a stable population, and the probability that any specified observation (event) will have a certain characteristic (outcome) is the same for all observations. Only when we know one or more of the observations are the probabilities changed. This is in contrast with a situation in which the probabilities are different for different observations, but not in ways related to how the other observations have turned out. An example of the latter is the sequence G S G S G   S G S G S G S G S G   S G S G S   G S G S G   S G S G S   G S S S, which records the winners of successive games of the first set of a tennis match between Gonzales and Schroeder, which Schroeder won, 18–16. The explanation for the alternating pattern is that, in tennis, the serve changes with each game and a player's probability of winning is usually much higher when he is serving than when he is not. This sequence could be regarded as a mixture of two sequences of random independent events, the odd-numbered observations from a population in which the probability of G was nearly 1, the even-numbered observations from a population in which the probability of an S was nearly 1. While the probabilities for any observation depend strongly on which population it is from, they do not (presumably) depend on how previous observations from either population have turned out.

In the sampling demonstration, the color of the bead in a certain position for one sample was independent of the color of the bead in that position—or in any other position—for any other sample. Similarly in tossing the ten-sided die, the digit on any one toss was independent of the digit on any other toss.

## 10.5
## PROBABILITY AND PREDICTABILITY

Randomness implies the approximate predictability of relative frequencies, the degree of approximation depending, by the Law of Large Numbers, on the number of observations. It also implies the unpredictability of individual observations. As the insurance companies say, "We don't know *who* will die, but we know how many." Actually, they don't know *exactly* how many, either, but only approximately.

Suppose that with the ten-sided die the probability of a "3" is 0.1, as it should be. Then if we predict a 3 before every toss, we will be right about ten percent of the time in a very long sequence of tosses. (In the 50 tosses reported above, there were three 3's or only six percent.) Now suppose we want to improve this record, to be right on more than ten percent of our guesses of 3 in a long sequence of tosses. For example, we might predict a 3 only when we have a strong hunch that it will appear, or only when a 3 has not occurred in the previous nine tosses. No ruse whatsoever will increase—or, for that matter, decrease—the probability that a prediction of 3 will actually be followed by a 3.

When observations are random but not independent, unpredictability applies slightly differently. Suppose the operation of our ten-sided die for some reason were such that a 3 has only half as large a chance of occurring if the preceding digit is a 3. Then we can predict the relative frequency with which 3's will be followed by 3's, but we cannot say which 3's will be followed by 3's.

Sequences of independent, random events often appear to be clustered. In discussing Fig. 109, for example, we called attention to a run of six consecutive 2's. As another example, consider the following 125 digits which were obtained from an electronic analogue

of the ten-sided die. Each 3 is shown in boldface. (The division into
5 x 5 blocks is simply for appearance.)

TABLE 321

125 Random Digits

| | | | | |
|---|---|---|---|---|
| 1 0 0 9 7 | **3** 2 5 **3 3** | 7 6 5 2 0 | 1 **3** 5 8 6 | **3** 4 6 7 **3** |
| **3** 7 5 4 2 | 0 4 8 0 5 | 6 4 8 9 4 | 7 4 2 9 6 | 2 4 8 0 5 |
| 0 8 4 2 2 | 6 8 9 5 **3** | 1 9 6 4 5 | 0 9 **3** 0 **3** | 2 **3** 2 0 9 |
| 9 9 0 1 9 | 0 2 5 2 9 | 0 9 **3** 7 6 | 7 0 7 1 5 | **3** 8 **3** 1 1 |
| 1 2 8 0 7 | 9 9 9 7 0 | 8 0 1 5 7 | **3** 6 1 4 7 | 6 4 0 **3** 2 |

*Source:* Table 632, first 25 columns of first 5 rows.

A uniform division of the 16 3's among the five blocks would mean
about three to a block. Actually, six are "clustered" in one block,
and four in each of two other blocks. Randomness does not produce
perfect uniformity, it produces chaotic irregularity—but regularity
in long-run relative frequency.

As another example, consider the following record of 15 tosses
of a coin:

$$H H H T T \quad H T T H T \quad H T T H T$$

Again, there seems to be a tendency to cluster. That this tendency
is almost inevitable can be seen by noting that of all possible sequences
of 15 tosses with 8 tails and 7 heads, only the following shows no
tendency at all to cluster:

$$T H T H T \quad H T H T H \quad T H T H T$$

Such perfect alternation would occur very rarely if 8 *T*'s and 7 *H*'s
were arranged in a random sequence, indeed only about once in
6,435 sequences. Furthermore, most sequences of 15 tosses would not
show exactly 8 *T*'s and 7 *H*'s, or 7 *T*'s and 8 *H*'s, and then some
clustering would be inevitable.

Many events in the real world seem to come in clusters: airplane
accidents, deaths of famous people, personal problems, etc. But if
these events occurred randomly and independently that is just what
one would expect. In fact, however, some of these events may not
behave strictly as sequences of random, independent events with
constant probability. The probability of airplane accidents may be
higher in winter than in summer, or the occurrence of a riot in one
prison may encourage convicts in another prison to riot. Or there
may be a tendency for clusters of events to be played up by the newspapers, while single occurrences are ignored. In any of these instances,

there would be, or appear to be, more clustering than would occur in a sequence of random, independent events with constant probability.

EXAMPLE 322   DO JESUITS DIE IN THREE'S?

A belief that Jesuits commonly die in three's was examined through a statistical study of historical records. The conclusion was that

> the persuasion in question, a supposed superstition, that Jesuits die in three's is confirmed by the series observed and, furthermore, has been shown to be the recognition of an important trait of all random historical sequences.[5]

It might seem that clustering affords a basis for forecasting the observations in a random sequence. In tossing coins, for example, you might forecast that the next toss will be the same as the last one. Such forecasts, however, will be right only half the time, on the average, if the coin is a fair one; this is the same proportion correct that will be achieved by any forecasts. The half of the time that an outcome is like the preceding one is sufficient to give the appearance of clustering.

Probably as common as misunderstandings of random clustering are misunderstandings to the effect that randomness requires compensating outcomes—for example, that "good luck" is inevitably punished by "bad luck," and vice versa. A gambler with a streak of "bad luck" may keep playing in the belief that the law of averages will bring an offsetting streak of "good luck." A baseball player in a batting slump is spoken of as being "due for a hit."[6] These attitudes are fallacious so long as the sequence of events referred to really results from an independent random process. Assume that the probability of a head on the toss of a coin is 0.50, that the tosses are independent, and that 10 tosses are heads. What is the probability of a head on the eleventh toss? Since we have *assumed* the probability to be 0.50, the probability is still 0.50. There is no reason, therefore, to expect a compensatory excess of ten tails to offset the ten consecutive heads. If more tosses are made, the proportion of heads would be expected to be very close to 0.50. The ten consecutive heads would be "swamped" (as Tippett aptly describes the effect) by a large number of subsequent observations. The discussion of the Law of Large Numbers in Sec. 4.7, including Example 122, should be reviewed in this connection.

Care must be taken to distinguish situations in which the Law of Large Numbers actually applies from those in which it does not, or

5. J. Solterer, "A Sequence of Historical Random Events: Do Jesuits Die in Three's?" *Journal of the American Statistical Association*, Vol. 36 (1941), pp. 477–484.

6. A famous baseball pitcher, Ted Lyons, was once removed for a pinch hitter on his fifth time at bat after having made a base hit on each of his first four times at bat. The reason given was that it was virtually impossible for a pitcher to get five hits in a row! (Lyons was a very good batter during his entire major league career.)

in which its applicability is uncertain. Basically the law applies only to random sequences in which the probability of each possible outcome remains constant. Suppose, for example, there are two baseball teams of approximately equal skill. One team is made up wholly of men of age 20; the other, of men of age 35. The two teams agree to play each other daily during the baseball season for the next ten years. In the first year, the young men win 80 and lose 70 games. Will the law of averages tend to swamp this difference? Even if the probability of either team winning were one-half at the start, the probability of the younger team winning would increase with time.

In applying the Law of Large Numbers to predicting chance events, it must also be remembered that, in practice, the probabilities themselves are unknown. Suppose you are tossing pennies with an acquaintance, using a supposedly fair coin that he has produced, and the first ten tosses are all heads. You would then have grounds for suspecting that the subsequent tosses would *not* swamp the first ten heads. That is, you would have made an inference from the sample of ten that the population proportion was not one-half.

Even though accurate predictions of individual chance events are not possible, predictions of relative frequencies may be very useful. A gambler with "loaded" dice may profit handsomely by his knowledge that the probability of each side coming up is more or less than one-sixth. For, in contrast to a popular conception, dishonest gambling devices *may* produce random sequences of observations, the dishonesty consisting simply in an alteration of the probabilities from those that would commonly be expected, as in the case of loaded dice. The basic notion of randomness does not require equal probabilities for both sides of a coin, all sides of a die, all numbers on a roulette wheel, or all cards in a deck. A more important example of the practical usefulness of a knowledge of probabilities is afforded by insurance businesses. By an approximate knowledge of the probability of death, of fire, or of accident for a certain group, an insurance company is able to make "bets" with given individuals that reduce the financial uncertainties to individuals of being unable to predict how chance will affect them as individuals, but that produce reliable long-run profits for the companies.

## 10.6
## PROBABILITY CALCULATIONS

The following two rules are often applied in the discussion of statistical inference, and other applications of probabilities.

## 10.6.1 The Addition Rule for Mutually Exclusive Occurrences

Mutually exclusive occurrences, or outcomes of events, are occurrences which cannot both happen. A bead cannot be both red and not red, a person both literate and illiterate, a college woman both beautiful and homely, or a person both 20 to 24 years of age and the son of a man of 30. In contrast, a college woman can be both an undergraduate and a junior, or both a senior and homely, a person can be both literate and unschooled, or both schooled and illiterate; so these occurrences are not mutually exclusive.

The *addition rule* for probabilities is: The probability that one or the other of two mutually exclusive occurrences will happen is the sum of their separate probabilities.

Consider the ten-sided die. Let $Pr(2)$ denote the probability of a 2 and $Pr(5)$, the probability of a 5. Assume $Pr(2) = Pr(5) = 0.1$. Then the probability of a 2 *or* a 5 on a single toss, $Pr(2 \text{ or } 5)$, is, by the addition rule,

$$Pr(2 \text{ or } 5) = Pr(2) + Pr(5)$$
$$= 0.10 + 0.10$$
$$= 0.20.$$

When we ask for the probability of a 2 or a 5, we are simply broadening the class of outcomes whose probability we seek. Since no outcomes that were included in the "2" class can also be in the "5" class, we add to the total number of occurrences of 2's all the occurrences of 5's. The rule would not apply if some outcomes were both 2 and 5, for then some of the 5's would already have been counted with the 2's.

## 10.6.2 The Multiplication Rule for Independent Occurrences

As we saw in Sec. 10.4, random events are independent if the probabilities of the various possible outcomes of each are the same whatever the outcome of the other. Several examples were given in Sec. 10.4.

The *multiplication rule* for probabilities is: The probability that two independent events will have specified outcomes is the product of the probabilities of these outcomes.

Suppose that we decide to toss the ten-sided die twice. What is the probability we will first get a 2, then a 5—$Pr(2,5)$? According to the multiplication rule, this will be

$$Pr(2,5) = Pr(2) \times Pr(5)$$
$$= 0.1 \times 0.1$$
$$= 0.01.$$

To see the reason for this rule, think of all outcomes that are possible in two tosses of the die:

| 0,0 | 1,0 | 2,0 | 3,0 | 4,0 | 5,0 | 6,0 | 7,0 | 8,0 | 9,0 |
|-----|-----|-----|-----|-----|-----|-----|-----|-----|-----|
| 0,1 | 1,1 | 2,1 | 3,1 | 4,1 | 5,1 | 6,1 | 7,1 | 8,1 | 9,1 |
| 0,2 | 1,2 | 2,2 | 3,2 | 4,2 | 5,2 | 6,2 | 7,2 | 8,2 | 9,2 |
| 0,3 | 1,3 | 2,3 | 3,3 | 4,3 | 5,3 | 6,3 | 7,3 | 8,3 | 9,3 |
| 0,4 | 1,4 | 2,4 | 3,4 | 4,4 | 5,4 | 6,4 | 7,4 | 8,4 | 9,4 |
| 0,5 | 1,5 | 2,5 | 3,5 | 4,5 | 5,5 | 6,5 | 7,5 | 8,5 | 9,5 |
| 0,6 | 1,6 | 2,6 | 3,6 | 4,6 | 5,6 | 6,6 | 7,6 | 8,6 | 9,6 |
| 0,7 | 1,7 | 2,7 | 3,7 | 4,7 | 5,7 | 6,7 | 7,7 | 8,7 | 9,7 |
| 0,8 | 1,8 | 2,8 | 3,8 | 4,8 | 5,8 | 6,8 | 7,8 | 8,8 | 9,8 |
| 0,9 | 1,9 | 2,9 | 3,9 | 4,9 | 5,9 | 6,9 | 7,9 | 8,9 | 9,9 |

There are 100 possible outcomes, all of which would be equally frequent in an infinite number of pairs of tosses. Only one of these 100 is the prescribed one, 2,5. Note that 5,2 is another one of the 100 possible outcomes, but that it is not the same as 2,5; the multiplication rule applies to a prescribed order.

### 10.6.3 Illustrative Computations and Generalization of the Rules

A few illustrations will help clarify these rules. Consider again our ten-sided die.

(i) What is the probability of an even number on a single toss?

$$Pr(\text{even}) = Pr(0) + Pr(2) + Pr(4) + Pr(6) + Pr(8)$$
$$= 0.1 + 0.1 + 0.1 + 0.1 + 0.1$$
$$= 0.5.$$

(ii) What is the probability of a prime number (including 1) on a single toss?

$$Pr(\text{prime}) = Pr(1) + Pr(2) + Pr(3) + Pr(5) + Pr(7)$$
$$= 0.1 + 0.1 + 0.1 + 0.1 + 0.1$$
$$= 0.5.$$

(iii) What is the probability, on two tosses, of a 2,5 or a 5,2?

$$Pr(2,5) = Pr(2) \times Pr(5)$$
$$= (0.1)\,(0.1)$$
$$= 0.01.$$

*Randomness and Probability*

Similarly,

$$Pr(5,2) = Pr(5) \times Pr(2) = 0.01.$$

Then,

$$Pr(2,5 \text{ or } 5,2) = Pr(2,5) + Pr(5,2)$$
$$= 0.01 + 0.01$$
$$= 0.02.$$

(iv) Suppose one throws the die and reads the side his thumb is pressing and the side to the right. What is the probability of getting a 4 and a 7? The probability that the thumb is pressing the 4 is 0.1. The probability is also 0.1 that the side to the right of the thumb is 7. But the desired probability is

$$\text{neither} \quad 0.1 \times 0.1 = 0.01$$
$$\text{nor} \quad 0.1 + 0.1 = 0.2.$$

The first result is wrong because the events are not independent. In fact, the die illustrated in Fig. 318 happens to be constructed so that 7 is always to the right of 4. The second result is wrong because the outcome is not "either a 4 or a 7," but the result 4, 7 obtained in the way just described. The result 4,7 is obtained when, and only when, the die comes down so that the thumb presses on 4, and the probability that this will happen is 0.1, not 0.01 or 0.2. Similarly, the probability of 4,2 would be 0, since for this die the digit to right of 4 is never 2 but always 7.

*When you apply the probability rules, be sure the rule fits the situation.*

(v) What is the probability of an even number or a number greater than 2?

The addition rule, blindly applied, would give

$$Pr \text{ (even or } > 2) = 0.5 + 0.7 = 1.2,$$

but this is obviously wrong. It is fortunate that the absurdity of the answer makes its incorrectness obvious; it might have been wrong but not obviously wrong, and the error might have passed unnoticed. The error is that the addition rule refers to *mutually exclusive* outcomes, and 4, 6, and 8 are both greater than 2 and even. In the erroneous calculation, these outcomes have each been counted twice.

You can see from this how to modify Rule 1 so that it will apply whether or not outcomes are mutually exclusive: the probability that either of two outcomes will occur is the sum of their probabilities *minus* the probability that both will occur together. Thus, in the example just cited,

$$Pr \text{ (even or } >2) = Pr \text{ (even)} + Pr(>2) - Pr \text{ (both even and } >2)$$
$$= (0.5) + (0.7) - (0.3)$$
$$= 0.9.$$

That is, since we originally counted the three outcomes twice, both as even and as greater than 2, we have to subtract 0.3 to compensate for this.

Although the addition rule is stated for two mutually exclusive outcomes of an event, and the multiplication rule for the outcomes of two independent events, both rules extend easily to more than two cases:

*Addition:* The probability that one of several mutually exclusive outcomes will occur is the sum of the individual probabilities.

*Multiplication:* The probability that each of several independent outcomes will occur is the product of the individual probabilities.

For example, if the probability of a 3 is 0.1, of a 7, 0.1, and of a 9, 0.1, then the probability of obtaining a 3 *or* a 7 *or* a 9 on a single observation is $0.1 + 0.1 + 0.1 = 0.3$. The probability of 3, then 7, then 9 in *three* observations is $0.1 \times 0.1 \times 0.1 = 0.001$.

## 10.7
## CONDITIONAL PROBABILITY

In the preceding section we saw how to compute probabilities for outcomes even if they are not mutually exclusive, by adding the probabilities and subtracting the amount of overlapping due to joint occurrences. To compute probabilities for events that are not independent, we have to use the idea of conditional probability.

Conditional probability simply refers to the probability that an event of a certain class will have a given outcome under the condition that it belongs to a specified subclass of the whole class. Actually, we have introduced a number of examples already: the probability that a United States adult is a male, subject to the condition that it is the particular adult Beryl Sprinkel or, alternatively, to the condition that the first name is Beryl, etc.; the probability that a member of the United States population of 1950 is from 20 to 24 years of age, subject to the condition that his father's age is 30; the probability that a college woman is beautiful, subject to the condition that she is a senior; the probability that the bead in a certain position of the sampling panel is red, subject to the condition that the bead in another position is red or, alternatively, subject to the condition that a certain number of all the other 19 beads in the panel are red.

*Randomness and Probability*

Our definition of independence could, in fact, have been stated in terms of conditional probability. Two events are independent if the conditional probabilities of the possible outcomes of one, given the outcome of the other, are the same as the unconditional probabilities.

As an example of the use of conditional probabilities, let us find the probability of getting four aces in four draws from an ordinary deck of cards. The probability of drawing an ace of spades from a deck of cards is $1/52$. The meaning attached to this statement can be expressed in terms of an indefinitely long series analogous to the one for the die; indeed, drawing a card from a deck is equivalent to one toss of a 52-sided die. There are also three other aces in the deck. Hence, applying the addition rule, we see that the probability of drawing one of the four aces on the first draw is $4/52$. If the first draw is an ace, the next draw comes from a different population, in which there are 51 cards of which three are aces, so the probability of an ace on the second draw is $3/51$; etc. In fact, we can view the problem this way: Suppose there are four decks of cards of sizes 52, 51, 50, and 49, with numbers of aces 4, 3, 2, and 1, respectively. If we draw one card from each deck, what is the probability that all four will be aces? This is now a problem to which the multiplication rule applies, for the four draws are independent. The probability is, therefore,

$$Pr(4 \text{ aces}) = \frac{4}{52} \times \frac{3}{51} \times \frac{2}{50} \times \frac{1}{49} = \frac{1}{270,725}$$

We can generalize this result by saying that *the probability that each of a specified sequence of outcomes will occur is the product of their conditional probabilities*, the conditions being that the preceding events in the sequence have had the specified outcomes.

EXAMPLE 328   CANCER DIAGNOSIS

Suppose, contrary to fact, that there were a simple diagnosis of high reliability for every kind of cancer. Let us specify high reliability as meaning that if the test is applied to people without the disease, 95 percent of the reactions are negative and only five percent are positive, and that if it is applied to people with the disease, 95 percent of the reactions are positive and five percent negative. Suppose that the probability that a person tested has cancer is 0.005. Then what is the probability that a person giving a positive reaction has cancer?

The unconditional probability of a positive reaction is obtained by adding the probabilities of the following two mutually exclusive outcomes: (1) has cancer and reacts positively, (2) does not have cancer but reacts positively.

### 10.7 Conditional Probability

Each of these in turn can be calculated from the multiplication rule. The probability of the first, has cancer and reacts positively, is

$$0.005 \times 0.95 = 0.00475,$$

while the probability of the second, does not have cancer but reacts positively, is

$$0.995 \times 0.05 = 0.04975.$$

Adding these two probabilities to obtain $Pr(+)$, the probability of a positive reaction, we have

$$Pr(+) = 0.00475 + 0.04975 = 0.05450.$$

Of these positive responses, the proportion who actually have cancer is, letting $Pr(c \mid +)$ be the probability of having cancer subject to the condition of having reacted positively to the test,

$$Pr(c \mid +) = \frac{0.00475}{0.05450} = 0.087,$$

that is, only one in 11 or 12 of those who give positive reactions will actually have cancer, even with so reliable a test. Since 0.005 is not an unrealistic figure for the incidence of cancer per year at the middle ages, these calculations reveal a difficulty facing mass screening projects, such as have been used for tuberculosis, even if a test of the reliability indicated could be developed.[7]

Conditional probabilities are extensively used in the insurance business. When a rate is quoted on, say, passenger car liability insurance, it is not based on the probability of loss for all cars in the country. It is based on the probability for that locality, age and occupation of driver, type of usage, etc.—on the conditional probability, in other words. If there were some reasonably numerous and easily identifiable group with an exceptionally low accident rate, it would pay special companies to serve this group at less than standard rates, if general companies did not discriminate in their rates in favor of these low-risk groups. This would draw some of the low-risk individuals from the general companies, forcing them to raise rates to their remaining customers, and the general companies would tend to become special companies for high-risk groups. Similar mechanisms result in the use of quite refined and specialized conditional probabilities in fire, life, and other forms of insurance.

---

7. John E. Dunn, Jr., M.D., and Samuel W. Greenhouse, *Cancer Diagnostic Tests: Principles and Criteria for Development and Evaluation*, Public Health Service Publication No. 9 (Washington: Government Printing Office, 1950), pp. 9–20.

## 10.8
## PITFALLS IN CALCULATING PROBABILITIES

We have already given, in Sec. 10.6.3, computations (iv) and (v), two examples of the danger of misapplying the addition and multiplication rules. In this section we give three examples of the dangers in calculating relative frequencies, another example of misuse of the addition rule, and a paradoxical but true result.

EXAMPLE 330A  NORTH- AND SOUTHBOUND TRAINS

A young man has two girl friends. To visit one he takes a southbound subway train, to visit the other a northbound train. He knows that trains in each direction run equally often, and since he can never make up his mind which girl to visit, he lets chance decide. He goes to the station and takes the first train that comes. In the long run, he reasons, this plan should result in his seeing each girl equally often. But he discovers that he sees Miss South about 80 percent of the time, Miss North only 20 percent. What is wrong with his probability calculation?

The young man thought that he was getting a random sample of trains by taking a random sample of times. Suppose that the train schedules are like this:

| Northbound: | 7:30 | 7:35 | 7:40 | etc. |
| Southbound: | 7:34 | 7:39 | 7:44 | etc. |

On four trips out of five, on the average, a randomly selected time will put him on a southbound train—any time from 7:30 to 7:34, 7:35 to 7:39, etc.

His mistake is in some ways similar to those discussed in connection with sampling. In Example 72B, for instance, the error was in thinking that a random sample of children gave a random sample of families. In Example 72C, it was in thinking that a random sample of wage earners gave a random sample of families.

EXAMPLE 330B  FIRST DIGITS OF CAR LICENSES

Take as a population all of the private passenger cars registered in Illinois, one of the few states with a large number of cars which still number license plates serially beginning with 1. What is the probability that the first digit on the license of a car selected at random will be 1? Assume that there are three million private passenger cars registered in Illinois.

This is a case where snap judgment would very likely underestimate the true probability as $\frac{1}{9}$, since there are 9 digits that may be first. This is wrong. We can get the true probability by enumeration, as in Table 331A:

## 10.8 Pitfalls in Probabilities

TABLE 331A

LICENSE PLATES WITH FIRST DIGIT 1

| Interval | Number of Licenses with First Digit 1 |
|---|---|
| 1 | 1 |
| 10–19 | 10 |
| 100–199 | 100 |
| 1,000–1,999 | 1,000 |
| 10,000–19,999 | 10,000 |
| 100,000–199,999 | 100,000 |
| 1,000,000–1,999,999 | 1,000,000 |
| Total | 1,111,111 |

The probability we are looking for is, therefore, 1,111,111 divided by 3,000,000, or 0.37—about $\frac{3}{8}$.

Notice that the snap answer, $\frac{1}{9}$, would be exactly right if the total number of cars registered were 9, or 99, or 999, ..., or 999,999. From 1,000,000 to 1,999,999, however, there are as many numbers beginning with 1 as there are numbers below 1,000,000 beginning with the digits 1 to 9. Thus, when the number of private cars registered in Illinois was only two million, the probability was more than one-half—$\frac{5}{9}$, in fact—that an initial digit would be 1.

EXAMPLE 331  FIRST DIGITS OF STATISTICAL TABLES

The previous example may make it a little easier to understand the following parlor game (or swindle). Get a *World Almanac*, *Statistical Abstract*, or any book that contains many statistical tables. Open the book haphazardly and read the numbers in a column. For each number whose first digit is 1, 2, 3, or 4, Abercrombie pays Fitch a dime; for each number whose first digit is 5, 6, 7, 8, or 9, Fitch pays Abercrombie a dime. Fitch will get rich.

It looks as if Abercrombie were favored with odds (the ratio of the probability for him to the probability against him) of 5 to 4, but let's try it. The 122 first digits tabulated in Table 331B are from the 122 entries in the first column, headed "total number of returns," of a table showing the number of corporation income tax returns filed, by industrial group, for 1950.[8]

TABLE 331B

DISTRIBUTION OF FIRST DIGITS IN A STATISTICAL TABLE

| Digit: | 1 | 2 | 3 | 4 | 5 | 6 | 7 | 8 | 9 | Total |
|---|---|---|---|---|---|---|---|---|---|---|
| Frequency: | 42 | 25 | 10 | 9 | 9 | 11 | 8 | 5 | 3 | 122 |

In this instance, 86 of the 122 first digits, about seven-tenths of them, are 1, 2, 3, or 4. This is about the average proportion, although there is a good deal of variation from one table to another.

---

8. *Statistical Abstract: 1954*, Table 416, pp. 384–387.

*Randomness and Probability*

We will not explain this phenomenon, except for two remarks: (1) The previous example illustrates the mechanism; (2) Some people interpret small first digits as representing small numbers, others interpret them as representing large numbers.

### EXAMPLE 332 CHUCK-A-LUCK

The following game, sometimes called "chuck-a-luck," is often played at small carnivals. The player pays a nickel to play. Three dice[9] are rolled. If any 6's appear, the player gets back his nickel, plus one nickel for each 6 that appears. Players frequently calculate that this game offers them an advantage. They say (correctly) that:

$$Pr(6 \text{ on first die}) = \tfrac{1}{6}$$
$$Pr(6 \text{ on second die}) = \tfrac{1}{6}$$
$$Pr(6 \text{ on third die}) = \tfrac{1}{6}$$

and then conclude (incorrectly) that the probability of winning is $\tfrac{3}{6} = \tfrac{1}{2}$. The player will win a nickel as often as he loses one, this erroneous argument runs, and in addition will get an extra nickel on those occasions when two 6's appear, and two extra nickels when three 6's appear.

The addition rule does not apply, however, because the three outcomes are not mutually exclusive. A six on the first die does not prevent a six on the second or third die.

A correct analysis is as follows: There are three outcomes that lead to a win: (a) three 6's, (b) exactly two 6's, (c) exactly one 6. The probability of each of these can be obtained by the multiplication rule. Letting $\cancel{6}$ represent any outcome other than 6, we have

(a) $Pr$ (three 6's) $= Pr\,(6, 6, 6) = \tfrac{1}{6} \times \tfrac{1}{6} \times \tfrac{1}{6} = \tfrac{1}{216}$.

(b) $Pr$ (exactly two 6's) $= Pr\,(6, 6, \cancel{6}) + Pr\,(\cancel{6}, 6, 6) + Pr\,(6, \cancel{6}, 6)$, since these are three mutually exclusive sequences that produce exactly two 6's. But

$$Pr\,(6, 6, \cancel{6}) = \tfrac{1}{6} \times \tfrac{1}{6} \times \tfrac{5}{6} = \tfrac{5}{216}$$

and the same for $Pr\,(\cancel{6}, 6, 6)$ and $Pr\,(6, \cancel{6}, 6)$. So $Pr$ (exactly two 6's) $= \tfrac{15}{216}$.

(c) $Pr$ (exactly one 6) $= Pr\,(6, \cancel{6}, \cancel{6}) + Pr\,(\cancel{6}, 6, \cancel{6}) + Pr\,(\cancel{6}, \cancel{6}, 6)$. The value of each of these probabilities is

$$\tfrac{1}{6} \times \tfrac{5}{6} \times \tfrac{5}{6} = \tfrac{25}{216}, \quad \text{so} \quad Pr \text{ (exactly one 6)} = \tfrac{75}{216}.$$

Thus, the probability of winning is not one-half but

$$\frac{1 + 15 + 75}{216} = \frac{91}{216}.$$

To calculate the value of the right which the player buys for 5 cents we see that his net income (receipts minus the 5 cents he pays to play) always has

---

9. Whenever we mean ten-sided dice we will say so; otherwise, as here, we mean six-sided dice.

*10.8 Pitfalls in Probabilities*

one of four values: $-5\cent$, $+5\cent$, $+10\cent$, or $+15\cent$. To get his average income we calculate a weighted mean (as explained in Sec. 7.4.2), using as weights the respective probabilities. The probabilities of the positive incomes have already been found. The probability of losing is $Pr\ (6,\ 6,\ 6) = \frac{5}{6} \times \frac{5}{6} \times \frac{5}{6} = \frac{125}{216}$. Letting $V$ be the mean income, or expected value, of a play, we find

$$V = \frac{125 \times (-5) + 75 \times 5 + 15 \times 10 + 1 \times 15}{216} \text{ cents}$$

$$= \frac{-85}{216} \text{ cents} = -0.4 \text{ cents.}$$

Thus, on the average, the player loses $\frac{2}{5}$ of a cent per play, or 8 percent of his payments of $5\cent$ per play.

The technical term for the $-0.4$ cents is *mathematical expectation*, or just *expectation*. A table of life expectations, or life expectancies, for example, is computed by multiplying each possible length of life in years by its probability for the group to which the table applies, and summing the products.

EXAMPLE 333   PARADOXICAL PROBABILITIES

Consider three dice, die $A$ "loaded" so that "3" always appears, die $B$ loaded so that "2" appears $\frac{2}{3}$ of the time and "5" appears $\frac{1}{3}$ of the time, and die $C$ loaded so that "1" appears $\frac{1}{3}$ of the time and "4" appears $\frac{2}{3}$ of the time. Then $A$ will be larger than $B$ most of the time, $B$ will be larger than $C$ most of the time, and $C$ will be larger than $A$ most of the time.

The probability that $A$ will be larger than $B$ is simply the probability that $B$ will be "2", namely $\frac{2}{3}$.

The probability that $B$ will be larger than $C$ is the probability that $B$ will be "2" and $C$ will be "1", plus the probability that $B$ will be "5"; that is,

$$(\tfrac{2}{3} \times \tfrac{1}{3}) + \tfrac{1}{3} = \tfrac{5}{9}.$$

The probability that $C$ will be larger than $A$ is the probability that $C$ will be "4", or $\frac{2}{3}$.

Similarly paradoxical results can hold with measures of correlation among three variables. Variable $X$ can be positively correlated with both $Y$ and $Z$, in the sense that $X$ increases, on the average, when $Y$ and $Z$ increase, yet $Y$ and $Z$ can be negatively correlated, in the sense that $Y$ decreases, on the average, when $Z$ increases. Again, if several universities are ranked in order of preference by a number of people, it may be that most prefer Emory to Duke, most prefer Duke to Vanderbilt, and most prefer Vanderbilt to Emory.

## 10.9
## SIMPLE RANDOM SAMPLING

### 10.9.1 General Method

We have talked about randomness and random sampling. But—unless you are interested in drawing beads from a box—we have not told you how to draw a random sample. We shall now turn to this subject, and its relation to the techniques of statistical inference that we shall take up in subsequent chapters.

The method of drawing a random sample amounts to assigning a number to each member of the population, selecting numbers at random by the use of a ten-sided die, and using as the sample those members of the population whose numbers were selected. This method not only gives each set of *n* members of the population the same chance of constituting the sample, as required by the definition of randomness, but, as we shall see, it goes further and gives each member the same chance of being any particular observation drawn—say the third, or the twentieth.

Tables of random digits have been published. These not only save the time involved in drawing random digits by devices such as shuffling cards or tossing a ten-sided die, but they are more nearly random. Almost any device and method of using it shows at best slight departures from the intended random pattern, but in the best of the published tables these biases are minute. Each page of such a table contains a large number of random digits. These are usually divided into blocks of five for convenience. The digits can be read in groups of two, three, or more, to produce two, three, or more digit numbers. The starting point should be selected haphazardly. Table 632 in this book provides ten thousand random digits.

### 10.9.2 A Detailed Example

Suppose you want a random sample of (the lengths of service of) the 781 faculty members of a certain university.

The first step is to draw up a list of these faculty members and assign them consecutive numbers, 001, 002, 003, . . ., 781. A faculty directory might be available, in which case all that is necessary is to number the names. Lists that are available are often not entirely complete. Since the population actually sampled will be that listed, this should coincide as nearly as possible with the target population.

*10.9 Simple Random Sampling*

The next step is to select random digits from a table of random numbers. The digits listed in Sec. 10.3 that were produced by our ten-sided die can be used for illustration. The first three are 2, 2, and 8. Therefore, faculty member number 228 would be included in the sample. The next three digits are 3, 1, and 5. Hence faculty member number 315 would be included in the sample. This method is continued until the required number of faculty members is obtained. If, for example, the size of the sample is to be 100, the selection would continue until 100 faculty members had been chosen. More than 50 digits would, of course, be required in order to obtain 100 observations. These numbers would be obtained by continuing to select digits consecutively in the table of random digits, continuing from left to right, moving down the page, just as in reading a book.

Numbers like 799 and 906 (the eleventh and twelfth three-digit numbers formed from the 50 digits produced by our ten-sided die), which do not correspond to any number on the list, are ignored.

A number may appear a second time before the sample is complete. In that case, the second appearance is ignored. In other words, each time an observation from the population has been drawn, the population left to be sampled is smaller by one.

To see that on each draw every observation in the population actually has an equal probability of selection—the second definition of random sampling in Sec. 10.9.1—we apply the rules for calculating probabilities. In this case, the problem is to find the probability of selection of each three-digit number from 001 to 781. Each individual digit from 0 through 9 has a probability of $1/10$. The probability of obtaining any particular sequence of three digits is, by the multiplication rule, $1/10 \times 1/10 \times 1/10 = 1/1000$. But each faculty member corresponds with one and only one three-digit number. Therefore, each faculty member has a probability of $1/1000$ of inclusion on the first draw. Faculty member 228, who was actually chosen on the first draw, had a probability of $1/1000$ of selection on the first draw, just as did the remaining 780 faculty members. The total probability that some faculty member will be selected on the first draw is $781/1000$; there is a probability of $219/1000$ that none will be selected. The conditional probability that *if* a faculty member is selected it will be a particular one is $1/781$ for each faculty member.

Once the number 228 is used, the universe "shrinks." We have 999 numbers, 780 of which correspond with faculty members. By the same kind of reasoning, each of the remaining faculty members has a probability of $1/780$ of being the second member selected. Similarly, the 779 faculty members remaining after the second member is selected each has a probability of 1/779 of being the third member selected, etc.

The probability that a given faculty member will appear in the sample, if 100 are chosen, is 100/781. To see this, consider some particular faculty member and calculate the probabilities of the following 100 mutually exclusive outcomes: (1) that he is the first faculty member selected; (2) that he is the second selected; ... to (100) that he is the 100th selected. These outcomes are mutually exclusive because an individual is removed from the population once he has been selected, and therefore cannot be selected twice. When we have these 100 probabilities we will add them to get the total probability that an individual will be selected. Taking the events in order we have:

(1) The probability that the particular member will be the first selected is the probability, when none have been selected, that *he* will be, namely 1/781, as we saw before.

(2) The probability that he will be the second selected is the probability that the first faculty member selected will not be he, 780/781, times the probability that he will be the member selected from the remaining 780; that is,

$$\frac{780}{781} \times \frac{1}{780} = \frac{1}{781}.$$

(3) Similarly, his chance of being the third member selected is

$$\frac{780}{781} \times \frac{779}{780} \times \frac{1}{779} = \frac{1}{781}.$$

Proceeding in this way, each of the 100 probabilities is found to be 1/781.

Another method of drawing the sample, which would be more useful with a smaller population than with one of 781, is to write a random number opposite each name on the list, then take those with the 100 lowest numbers as the sample. If two random numbers are the same, rank them by the digits to the right in the table of random numbers. This method is particularly useful for such problems as randomly dividing a group of, say, 24, into two groups of specified sizes. Selecting a sample of 100 from a list of 781 amounts to dividing the list into two parts at random, one part to serve as the sample.

Suppose there had been only 240 faculty members. Then 76 percent of the three-digit random numbers would be "duds"—fail to correspond with any member of the population. We could improve the yield by letting the numbers 001, 251, 501, and 751 all represent the first member on the list, 002, 252, 502, and 752 represent the second member, and so on, with 240, 490, 740 and 990 representing the 240th member. Numbers from 241 to 250, 491 to 500, 741 to 750, and 991 to 000 would not correspond with any member of the list; there are only 40 such numbers, not enough to assign one to each member on the list, and assigning them to some but not other members would give those members a higher probability than the others of inclusion in the sample. With this method, only 4 percent of the numbers are "duds" for this example.

Sometimes the members of the population are not actually numbered, but numbers are assigned to them by some feature such as their location. If the 781 faculty members are listed on 52 pages, exactly 15 to a page, but only one on the last page, a two-digit random number could select the page and another two-digit random number the name on the page. Numbers above 1 for the last page would be ignored. If the number of names on a page varies, say up to 20, it would be necessary after selecting the page to select an entry number from 1 to 20. If the entry number exceeded the number of names on the particular page, it would be ignored, and a new page and entry number selected. It would *not* do to select a page, count the number of names, and then select one at random—for example by one random digit if there were only ten names. If that were done, the probability of a name being used would be inversely proportional to the number of names on the page. For example, a name on a page containing eight names would have a probability of selection, at any one selection, of

$$\frac{1}{53} \times \frac{1}{8} = \frac{1}{424} = 0.0024,$$

whereas a name on a page containing 20 names would have a proba·bility of

$$\frac{1}{53} \times \frac{1}{20} = \frac{1}{1060} = 0.0009.$$

## 10.10
## MISCONCEPTIONS ABOUT RANDOMNESS

(1) "Random" is not the same as "haphazard," "aimless," or "hit-or-miss." While randomness may be roughly described as "haphazard," it is not true that "haphazard" procedures are often, or ever, random. Randomness is usually attained by some mechanical or electronic process which has been carefully tested. Aimless or haphazard methods do not often lead to results that can be called random.

*Randomness and Probability*

If, for example, you were to attempt to write down a sequence of 300 digits in an "aimless" manner, you ought not to be surprised to find evidences of regularity that would be exceedingly difficult to reconcile with the hypothesis of randomness. (See Example 141B, Imaginary Coin Tosses.) Similarly, you cannot expect to proceed in an "aimless" or "hit-or-miss" fashion to select a random sample. Hit-or-miss selection, where someone picks cases that he *thinks* are "typical" or "representative," is even further from randomization. Human beings have all sorts of subtle, often unconscious, biases. You would not be able to pick cards deliberately so that every player had a fair share of good and bad cards. You might be overconscientious and deprive yourself; you might favor the weakest player a little; you might deal hands that one person could have played well to a person who did not see how to use them effectively; you might deal monotonously "typical" hands. It is best to leave the dealing to chance: then everyone knows what kind of process is operating, and has a basis for inferring from the cards he sees what the cards are that are hidden from him. In duplicate bridge, for example, where the hands are duplicated and played by different players, it is always insisted that the hands be formed originally by the usual methods of shuffling, cutting, and dealing.

(2) Often it is not appreciated how difficult it is to achieve randomness.

Example 338  1940 Draft Lottery

A classic illustration of inadequate randomization on a grand scale is the lottery used in establishing order numbers for the draft in 1940. Ten thousand numbers were written on slips of paper, the slips were put in capsules, and the capsules were put into a bowl and mixed. The capsules were then drawn by various blindfolded dignitaries in a public ceremony. The results showed marked departures from randomness. Apparently the difficulties of adequate mixing were not understood.[10]

(3) Another misconception is that a random sample is necessarily a "representative sample" or a "true cross section." Unless enough is known about a population to make sampling unnecessary, one cannot guarantee that *any* sampling method, random or other, will produce a "representative sample."

It might be that only 100 of the 781 faculty members at a university had served more than 15 years, yet it is *possible* that all 100 of these might be included in the simple random sample of size 100.

---

10. See statement by Samuel A. Stouffer and Walter Bartky, *Chicago Tribune*, November 2, 1940, p. 4.

The probability of this happening is fortunately extremely small. But if it did happen, and if the sampling process were "known" to be random, it would have to be acknowledged that the sample was in fact a random but a most "unrepresentative" sample. Whenever the term "representative" is used to describe a sample, it is necessary to examine the statement carefully to see what is meant, since it is impossible to assure the selection of a sample that will be representative of the population in regard to characteristics not known in advance of sampling.

(4) Still another common misconception is that randomness is less important in large samples than in small ones. Unless the sample is so large as to contain nearly the whole of the population, this is groundless. Indeed, departures from randomness that in small samples would be disregarded as possibly sampling error will, because of the law of large numbers, be unmistakable in large samples. Had the apparent relation between smoking and cancer (Example 287) been observed in about 180 men instead of about 180,000, there would have been little interest. It is only those errors or fluctuations due to randomness in itself that are reduced in large samples; errors due to nonrandomness have just as much effect on large as on small samples. The *Literary Digest* presidential poll based on a sample of over two million (Example 74A) is a case in point.

## 10.11
## OTHER PROBABILITY SAMPLING METHODS

Simple random sampling is only one of the methods of assigning a known, nonzero probability to every element in a population. There are many other probability sampling processes that accomplish the same thing, and some of these are more useful in certain sampling problems than is simple random sampling. Some of the more interesting and useful of these processes are discussed in Chap. 15. The following brief descriptions, however, hint at some of the possibilities.

(1) *Stratified Sampling.* In this type of sampling, already referred to in Sec. 4.6.3 and Sec. 9.6.1, the population is divided into groups according to some relevant characteristic, and a simple random sample is taken from each group.

(2) *Cluster Sampling.* This is best illustrated by an example. To draw a sample of people from a given state, a random sample of the counties might be selected, from the counties random samples of city blocks and open country areas, and from these simple random samples (or sometimes 100 percent samples) of the people living in them.

*Randomness and Probability*

(3) *Systematic Sampling.* A starting point in the first *i* observations is chosen at random, then every *i*th observation thereafter is chosen. Here every observation has the same probability of being included, as in simple random sampling, but the probabilities are not independent.

For the main development of the principles of statistical inference, however, we shall deal only with simple random sampling. There are three reasons for this. In the first place, all probability sampling designs are built up from simple random sampling. Before this process of "building up" can be understood, simple random sampling and the techniques directly based upon it must be understood. Second, simple random sampling frequently does apply directly to practical problems. Third, the heart of the ideas of statistical inference can be grasped most easily in connection with simple random samples.

## 10.12
## CONCLUSION

The uncertainty which always attaches to a conclusion about a population based on a sample can be measured in terms of probability. By the probability that an event will have a certain outcome we mean the relative frequency of occurrence of that outcome in independent repetitions of the event. The conditional probability of an outcome is the probability of that outcome under special conditions, such as that another event has had a certain outcome. Two events are independent if the probabilities of their various outcomes are the same no matter how the other turns out.

In calculating the probabilities of complex outcomes from the probabilities of simple ones, two rules are especially useful:

(1) The probability that one or another of several mutually exclusive outcomes will occur (that is, outcomes of which only one can occur) is the sum of their separate probabilities. If the outcomes are not mutually exclusive, this rule can be adjusted; for example, if not more than two of the outcomes can occur simultaneously, the sum of the probabilities for all pairs of simultaneous outcomes is subtracted from the sum of the probabilities of individual outcomes.

(2) The probability that each of a sequence of outcomes of independent events will occur is the product of their probabilities. If the events are not independent, the probability is given by the product of the conditional probabilities, the conditions for each being the occurrences of the preceding outcomes of the sequence.

Randomness means, essentially, selection in such a way that the laws of probability can be applied. This can be achieved by assuring each group of *n* members of a population the same chance of constituting the sample. The methods of achieving this amount to assigning a number to each member of the population, selecting some of the numbers by using published tables of random digits, and taking as the sample those members of the population whose numbers were selected. This is called simple random sampling. More complicated probability sampling methods—methods that assign known, nonzero, but not necessarily equal probabilities of selection to each member of the population—rest ultimately on simple random samples.

Random observations are unpredictable individually even if the population is known, but predictable in the mass. That is, the relative frequency of occurrence of the various results converges, as sample size increases, to the population relative frequency, and in this sense it is predictable. Thus, we know quite accurately what proportion of newborn infants will die within a year, but we cannot predict which ones.

With these ideas of probability, conditional probability, randomness, independence, unpredictability of individual events and predictability of relative frequencies, and with the addition and multiplication rules for calculating probabilities, we are now ready to construct sampling distributions—those patterns of sampling variability which, we showed in Chap. 4, are the background against which any particular sample is interpreted. This is the task of the next chapter.

## DO IT YOURSELF

EXAMPLE 341A

Give several illustrations of populations, specifying each one as carefully as you can. Which are finite and which are infinite? Explain.

EXAMPLE 341B

Comment critically on the following quotation:

> Miss Deanne Skinner of Monrovia, California, asks: Can the Wizard tell me what the odds are of the next President of the United States being a Democrat? . . . Without considering the candidates, the odds would be 2 to 1 in favor of a Republican because since 1861 when that party was founded, there have been 12 Republican Presidents and only 7 Democrats.[11]

---

11. Leo Guild ("The Wizard of Odds") *What Are the Odds?* (New York: Pocket Books, Inc., 1949), p. 202. Originally published under the title *You Bet Your Life* by the Marcel Rodd Company, 1946.

EXAMPLE 342A

The following questions are based on Table 280:
What is the probability that one of the 643 women selected at random will be:
  (1) a senior?
  (2) a plain senior?
  (3) a homely undergraduate?
  (4) a good-looking graduate student?
  (5) beautiful and homely?

EXAMPLE 342B

(1) Without looking at the digits in Table 321, devise an unambiguous forecasting rule for predicting each successive observation *solely on the basis of observations which have already occurred*. Write down your rule explicitly, so that it leaves no room whatever for discretion once you start to apply it. Make your rule as elaborate as you wish, but keep it unambiguous. Assume that the observations in Table 321 occur in sequence from left to right across each line and then down to the next one. Apply your rule to the entire table and keep score on your proportion of successes. What do you conclude about your forecasting method?

(2) Study Table 321 to see if you can devise another unambiguous rule which will improve your forecasting performance, and see what proportion of correct forecasts it yields.

(3) Try out the rule evolved in Step (2) for a series of predictions beginning with a haphazard starting point in Table 632. What is your proportion of successes?

(4) Did you get a higher proportion of successes in Step (2) or Step (3)? Would you have expected to get a higher or lower, or the same, proportion of successes? Explain.

EXAMPLE 342C

Comment critically on the following quotation:

> ... Nothing ... is more difficult than to convince the merely general reader that the fact of sixes having been thrown twice in succession by a player at dice, is sufficient cause for betting the largest odds that sixes will not be thrown in the third attempt. A suggestion to this effect is usually rejected by the intellect at once. It does not appear that the two throws which have been completed, and which lie now absolutely in the Past, can have influence upon the throw which exists only in the Future. The chance for throwing sixes seems to be precisely as it was at any ordinary time. ... And this is a reflection which appears so exceedingly obvious that attempts to controvert it are received more frequently with a derisive smile than with anything like respectful attention. The error here involved—a gross error redolent of mischief—I cannot pretend to

*Do It Yourself*

expose within the limits assigned me at present; and with the philosophical it needs no exposure. . . .[12]

EXAMPLE 343A

In a certain small part of Iowa, crop-devastating hailstorms have occurred in the past about once every 17 years, on the average. A student whose family owned a farm in that area, and carried no insurance against crop devastation by hailstorms, asked us whether, since it had been nearly 17 years since the last devastating hailstorm, it was an especially favorable time to buy the insurance. What answer should we have given?

Suppose, contrary to fact, that the question concerned drought insurance. Should the answer be modified? How about 17-year locusts?

EXAMPLE 343B

The following questions all require probability calculations for the ten-sided die. What is the probability of:

(1) an odd number?
(2) an even number?
(3) an odd or an even number?
(4) an even number or a number in excess of 7?
(5) a 4, then a 7?
(6) a 7, then a 4?
(7) not getting a 4 in two spins?
(8) not getting a 4 in five spins?
(9) 4 4 4 4 4?
(10) 4 7 4 7 4    7 4 7 4 7?
(11) 3 1 4 1 5    9 2 6 5 3?
(12) 8 9 0 2 4    3 2 0 5 4?

Show all your calculations.

EXAMPLE 343C

Analyze Example 96A in the light of the rules for probability calculation given in this chapter.

Show and explain any necessary calculations.

EXAMPLE 343D

In the following exercises, show all calculations and explain your reasoning where it is not obvious from the calculations.

(1) What is the probability of drawing two hearts successively from a deck of well shuffled cards?

---

12. Edgar Allan Poe, "The Mystery of Marie Rogêt," final paragraph, in *Murders in the Rue Morgue.*

(2) What is the probability of a royal flush in five-card stud poker? That is, what is the probability of one of the four following hands?

| | | | | | |
|---|---|---|---|---|---|
| A, | K, | Q, | J, | 10 | (All hearts) |
| A, | K, | Q, | J, | 10 | (All spades) |
| A, | K, | Q, | J, | 10 | (All diamonds) |
| A, | K, | Q, | J, | 10 | (All clubs) |

(3) What is the probability of obtaining a total of 7 in one roll of two six-sided dice? In two rolls of one six-sided die?

(4) The probability of a white male aged 60 dying within one year is (as of 1951) .023, and the probability of a white female aged 55 dying within one year is .008. If a man and his wife are 60 and 55 respectively, what is the probability of their both living a year? Of at least one of them dying within a year? Of at least one of them living a year? How do you interpret the meaning of these probabilities for a particular couple of these ages?

EXAMPLE 344A

Use Table 632 to select a simple random sample of $n = 10$ from the following populations:

(1) A list of 30 of your friends and acquaintances.
(2) The pages of this book.
(3) The lines of this book, omitting tables and charts.

Record your results. What apparently "non-random" features do you observe?

EXAMPLE 344B

In Sec. 10.11 we gave this definition of systematic sampling: "A starting point in the first $i$ observations is chosen at random, then every $i$th observation thereafter is chosen." Suppose $i = 5$ and that there are 20 observations in the population. What is the probability that the first observation in the population will be included in the sample? The 17th? The 19th?

What is the probability that the following observations will be included:

(1) 2nd, 7th, 12th, 17th?
(2) 2nd, 6th, 12th, 17th?
(3) 1st, 2nd, 6th, 11th, 16th?

EXAMPLE 344C

Suppose that a person asserts that by spinning a silver dollar a certain way, he can get heads most of the time. He gets 18 heads consecutively. Which side of an even money bet would you prefer on the 19th spin, if you would have a preference? Explain.

# Chapter 11

# Sampling Distributions and the Normal Distribution

## 11.1
### THE NATURE OF A SAMPLING DISTRIBUTION

In Sec. 4.3 we described a demonstration in which a sample of 20 beads was drawn at random from a large population. This first sample contained 13, or 65 percent, red beads. Before inferences could be drawn about the population, except such trivial ones as that some but not all of the beads in the population are red, it was necessary, we saw, to know the pattern of the various results that might equally well have occurred in random sampling.

Suppose, to be concrete, that someone had claimed that the proportion, $P$, of red beads in the box was more than 0.37, and the sample had been drawn to test this claim. The reasoning by which we reach a conclusion about the truth of the claim is a bit reminiscent of a type of reasoning called *reductio ad absurdum* that you may remember from plane geometry. This type of reasoning starts by *assuming* tentatively that the proposition under investigation is false. It then examines the implications of assuming it false, with a view to showing that these implications are absurd—that they fly in the face of known facts. If such a contradiction can be shown, the assumption is proved to be wrong; hence the original proposition is proved true. On the

**345**

*Sampling Distributions and the Normal Distribution*

other hand, if the assumption is not proved wrong, that does not prove that the original proposition is true.

Here, too, we start by *assuming* that it is not true that $P$ exceeds 0.37. We assume that $P$ is exactly 0.37. Then we examine the implications of this assumption. If it is true that $P$ is 0.37, what value will $p$ have in a random sample of 20? Here things become more complicated than in plane geometry. We cannot say precisely what value $p$ will have. It *might* be 0, it *might* be 1, or it *might* be any whole multiple of 0.05 between 0 and 1. But while none of these values is impossible, some of them are improbable if $P$ is 0.37. If one of the improbable results does in fact occur, we say that the assumption flies in the face of this fact. We say so not with certainty, as in plane geometry, but with a degree of confidence that is higher, the less the probability of the observed result. If we reject the assumption that $P$ is 0.37, as inconsistent with the facts, we accept the claim that $P$ exceeds 0.37.

To apply this method, which might be described as "reduction to improbability," we have to be able to deduce from the assumption that $P$ is 0.37 the probabilities of the various possible values of $p$. Such a set of probabilities, deduced from an assumption about the population, is a *sampling distribution*. The principles of probability explained in Chap. 10 make it possible to deduce the sampling distribution of a statistic from an assumption about the population.

In this way we can test whether some specific assumption about the population parameter is valid. We can also find a range within which the population parameter may be presumed to lie in view of the facts observed in a sample, by simply taking all values of the parameter which lead to sampling distributions for which the observed sample is not too improbable. The first of these two types of inference, where we start with a specific idea about the population, is called *testing hypotheses*, or sometimes just *testing*. The second type, where we start from the observations and derive from them an idea about the population, is called *estimation*.

A sampling distribution, it is important to note, is deduced from assumptions made for the purpose of testing their consistency with the observed facts. We assumed, in the case we have been discussing, that the sample of 20 was drawn at random from the population. And we assumed that the proportion red in the population was 0.37. There is no limit to the number of sampling distributions we could have for this situation; every value we might assume for the population proportion leads to a different sampling distribution. If we assume 63 percent red beads in the population, for example, we obtain a sampling distribution for samples of 20 which indicates that

## 11.1 Nature of a Sampling Distribution

65 percent is a very likely observation, hence entirely consistent with the idea that the population percent is 63.

A second thing that it is important to remember about sampling distributions, besides the fact that they are deduced from assumptions about the population and the method of sampling, is that they relate to statistics. For each possible value of some statistic, a sampling distribution shows the probability of a sample in which the statistic has that value. With the same assumptions about the population, the sampling distribution will be different according to what statistic we are interested in—the mean or proportion, the standard deviation, the median, the range, etc.

Thus, when we speak of a sampling distribution, we have to specify the population, the sample size, and the statistic. (We do not ordinarily specify randomness, since random samples are the only ones for which sampling distributions can be found—that is the reason randomness is essential—but if some more complicated probability sampling methods are involved we must specify them.) For the present example, then, we would discuss the distribution of the sample proportion in samples of 20 drawn from a population in which the proportion is 0.37.

Such a population, incidentally, in which the observations are classified simply as having or lacking a certain characteristic (dichotomous) is called a *binomial population*, and a sampling distribution derived from it is called a *binomial distribution*. There are innumerable different binomial populations, as many as there are different sample sizes and different values of the population parameter, the proportion having the characteristic.

The particular distribution we are discussing would be called "a binomial distribution with parameter 0.37 and sample size 20," or "with $P = 0.37$ and $n = 20$."

A sampling distribution is, then, essentially a description of one population which is derived from another, or parent, population. It describes a population in which each measurement is a statistic computed from a sample from the parent population. The parent population in our example consists of a large, but finite, set of numbers, of which 37 percent are 1's and 63 percent are 0's. From this we derive another, and infinite, population by drawing samples of 20, finding the sample proportion, and returning the sample to the parent population. In this derived population, or population of sample proportions, there appear not just two but 21 different numbers, 0, 0.05, 0.10, . . . , 0.95, and 1. When we find the proportions in which these 21 numbers appear in the population of sample proportions, we call the result the sampling distribution of the sample proportion in samples of 20 from a binomial population with parameter 0.37—or, more compactly, the binomial distribution for $n = 20$ and $P = 0.37$.

In this chapter we will consider a number of sampling distributions. We will show, however, that one special distribution, the standard normal distribution, will serve in many and varied practical situations; it is a veritable boy scout knife of a distribution. We shall, therefore, study in some detail exactly how to use the standard normal distribution. It is the only sampling distribution used in the rest of the book. Since it is used innumerable times in a variety of situations, its mechanics must be mastered; fortunately, they are not complicated.

## 11.2
## HOW SAMPLING DISTRIBUTIONS ARE DEDUCED

### 11.2.1 An Illustrative Calculation for a Binomial Population

One of the important accomplishments of mathematical statisticians is the deduction of sampling distributions from the principles of mathematical probability. In using or interpreting statistics, it is not necessary to know the details of these derivations, for the results are available in convenient tables; but it is essential to have some idea of the nature of the principles underlying the tables.

The procedure in deriving a sampling distribution is as follows:

(1) We start with a population which we assume to be completely known. You must fix this fact in mind through the entire discussion which follows. We assume the population to be known in order to see what would happen if this were the true population. Thus, the procedure is deductive. Later, comparison of the deductive results with the empirical results ("the facts") will be the basis for our inference about the actual population.

(2) We assume simple random sampling. For theoretical simplicity only, it will be assumed also that the population is so large relative to the sample that we can ignore the changes in the population distribution that occur as the sample is drawn. Strictly speaking, when a particular observation is drawn, there is less probability of a similar observation on subsequent draws, since this type of observation is now relatively less frequent in the remainder of the population; but we shall assume here that the reduction is negligible. Later (Sec.11.4.3) we will see that this assumption is ordinarily correct.

## 11.2 How Sampling Distributions Are Deduced

(3) From these assumptions about the population and about the nature of the process of sampling from the population, together with our knowledge of probability theory, we deduce the sampling distribution of a statistic computed from samples drawn from this population. The theoretical sampling distribution is also called a *probability distribution* because it distributes the total probability (1) among the different possible sample outcomes.

It happens that the sampling distribution of the sample proportion, $p$, is the easiest to deduce, so it will be used to illustrate the procedure. Here are the assumptions which we shall use for an illustrative example.

(1) The population consists of 3,000,000 Illinois private passenger car license numbers, of which 1,111,111 begin with "1" (see Example 330B). Thus the true proportion, $P$, of numbers beginning with 1 is 0.37, to two decimals. As we saw in Sec. 4.3.1, such a population is formally identical with many populations of practical interest, such as a production lot of 1,000 items, some of which are defective and others non-defective, or a group of people inoculated against a disease, some of whom have contracted the disease and some of whom have not.

(2) The sample size is 5.

(3) The sample statistic is the proportion, $p$, of licenses beginning with 1.

Given this population and this sample size, the only possible values of $p$ are 0, 0.2, 0.4, 0.6, 0.8, and 1, corresponding with 0, 1, 2, 3, 4, and 5 initial 1's in the sample. To find the sampling distribution, we apply the multiplication and addition rules of Sec. 10.6 to find the probabilities of samples leading to each possible value of $p$.

*For $p = 0$:* The only way for this to happen is for a digit other than 1, which we will denote by 0 (representing the number of 1's in the first position of the license number), to occur all five times when sample observations are drawn. The probability of this is $1 - 0.37 = 0.63$ on each draw, and the probability of its happening five times is, by the multiplication rule,

$$0.63 \times 0.63 \times 0.63 \times 0.63 \times 0.63 = 0.099,$$

or just under 1 chance in 10.

*For $p = 0.2$:* There are five mutually exclusive ways for this to happen, namely any one of the following sequences of observations:

$$10000 \quad 01000 \quad 00100 \quad 00010 \quad 00001.$$

The probability of each of these is, by the multiplication rule,

$$0.37 \times 0.63 \times 0.63 \times 0.63 \times 0.63 = 0.05829.$$

Then the probability that some one of the five samples will occur is five times this—that is, a sum of five terms, each equal to 0.05829 —or 0.291.

*For p = 0.4:* There are ten mutually exclusive ways for this to happen, namely any one of the following sequences:

$$11000 \quad 10100 \quad 10010 \quad 10001 \quad 01100$$
$$01010 \quad 01001 \quad 00110 \quad 00101 \quad 00011.$$

Each has probability

$$0.37 \times 0.37 \times 0.63 \times 0.63 \times 0.63 = 0.03423.$$

The sum of the probabilities for the ten sequences is then 0.342.

*For p = 0.6:* We can list the possibilities here by simply interchanging 0's and 1's in the list for $p = 0.4$:

$$00111 \quad 01011 \quad 01101 \quad 01110 \quad 10011$$
$$10101 \quad 10110 \quad 11001 \quad 11010 \quad 11100.$$

Each probability is

$$0.37 \times 0.37 \times 0.37 \times 0.63 \times 0.63 = 0.02010,$$

and the sum is 0.201.

*For p = 0.8:* The possibilities are obtained by interchanging 0's and 1's in the list for $p = 0.2$:

$$01111 \quad 10111 \quad 11011 \quad 11101 \quad 11110.$$

Each probability is

$$0.37 \times 0.37 \times 0.37 \times 0.37 \times 0.63 = 0.01181,$$

and the sum is 0.059.

*For p = 1:* The only sample leading to this result is 11111, for which the probability is

$$0.37 \times 0.37 \times 0.37 \times 0.37 \times 0.37 = 0.007.$$

These results are summarized in Table 351.

The use of such a sampling distribution is a subject for later chapters. Here we will simply remark that to use it, we must make some decision as to the boundary between "reasonably probable" and "improbable." For example, if we consider improbable those events that would occur only one time in a hundred, or less frequently, then the only samples of five that can be regarded as inconsistent with the notion that the population proportion is 0.37 would be those in which all the license numbers begin with 1. If we require a rarity of one in a thousand, no sample of five can lead us to reject the notion that $P = 0.37$. On the other hand, if we regard an event which

## 11.2 How Sampling Distributions Are Deduced

TABLE 351

DISTRIBUTION OF THE SAMPLE PROPORTION IN SAMPLES OF 5
FROM A BINOMIAL POPULATION WITH PARAMETER 0.37

| Sample Proportion $p$ | Probability | |
|:---:|:---:|:---:|
| | Specific | Cumulated[a] |
| 0 | 0.099 | 0.0992 |
| 0.2 | 0.291 | 0.3907 |
| 0.4 | 0.342 | 0.7330 |
| 0.6 | 0.201 | 0.9340 |
| 0.8 | 0.059 | 0.9931 |
| 1 | 0.007 | 1.0000 |
| Total | 1.000[a] | |

[a] For purposes of comparisons to be made later, these figures have been cumulated from figures computed to more decimal places than those shown in the "Specific" column.

occurs one time in ten, or less often, as improbable, we can regard values of $p = 0$ as inconsistent with the idea that $P = 0.37$.

The last level of improbability, 0.1, brings up a new point. We could *not* reject the notion that $P = 0.37$ if *any* result having a probability less than 0.1 occurs. In this case, such results have a total probability of $0.099 + 0.059 + 0.007 = 0.165$, so we would find "improbable" outcomes occurring about one time in six, not one in ten, if we included them all. If we work at a one in ten level of improbability (called, as we shall see in Chap. 12, a 0.1 level of significance), the whole group of outcomes to be regarded as improbable must have a total probability not more than 0.1. Then if $P$ really is 0.37, there will be only a 10 percent risk that sampling error will lead us to reject that value. As a matter of fact, there are typically many ways to select a group of results having a total probability of only 0.10 if the assumption is true; and in Chap. 12 we will have to consider how to choose among them.

To find a sampling distribution like that of Table 351, it is not necessary to make the calculations we have presented for illustrative purposes. Instead, published tables can be used.[1] Furthermore, as

1. For values of $n$ up to 150, the most detailed table published so far is Office of Chief of Ordnance, *Tables of the Cumulative Binomial Probabilities* (Ordnance Corps Pamphlet, No. ORDP 20-1) (Washington, 1952). This covers values of $P$ from 0 to 1 by steps of 0.01.

Somewhat less detail for values of $n$ up to 150, plus coverage for selected values of $n$ up to 1,000, is given by Staff of the Computation Laboratory of Harvard University, *Tables of the Cumulative Binomial Probability Distribution* (Cambridge, Mass.: Harvard University Press, 1955). This includes a number of common fractions that are not whole multiples of 0.01, namely, all whole multiples of $\frac{1}{12}$ and of $\frac{1}{16}$.

Another useful table is National Bureau of Standards, *Tables of the Binomial Probability Distribution*, which covers values of $P$ from 0 to 1 by steps of 0.01, for $n$ up to 49.

we shall see in Sec.11.5, even these tables can usually be replaced by a single table (the standard normal distribution) by means of simple approximations.

11.2.1.1 *Effect of Varying the Parameter.* To show how the sampling distribution for samples of five from a binomial population depends on the parameter $P$, we have prepared Fig. 352. This shows the

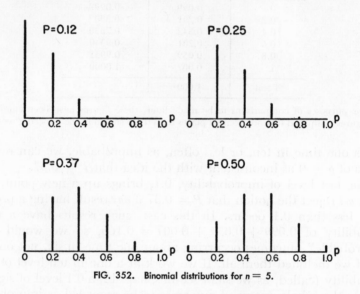

FIG. 352.  Binomial distributions for $n = 5$.

sampling distribution of $p$ for $n$ of five, and four values of $P$, 0.12, 0.25, 0.37, and 0.50. At each value of $p$, the vertical line is proportional to the probability of that value of $p$.

For $P = 0.12$, the distribution is concentrated about the small values of $p$, 0, 0.2, and 0.4, with very small probabilities for larger values of $p$. The mode is at $p = 0$, and the distribution is of a general shape described as "skewed to the right."

For $P = 0.25$, the distribution shows wider dispersion, the mode is 0.2, and the distribution is still skewed to the right.

For $P = 0.37$ (the distribution shown in Table 351), there is still more dispersion. The mode is at 0.4, and there is little skewness.

Finally, for $P = 0.50$, the dispersion is greatest. The distribution is symmetrical, with modes at both 0.4 and 0.6.

Let us follow a particular value of $p$, say 0.4, from one distribution to the next. As $P$ increases, the probability that $p$ will be 0.4— $Pr(p = 0.4 \mid P)$—increases, reaching its highest value among these four distributions when $P = 0.37$; thereafter, the probability that

### 11.2 How Sampling Distributions Are Deduced

$p = 0.4$ declines as $P$ increases. Had we shown the distribution for $P = 0.4$, the probability that $p = 0.4$ would have been still larger than when $P = 0.37$, and in fact larger than for any other value of

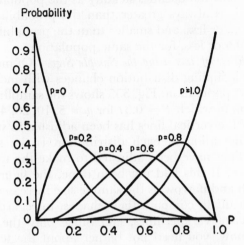

FIG. 353A. Probability that $p = 0, 0.2, 0.4, 0.6, 0.8, 1.0$ in a sample of five as a function of $P$.

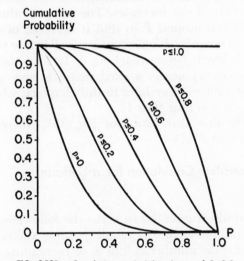

FIG. 353B. Cumulative probability that $p \leq 0, 0.2, 0.4, 0.6, 0.8, 1.0$ in a sample of five as a function of $P$.

$P$. Fig. 353A shows how the probabilities of various values of $p$ vary with $P$. Each value of $p$ attains its highest probability under the assumption that $P = p$. Fig. 353B shows the same information as

Fig. 353A, but for the cumulated probabilities. It is somewhat easier to read because the curves slope in the same direction all the way and do not cross one another. Thus, the probability of a sample proportion of 0.4 *or less* declines steadily as the population proportion increases; but it is always greater than the probability of a sample proportion of 0.2 or less, and smaller than the probability of a sample proportion of 0.6 or less, for the same population.

11.2.1.2   *Effect of Increasing the Sample Size.*   Of more importance is the way the sampling distribution changes with increasing sample size, for a fixed population. Fig. 355 shows the sampling distributions for a population in which $P = 0.37$ for $n = 5, 10, 20, 40, 80,$ and 160. The spacing of the vertical lines has been adjusted in such a way that the standard deviation of each distribution is the same, and the heights of the lines have been adjusted to make the total area of the figures the same. If this had not been done, the figures would have become so high and narrow as the sample size increased that it would have been difficult to compare their shapes, in regard to symmetry, relative heights of various lines, etc. Since only the shapes are of interest to us now, you need not bother about the technical details of the scales.[2]

The asymmetry which is noticeable for the small samples disappears as the sample size increases. The sample values of $p$ become sufficiently clustered around $P$ so that it becomes of no importance that they *could* go considerably further above $P$ before reaching 1 than they could go below it before reaching 0. In fact, the sampling distribution takes on very nearly a fixed shape, known as a *normal distribution* (of which much more later in this chapter), when $n$ is not too small and $P$ is not too near 0 or 1.

Note that the first distribution of Fig. 355, where $n = 5$, is the one shown in Table 351.

## 11.2.2   An Illustrative Calculation for a Uniform Population

For a second illustration, let us make the following assumptions:

(1) The population consists of all possible digits produced by tossing a ten-sided die numbered 0 to 9, or by reading a digit from a

---

2. A technical note on the scales of Figs. 355, 356, and 358: The unit of length used on the horizontal scale is the standard deviation of the particular distribution, and the scales are placed so that their means, at $P = 0.37$, are aligned vertically. The vertical scale is adjusted so that the total area would be 1 if the vertical lines were replaced by bars centered on the lines and of width equal to the distance between possible values of the variable; this adjustment is made by multiplying the actual probability by $n$ times the standard deviation of the distribution.

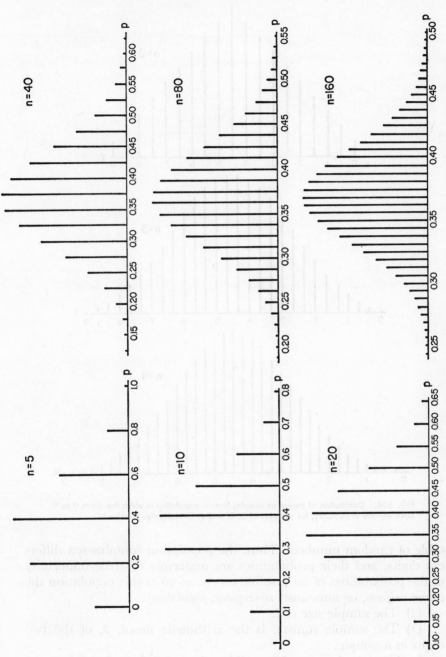

FIG. 355.   Distribution of sample proportions, p, from a binomial population with P = 0.37 for samples of 5, 10, 20, 40, 80, and 160.

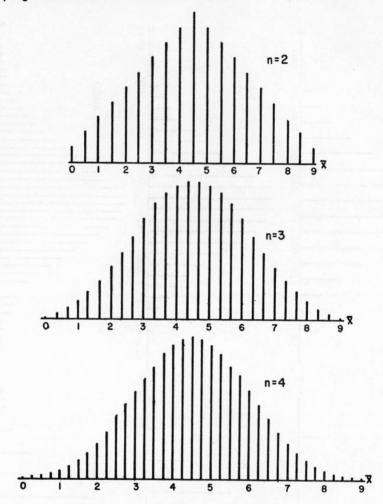

FIG. 356. Distribution of means of samples from a population in which the digits 0 to 9 have uniform probability, for samples of 2, 3, and 4. (Standardized scales.)

table of random numbers. Thus, the population contains ten different digits, and their probabilities are uniformly 0.1 (the uniformity of the probabilities of all possible outcomes gives this population the name *uniform*, or sometimes *rectangular*, *population*).

(2) The sample size is 2.

(3) The sample statistic is the arithmetic mean, $\bar{x}$, of the two digits in a sample.

In this case, there are 19 possible values of $\bar{x}$, from 0 to 9 by steps of $\frac{1}{2}$. As an example of the calculations involved in obtaining the sampling distribution, consider the probability that $\bar{x} = 2$. This means that the sum of the two observations must be 4. For this to happen, one of the following pairs of observations must occur:

$$0,4 \qquad 1,3 \qquad 2,2 \qquad 3,1 \qquad 4,0.$$

Since each digit has probability 0.1, the probability of each pair is 0.01, by the multiplication rule. Since the outcomes are mutually exclusive, the probability that one of them will occur is the sum of their individual probabilities, or 0.05.

Again, for a mean of 6, one of the following samples must occur:

$$3,9 \qquad 4,8 \qquad 5,7 \qquad 6,6 \qquad 7,5 \qquad 8,4 \qquad 9,3$$

and the probability is 0.07.

The whole sampling distribution is shown (on a standardized scale) at the top of Fig. 356; it is triangular.

11.2.2.1 *Effect of Increasing Sample Size.* Fig. 356 also shows the sampling distributions for the same population but samples[3] of sizes 3 and 4. As in Fig. 355, the horizontal scales have been adjusted to keep the standard deviations the same, and the heights adjusted correspondingly to keep the areas of the figures the same, thus making it possible to compare shapes. Again, we see that as the sample size increases the sampling distributions approach a fixed shape, and again it is the normal distribution. Indeed, the difference between a normal distribution and a curve through the tops of the lines in the diagram for $n = 4$ is too small to show on the scale of Fig. 356; the maximum difference occurs at the peak, and is only 3.6 percent of the height of the center line. Even for $n = 3$, the maximum difference is only 6.9 percent of the height of the center line. You may, in fact, keep in mind a smooth curve through the tops of the lines for $n = 4$ as a very good picture of the shape we mean when we talk about a normal population.

## 11.3
## THE NORMAL DISTRIBUTION

### 11.3.1 The Central Limit Theorem

The tendency illustrated in Sec. 11.2 for sampling distributions to take on a common shape as larger samples are considered is a prevalent one. This approach to normality occurs in the sampling

---

3. This sampling distribution for $n = 3$, but for six-sided dice, was first obtained by Galileo (1564–1642) in the earliest known publication on probability. See F. N. David, "Dicing and Gaming (A Note on the History of Probability)," *Biometrika*, Vol. 42 (1955), pp. 1–15, especially pp. 11–13.

### Sampling Distributions and the Normal Distribution

distributions of many of the statistics which are of practical importance. There is, in fact, a general law that, almost regardless of the shape of the original population, the shape of sampling distributions derived from it by considering the statistics commonly computed from samples will be approximately normal. This law can be proved mathematically, and the conditions under which it holds stated more precisely. It is known as the *central limit theorem*. Some statistics, however, are not subject to it; the range is an important example.

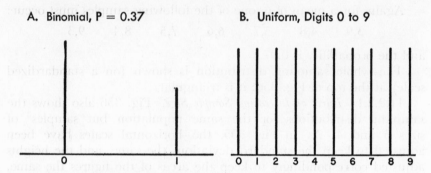

FIG. 358. Binomial and uniform populations. (Standardized scales.)

If the original population were itself normal, the sampling distribution of the mean would be exactly normal, no matter how small the sample. The sampling distribution of the standard deviation, however, would not be exactly normal, even if the original population were exactly normal; but it would rapidly approach normality as the sample size increased. Even when the original population is far from normal, however, the distributions derived from it ordinarily approach normality. Indeed, the original populations considered in Secs. 11.2.1 and 11.2.2 were not at all like normal populations; see the diagrams of them in Fig. 358, which uses the same standardized scales used for Figs. 355 and 356. Nevertheless, the distribution of means of samples drawn from them is approximately normal if the samples are large enough, as Figs. 355 and 356 show. (Remember that a sample proportion is a special case of a mean.)

The important fact stated by the central limit theorem, that the sampling distributions of common statistics tend to be approximately normal, almost regardless of the shape of the original population, results in enormous simplifications. It means that a wide class of important problems can be solved, to satisfactory practical approximations, by this single pattern of sampling variability.

### 11.3 Normal Distribution

The normal distribution was first discovered by the English mathematician DeMoivre (1667–1754). It was later rediscovered and applied in science (both natural and social) and in practical affairs by the French mathematician Laplace (1749–1827). It was also extensively developed and utilized by the German mathematician, physicist, and astronomer, Gauss (1777–1855). One of the first to make extensive use of the normal distribution in social statistics was the Belgian astronomer and statistician Quetelet (1796–1874). A pioneer in its application to biological data was the English anthropologist, biometrician, criminologist, geneticist, meteorologist, psychologist, and statistician, Sir Francis Galton (1822–1911)—a cousin of Charles Darwin. Galton's boundless admiration for the normal distribution was expressed with amusing Victorian enthusiasm:

> I know of scarcely anything so apt to impress the imagination as the wonderful form of cosmic order expressed by the "Law of Frequency of Error." The law would have been personified by the Greeks and deified, if they had known it. It reigns with serenity and in complete self-effacement amidst the wildest confusion. The huger the mob and the greater the apparent anarchy, the more perfect its sway. It is the supreme law of Unreason. Whenever a large sample of chaotic elements are taken in hand and marshalled in the order of their magnitude, an unsuspected and most beautiful form of regularity proves to have been latent all along.[4]

A contemporary statistician, W. J. Youden, whose hobby is typography, expresses his admiration this way:

THE
NORMAL
LAW OF ERROR
STANDS OUT IN THE
EXPERIENCE OF MANKIND
AS ONE OF THE BROADEST
GENERALIZATIONS OF NATURAL
PHILOSOPHY ◆ IT SERVES AS THE
GUIDING INSTRUMENT IN RESEARCHES
IN THE PHYSICAL AND SOCIAL SCIENCES AND
IN MEDICINE AGRICULTURE AND ENGINEERING ◆
IT IS AN INDISPENSABLE TOOL FOR THE ANALYSIS AND THE
INTERPRETATION OF THE BASIC DATA OBTAINED BY OBSERVATION AND EXPERIMENT

Although modern statisticians are aware of far more applications of the normal distribution (still sometimes called the Law of Error) than Galton was, it is now realized that he overstated its universality for distributions of basic data. A wide class of phenomena turn out to be roughly normally distributed if a frequency distribution is made from a large enough number of cases; the weights and dimensions of

---

4. Quoted in Helen M. Walker, *Elementary Statistical Methods* (New York: Henry Holt and Company, 1943), p. 166.

plants or animals of a given species or of objects produced under similar conditions, the number of words on printed pages of a given size and typography, intelligence quotients, and baseball batting or fielding averages are examples. The characteristic shape is a humping up in the middle, falling off in either direction, at first with increasing steepness and then with decreasing steepness. The quotation from Youden, or the bottom part of Fig. 356, conveys the picture. On the other hand, many kinds of data depart widely from normal: the incomes of families or individuals, or the number of cars per family, or the number of wheels per car are examples.

The fundamental importance of the normal distribution in statistics, however, arises from the fact that the *measures computed from samples usually tend to be normally distributed whether or not the original data are normally distributed.*

There are, of course, exceptions. Populations are conceivable for which the distribution of certain common statistics cannot be well approximated by a normal distribution no matter how large the sample. Such populations rarely occur in practice, but populations do sometimes occur for which the approach to normality is too slow to give good approximations for samples of practicable sizes. Such is the case with binomial populations having parameters near 0 or 1. Some statistics—the range is an example—do not take on normal distributions even for large samples from normal populations. For our purposes, of explaining the principles of statistical reasoning, however, we shall make out well enough with the normal distribution.

Since the normal distribution is so useful, it will be worth our while to examine it in some detail before we turn to its application to the sampling distributions of statistics.

### 11.3.2 Characteristics of the Normal Distribution

The preceding discussion may have created an impression that "the" normal distribution possesses one inflexible form. It is true that only one table is needed for the normal distribution, but this is because all normal distributions can easily be adjusted to a standard form. There are, in fact, many normal distributions varying in location and in dispersion. Fig. 361A shows two normal distributions which differ only in location, and Fig. 361B shows three which differ only in dispersion.

The *location* of a normal distribution is measured by its arithmetic mean, $M$. This is the same as its median, because of symmetry, and it is also the same as its mode. The standard normal distribution has

*11.3 Normal Distribution*

a mean of zero. Any actual normal distribution can be converted to one of mean zero by shifting it right or left an amount equal to its mean. This is done by subtracting $M$ from each observation in the population, thereby obtaining new observations whose mean is 0.

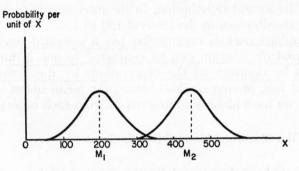

FIG. 361A.   Normal curves with two different means.
($M_1 = 200, M_2 = 450; \sigma_1 = \sigma_2 = 50.$)

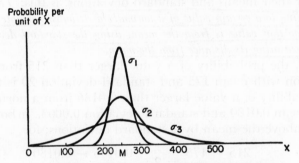

FIG. 361B.   Normal curves with three different dispersions.
($\sigma_1 = 25, \sigma_2 = 50, \sigma_3 = 100; M_1 = M_2 = M_3 = 250.$)

Thus, if we want to know the probability of a value larger than 215 in a normal distribution with mean of 175, we may ask instead for the probability of a value larger than 40 in a normal distribution with mean of 0 and the same dispersion.

The *dispersion* of a normal distribution is measured by its standard deviation, $\sigma$. As the curve of a normal distribution declines in either direction from its peak, it declines gradually at first, but gets continually steeper. At a certain point, called the *point of inflection*, it stops getting steeper, and begins to level off. The distance of either point of inflection from the mean is equal to the standard deviation. About two-thirds of the observations in any normal distribution are between the two points of inflection—that is, within one standard deviation of the mean. About ninety-five percent are within two standard deviations, and practically all (99.73 percent) are within three standard deviations. For example, in Fig. 361B, two-thirds of the observations for the first distribution are in the interval 225 to 275; for the second distribution, in the interval 200 to 300; and for the third distribution, in the interval 150 to 350.

A standard normal distribution has a standard deviation of 1. Any normal distribution can be converted to one of unit standard deviation by dividing all the observations by the actual standard deviation; this, of course, also changes the mean unless it is 0, as it will be if we have already subtracted $M$ from each observation.

### 11.3.3 The Standard Normal Distribution

The fact about the normal distribution that renders a single standard normal distribution sufficient to cover all normal distributions, whatever their means and standard deviations, is this: *The probability corresponding to a certain value of a normally distributed variable depends only on how far that value is from the mean, using the standard deviation as the unit for measuring its distance from the mean.*

Thus, the probability of a value larger than 215 from a normal population with mean 175 and standard deviation 20 is the same as the probability of a value larger than 0.0186 from a normal population of mean 0.0180 and standard deviation 0.0003. In both cases, the value is above the mean by 2 standard deviations; for

$$\frac{215 - 175}{20} = 2 = \frac{0.0186 - 0.0180}{0.0003}$$

In general, probabilities from normal distributions are found as follows:

*First*, subtract the mean, $M$, from the value whose probability is wanted. This produces a variable that is normally distributed with mean of 0, and the original standard deviation.

*Second*, divide the difference from the mean by the standard deviation, $\sigma$. This produces a variable that is normally distributed with standard deviation of 1; its mean still has the value 0 given it in step (1), for dividing 0 by $\sigma$ leaves 0.

These two steps can be summarized in a simple formula. Let $x$ be a value of a variable that is normally distributed with mean $M$ and standard deviation $\sigma$. Then

$$K = \frac{x - M}{\sigma}$$

where $K$ is the *standard normal variable*, often called a *unit normal deviate*, a *normal deviate*, or a *standardized score*. Like $x$, $K$ is normally distributed, but its mean and standard deviation are 0 and 1 no matter what $M$ and $\sigma$ are.

In going from $x$ to $K$ we have, in effect, converted from one measuring scale to another. Sometimes we want to go from the $K$-scale to the $x$-scale. In that case we use the formula

$$x = M + K\sigma.$$

To illustrate, let us assume that heights of men are normally distributed with $M = 68''$ and $\sigma = 3''$. One of the authors is $70''$ tall. To convert this $x$-measurement into $K$-units, we substitute into the formula:

$$K = \frac{x - M}{\sigma} = \frac{70'' - 68''}{3''} = +0.67.$$

This author is, therefore, two-thirds of a standard deviation above the mean. The other author's height, in units of $K$, is $+2.67$. To find his height in inches, we use the second version of the formula:

$$x = M + K\sigma = 68'' + (2.67 \times 3'') = 76''.$$

### 11.3.4 Tables of the Standard Normal Distribution

Once we have a value of $K$, we can refer to a table of the standard, or unit, normal distribution to find the required probability.

Table 365 is a compact table of the normal distribution. It shows the probability of a value larger than any particular $K$, for values of $K$ from 0 to $+3.09$ by steps of 0.01. For $K = 0.67$, for example, we look in the row headed 0.6 and the column headed 7 and find 0.2514. This means that the probability is 0.2514 (25.14 percent) that a normally distributed variable will exceed the mean of the population by 0.67 standard deviations, or more.

To illustrate, let us assume again that the heights of men are normally distributed with mean $68''$ and standard deviation $3''$. On these assumptions, what proportion of men are taller than each of the authors, whose heights are $70''$ and $76''$?

We have already seen that the standardized normal scores are $K = 0.67$ and $K = 2.67$. Referring to Table 365, we find that 25.14 percent of a normal distribution exceeds $K = 0.67$, that is, 70″. Similarly, in the line headed 2.6 and the column headed 7 we find 0.0038, indicating that 0.38 percent[5] are taller than 76″.

Suppose the question had been, What proportion of men are between the two authors in height? All of those above the shorter author are between the two authors, except those above the taller author. We therefore subtract 0.38 percent from 25.14 percent, obtaining 24.76 percent.

Possibly it has already occurred to you that negative values of $K$ are unnecessary in Table 365 because of the symmetry of the normal distribution. For example, let us ask what proportion of men are taller than 67″, under our previous assumptions. Now

$$K = \frac{67'' - 68''}{3''} = -0.33.$$

In view of the minus sign on $K$ we interpret this as showing that 37 percent are below 67″; hence 63 percent are above.

In general, it is possible to obtain from Table 365 any probabilities for normally distributed variables by making use of two facts:

(1) the normal distribution is symmetrical;

(2) the total probability is 1.

Since you may encounter other tables of the normal distribution, we must warn you that, while the tables all show the same thing, they do not all show it in the same way. What our table shows is called an "upper-tail" probability. Some tables show the probabilities for two tails, which are double ours. Some show the probability between $K$ and the mean. These and other possibilities are represented by the shaded areas of the following diagram.

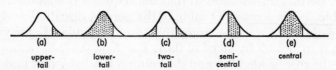

| (a) | (b) | (c) | (d) | (e) |
|-----|-----|-----|-----|-----|
| upper-tail | lower-tail | two-tail | semi-central | central |

Unfortunately, many tables fail to specify what probabilities are tabulated; and some that specify it do so in mathematical symbols

---

5. Actually, for the first probability we might have *interpolated*, that is, taken the probability as $\frac{2}{3}$ of the way from 0.2546 (shown for $K = 0.66$) to 0.2514 (shown for $K = 0.67$). The reason for doing this is that $K$ was actually 0.666 . . . , not exactly 0.67. From 0.2546 to 0.2514 is −0.32, and $\frac{2}{3}$ of this is −0.21, so we would take 0.2546 − 0.0021, or 0.2525. Ordinarily, instead of interpolating for greater accuracy, we will be rounding for less accuracy—for example, rounding the 0.2514 shown for $K = 0.67$ to 0.25—since two decimals are ordinarily sufficient for practical purposes.

## 11.3 Normal Distribution

TABLE 365

PROBABILITIES THAT GIVEN STANDARD NORMAL VARIABLES
WILL BE EXCEEDED

The probabilities shown are for the upper-tail.

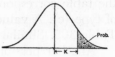

| Normal Variable | 0 | 1 | 2 | 3 | 4 | 5 | 6 | 7 | 8 | 9 |
|---|---|---|---|---|---|---|---|---|---|---|
| 0.0 | .5000 | .4960 | .4920 | .4880 | .4840 | .4801 | .4761 | .4721 | .4681 | .4641 |
| 0.1 | .4602 | .4562 | .4522 | .4483 | .4443 | .4404 | .4364 | .4325 | .4286 | .4247 |
| 0.2 | .4207 | .4168 | .4129 | .4090 | .4052 | .4013 | .3974 | .3936 | .3897 | .3859 |
| 0.3 | .3821 | .3783 | .3745 | .3707 | .3669 | .3632 | .3594 | .3557 | .3520 | .3483 |
| 0.4 | .3446 | .3409 | .3372 | .3336 | .3300 | .3264 | .3228 | .3192 | .3156 | .3121 |
| 0.5 | .3085 | .3050 | .3015 | .2981 | .2946 | .2912 | .2877 | .2843 | .2810 | .2776 |
| 0.6 | .2743 | .2709 | .2676 | .2643 | .2611 | .2578 | .2546 | .2514 | .2483 | .2451 |
| 0.7 | .2420 | .2389 | .2358 | .2327 | .2296 | .2266 | .2236 | .2206 | .2177 | .2148 |
| 0.8 | .2119 | .2090 | .2061 | .2033 | .2005 | .1977 | .1949 | .1922 | .1894 | .1867 |
| 0.9 | .1841 | .1814 | .1788 | .1762 | .1736 | .1711 | .1685 | .1660 | .1635 | .1611 |
| 1.0 | .1587 | .1562 | .1539 | .1515 | .1492 | .1469 | .1446 | .1423 | .1401 | .1379 |
| 1.1 | .1357 | .1335 | .1314 | .1292 | .1271 | .1251 | .1230 | .1210 | .1190 | .1170 |
| 1.2 | .1151 | .1131 | .1112 | .1093 | .1075 | .1056 | .1038 | .1020 | .1003 | .0985 |
| 1.3 | .0968 | .0951 | .0934 | .0918 | .0901 | .0885 | .0869 | .0853 | .0838 | .0823 |
| 1.4 | .0808 | .0793 | .0778 | .0764 | .0749 | .0735 | .0721 | .0708 | .0694 | .0681 |
| 1.5 | .0668 | .0655 | .0643 | .0630 | .0618 | .0606 | .0594 | .0582 | .0571 | .0559 |
| 1.6 | .0548 | .0537 | .0526 | .0516 | .0505 | .0495 | .0485 | .0475 | .0465 | .0455 |
| 1.7 | .0446 | .0436 | .0427 | .0418 | .0409 | .0401 | .0392 | .0384 | .0375 | .0367 |
| 1.8 | .0359 | .0351 | .0344 | .0336 | .0329 | .0322 | .0314 | .0307 | .0301 | .0294 |
| 1.9 | .0287 | .0281 | .0274 | .0268 | .0262 | .0256 | .0250 | .0244 | .0239 | .0233 |
| 2.0 | .0228 | .0222 | .0217 | .0212 | .0207 | .0202 | .0197 | .0192 | .0188 | .0183 |
| 2.1 | .0179 | .0174 | .0170 | .0166 | .0162 | .0158 | .0154 | .0150 | .0146 | .0143 |
| 2.2 | .0139 | .0136 | .0132 | .0129 | .0125 | .0122 | .0119 | .0116 | .0113 | .0110 |
| 2.3 | .0107 | .0104 | .0102 | .0099 | .0096 | .0094 | .0091 | .0089 | .0087 | .0084 |
| 2.4 | .0082 | .0080 | .0078 | .0075 | .0073 | .0071 | .0069 | .0068 | .0066 | .0064 |
| 2.5 | .0062 | .0060 | .0059 | .0057 | .0055 | .0054 | .0052 | .0051 | .0049 | .0048 |
| 2.6 | .0047 | .0045 | .0044 | .0043 | .0041 | .0040 | .0039 | .0038 | .0037 | .0036 |
| 2.7 | .0035 | .0034 | .0033 | .0032 | .0031 | .0030 | .0029 | .0028 | .0027 | .0026 |
| 2.8 | .0026 | .0025 | .0024 | .0023 | .0023 | .0022 | .0021 | .0021 | .0020 | .0019 |
| 2.9 | .0019 | .0018 | .0018 | .0017 | .0016 | .0016 | .0015 | .0015 | .0014 | .0014 |
| 3.0 | .0013 | .0013 | .0013 | .0012 | .0012 | .0011 | .0011 | .0011 | .0010 | .0010 |

The digits heading the columns are additional digits for the values of the normal variable shown in the first column. Thus, the probability corresponding with the standard normal variable 1.32 is found in the row in which "1.3" appears at the left and the column in which "2" appears at the top. The probability is 0.0934.

that may not be intelligible to you. The best way to find out what is tabulated is to remember two or three key probabilities that are useful for many other purposes as well. As we have already mentioned, the probability is approximately 95 percent that a normally distributed variable is within two standard deviations of the mean. The probability is exactly 95 percent for $K = 1.96$. Remembering that, you can look up the probability shown for $K = 1.96$. If 0.95 is shown, the table corresponds with (e) on our diagram. If 0.05, the table is of type (c). A value of 0.025 corresponds with type (a)—the type of Table 365. Similarly, 0.475 indicates type (d), and 0.975 type (b).

Another possible confusion is that some tables show both probabilities (often labeled "areas," since the area of a normal curve between two values of the variable measures the probability) and ordinates, that is, heights for plotting a normal curve; and ordinates are sometimes mistaken for probabilities.

Finally, there are some tables of the normal distribution that refer to a distribution with standard deviation $\sqrt{2}$ instead of 1; such tables are no longer common. Our $K$ would have to be multiplied by $\sqrt{2} = 1.4142$ to enter these tables.

There is another form of normal table that we shall introduce later, in which values of the probability are given and the corresponding values of $K$ are shown in the body of the table (Table 391). This is simply an inversion of tables of the kind we have been discussing, which tabulate probabilities for given normal deviates. The probabilities in these "percentage point" tables may be any of the types we have mentioned before, but almost always are either the upper-tail type (a) or the two-tail type (c).

## 11.4
## THE SAMPLING DISTRIBUTION OF THE MEAN

We have now seen that the sampling distribution of the mean is ordinarily given to a satisfactory approximation by a normal distribution having the same mean and standard deviation as the actual sampling distribution. This means that instead of laborious calculations like those shown in Sec. 11.2.1 or those illustrated in Sec. 11.2.2, all we need is to find the mean and standard deviation of the sampling distribution, compute a normal deviate, $K$, and use the tables of the standard normal distribution in the manner described in Secs. 11.3.3 and 11.3.4.

### 11.4.1 The Mean of the Sample Means

It is possible to prove mathematically that *the mean of the sampling distribution of the mean equals the mean of the original population from which the samples are drawn*.[6] This is perhaps a mean sentence to interpret,

6. This sentence is true only for simple random samples and some other probability samples. It is almost but not quite correct for some more complex kinds of probability sampling. The sampling distribution of the mean for non-probability sampling methods cannot be determined except, perhaps, approximately by repeated sampling.

but the idea, far from being complicated, is just what common sense would lead you to expect; so let us say it over again less technically:

If a random sampling process is repeated a "very large" (infinite) number of times, and the mean of each sample is computed, the average of all the sample means will equal the population mean. Thus, the average result in repeated sampling tends to equal the correct value. For that reason, we speak of the sample mean, $\bar{x}$, of a simple random sample as an *unbiased* estimate of the mean of the population from which the sample was drawn. For example, if we were to draw a large number of random samples of size $n$ from the population of heights of men and compute the mean for each sample, the average of these sample means would be very close to 68 inches, if that is the true mean of the population.

If (contrary to fact) the mean of the sampling distribution of $\bar{x}$ were, say, $M + 17$, we would call $\bar{x}$ a *biased* estimate of $M$; the bias, in this example, would be 17.

Two reminders are in order here. (1) It is *not* possible, when sampling from an unknown population, to deduce an exact numerical value of the mean of the sampling distribution of the mean. What the italicized statement above says is that the mean of the sampling distribution of the mean will equal the mean of the population, even though it may be unknown. We apply this result by *assuming* various values of the population mean, $M$, then determining whether, in the light of the sampling distribution that would follow, an observed $\bar{x}$ is consistent with the assumption. (2) The fact that $\bar{x}$ averages to $M$ in repeated sampling does not mean that $\bar{x}$ in any one sample will equal, or even be close to, $M$.

## 11.4.2  The Standard Error of a Mean

The standard deviation of the sampling distribution of the mean is called the *standard error of the mean*. Conceptually, it would be obtained by treating the collection of sample means as individual observations and computing their standard deviation just as for any other population of observations. There is, however, a simple relation between the standard error of the mean and the standard deviation of the population from which the samples making up the sampling distributions were obtained:

$$\sigma_{\bar{x}} = \frac{\sigma}{\sqrt{n}},$$

where $n$ is the size of the sample and $\sigma$ is the standard deviation of the population from which the sample is drawn. In words: the standard error of the sample mean in random samples of size $n$ is equal to the standard deviation of the population divided by the square root of the sample size. Thus, for a random sample of size 625 from the population of heights of men (where the standard deviation of the population is assumed to be 3 inches), the standard deviation of the mean, $\sigma_{\bar{x}}$, would be equal to

$$\sigma_{\bar{x}} = \frac{\sigma}{\sqrt{n}} = \frac{3}{\sqrt{625}} = \frac{3}{25} = 0.12 \text{ inches.}$$

This relationship is deduced from the basic rules of probability theory, but we shall not attempt to give even an intuitive argument for its correctness. The only condition that must be met is that the observations be independent.

As the sample size increases, the standard error of the mean diminishes. It diminishes, however, at a diminishing rate; that is, the first few observations bring about substantial reductions in the standard error, but later it takes many observations to bring about similar reductions. The second observation, for example, reduces the standard error by 29 percent from what it would have been for only one observation; the third brings about a further reduction of 18 percent, and the fourth of 13 percent. The tenth observation, however, reduces the standard error by only 5 percent from what it would be for nine observations, the fiftieth observation by only 1 percent, and the hundredth by only 0.5 percent. The standard error of the mean is inversely proportional to the square root of the sample size. To cut $\sigma_{\bar{x}}$ in half, it is necessary to quadruple the size of the sample.

Another thing to notice about the formula for the standard error of a mean is that the greater the variability in the population, as measured by $\sigma$, the larger the standard error of the mean. Common sense suggests that this must be so. For example, if all American males were exactly the same height, say 68 inches, every sample would have exactly the same mean and the standard error of the mean would be zero. On the other hand, if men's heights were frequently as small as three feet and as large as ten feet, there would be a high degree of variability from one sample mean to the next.

### 11.4.3 The Effect of Population Size on the Standard Error of a Mean

In one respect, the formula for the standard error of the mean does not agree with common sense. The formula says nothing about the *fraction* of the total population that is included in the sample. Common sense suggests (erroneously, for the most part, as we shall see) that the variability of samples of $n$ depends on how large the population is. But the formula says that a sample of given size has a standard error that depends on the variability, $\sigma$, of the population, and on the number of observations in the sample, $n$, but not on the number of observations in the population. This implies, for example, that a random sample of 100 observations from a small city would be no better for drawing inferences about that city than a random sample of 100 observations from New York would be for drawing inferences about New York, assuming the variability within both cities to be the same. Which is right, then, intuition or the formula?

The fraction of the population included in the sample does have a mild effect on the standard error of the mean in addition to that already noted for absolute sample size. Here is the complete formula:

$$\sigma_{\bar{x}} = \sqrt{\frac{N-n}{N-1}} \cdot \frac{\sigma}{\sqrt{n}},$$

where $N$ is the number of observations in the population. This formula is simply the earlier formula, $\sigma/\sqrt{n}$, multiplied by $\sqrt{\dfrac{N-n}{N-1}}$, which is known as the "finite population factor."

The finite population factor can be rewritten as follows, where "$\simeq$" means "is approximately equal to,"

$$\sqrt{\frac{N-n}{N-1}} \simeq \sqrt{1 - \frac{n}{N}} \simeq 1 - \frac{n}{2N}.$$

Now this factor will always be less than 1 unless $N$ is infinite or $n$ is 1. For example, if $n/N$, the fraction of the population included in the sample, were 20 percent or 0.20, the finite population factor would be approximately

$$1 - \frac{0.20}{2} = 0.9.$$

Hence, for a 20 percent sample the standard error of the mean is obtained by multiplying $\sigma/\sqrt{n}$ by 0.9. That is, the exact standard error of the mean in this case is about 10 percent less than is given by the formula not allowing for population size.

Thus, even with so large a fraction of the population in the sample as 20 percent, the reduction of the standard error of the mean is only 10 percent. This point may be developed further by an example.

Suppose a sample of 100 observations is drawn from a population for which the standard deviation is 100. Then the standard error of the mean is given by the formula that does not allow for population size as

$$\sigma_{\bar{x}} = \frac{\sigma}{\sqrt{n}} = \frac{100}{\sqrt{100}} = 10.$$

Now assume that the sample size and standard deviation are both the same as before, and that the population consists of 1,000,000 observations ($N = 1,000,000$ and $n/N = 0.0001$). The standard error of the mean is now computed by multiplying the previous result, 10, by

$$\sqrt{\frac{1,000,000 - 100}{1,000,000 - 1}} \simeq 1 - \frac{0.0001}{2} = 0.99995.$$

Thus, the standard error is 9.9995. Next, assume that everything remains the same as before except that the population consists of only 1,000 observations ($N = 1,000$ and $n/N = 0.1$). Then the finite population factor is approximately

$$1 - \frac{0.1}{2} = 0.95$$

and

$$\sigma_{\bar{x}} = 9.5.$$

Finally, assume there are only 200 observations in the population. Here we use the exact finite population factor, since the approximation loses accuracy for large values of $n/N$:

$$\sqrt{\frac{200 - 100}{200 - 1}} = 0.709.$$

Hence

$$\sigma_{\bar{x}} = 7.09.$$

By examining these four examples, the following conclusions may be drawn.

(1) The standard error of the mean of a sample from a finite population is smaller than for the same size sample from an infinite population of the same $\sigma$.

(2) For all practical purposes, however, the reduction in the standard error due to the limited size of a population is negligible unless the sample contains a large proportion of the total population,

as in the last of the series of examples just given. The ordinary statistical practice is to ignore the finite population factor unless the sample contains more than 20 percent (some say ten percent) of the total population. This practice may result in a small overstatement of the standard error—not more than about 11.8 percent (5.4 percent if the ten percent rule is used). In some instances, particularly when sampling from small populations, the sample size will be more than 20 percent of the population. Then it is important to take the correction into account.

The conclusion that the size of a population usually has little to do with the standard error, and hence with the reliability of a sample, is the opposite of what common sense usually leads people to expect. Nevertheless, it is true.

Try to answer the following question without looking back or looking ahead: It is agreed that a sample of 1,000 will give a satisfactory standard error for an estimate of mean family income in Rockville, a city with 30,000 families. Assume that the standard deviations of family income in Rockville and Chicago are the same. The objective is to obtain the same standard error of the mean for Chicago, where there are about 1,000,000 families. Assuming simple random sampling in each city, how large a sample is needed in Chicago? The answer is given in the footnote* at the bottom of page 374.

Frequently it is said that a sample must include a certain proportion of the total population in order to assure satisfactory results. One person may contend that a 20 percent sample is needed for "reliable" results; another may be willing to settle for three percent. Engineers commonly believe in ten percent. In general, all such assertions are wrong, for the standard error of a sample estimate depends almost entirely on the actual number of observations, and scarcely at all on the relation of the sample size to the population size. Frequently statisticians are asked, "What percentage of the population should I include in my sample?" This is like asking an expert cook, "What percentage of the flour in the bin should I put in my cake?" The answer in both cases happens to be the same, namely that you need a certain amount, and how much you need depends on what you are going to do with it, regardless of the size of the supply from which you take it.

The formulas $\dfrac{\sigma}{\sqrt{n}}$ and $\sqrt{\dfrac{N-n}{N-1}} \cdot \dfrac{\sigma}{\sqrt{n}}$ apply only to simple random sampling. Analogous, but more complicated, formulas apply to other probability sampling designs.

## 11.4.4 An Example

The mean of a uniform population in which the integers 0 to 9 are equally frequent is 4.5 and the standard deviation is 2.8723. You can verify these figures by taking any group of numbers in which the digits 0 to 9 are equally frequent—10 numbers, with each digit appearing once, is simplest—and calculating

$$M = \frac{\sum x}{n}$$

$$\sigma = \sqrt{\frac{\sum (x - M)^2}{n}} = \sqrt{\frac{\sum x^2}{n} - \left(\frac{\sum x}{n}\right)^2}.$$

This formula for the population standard deviation differs from the one on page 252 for $s$, the sample standard deviation, because the deviations are now being taken from $M$, the population mean, rather than from $\bar{x}$, a sample mean. Whenever $M$ is used in a standard deviation instead of $\bar{x}$, we divide by $n$ instead of $n - 1$. For the integers 0 to 9, $\sum x^2 = 285$ and $\sum x = 45$, so $\sigma = \sqrt{8.25} = 2.8723$, as asserted.

Suppose a random sample of four is taken from such a population. What is the probability that its mean will be 2 or less?

Actually, the authors happen to know that the exact answer is 0.0495, to four decimals, because we went to considerable trouble to calculate the exact sampling distribution in order to picture it in Fig. 356 and show you that the normal distribution would have done practically as well. We reinforce that point now by calculating an approximate answer to this problem by the normal distribution.

From the central limit theorem, we know that the sampling distribution of means of 4 will be approximately normal. Since the original population has a mean of 4.5, so will the sampling distribution. Since the original population has a standard deviation of 2.8723, the sampling distribution will have a standard error of

$$\sigma_{\bar{x}} = \frac{2.8723}{\sqrt{4}} = 1.436.$$

Hence our problem becomes: What is the probability that a variable that is normally distributed with mean 4.5 and standard deviation 1.436 will be 2 or less?

11.4.4.1 *Continuity Adjustment.* At this point we introduce a special wrinkle which, while unimportant for large samples, is often just the trick needed to make the normal approximation adequate even for fairly small samples. We notice that the exact sampling distribution of the mean is a discrete distribution in this case because the original population was discrete. The possible observations are the consecutive integers 0 to 9, so the possible sums of 4 observations are the consecutive integers 0 to 36. When these sums are divided by 4 to get means, only the 37 values 0, 0.25, 0.50, etc., to 9.00—whole multiples of 0.25—can occur. Thus, from the exact distribution the answer will be the same if we ask for the probability of a value of 2 or less, of 2.1 or less, of 2.17 or less, or of any value less than 2.25 but not less than 2; for the only values less than numbers between 2 and 2.25 that can occur are those that are 2 or less. We are approximating this discrete distribution with a continuous one that gives different answers for all these values from 2 to 2.25, for which the exact answer is the same. Which value from the continuous distribution shall we take? It usually works well to take the value at the midpoint of the interval in which the exact answer is constant. This is called a *continuity adjustment* or *continuity correction*.

Thus, in this example, where the exact probability we seek would apply to any value from 2 to 2.25, we will use 2.125 in applying the normal distribution to find the probability of a mean of 2 or less.

With the continuity adjustment, the problem becomes, What is the probability that a variable that is normally distributed with mean 4.5 and standard deviation 1.436 will be 2.125 or less? To answer this we standardize the normal variable, by computing

$$K = \frac{2.125 - 4.5}{1.436} = -1.65.$$

Now we refer to Table 365, where, in the line for 1.6 and column for 5, we find 0.0495. Since the value of $K$ is *negative*, this is the probability of a value *below* $-1.65$, and is the probability we seek. Had we computed $K$ to more decimals it would have been $-1.6537$, so we might have taken a probability 37 percent of the way from 0.0495 to 0.0485, the values shown in Table 365 for 1.65 and 1.66, obtaining 0.0491 as a more exact normal probability. This approximate result thus agrees to 3 decimals with the exact answer. While the agreement is usually close, it is seldom this close, as Table 374 shows by making similar comparisons for all possible sample means from 0 to 4.5. (Probabilities for values above 4.5 can be found from symmetry considerations; thus the probability of a mean of 7 or more is the same as the probability of a mean of 2 or less.) The largest discrepancy shown in Table 374 is at $\bar{x} = 3.25$, where the normal distribution shows 0.2167 but the correct probability is 0.2240, a difference of

## Sampling Distributions and the Normal Distribution

TABLE 374

EXACT SAMPLING DISTRIBUTION AND NORMAL APPROXIMATION
FOR MEANS OF SAMPLES OF 4 FROM A UNIFORM POPULATION
OF DIGITS 0 TO 9

| Sample Mean | Normal Deviate[a] | Cumulative Probability[b] | | |
|---|---|---|---|---|
| | | Normal[c] | Exact[d] | Error |
| 0.00 | −3.046 | 0.0012 | 0.0001 | +0.0011 |
| .25 | −2.872 | 0.0020 | 0.0005 | +0.0015 |
| .50 | −2.698 | 0.0035 | 0.0015 | +0.0020 |
| .75 | −2.524 | 0.0058 | 0.0035 | +0.0023 |
| 1.00 | −2.350 | 0.0094 | 0.0070 | +0.0024 |
| .25 | −2.176 | 0.0148 | 0.0126 | +0.0022 |
| .50 | −2.002 | 0.0226 | 0.0210 | +0.0016 |
| .75 | −1.828 | 0.0338 | 0.0330 | +0.0008 |
| 2.00 | −1.654 | 0.0491 | 0.0495 | −0.0004 |
| .25 | −1.480 | 0.0694 | 0.0715 | −0.0021 |
| .50 | −1.306 | 0.0958 | 0.0997 | −0.0039 |
| .75 | −1.132 | 0.1288 | 0.1345 | −0.0057 |
| 3.00 | −0.9574 | 0.1692 | 0.1760 | −0.0068 |
| .25 | −0.7833 | 0.2167 | 0.2240 | −0.0073 |
| .50 | −0.6093 | 0.2712 | 0.2780 | −0.0068 |
| .75 | −0.4352 | 0.3317 | 0.3372 | −0.0055 |
| 4.00 | −0.2611 | 0.3970 | 0.4005 | −0.0035 |
| .25 | −0.0870 | 0.4653 | 0.4665 | −0.0012 |
| .50 | +0.0870 | 0.5347 | 0.5335 | +0.0012 |

[a] Computed with continuity adjustment; the formula for this case is

$$K = \frac{\bar{x} + \frac{1}{2n} - M}{\sigma_{\bar{x}}} = \frac{\bar{x} - 4.375}{1.43614}$$

where $\bar{x}$ is the sample mean, $n$ the sample size, $M$ the population mean, and $\sigma_{\bar{x}}$ the standard error of the mean.

[b] The probability of a sample mean as small as that shown or smaller.

[c] Obtained from the National Bureau of Standards *Tables of Normal Probability Functions* (Washington: Government Printing Office, 1953). The number of decimals to which values of $K$ are computed corresponds with the number to which $K$ is shown in that table.

[d] Computed by the methods indicated in Sec. 11.2.2.

less than 1 percentage point. Had we chosen for our illustration the case of samples of 3, the agreement would have been less good, but for samples of 5 it would have been better.

Suppose our problem had been to find the probability that the sample mean would be 2 or larger, instead of 2 or smaller. In that case, the exact answer would be the same as if we had asked about 1.78 or larger, 1.92 or larger, or any other value between 1.75 and 2;

---

* The main point is that about 1,000 interviews would be needed in each city, despite the disparity in the number of families in the two cities. Actually, 1,033 interviews would be needed in Chicago.

for no values larger than 1.75 can occur except those that are 2 or larger. Thus, the interval over which the exact answer is constant is now 1.75 to 2. As a continuity adjustment we would, in computing $K$, replace 2 by 1.875, the midpoint of the interval 1.75 to 2.00 over which the exact probability we want holds. Then $K = -1.828$. The probability of a value this small or smaller is shown by the normal distribution as 0.0338 (see Table 374). So, remembering that for a continuous distribution, such as the normal, the probability of exactly a certain value is 0 (see page 174), we see that the probability that $K$ will be $-1.828$ or larger is $1 - 0.0338$, or 0.9662.

Note that the probability of a sample mean of 1.75 or less (shown in Table 374 as 0.0338) plus the probability of a sample mean of 2 or more (just computed as 0.9662) totals 1.0000. Since these events are mutually exclusive, and between them cover all possibilities, this is as it should be.

The steps in using the continuity adjustment are, then,

*First*, find the possible values which the statistic can have in the immediate neighborhood of the observed value for which a probability is sought. When the possible values of the observations are consecutive integers, the possible values of the mean are spaced every $1/n$.

*Second*, decide in which interval between possible values of the statistic the exact probability sought will be constant.

*Third*, use the midpoint of this interval in calculating $K$.

Thus, when the possible values of the observations are consecutive integers, the continuity adjustment for a sample mean consists of (1) replacing $\bar{x}$ by $\bar{x} - (1/2n)$ if we seek (a) the probability of a value as large as $\bar{x}$ or larger, or (b) the probability of a value smaller than $\bar{x}$; (2) replacing $\bar{x}$ by $\bar{x} + (1/2n)$ if we seek (a) the probability of a value as small as $\bar{x}$ or smaller, or (b) the probability of a value larger than $\bar{x}$. Since a sample proportion is a mean of observations whose possible values are the consecutive integers 0 and 1, $\bar{x}$ may be replaced by $p$ in this paragraph.

## 11.5
## THE SAMPLING DISTRIBUTION OF A PROPORTION

Since a proportion is a special case of a mean, in which the observations averaged are all 0's or 1's, our results about the sampling distribution of means apply to proportions also. You should review Sec. 7.4.3 on this point before you proceed with the rest of this section.

### 11.5.1 The Mean of the Sample Proportions

The mean of the sampling distribution of proportions is the population "mean," in other words, the population proportion, $P$. If $P = 0.37$, as in Sec. 11.2.1, and if an indefinitely large number of samples is drawn, the mean of the sample proportions, $p$, will tend to $P$. Thus, $p$ is unbiased, in the sense in which "unbiased" was used in Sec. 11.4.1.

### 11.5.2 The Standard Error of a Proportion

The standard error of a proportion, like that of a mean, is the standard deviation of the population sampled, divided by the square root of the sample size:

$$\sigma_p = \frac{\sigma}{\sqrt{n}}.$$

In the case of proportions, however, the standard deviation of the population is related to the mean of the population—that is, to the population proportion, $P$—by a simple formula,

$$\sigma = \sqrt{P(1 - P)},$$

often written

$$\sigma = \sqrt{PQ},$$

where $P$ is the probability that an item has the characteristic under consideration and $Q = 1 - P$ is the probability that it does not.

For the population in which $P = 0.37$, for example,

$$\sigma = \sqrt{0.37 \times 0.63} = \sqrt{0.2331} = 0.4828.$$

You can verify this by considering any group of numbers of which 37 percent are 1's and 63 percent are 0's, and applying a formula equivalent to that used in Sec. 11.4.4,

$$\sigma = \sqrt{\frac{\sum(x - P)^2}{n}} = \sqrt{\frac{\sum x^2}{n} - \left(\frac{\sum x}{n}\right)^2}.$$

The smallest group of observations that can have exactly 37 percent 1's is one of a 100, with 37 1's and 63 0's. Then $\sum x = 37$. Since $x$ is always 0 or 1, and since $0^2 = 0$ and $1^2 = 1$, $\sum x^2 = 37$ also. Hence

$$\sigma = \sqrt{\frac{37}{100} - \left(\frac{37}{100}\right)^2}$$

$$= \sqrt{\frac{37}{100}\left(1 - \frac{37}{100}\right)}$$

$$= \sqrt{\frac{37}{100} \times \frac{63}{100}} = 0.4828.$$

Perhaps you can see from this calculation why $\sigma$ will always come out $\sqrt{PQ}$, whatever value of $P$ we start with.

Thus, the standard error of a proportion is

$$\sigma_p = \frac{\sigma}{\sqrt{n}} = \sqrt{\frac{P(1 - P)}{n}} = \sqrt{\frac{PQ}{n}},$$

where $Q = 1 - P$. If, for example, $n = 5$ and $P = 0.37$, we have

$$\sigma_p = \sqrt{\frac{0.37 \times 0.63}{5}} = \sqrt{0.04662} = 0.2159.$$

Exactly the same finite population factor is appropriate for the standard error of a proportion as for the standard error of a mean. In other words, a more accurate (but usually only slightly more accurate) formula for $\sigma_p$ is

$$\sigma_p = \sqrt{\frac{N - n}{N - 1}} \times \sqrt{\frac{PQ}{n}},$$

where $N$ represents the number of items in the population. The conclusion of Sec. 11.4.3 thus applies here, too, that the standard error of a proportion depends almost entirely on the actual number of observations in the sample, and scarcely at all on what fraction they are of the population. The only exceptions to this are when more than 20 percent of the population is included in the sample; otherwise, the standard error given by the simpler formula will be reduced less than 11 percent by taking account of the finite population factor.

### 11.5.3 An Example

Suppose a sample of five is selected from a binomial population in which $P = 0.37$. What is the probability that the sample proportion, $p$, will be 0.2 or smaller?

The appropriate normal distribution for approximating the sampling distribution of $p$ is one with mean equal to $P$, or 0.37, and standard deviation equal to $\sigma_p$, 0.2159 according to the calculations just made.

### Sampling Distributions and the Normal Distribution

The variable is discrete and can take only the 6 values 0, 0.2, 0.4, 0.6, 0.8, and 1. The exact value of the probability sought is the same for all values between 0.2 and 0.4, so we work with 0.3, the midpoint of this interval.

The question then is, What is the probability of a value of 0.3 or less for a normally distributed variable with mean of 0.37 and standard deviation of 0.2159? Compute a standardized normal variable,

$$K = \frac{0.3 - 0.37}{0.2159} = -0.32.$$

Entering Table 365, in the line for 0.3 and the column for 2, we find 0.37. This is accurate enough for practical purposes, but for more precision in comparing our result with the exact value, we compute $K$ more accurately as $-0.3242$ and interpolate 42 percent of the way from 0.3745 to 0.3707, shown in Table 365 for $K = 0.32$ and $K = 0.33$, thereby getting a probability of 0.3729. Since $K$ is negative, 0.3729 is the probability of a value this small or smaller, which is what we required. The exact value, shown in Table 351, is 0.3907. Thus, the normal distribution has provided a simple but satisfactory approximation to the exact probability, even though the sample size is small and we deliberately chose the value of $p$ for which there is

TABLE 378

EXACT SAMPLING DISTRIBUTION AND NORMAL APPROXIMATION
FOR PROPORTIONS IN SAMPLES OF 5 FROM A BINOMIAL POPULATION
WITH PARAMETER 0.37

| Sample Proportion | Normal Deviate[a] | Cumulative Probability[b] | | |
|---|---|---|---|---|
| | | Normal[c] | Exact[d] | Error |
| 0.0 | −1.250 | 0.1056 | 0.0992 | +0.0064 |
| 0.2 | −0.3242 | 0.3729 | 0.3907 | −0.0178 |
| 0.4 | +0.6021 | 0.7264 | 0.7330 | −0.0066 |
| 0.6 | +1.528 | 0.9367 | 0.9340 | +0.0027 |
| 0.8 | +2.455 | 0.9930 | 0.9931 | −0.0001 |
| 1.0 | +3.381 | 0.9996 | 1.0000 | −0.0004 |

[a] Computed with continuity adjustment; the formula in this case is

$$K = \frac{p + \frac{1}{2n} - P}{\sigma_p} = \frac{p - 0.27}{0.2159}.$$

[b] The probability of a sample proportion as small as that shown or smaller.

[c] Obtained from the National Bureau of Standards *Tables of Normal Probability Functions* (Washington: Government Printing Office, 1953). The number of decimals to which $K$ is computed corresponds with the number to which $K$ is shown in that table.

[d] See Sec. 11.2.1. and Table 351.

most disagreement between the approximate and exact probabilities. Table 378 presents similar comparisons for other possible values of $p$, when $P = 0.37$ and $n = 5$.

We would have gotten poorer agreement between the normal and the exact distributions for a smaller sample size, better agreement for a larger sample size. We would also have gotten poorer agreement for a value of $P$ further from 0.5, and better agreement for a value of $P$ nearer to 0.5.

A common rule is that if both $nP$ and $n(1 - P)$ are 5 or more, the normal distribution provides a satisfactory approximation to the sampling distribution of a proportion. As we see in our example, where $np = 5 \times 0.37 = 1.85$, the normal distribution may be satisfactory even when the condition $nP \geq 5$ and $n(1 - P) \geq 5$ is not met. However, there are cases—for example, $P = 0.01$, and $n = 200$, or $P = 0.001$ and $n = 2,000$—where the normal distribution does not provide a good approximation even for large samples. But the condition $nP \geq 5$ is sufficient; if it is met the normal approximation will be satisfactory for most practical purposes.

# 11.6
# CONCLUSION

In order to draw reliable conclusions about a population from a sample, it is necessary to distinguish as well as possible between the purely fortuitous aspects of the sample, which result from the role of chance in selecting one set of observations from the population rather than another for the sample, and the aspects of the sample that are true reflections of the population. That is, any particular statistic computed from a sample, such as a mean, proportion, or standard deviation, must be interpreted in the light of its sampling distribution—the range of values it might have taken in other random samples and the probabilities of its taking each of them.

Sampling distributions can be derived from original populations by the methods of computing probabilities presented in Chap. 10, and more complicated methods. The statistician's kit of tools contains a good many sampling distributions which enable him to deal rather precisely with even small samples. (This does not mean that he can always learn a great deal from a small sample, but only that he can wring out whatever information it does contain *and* can state fairly precisely the degree of confidence or uncertainty attached to whatever information he does get.) The principal sampling distributions in his kit go by the names normal, binomial, Poisson, exponential, Student's, chi-square, variance ratio, noncentral-$t$, and so forth.

Many of the most common problems of statistical inference, however, can be handled reasonably accurately and quite simply by a single distribution, the standard normal distribution. Although this distribution often serves quite well as a description of a basic population of observations, its fundamental importance in statistics is due to its versatility as an approximate sampling distribution. The fact that the distributions of statistics computed from samples tend to be normal, almost regardless of the shape of the basic population from which the observations come, is known as the central limit theorem. There are exceptions: certain statistics, certain populations, and especially certain (small) sample sizes may not lead to distributions that are satisfactorily approximated by the normal distribution. But for purposes of explaining the principles of statistical reasoning, and indicating how to handle many important problems, the normal distribution suffices.

The normal distribution involves only two parameters, its mean and its standard deviation. Any normal distribution can be converted to a standard normal distribution of zero mean and unit standard deviation by simply subtracting its mean from each observation, then dividing by the standard deviation. Probabilities for this standard normal distribution have been extensively tabled.

To use the normal distribution as an approximate sampling distribution, all we need know about the exact sampling distribution is its mean and standard deviation, called the standard error. For many statistics in common use, such as the mean and proportion, the mean of the sampling distribution is the same as the mean of the population from which the samples come, and the standard error is related to the standard deviation of the population by a simple formula. In the case of a mean or proportion this formula is that the standard error is the standard deviation of the population divided by the square root of the sample size. For proportions, the standard deviation of the population is simply $\sqrt{P(1-P)} = \sqrt{PQ}$.

Strictly speaking, when the population is of limited size, the formula relating the standard error of a mean or proportion to the standard deviation of the population should include a finite population factor to allow for the relation of the sample size to the population size. The effect of this factor is to reduce the calculated standard error by a percentage equal to approximately half the percentage of the population included in the sample. Thus, if less than 20 percent of the population is included in the sample, the standard error is reduced by less than 11 percent. In general, the standard error of a sample statistic depends almost entirely on the actual number of observations in the sample, and hardly at all on the proportion of the population in the sample.

Use of the normal distribution as an approximate sampling distribution when the sample size is small, and the exact distribution is discrete, is usually improved by a continuity adjustment. The discreteness of the distribution implies that the exact probability of a value more than a certain amount, or less than a certain amount, does not change in the interval between possible values. In applying the normal approximation we use the midpoint of the interval for which the exact probability we want would be constant. Thus, if we want the probability of a certain value or more, we use an adjusted value halfway back to the next possible lower value. Similarly, for the probability of a certain value or less, we use an adjusted value halfway up to the next higher possible value.

You are now equipped with the basic information needed to deal with such problems as testing whether an assumption about a population is supported or contradicted by the data, and estimating from data what the characteristics of the population probably are. We turn to testing in the next two chapters, then to estimation in Chap. 14.

## DO IT YOURSELF

### Example 381A

Calculate the binomial distributions for $n = 5$, $P = 0.12, 0.25$, and $0.50$. Compare your results with Fig. 352.

### Example 381B

Reread Example 141B (Imaginary Coin Tosses). Explain how the expected frequencies can be computed.

### Example 381C

Compute the binomial distribution for $n = 6$, $P = 0.50$.

### Example 381D

Consider the population consisting of the values 1, 2, and 3, each very numerous but equally frequent, and samples of $n = 3$ from this population.

(1) Compute the sampling distribution of $\bar{x}$. [Hint: List all 27 possible samples of 3.] Compute the mean and standard deviation of this distribution.

(2) What are $M$ and $\sigma$ for this population? How do these numbers relate to the mean and standard deviation computed in (1)?

*Sampling Distributions and the Normal Distribution*

EXAMPLE 382A

Using Table 365, compute the following probabilities for the standard normal distribution. Make a rough sketch with each computation to show the area of a normal curve corresponding with the probability.

(1) $Pr(-1 < K < 1)$ (Read: "Probability that $K$ is between $-1$ and $+1$.")
(2) $Pr(-2 < K < 2)$
(3) $Pr(-3 < K < 3)$
(4) $Pr(K > 1.28)$
(5) $Pr(K < -1.64)$
(6) $Pr(K > 1.96)$
(7) $Pr(K > 1.28$ or $K < -1.28)$
(8) $Pr(K > 3.09)$
(9) $Pr(K > 1.96$ or $K < -1.96)$
(10) $Pr(-1.28 < K < 1.64)$

EXAMPLE 382B

By using the normal approximation, find the probability that $p = 0.65$ for a sample of 20 from a population in which $P = 0.37$.

EXAMPLE 382C

What is the approximate probability $p \geq 0.57$ if $n = 100$ and $P = 0.50$? That $p \leq 0.43$?

EXAMPLE 382D

What is the approximate probability that $p \leq 0.56$ if $n = 100$ and $P = 0.60$?

EXAMPLE 382E

Verify the assertion of the footnote of page 374.

EXAMPLE 382F

Compute the binomial distribution for $n = 5$, $P = 0.50$.
(1) Compute the mean and standard deviation of this distribution.
(2) Compute $M$ and $\sigma$ of the population specified by this problem. Do your answers to (1) and (2) agree with the formulas

$$M_p = M \qquad \text{and} \qquad \sigma_p = \frac{\sigma}{\sqrt{n}}?$$

EXAMPLE 382G

(Sequel to Example 382F, which should be done first.) Suppose there are 10 people in a room, 5 men and 5 women. Five people are selected at random and the proportion $p$ of men is computed.
(1) Compute the sampling distribution of $p$. [Hint: The probability that $p = 0$ is $\frac{5}{10} \times \frac{4}{9} \times \frac{3}{8} \times \frac{2}{7} \times \frac{1}{6}$.]

*Do It Yourself*

(2) Compute the mean and standard deviation of this distribution.

(3) Compute $M$ and $\sigma$ of the population.

(4) What is the relation between your answers to (2) and (3)?

(5) Graph the distribution of step (1) and of that of step (1) in Example 382F. Which shows the smaller dispersion? Why? Is this an exception to the generalization that $n/N$ is irrelevant to sample accuracy? Explain.

EXAMPLE 383

Consider the 32 weights on page 171 as a population rather than a sample, and consider a sample consisting of the weights 198, 189, 148, . . ., 175, 191—that is, the first, third, fifth, . . ., twenty-ninth and thirty-first weights.

(1) What is the mean of the population?

(2) What is the mean of the sample?

(3) What is the standard deviation of the population?

(4) What is the standard error of the sample mean?

(5) What is the probability of a sample mean as small as or smaller than the actual sample mean?

(6) What is the probability of a sample mean at least as far from the population mean as the actual sample mean?

# Chapter 12

# Statistical Tests and

# Decision Procedures

## 12.1
### INTRODUCTION

Now we come to one of the main branches of analytical statistics, the branch called, variously, tests of hypothesis, tests of significance, and decision procedures.

The "testing" terminology is old and established. The "decision" terminology is new and has only begun to come into use during the past decade. The differences are more than term-deep, however; there are fundamental differences in viewpoint.

The testing point of view starts from a hypothesis ("null hypothesis") about the true state of affairs in the population. The hypothesis might be, for example, that the population proportion of heads when a certain coin is tossed in a certain way is $\frac{1}{2}$. The problem then is to determine whether the results of a sample are consistent with this hypothesis. The sample results are regarded as consistent with the hypothesis if they are within the bounds of "reasonable" sampling variation—reasonable in the sense that discrepancies as great as or greater than that observed would occur with more than some predetermined probability, say 0.05 or 0.01. We have briefly illustrated this approach in Sec. 11.1 and elsewhere.

A difficulty with this viewpoint is that it is often known that the hypothesis tested could not be precisely true. No coin, for example, has a probability of *precisely* $\frac{1}{2}$ of coming heads. The true probability

**384**

will always differ from $\frac{1}{2}$, even if it differs by only 0.000,000,000,1. Neither will any treatment cure *precisely* one-third of the patients in the population to which it might be applied, nor will the proportion of voters in a presidential election favoring one candidate be *precisely* $\frac{1}{2}$. Recognition of this leads to the notion of differences that are or are not of practical importance. "Practical importance" depends on the actions that are going to be taken on the basis of the data, and on the losses from taking certain actions when others would be more appropriate.

Thus, the focus is shifted to decisions: Would the same decision about practical action be appropriate if the coin produces heads 0.500,000,000,1 of the time as if it produces heads 0.5 of the time precisely? Does it matter whether the coin produces heads 0.5 of the time or 0.6 of the time, and if so does it matter enough to be worth the cost of the data needed to decide between the actions appropriate to these situations? Questions such as these carry us toward a comprehensive theory of rational action, in which the consequences of each possible action are weighed in the light of each possible state of reality. The value of a correct decision, or the costs of various degrees of error, are then balanced against the costs of reducing the risks of error by collecting further data. It is this viewpoint that underlies the definition of statistics given in the first sentence of this book.

Decision theory in this broad sense is not yet ready for practical application, except perhaps in special circumstances. The term "decision theory" has, of course, great appeal, for everyone has to make decisions, and everyone hopes for some magic formula by which to make them; as a result, more has sometimes been claimed for decision theory than it can yet produce in practice. Already, however, decision theory has resulted in clarification of the standard significance testing procedures. That is, it illuminates the nature, uses, and limitations of significance tests.

In this chapter we shall try to bring out the nature of significance tests by examining one special problem. Then in the next chapter we shall present the technical apparatus necessary for a variety of significance tests. It is essential not to confuse the statistical usage of "significant" with the everyday usage. In everyday usage, "significant" means "of practical importance," or simply "important." In statistical usage, "significant" means "signifying a characteristic of the population from which the sample is drawn," regardless of whether the characteristic is important.

## 12.2
## A DECISION PROBLEM

We shall illustrate the basic ideas of a test of significance by a hypothetical but realistic example. Suppose that a psychiatric hospital wants to decide whether to adopt a new method of therapy for a certain class of patients. Assume that past experience indicates that while some people recover, no existing therapy has any influence on the recovery rate. That is, as many recover without therapy as with it.

Ideally, the research staff of the hospital would like to conduct an experiment in which patients are divided randomly (perhaps with some stratification by length of hospitalization, etc.) into two groups, the first group to be treated according to current practice (the "control" group), the second group to be treated by the new therapy. Then the recovery rates in the two groups would be compared. Such an experimental arrangement is often desirable, but let us assume that it is not desirable in this problem. For example, the hospital staff may feel strongly that no promising treatment should be withheld from any patient.[1] Moreover, they feel that a control group can be dispensed with because past experience has shown the recovery rate to be remarkably stable, regardless of the hospital or type of therapy used.

Let us assume that of all people initially admitted to mental hospitals with this disorder, 50 percent recover within 5 years. Then the proper decision simply depends on whether the new therapy will result in a rate of recovery above 50 percent. In other words, if the population proportion of recoveries exceeds 50 percent, the new method should be adopted; if it does not, the old method should be retained.

This formulation of the problem to illustrate a statistical technique passes over a host of questions that should be raised in an actual investigation. Many of these questions are of the "common sense" kind considered in Part I. What is the evidence that the recovery rate is really stable and equal to 50 percent? How reliable is diagnosis of the disorder? Is it really possible to define "recovery" objectively, so that different psychiatrists would agree (in most cases, at least) that a given patient had recovered or that he had not? We assume that such questions can be answered satisfactorily, and we turn to the statistical problems raised by this experiment.

---

1. This attitude is common in the medical profession. The reasons for it are understandable and scarcely assailable. Unfortunately, however, it often leads to failure to find out whether a treatment is of any real value, because there is no sound bench mark for comparison.

## 12.3
## A DECISION PROCEDURE

A natural way to proceed would be to try the new therapy on a sample of patients, then make a decision in favor of the new therapy if more than half of the patients in the sample recover, or a decision against it if not more than half recover.

Suppose that a sample of 100 patients is treated by the new therapy. Then if 51 or more recover the new method would be adopted, but if 50 or fewer recover the old method would be retained.

## 12.4
## RISKS OF A WRONG DECISION

What are the risks of wrong decision with this procedure? If the true proportion, $P$, of recoveries with the new method is $\frac{1}{2}$ or less, so that the correct decision would be not to adopt the new therapy, what is the probability that the sample will show 51 or more recoveries, and thereby lead, erroneously, to adoption of the new method? And if $P$ exceeds $\frac{1}{2}$, so that the correct decision would be to adopt the new therapy, what is the probability that the sample will show 50 or fewer recoveries, and thereby mislead us into deciding against the new method? The possibilities for correct and incorrect decision may be presented as follows:

| Decision | Reality (unknown) | |
|---|---|---|
| | Old better | New better |
| Retain old | Correct decision | Error of Type II |
| Adopt new | Error of Type I | Correct decision |

The probability of a Type I error, adopting the new when it is not better, will depend on how much the value of $P$ falls short of $\frac{1}{2}$. The lower $P$ is, the less the chance that the sample proportion, $p$, will exceed $\frac{1}{2}$. Correspondingly, the better the new method, the less the chance that $p$ will fail to exceed $\frac{1}{2}$. These probabilities may, in fact, be calculated by the methods shown in Sec. 11.5.3. Suppose, for example, that the true value of $P$ is 0.4, so that the new therapy ought to be rejected. The probability of an erroneous decision is the probability that $p$ will equal or exceed 0.51. To find the probability for a sample of size 100, compute the usual standard normal variable

$$K = \frac{0.51 - 0.005 - 0.4}{\sqrt{\dfrac{0.4 \times 0.6}{100}}} = \frac{0.105}{0.049} = 2.14.$$

Table 365 shows the upper-tail probability as 0.016. Thus, if $P = 0.4$, the probability of erroneously adopting the new therapy is only about one in sixty.

For a value of $P$ larger than $\frac{1}{2}$, say $P = 0.55$, an erroneous decision is made if $p$ is 0.50 or less. To find this probability, compute

$$K = \frac{0.50 + 0.005 - 0.55}{\sqrt{\dfrac{0.55 \times 0.45}{100}}} = \frac{-0.045}{0.050} = -0.90.$$

Table 365 shows the lower-tail probability as 0.184.

Proceeding in this way, a curve showing the probability of erroneous decision can be drawn, as in Fig. 389A.

## 12.5
## OPERATING-CHARACTERISTIC CURVE

Ordinarily, instead of a curve showing the probability of error, a curve called the *operating-characteristic curve*, usually abbreviated to "OC curve," is shown. The probability shown in an OC curve applies to the same decision for all values of $P$, whether that decision is the correct one or not. Fig. 389B shows the OC curve corresponding to Fig. 389A. It gives the probability of retaining the old method. For values of $P$ up to 0.5, this is the correct decision, so the height of the OC curve is 1 minus the height of the error curve. For values of $P$ above 0.5, retaining the old method is wrong, so the OC curve is the same as the error curve. The OC curve shows how the test or decision procedure operates, regardless of whether that operation is correct or incorrect. Obviously an error curve can easily be drawn from an OC curve, once it is decided which decision is appropriate to the possible values of the population proportion.

## 12.6
## ADJUSTING THE DECISION CRITERION

The risks of error shown in Figs. 389A and 389B might well be considered undesirable. If the new therapy really obtains only the same proportion of recoveries as would occur under the old practice,

### 12.6 Adjusting the Decision Criterion

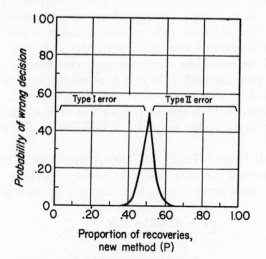

FIG. 389A. Probability of wrong decision for various values of *P*.

Decision procedure: Take sample of 100. Adopt new method if $p \geq 0.51$. Retain old method if $p \leq 0.50$.

Correct decisions: If $P > 0.5$, adopt new method. If $P \leq 0.5$, retain old method.

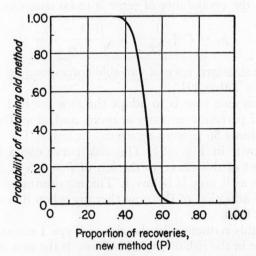

FIG. 389B. Operating-characteristic curve for test based on sample of 100, old method retained if sample shows 50 or fewer recoveries.

0.5, it has almost a 50 percent chance of being adopted. Typically there is more cost to making a change than to continuing an existing method. Perhaps, also, the new treatment involves some risk or discomfort for the patient not present in the old. Considerations of economy and conservatism might suggest, therefore, that the risk of adopting the new method, if in fact it results in only 50 percent recoveries, should be kept below a certain level, say 0.10. This may be achieved by requiring the sample proportion to be substantially larger than 50 percent in order to make a decision in favor of the new method.

How much larger? Table 391 shows that 1.28 is the value that a standard normal variable exceeds just ten percent of the time. Hence the decision level, say $p_r$, for adopting the new method, must be such that

$$\frac{p_r - 0.005 - 0.5}{\sqrt{\dfrac{0.5 \times 0.5}{100}}} = 1.28.$$

A little algebra shows that

$$p_r = 0.505 + 0.05 \times 1.28 = 0.569.$$

More generally, if $n$ is the sample size and $P$ the population proportion for which the probability of error is to be controlled,

$$p_r = P + \frac{1}{2n} + K \sqrt{\frac{P(1 - P)}{n}},$$

where $K$ is the standard normal variable corresponding with the risk, and is shown in Table 391.

The decision rule now is to adopt the new method if the sample of 100 shows 57 percent recoveries or more, and adopt the old method if the sample shows 56 percent recoveries or fewer. The OC curve for this rule is shown in Fig. 391. The risk curve could be drawn by putting the part of this curve to the left of $P = 0.5$ as far above the horizontal axis as it now is below 1. The maximum probability of a Type I error—adoption of the new therapy when it is really no improvement—is 0.10.

Offsetting this reduction in the risk of Type I errors, however, is a large increase in the risk of Type II errors. If the new method really achieves 60 percent recoveries, for example, the probability of its adoption is only 0.76. A 60 percent recovery rate means recovery for one-fifth of those who would not otherwise have recovered, a high enough figure to make it quite desirable under most circumstances to adopt the new method.

### 12.6 Adjusting the Decision Criterion

TABLE 391

STANDARD NORMAL VARIABLES EXCEEDED WITH GIVEN PROBABILITIES

Probabilities are upper-tail probabilities, and $K$ is the standard normal variable exceeded with the specified probability.

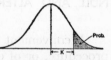

| Prob. | 0.00 | 0.0 | 0.1 | 0.2 | 0.3 | 0.4 |
|-------|------|-----|-----|-----|-----|-----|
| 0 |       |       | 1.282 | 0.842 | 0.524 | 0.253 |
| 1 | 3.090 | 2.326 | 1.227 | 0.806 | 0.496 | 0.228 |
| 2 | 2.878 | 2.054 | 1.175 | 0.772 | 0.468 | 0.202 |
| 3 | 2.748 | 1.881 | 1.126 | 0.739 | 0.440 | 0.176 |
| 4 | 2.652 | 1.751 | 1.080 | 0.706 | 0.412 | 0.151 |
| 5 | 2.576 | 1.645 | 1.036 | 0.674 | 0.385 | 0.126 |
| 6 | 2.512 | 1.555 | 0.994 | 0.643 | 0.358 | 0.100 |
| 7 | 2.457 | 1.476 | 0.954 | 0.613 | 0.332 | 0.075 |
| 8 | 2.409 | 1.405 | 0.915 | 0.583 | 0.305 | 0.050 |
| 9 | 2.366 | 1.341 | 0.878 | 0.553 | 0.279 | 0.025 |

For a probability of 0.025, $K = 1.960$.

For probabilities above 0.5, subtract the probability from 1, and attach a minus sign to the corresponding value of $K$.

The entries in the "Prob." column are additional digits for the probabilities shown in the "Prob." row. Thus, the entry 1.126 in column (0.1) and row (3) is the value of $K$ for a probability of 0.13; and $-1.126$ is the value of $K$ for a probability of $1 - 0.13 = 0.87$.

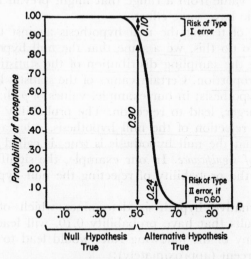

FIG. 391. Probability of accepting the null hypothesis as a function of P. ($n = 100$, $p_r = 0.57$.)

## 12.7
## NULL AND ALTERNATIVE HYPOTHESES

A convenient way of specifying what is required of a decision procedure, or of evaluating a proposed procedure, is to focus attention on two possible values of the parameter. In the preceding section, for example, we focused attention on the values $P = 0.5$ and $P = 0.6$. For values of $P$ below 0.5, it was assumed to be important that there be a high probability of deciding against the new therapy. For values of $P$ above 0.6, it was assumed to be important that there be a high probability of deciding in favor of the new therapy. For values of $P$ between 0.5 and 0.6, it is not of great importance which decision is made.

Such a pair of values of the parameter are called the *null hypothesis* and the *alternative hypothesis*. It is more or less arbitrary which is called which. Typically, however, the null hypothesis is precise (a treatment has no effect, a coin is fair) and corresponds with the absence of effects of the kind being studied—it "nullifies" the effect of the treatment, so to speak. In our example, the null hypothesis is that the therapy makes no difference at all. The alternative hypothesis may be less sharply determined; often, as with our value $P = 0.6$, it is simply one value from a range that might prevail if the methods under study do have some effect.

We speak of testing the null hypothesis against the alternative hypothesis. To do this, we assume that the null hypothesis is true, and calculate the sampling distribution of the statistic, in this case the sample proportion. Certain values of the statistic lead to *rejection* of the null hypothesis; in our example, values of $p$ that are as large as 0.57, or larger, lead to rejection.  The probability that a sample will result in rejection of the null hypothesis, calculated under the assumption that the null hypothesis is true, is called the *significance level* or *level of significance*. In our example, the significance level is 0.10. This is the probability of rejecting the null hypothesis if it is true.

The alternative hypothesis determines which of the possible groups of results that have probability 0.10, will lead to rejection. If $P = 0.5$, any of the following rules would lead to a significance level of 10 percent (approximately):

     (i) reject if $p \geq 0.57$,
     (ii) reject if $p \leq 0.43$,

$\qquad$ (iii) reject if $0.42 \leq p \leq 0.44$,
$\qquad$ (iv) reject if $p = 0.50$ or $p = 0.59$,
$\qquad$ (v) reject if $p \geq 0.59$ or $p \leq 0.41$,

and many others. All of these are equally satisfactory if the null hypothesis is true, for they all lead to rejection ten percent of the time. But if the alternative hypothesis is true, rule (i) is definitely the best, for then it has the greatest chance of leading to rejection of the null hypothesis. This is the best region of rejection (at a ten percent significance level) not only for the alternative hypothesis $P = 0.6$, but for any other alternative hypothesis in which $P$ exceeds 0.5. Such alternative hypotheses are called "one-sided," and they lead to "one-tail tests"—the upper-tail of the sampling distribution if the alternative hypothesis specifies a larger value than the null hypothesis, and the lower tail if the alternative hypothesis specifies a smaller value than the null hypothesis. These one-tail tests have higher probability of rejecting the null hypothesis when the alternative hypothesis is true than do any other decision rules with the required significance level.

Thus, the null hypothesis and significance level do not tell which is the best decision rule; account must be taken of the alternative hypothesis too.

The designation "error of the first kind," or "Type I error" is applied to the probability of rejecting the null hypothesis when it is true, and "error of the second kind," or "Type II error" to the probability of accepting the null hypothesis when the alternative hypothesis is true.

## 12.8
## BALANCING THE RISKS OF ERROR

Let us look back at Fig. 391, which gives the OC curve for $n = 100$, $p_r = 0.57$. Suppose we ask this question: "Keeping the sample size fixed at 100, can we adjust the rejection proportion so that we are better protected against the two kinds of errors?" This is equivalent to asking, "Have we properly balanced the risks of Type I and Type II errors?"

One way of answering the question is this. If the true $P$ were just a little better than 0.5, it would not be important that this be discovered. In fact, as we mentioned in Sec. 12.6, it might actually be preferable not to use the new method unless the gain is more than this. If, for example, $P$ were 0.51, then the use of the new therapy would lead to the recovery of only one person among each 50 who would not have recovered without it. Suppose, however, that $P$ were 0.6. The psychiatrists might feel that if the therapy were really that effective the risk should be very small of not discovering that it was better than the existing treatment. If $n = 100$ and $p_r = 0.57$, this risk, as we have seen, is 0.24. Since we are assuming the sample size fixed at 100, the only way to reduce the risk is to use a smaller rejection proportion. But while this decreases the risk of Type II error, it increases the risk of Type I error.

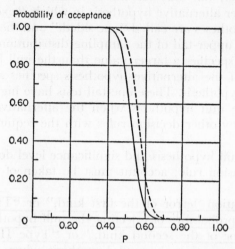

FIG. 394. Operating-characteristic curves for tests of the hypothesis that $P = 0.50$. ($p_r = 0.57$ and 0.53.)

Another way of saying this is that using a smaller rejection proportion moves the whole OC curve of Fig. 391 to the left. The horizontal displacement is the distance between the two rejection numbers. The two situations are shown in Fig. 394, where the dashed line is the same as the OC curve of Fig. 391, and the solid line is the OC curve for a rejection proportion of 0.53. The disadvantage of the new OC curve is that the risk of Type I error has been greatly increased. If the null hypothesis were true, about three samples in ten would falsely show an apparent superiority for the new therapy.

We might try the other expedient, that of increasing the rejection proportion, say from 0.57 to 0.59. This moves the OC curve to the right. The risk of Type I error is reduced from 0.1 to 0.045. But the risk of Type II error is increased. If $P$ were 0.6, for example, the risk of Type II error would be about 0.38.

### 12.8 Balancing the Risks of Error

This situation is a Scylla and Charybdis. As we get further from one danger, the other danger increases. So long as the sample size and sampling method remain the same, nothing can be done to avoid this uncomfortable situation. In practice the dilemma is all too often ignored by simply setting the risk of an error of the first kind at some conventional value, usually 0.05 or 0.01. These two numbers are so commonly used that sometimes a value found significant with a risk of Type I error of 0.05 is called "significant" while a value found significant with a risk of 0.01 is termed "highly significant," without explicit definitions. This solution is not satisfactory because it focuses attention on one of the two kinds of risk to the exclusion of the other. If the risk of Type I error were set at 0.01 in our problem, the risk of Type II error would be about 0.69 if $P$ were in fact 0.6.

A preferable solution is to make some rough evaluation of the consequences of each type of error. For example, if a Type I error would lead to serious practical consequences while a Type II error would not be so serious, the risk of an error of the first kind would be set much smaller than the risk of an error of the second kind. Each problem must be decided on its merits. On the one hand, the psychiatrists do not want to report that the therapy is not effective if it really is (Type II error). This might mean that a promising treatment would be discarded because of misleading chance variation in the sample of 100. On the other hand, the psychiatrists do not want to claim that their therapy is effective when it really is not. Such a claim would encourage reliance on a therapy that did not help people. It might also misdirect scientific work in the areas of psychiatry and clinical psychology, because scientists would try to formulate hypotheses that would account for the success of the therapy. These hypotheses might be incorrect, but their apparent consistency with the experimental results would encourage their acceptance into the body of psychological knowledge. Since an experiment of the kind we are discussing would be costly and therefore hard to duplicate, and the field is not one prone to objective experimentation, such an error might persist for years. These errors cannot be taken lightly, for there may be scores of other organizations working on similar problems, and even with good error control a considerable number of wrong answers—false leads—are going to be produced and have to be pursued to the bitter and costly end.

The two types of error must be weighed against each other in each problem. In scientific work, the consequences of erroneous claims of verification of new hypotheses are serious, and the risk of Type I error is usually set low: 0.01 and even 0.001 are not uncommon. For practical action, a higher risk of Type I error may be acceptable. But generalization is hazardous, for it is necessary to consider carefully in advance just how serious are the consequences of each kind of error in each specific instance. By moving the rejection number, tentatively, back and forth, different combinations of Type I and Type II errors—or, more fully, different OC curves—can be considered. The rejection number is then set where it gives the most satisfactory combination of the two risks, evaluated in terms of the practical consequences of each type of error, and judgment as to the possibility of various values of $P$. If the two kinds of error are equally serious, the rejection number may be set near the indifference point, that is, the recovery rate for which it makes no difference whether the new method is adopted or not.

Thus, we now see how the choice of the risk of Type I error is made. For any proposed Type I error, the associated Type II error is ascertained. The best available combination of the two risks is then chosen.

## 12.9
## ADJUSTING THE SAMPLE SIZE

After balancing the two risks relative to each other as well as can be done by adjusting the rejection number for a given sample size, we have to consider changes in the sample size, if we have any control over this. It may be that reducing the sample size will not increase the risks enough to offset the savings. In that case, a smaller sample is appropriate. On the other hand, it may be that increasing the number of observations will not increase the cost enough to offset the reduction in the risks. In that case, a larger sample is appropriate. The cost of information must be balanced against its value.

Suppose the null and alternative hypotheses are $P = 0.5$ and $P = 0.6$. Then the sampling distributions of $p$, under the two assumptions, are approximately as shown by the two solid normal curves of Fig. 397. If the rejection proportion is $p_r = 0.57$, the area of the curve for $P = 0.5$ to the right of 0.565 (0.57 minus a continuity adjustment of 0.005) represents the probability of an error of the first kind; similarly, the area of the curve for $P = 0.6$ to the left of 0.565 represents the probability of an error of the second kind.

The standard deviations of these two distributions are

$$\sigma_a = \frac{0.5}{\sqrt{n}} \qquad \text{and} \qquad \sigma_b = \frac{0.49}{\sqrt{n}},$$

*12.9 Adjusting the Sample Size*

where $n$ is 100 for the solid curves. If $n$ is increased, thereby reducing both standard deviations, the two shaded areas will be reduced in size, and both risks of error will be less. In Fig. 397, the dashed curves illustrate this. In technical terminology we would say that we have obtained a more *powerful* test than before, since the risks of error have been reduced. The OC curve would be steeper than before, indicating sharper discrimination, in the sense of a shorter interval be-

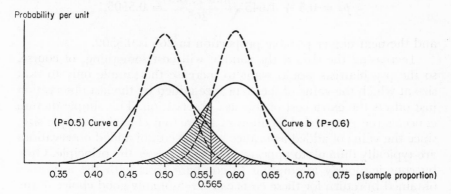

Probability per unit

(P=0.5) Curve a          Curve b (P=0.6)

0.35   0.40   0.45   0.50   0.55↓   0.60   0.65   0.70   0.75   p(sample proportion)
0.565

**FIG. 397.** Effect on the sampling distributions of increasing sample size.

tween values of $P$ for which the null hypothesis is almost always accepted and values of $P$ for which the alternative hypothesis is almost always accepted.

In the present example, we could reduce both risks—the risk of rejecting if $P = 0.5$ and the risk of accepting if $P = 0.6$—to about 0.05 by using a sample of 265, with a rejection proportion of 0.5509. A formula from which such sample sizes can be approximated is

$$n = \left( \frac{K_1\sqrt{P_1(1 - P_1)} + K_2\sqrt{P_2(1 - P_2)}}{P_2 - P_1} \right)^2,$$

where $K_1$ and $K_2$ are the standard normal variables (from Table 391) corresponding with the intended probabilities of errors of the first and second kinds. In the present example,

$$K_1 = K_2 = 1.645, \qquad P_1 = 0.5, \qquad P_2 = 0.6,$$

so

$$n = \left( \frac{1.645(\sqrt{0.5 \times 0.5} + \sqrt{0.6 \times 0.4})}{0.6 - 0.5} \right)^2$$
$$= 265.$$

Then the rejection proportion is

$$p_r = P_1 + K_1 \sqrt{\frac{P_1(1 - P_1)}{n}},$$

rounded up to the next higher proportion that is possible for the sample size. In this case

$$p_r = 0.5 + 1.645 \sqrt{\frac{0.5 \times 0.5}{265}} = 0.5505,$$

and the next higher possible proportion in 265 is 0.5509.

Increasing the size of the sample will cost something, of course, so the psychiatrists would want to increase the sample only to that size at which the value of the assurance added by the last observation just offsets the extra cost of this last observation. This simple dictum conceals the practical difficulties of the actual choice of sample size, since the value of added assurance and the cost of added observations are typically difficult to appraise; but it expresses the principle. Only by conscientiously attempting to compare costs with the assurance obtained in return for these costs can a reasonably good choice of the sample size be made. Each study must be examined separately on its merits. A statistician can calculate the various combinations of assurance against error that can be obtained with different possible sample sizes. The user of statistics must estimate how much any given combination of assurance is worth, and how much the data cost. Both of these depend on the special circumstances of each particular problem. For that reason, blanket assertions as to the sample size desirable for all applications are worthless.

Increasing the sample size reduces the risks of error because it reduces the standard error of the statistic upon which the test is based, $\sigma_p$ in this case. There are other ways of reducing the risks, and these should be explored carefully in the planning stage of any study. One way is to use a more elaborate design than simple random sampling. For the same outlay of resources, a smaller standard error for the test statistics might be obtained by using, for example, a stratified random sample in preference to a simple random sample[2] (see Chap. 15).

A second alternative is not to fix the sample size in advance, but rather to add one observation at a time until a decision can be made subject to the desired risks of error. This method is called *sequential*

---

2. If, however, the recovery rate from the psychological disorder of our example were really independent of any known variable, nothing would be gained by stratification.

*sampling* or *sequential analysis*. With this method it is not possible to predict in advance how large a sample will be needed before a decision is reached, but it is possible to calculate the mean sample size for repeated sampling, called the "expected sample size." For the same risks of error, the expected sample size with sequential sampling is typically a third to a half smaller than the sample size for a plan that fixes sample size in advance. The saving is less if the true situation is between the null and alternative hypotheses, more if it is outside their span. (Sequential sampling is discussed further in Secs. 16.3.3 and 16.3.4.)

A third method of reducing the risks without increasing the size of the sample is to improve the basic observational procedures so that more information is obtained from each individual. If, instead of classifying each patient merely as recovered or not, some scale measuring degree of recovery could be established, more information would be available in each observation, and fewer observations would be needed for given risks, or lower risks could be attained with the same number of observations. Similarly, if some other characteristic, psychological, physiological, or sociological, could be found which tended to be associated with recovery or nonrecovery, additional information could be obtained from each patient and the risks reduced thereby.

But once the best measuring devices available have been selected and the best type of sampling process decided, there will still remain the problem of determining the sample size, and the principles we have been discussing will then come into play.

## 12.10
## TYPES OF ALTERNATIVE HYPOTHESES

### 12.10.1 One-Sided and Two-Sided Alternatives

In the example we have been considering, the null and alternative hypotheses can be formulated as follows, where $P$ represents the true (but unknown) effectiveness of the therapy:

$$\text{Null:} \qquad P = 0.50,$$

$$\text{Alternative:} \quad P > 0.50.$$

For some purposes, a specific value of the alternative hypothesis must be selected, for example, to calculate the risk of an error of the second kind, or to calculate a sample size. But for some purposes it is suffi-

*Statistical Tests and Decision Procedures*

cient merely to know, as here, the relation of the alternative hypothesis to the null hypothesis. The decision rule, for example, depends only on this relation, not the specific value of the alternative hypothesis; this was pointed out in Sec. 12.7.

Suppose, now, that the situation had been different. Suppose that the therapy was not a new one but one that had been used before. Suppose that the users of the therapy had been somewhat careless about reporting results, and that no one really knew how effective the therapy had been. Suppose that a new development in psychological theory suggests that there is good reason to believe that the therapy may actually be doing harm rather than good. It is decided to draw a sample of case histories in which the therapy has been applied and study each case history to determine whether the patient had recovered at the end of five years. This might involve tracing down some patients whose progress was not completely recorded on the case histories. Hence the expense of carrying through the study might not be negligible, and a carefully worked-out sampling plan might be desirable. The way the problem is now formulated, the null and alternative hypotheses would be as follows:

$$\text{Null:} \qquad P = 0.50,$$

$$\text{Alternative:} \quad P < 0.50.$$

Consider now the sampling distribution of $p$ if the null hypothesis is really true. This sampling distribution is exactly the same as before. Now, however, *small* values of $p$ or $K$ tend to support the alternative

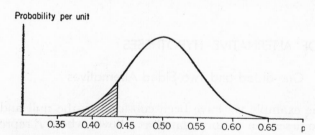

Probability per unit

| 0.35 | 0.40 | 0.45 | 0.50 | 0.55 | 0.60 | 0.65 | $p$ |

**FIG. 400. Rejection region when alternative hypothesis specifies lower values of the parameter.**

hypothesis, since the smaller the sample $p$, the more plausible the hypothesis that the treatment is really harmful. Thus the rejection number is now in the lower tail of the distribution. If the risk of a Type I error is to be 0.1, the rejection value of $K$ would be $-1.28$, corresponding with $p_r = 0.43$, as shown in Fig. 400. If this risk is to be 0.05, the rejection number would be $-1.64$, and so on. For-

### 12.10 Types of Alternative Hypotheses

mally, there is really nothing new in this situation. It is perfectly symmetrical with the old; all we have to do is turn the other example around. For instance, with the earlier formulation of the alternative hypothesis, the null hypothesis is rejected if $K$ or $p$ is *large*. With the second formulation, it is rejected if $K$ or $p$ is *small*.

In both of these cases, the alternative hypothesis is called a "one-sided alternative" because it specifies values only to one side of the null hypothesis, hence sample values which lead to rejection of the null hypothesis and acceptance of the alternative hypothesis are all in one tail of the sampling distribution. A test of significance based on a one-sided alternative is sometimes called a *one-tail* test; a one-tail test may be described more precisely as either a *lower-tail test* or an *upper-tail test*.

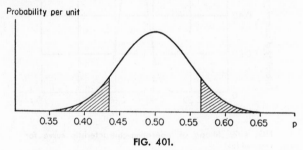

FIG. 401.

Another situation in psychiatric research might be as follows. Return to our earlier assumption that the method of therapy is a new one that has not yet been tried out. The psychiatrists are confident that the therapy could not possibly reduce the probability of recovery, but others in the field fear that it might, so it is agreed that both alternative hypotheses should be considered. In statistical terms, this situation can be represented as follows:

$$\text{Null:} \qquad P = 0.50,$$

$$\text{Alternative:} \qquad P \neq 0.50,$$

$$\text{that is, either} \quad P > 0.50,$$

$$\text{or} \qquad P < 0.50.$$

The sampling distribution of $p$ or $K$ under the null hypothesis is still the same as in the two previous cases, since the null hypothesis is still the same. Now, however, either high or low values of $p$ (or $K$) tend to support the alternative hypothesis, and thus to lead to rejection of the null hypothesis. We must, therefore, have *two* rejection numbers. In Fig. 401 these rejection numbers are set at $-1.28$ and

+1.28, corresponding with $p_r = 0.43$ and $p_r = 0.57$, hence the null hypothesis is rejected whenever $K$ is *either* smaller than $-1.28$ *or* greater than $+1.28$. The sum of the two shaded areas in the graph corresponds with the risk of an error of the first kind, which is $0.1 + 0.1 = 0.2$ (by the addition rule of Sec. 10.6.1).

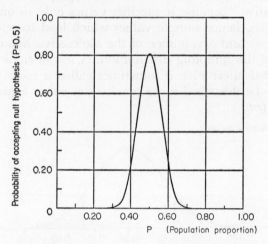

FIG. 402. Shape of operating-characteristic curve for two-tail test.

This formulation of the alternative hypothesis is called a *two-sided* alternative and the test is called a *two-tail* test. The outcome of the experiment will lead to the conclusion that the treatment is beneficial, that it has no effect, or that it is harmful. The OC curve, showing the probability of accepting the null hypothesis, is then high in the middle and low on both ends, as in Fig. 402.

### 12.10.2 Choice between One-Sided and Two-Sided Alternatives

If a one-sided alternative hypothesis is used and the rejection value of $K$ set either at $+1.28$ or at $-1.28$, as the case may be, the risk of an error of the first kind is 0.1, as we have seen. If, on the other hand, a two-sided alternative is under consideration, the rejection numbers would have to be $\pm 1.64$ in order to obtain the same risk of Type I error, 0.1. If the rejection numbers $\pm 1.28$ were used, as in the previous example, the risk of an error of the first kind would be 0.2.

### 12.10 Types of Alternative Hypotheses

Consider a two-sided alternative with a 0.1 risk of Type I error, that is, with rejection numbers ±1.64. A sample is drawn. The sample $p$ leads to a $K$ of 1.37, and the null hypothesis is accepted. With exactly the same risk of Type I error but with a one-sided alternative, the rejection number would be +1.28. In this case the sample $K$ of 1.37 would indicate rejection of the null hypothesis. The same sample result thus leads either to acceptance or rejection of the null hypothesis, depending on the alternative hypothesis. The reason is that the alternative actions between which a choice is to be made are different, and more numerous. It is important to be completely clear whether a one-sided or two-sided alternative is appropriate to the problem at hand. This admonition, of course, is just another expression of a fundamental rule of research—to specify as clearly and unambiguously as possible what the research is trying to find out.

The best way to avoid confusion is to think in terms of the practical decisions to be made. If the only question is whether or not the new therapy is to be used in mental hospitals, the one-sided alternative is probably appropriate. Presumably no one would want to institute a new system of therapy unless it were really more effective, possibly taking some account of costs in measuring effectiveness, than the existing therapy. On the other hand, if the new therapy were definitely less effective, this fact might have important implications for psychological and psychiatric theory—it might suggest, for example, how the disorder came about in the first place. Then the two-sided alternative would be appropriate.

Here again, as in the consideration of errors of the first and second kinds, conventional statistical procedures are often stereotyped and tend to obscure the real issues. Often people use a two-sided alternative when a one-sided alternative would be more appropriate to the question they want to answer and the practical actions they are contemplating. Sometimes the reason is simply that they have learned a computational routine for two-sided alternatives, or even just that they confuse one- and two-tail tables of sampling distributions (see Sec. 11.3.4).

In using Table 391 to get rejection numbers, remember that these are rejection numbers for one-sided tests. Their signs should be taken as plus for upper-tail tests, minus for lower-tail tests. To get rejection numbers for two-sided tests from Table 391 it is necessary to enter the table with *half* the value of the intended probability of an error of the first kind. For example, the figure 2.576 shown for a probability of 0.005 is the rejection number for an upper-tail test at significance level 0.005; −2.576 is the rejection number for a lower-tail test at significance level 0.005; and $+2.576$ and $-2.576$ are the rejection numbers for a two-tail test at significance level 0.01.

## 12.11
## FORMULATING HYPOTHESES

Everything said up to now has assumed that the hypotheses being tested are formulated in advance of collection, or at least of inspection, of the data to be used in their testing. If this is done, then the uncertainty about the inferences which are drawn can be expressed in terms of risks of Type I and Type II errors—or, better, the whole OC curve. Often in scientific work this requirement cannot be, or at least is not, met. Suppose that it is ignored altogether. Suppose a researcher collects a large amount of data without definite hypotheses in mind. He looks through the data and his attention is attracted by certain indications of interesting and important conclusions. Then, after having been led to the formulation of his hypothesis by examining the sample results, he tests the hypothesis in the ordinary way, using the very data which suggested the hypothesis in the first place.

The objection to this is that any body of data has "accidental" extremes that are likely to catch the attention. If features of the data are chosen for consideration because they look unusual, naturally the consideration leads to the conclusion that they are unusual. The risk of Type I error is indeterminately higher than if the hypotheses had been formulated in advance. The risk of Type I error is computed as the probability that if the null hypothesis were true a *randomly* selected result would be of a certain kind. In cases where the hypothesis was formulated from the same data being used to test it, the risk of error of Type I should be computed as the probability that if the null hypothesis were true, a result selected for its unusualness would be of a certain kind. No way to do this has yet been formulated, except in those cases where it may have been decided before examining the data to consider the most unusual—for example, the largest—result, defined in a clear-cut manner. Calling your shots in advance makes all the difference in the world in supporting a claim of marksmanship.

This does not mean that one must set up all the hypotheses in advance and try to anticipate every aspect of the results. That would usually be impossible. The following suggestions may be valuable for guidance:

(1) Specify as many hypotheses in advance as are relevant in the light of existing knowledge, and design tests of significance for them before the data are even collected.

## 12.11 Formulating Hypotheses

(2) After the data are available and the hypotheses specified in advance have been tested, look through the results carefully to find hypotheses which had not been anticipated in advance. While such hypotheses usually cannot be given a rigorous test by the data which suggested them, they may be regarded as new hypotheses to be tested in future studies, or by other samples collected independently. Occasionally an investigator finds results that would have been significant with a risk of error of the first kind as low as 0.000,01 or 0.000,000,001. Even though he has deliberately "hunted" for extreme results, he is, as a practical matter, likely to take such findings quite seriously. But he must be much more cautious than if he had specified the hypothesis in advance. (Also, he should check his calculations, as errors in arithmetic are the source of many such results.)

These suggestions avoid two faulty procedures: (1) Failure to specify any hypotheses in advance, which almost amounts to failure to plan a study in advance. (2) Failure to take advantage of all the information which has been collected—ignoring valuable new hypotheses which may be suggested by the data. It is important in scientific work both to test existing hypotheses and to formulate new ones. Neither objective should be neglected, and most investigations give an opportunity to do both.

The fallacy of singling out the extremes of a given set of data occurs in so many forms and pervades popular discussion to such an extent that further examples may be instructive.

EXAMPLE 405   DRABIK GIRLS

> *Seventh Girl in a Row Born to Woman, 28.*—By Rita Fitzpatrick.
> The odds against it were almost astronomical.
> Practically all Chicago was betting against the chance that Mrs. Henry Drabik, 28, of 7609 St. Lawrence Ave., would have a seventh daughter. But she did.
> The Drabiks are the couple who, when they married in 1941, told everyone they wanted 10 girls. They did fine. In fact, they had six girls in a row, one after the other, until May, 1951. All were named Mary.
> There are Marybeth, 10; Marykay, $8\frac{1}{2}$; Marysue, $7\frac{1}{2}$; Marylynn, 6; Maryjan, $3\frac{1}{2}$; and Marypat, 20 months. Now there is Maryrose.
> The youngest of the Drabik girls, seven pounds nine ounces, was born at 10 a.m. Saturday in Little Company of Mary Hospital.
> "I knew it was going to be a little girl, although my doctor said he thought I would have a boy," Mrs. Drabik said happily yesterday. "I would like to have all the girls in Chicago, if it were possible, and so would my husband."
> Relatives offered to bet the Drabiks 10 to 1 that the next child would be a boy. Mathematicians in 1951 said the odds against the remaining three being girls are reduced to 8. They go down then successively to 4 and to 2.

However, these odds apply only to the whole 10. For each individual birth, the chances are even against a girl. To reach such an answer, the mathematicians said that it must be assumed that the birth of a boy or girl is equally likely. Multiple births must also be ruled out.[3]

This example illustrates vividly the point we have just been discussing. The Drabiks would understandably reject the hypothesis that they are equally likely to have a boy or a girl. But would we reject it? To answer the question we would want to look not just at this one case, selected because of its unusualness, but at *all* seven-child families in Chicago. If all parents were equally likely to have a boy or a girl, we would expect about one in 128 families of seven children to have all girls and another one in 128 to have all boys, just by chance. (These results are obtained by the multiplication rule of Sec. 10.6.2.) If it turned out that many (significantly) more than two families per 128 of the seven-child families had children all of one sex, we would then have a real ground for doubting the hypothesis that all parents (and the Drabiks in particular) are equally likely to have a child of either sex.[4]

This illustrates one possible means of allowing for the fact that we initially picked an extremely unusual occurrence which attracted our attention. Often, however, it is hard to enumerate all the equally unusual events that might have attracted our attention.

### EXAMPLE 406A BUSINESS FORECASTS

One of many economists forecasts a decline in general business conditions and the decline materializes. The successful economist is singled out for his unusual foresight. He may deserve such acclaim, but first one would want to see his entire record of economic predictions. One economist who achieved a reputation in Washington by forecasting the sharp decline of 1937, had actually been forecasting such a decline ever since the preceding one, in 1934.

### EXAMPLE 406B COMMODITY SPECULATIONS

Every so often, prices on the grain market, or some other commodity market, fall sharply. A few speculators profit greatly because they have "sold short" before the decline. Several Congressmen then demand an investigation to see if these speculators somehow had manipulated the decline. Speculators often sell short, however, so there are some in short positions any time a market breaks. This does not prove, of course, that none of those in short positions manipulated the market, but it does suggest

---

3. *Chicago Tribune*, February 3, 1953. If the probability of an event is $P$, the odds for it are $P/(1 - P)$ and the odds against it are $(1 - P)/P$.

4. The *Chicago Tribune* of December 28, 1953, reported the birth of a boy to the Drabiks.

*12.11 Formulating Hypotheses*

restraint in concluding that a short position is evidence of manipulation—or even of business acumen, the source to which those involved attribute it.

EXAMPLE 407A   CORRELATED SEQUENCES

One of the most instructive examples of the dangers of "hunting around" is given by economic and business "time series." In the desire to predict future movements of the stock market, over-all business conditions, etc., many people have looked with great persistence for other time series which have in the past been closely correlated with the immediate future of the series of primary interest. Looking back over the past, it is often possible to find two time series that are closely correlated. Frequently when such time series are found, it is discovered that the correlation is of little or no use in predicting future changes in one time series from the changes in the other. The correlation observed in the past was probably an "accidental" random fluctuation. The following example is instructive.

> Suppose we wish to find a formula for estimating the digits on Line 28 of the Table of 105,000 Random Decimal Digits published by the Interstate Commerce Commission, Bureau of Transport Economics and Statistics. The first ten digits on this line are 0, 0, 7, 4, 2, 5, 7, 3, 9, 2. These digits are closely correlated with the first ten digits on line 32, which are 0, 0, 7, 2, 5, 6, 9, 8, 8, 4. The question arises, is line 32 useful for forecasting line 28? Proceeding in the usual manner . . . [we find] . . . a correlation of .796. . . . The probability of so large a [correlation] . . . is less than .01. An unsuspecting individual might therefore conclude that the answer to our question is Yes.
>
> The reason for the unusually large correlation is that I deliberately looked for two lines that appeared to be highly correlated . . . .
>
> I receive letters from time to time about methods of forecasting the stock market, pointing out how wonderfully well the method would have worked for some past period. I suspect that in nearly all cases this period coincides with the period used to arrive at the forecasting procedure.[5]

EXAMPLE 407B   SLIPPERY ROCK VS. NOTRE DAME

Often the sports pages of newspapers contain comparisons of the following kind. Slippery Rock State Teachers College beat team B by 13 points, B beat C by 1 point, C beat D by 6 points, D beat E by 20 points, E beat F by 3 points, F beat G by 10 points, and G beat Notre Dame by 1 point. Therefore Slippery Rock could beat Notre Dame by 13 + 1 + 6 + 20 + 3 + 10 + 1 = 54 points. Actually, of course, it may be that no team in the chain would ever win in another game against the same opponent, much less win by the same margin. The point here, however, is that this may be the only one of many chains of the same kind leading from Slippery Rock to Notre Dame that does not "show" a wide margin in favor of Notre Dame.

---

5. From an unpublished statement by Howard L. Jones, Illinois Bell Telephone Company, June 20, 1949.

## 12.12
## CONCLUSION

This chapter has gone into a relatively large amount of technical detail. It will be helpful, perhaps, to summarize its highlights by suggesting how to go about the interpretation of the results of tests of significance encountered in reports of statistical investigations.

First, suppose that an author concludes that one of his findings is "significant."

(1) Does he mean "statistically significant," or does he mean "practically significant?" If he means "practically significant," you should check to see if there is any way to discover if the finding is statistically significant. It is so interesting to speculate on the "practical significance" of a finding that a check on statistical significance seems pedantic and tedious; nonetheless, it is futile to speculate on practical significance unless the finding is statistically significant. If the author means, however, that his finding is statistically significant, then various qualifications are necessary to give this term concrete meaning. Hence we come to a second question.

(2) Did the author use a one-sided or two-sided alternative, and was this appropriate to his problem? If the choice was inappropriate, the effect of this error must be examined.

(3) Since the author concluded that the finding was significant, the only kind of error he might have made is an error of the first kind. It is important, therefore, to know the risk of Type I error inherent in his testing procedure. Sometimes this risk will be stated explicitly; for example, "the finding was significant at the one percent level of significance." This is equivalent to saying, "The finding was significant subject to a risk of Type I error of 0.01." That is, the terms "level of significance" and "risk of Type I error" are equivalent. Again, the statement is often encountered that a finding is "significant at the $2\sigma$" (or "two standard deviation") level. This means that the rejection level of the statistic corresponded with a standard normal variable of 2, giving a risk of Type I error of about 0.05 if the test is two-tail, or 0.025, if it is one-tail. At any rate, if the reference to risk of Type I error is made, it is necessary to ask if the author's choice of this risk was really appropriate.

(4) One should always try to determine whether the author formulated his hypotheses in advance of looking at the data, or whether the hypotheses were in some measure suggested by the same data used for the test. This is often difficult to ascertain from the report, but it is always a vital question in evaluating the results.

(5) Finally, if it appears that the finding is really statistically significant, the question of its *practical* significance should be considered. Sometimes authors are so intrigued by tests of significance that they fail even to state the actual *amount* of the effect, much less to appraise its practical importance.

If, on the other hand, the author asserts that a particular finding is not significant, an analogous series of questions should be asked. It is not difficult to see the necessary modifications in the above sequence. Only one comment seems worth emphasizing explicitly. If a finding is "insignificant" it is necessary to examine the risk of Type II error which was involved in the test procedure. The best way to examine this is through the operating-characteristic curve. This is not always easy or even possible from the evidence presented in the report. The importance of raising the question is, however, always great. If the risk of Type II error is large, as it is likely to be if the sample is small and the risk of Type I error low, the finding of "no significant difference" must be viewed with great reserve. The sample size may be so small and the risk of Type I error so small that even a fairly substantial effect would have little chance of being detected. Under such circumstances, failure to find an effect is not particularly indicative that there is none.

Finally, the ideas of significance tests apply to much more complicated examples than the one we have considered. The questions to be considered in these more complicated cases are almost identical with the ones we have discussed. The main difference is that it may be necessary to know a good deal about technical statistics, or to consult a statistician about them.

## DO IT YOURSELF

Example 409A

It is reported in a newspaper that 37 of 43 births in a particular hospital for a specified period of time are males. Is this consistent with the hypothesis that the true probability of a male birth in this hospital is $\frac{1}{2}$?

Example 409B

The government, in buying combat boots from a shoe manufacturer, specifies that no more than one percent of an order of 100,000 pairs is to be "defective" according to an agreed-upon standard. In a sample of 1,000 pairs, chosen at random from the 100,000, 15 pairs of boots are defective. What can the government conclude about the acceptability of the entire 100,000?

*Statistical Tests and Decision Procedures*

EXAMPLE 410A

A physical anthropologist obtains a certain skull measurement for 25 members of a primitive tribe. He obtains the same measurement for 25 members of a second tribe. Pairing measurements at random, he finds that in 21 cases the member of the first tribe had the larger measurement. What inference can he draw in relation to a hypothesis that the first tribe, on the average, has the greater measurements?

EXAMPLE 410B

Look back to Table 279. Were actual purchases significantly greater than planned purchases?

EXAMPLE 410C

The following experiment was conducted by a mail-order company. Two catalogs, *A* and *B*, identical except that *A* used more color illustrations, were mailed to two groups of customers. The two groups of customers were not selected at random, the group receiving catalog *A* being made up of customers for whom sales per capita had been high in the past. For each advertisement in the catalog, the following ratio was computed: sales from catalog *A* divided by sales from catalog *B*. This ratio was then converted to a percentage figure. Each of the departments shown in Table 410 had some advertisements with monotone illustrations in both catalog *A* and catalog *B*, and other advertisements with four-color illustrations in *A* but monotone in *B*. Did four-color advertising increase sales significantly? Do you have any qualifications about your conclusion?

TABLE 410

SALES FROM CATALOG *A* AS PERCENTAGE OF SALES FROM CATALOG *B*

| Department | Ads Four-Color in *A*, Monotone in *B* | Ads Monotone in *A* and *B* |
|---|---|---|
| 1 | 568 | 391 |
| 2 | 665 | 510 |
| 3 | 413 | 550 |
| 4 | 681 | 412 |
| 5 | 1,129 | 464 |
| 6 | 814 | 380 |
| 7 | 598 | 424 |
| 8 | 576 | 456 |
| 9 | 923 | 425 |
| 10 | 623 | 449 |

EXAMPLE 410D

Reread the paragraph starting on the middle of page 172. Can you now supply the calculations on which the conclusion of that paragraph was based?

EXAMPLE 410E

A study of "leading admirals" revealed that 14 out of 31 had come from the upper quarter of their classes at Annapolis. Can one conclude from this

*Do It Yourself*

that scholastic standing at Annapolis has something to do with becoming a "leading admiral"?

EXAMPLE 411A

In Table 597A does stress have a significant influence on tensile strength?

EXAMPLE 411B

A statistician talked with one of the authors about analyzing some data that were supposed to show whether male and female employees differ in their attitudes toward a certain company. He had questions about *t* tests, analysis of variance, homogeneity of variances, and so on. The author didn't know the answers to all of his questions, but he could see that there definitely was a difference between the sexes.

Each sex had been divided into six groups, three length-of-service categories for each of two races. The Negro short-term males were less favorable to the company than were the Negro short-term females. The white short-term males were also less favorable than the white short-term females. In fact, all six pairs showed the males less favorably disposed toward the company than the corresponding group of females.[6]

How often would this happen by chance if the two sexes were equally favorable? Would you use a one-sided or a two-sided alternative?

EXAMPLE 411C

The median income of families in a certain city is claimed to be $7,000 by the Chamber of Commerce. You draw a random sample of 100 from the families in this city and find only 23 of their incomes above $7,000. Do you believe the Chamber of Commerce claim?

EXAMPLE 411D

In testing a new rifle, the new rifle and a standard rifle are fired a large and equal number of times at a target, under as nearly equal conditions as possible (fixed mount, same time of day, etc.). The new rifle makes 53 hits, the old, 47. Is the new rifle more accurate?

EXAMPLE 411E

An educator wishes to find out the relative effectiveness of two methods of teaching arithmetic, *A* and *B*. He finds in the literature that 22 apparently well-designed experiments have been conducted independently over the years. Thirteen of these failed to show a statistically significant difference, though in ten of them *A* did better than *B*. The remaining nine experiments showed differences which were significant at levels of significance of five percent or less. Eight of these showed *A* better than *B*. What can the educator conclude?

6. W. Allen Wallis, "Rough and Ready Statistical Tests," *Industrial Quality Control*, Vol. 8, No. 5 (1952), p. 35.

# Chapter 13

# Further Test Procedures

## 13.1
### INTRODUCTION

Chap. 10 presented some elementary principles of probability, Chap. 11 showed how sampling distributions can be deduced through probability principles, and Chap. 12 showed how statistical decisions or tests of significance can be based on sampling distributions. This chapter will present a series of actual test procedures, and show how to use them.

The tests included in Chap. 12 were there to illustrate principles. Now that the principles of significance testing have been established, we will present a series of tests for several problems of types that arise frequently. Our chief aim in presenting these is to convey the nature of the details embraced by the principles, not to make you proficient in these details. Should you meet problems like these in your own work, you will be wise to obtain statistical advice, to learn more about technical statistics than an introductory book conveys, or to turn to Chap. 19. In Chap. 19 there are shortcut methods for tackling nearly all of the problems discussed in this chapter. These shortcut methods are easier and safer for the beginner. Proper understanding of them depends on understanding the ideas presented in this chapter, but not necessarily the technical details. The technical details of this chapter should enable you to make a few calculations to be sure you understand the ideas. They should also help you in reading statistical materials, since the techniques presented here are widely used.

First, we will consider a group of problems relating to means. Then, we will consider the same problems for proportions. For each of the two types of statistic we will consider (1) how to test an assumption about the corresponding population parameter, (2) how to decide whether two population parameters are equal, and (3) how to decide whether a set of population parameters are equal.

While the normal distribution provides an approximate sampling distribution for most of these six situations, some of them are handled by statisticians through special distributions, known as Student's, the chi-square, and the variance ratio distributions. We will not explicitly introduce these distributions and the special tables needed for them, but instead will show how the problems can be handled sufficiently well for our purposes by means of the normal distribution.

This chapter is not really essential to the general stream of ideas in the book; and you may even want to bypass it on a first reading. But once you have finished reading the book, if you refer to it again for practical guidance, it will often be this chapter and Chap. 19 to which you refer.

## 13.2
## TESTS OF MEANS

### 13.2.1  Testing an Assumption About a Population Mean

Suppose we wish to test the assumption that the mean, $M$, of a large number of weights is 170 pounds. We draw a random sample of size $n = 32$ and find $\sum x = 5,629$ and $\sum x^2 = 1,003,483$. From these we compute $\bar{x} = 175.91$ and $s = 20.718$. These are, in fact, the data and computations shown in Secs. 7.4.1 and 8.5.2.

From the central limit theorem (Sec. 11.3.1) we know that if repeated samples were drawn from the same population the variation in the values of $\bar{x}$ would be described well by a normal distribution. The mean of this normal distribution, if the population mean is 170, will also be 170 (Sec. 11.4.1). The standard deviation, or standard error, of this normal distribution of means will be $\sigma/\sqrt{n} = \sigma/\sqrt{32}$, where $\sigma$ is the standard deviation of the population of individual weights (Sec. 11.4.2).

We do not know $\sigma$, the population standard deviation, but we will use $s$, the sample standard deviation, as though it were $\sigma$. This causes some lack of precision, for instead of the exact, stable unit of measure, $\sigma$, appropriate to measuring sampling fluctuations in the mean, we are using a unit of measure, $s$, which is itself subject to sampling fluctuations. Thus, a value of $\bar{x}$ may appear to be relatively far from $M$ simply because the sample has yielded a value of $s$, in terms of which we measure $\bar{x} - M$, that is less than $\sigma$; or $\bar{x}$ may appear relatively close to $M$ if we happen to have a value of $s$ that exceeds $\sigma$. For samples as large as, say, 10 (or anyway 20) the error introduced from this source is unimportant for most practical purposes; and we will indicate later a refinement that virtually eliminates the error even when $n$ is smaller than 10.

We have, then, in our example

$$s_{\bar{x}} = \frac{s}{\sqrt{n}} = \frac{20.718}{5.657} = 3.662,$$

from which the standard normal variable is

$$K = \frac{\bar{x} - M}{s_{\bar{x}}} = \frac{175.91 - 170}{3.662} = 1.614.$$

Referring to Table 365 we find a probability of 0.053.

The decision to be made on the basis of this probability of 0.053 depends on the alternative hypothesis and the level of significance, or risk of error of the first kind. Suppose that the data referred to a random sample of airplane passengers, and that the test were being made to decide whether 170 lbs. is a safe average to use in figuring airplane loads without weighing individual passengers. We would want to reject the null hypothesis that $M$ is 170 only if $M$ exceeds 170. A one-sided alternative hypothesis is involved, so an upper-tail probability is appropriate. Thus, for any level of significance 0.053 or more we would reject the null hypothesis and decide that $M$ exceeds 170. At a ten percent level, for example, we decide that 170 is not an acceptable average to use for this purpose.

If maximum loading rather than safety were the only consideration, the alternative hypothesis would be that $M$ is less than 170, for only in that case would we be unwilling to work with 170 as the mean. This would make a lower-tail probability appropriate, and we find a probability of 0.947; this leads to acceptance of the null hypothesis, since a significance level as high as 0.947 (or even as high as 0.5) would rarely, if ever, be used.

If it is important to guard against *either* overloading or underloading, and this is the most realistic problem, the alternative hypothesis is two-sided, and a two-tail probability is relevant. The probability guiding our decision then becomes twice 0.053 or 0.106, the probability of being as far from 170 as $\pm(\bar{x} - M) = \pm(175.91 -$

170) $= \pm 5.91$, and we would reject the null hypothesis only if we were accepting more than a 0.106 risk of an error of the first kind. At a ten percent level, for example, 170 would be accepted as a working mean.

So far we have been able to discuss these tests by simply indicating which risks of error of the first kind would lead to acceptance or rejection for the sample actually observed. To compute an OC curve, however, we must consider all the samples that could arise, and which of them would lead to rejection of the null hypothesis. This requires us to select some specific value of the risk of rejecting the null hypothesis if it is correct. For illustration, take this risk, or significance level, as 0.1. Then the upper-tail test rejects if $K$ equals or exceeds 1.282, the lower-tail test if $K$ equals or is less than $-1.282$. (These values are from Table 391.) Now with a null hypothesis of $M = 170$ and a standard error of the mean of 3.662, sample means will lead to acceptance by the upper-tail test if

$$K = \frac{\bar{x} - 170}{3.662} < 1.282,$$

that is, if

$$\bar{x} < (1.282 \times 3.662) + 170 = 174.69.$$

Thus, 174.69 is the rejection number for $\bar{x}$ for the upper-tail test. Similarly, the lower-tail test will lead to acceptance for those samples in which $\bar{x}$ exceeds 165.31.

A two-tail test with significance level 0.1 rejects if $K$ equals or exceeds 1.645 or $K$ equals or falls short of $-1.645$ (see Table 391), and by the type of computation just illustrated, a two-tail test will lead to acceptance for those samples in which $163.98 < \bar{x} < 176.02$ (that is, $\bar{x}$ is between 163.98 and 176.02).

An OC curve for each of these three alternative tests at the 0.1 significance level is computed by assuming various values of $M$ in turn, and for each asking, If this were the population mean, what would be the probability that a sample of 32 would have a sample mean leading to acceptance of the 170 figure? Such probabilities can be found from Table 365.

To illustrate the calculations for an OC curve, suppose the true population mean is 177. Then for the upper-tail test, the probability of accepting the null hypothesis is the probability of getting a sample with a mean less than 174.69. To find this probability we compute

$$K = \frac{174.69 - 177}{3.662} = -0.631.$$

From Table 365 we find that the probability of a standard normal variable being less than −0.631 is 0.264. In other words, the probability of accepting 170 as a basis for calculations of passenger weights would be a little more than one-fourth if the true mean were 177. By making similar computations for $M = 164, 166, 168, \ldots, 184, 186$, we obtained the OC curve for the upper-tail test plotted in Fig. 417.

Computations for the lower-tail test are similar. For example, if $M$ is 177, the probability of a sample in which $\bar{x}$ exceeds 165.31, leading to acceptance of the null hypothesis, is found by computing

$$K = \frac{165.31 - 177}{3.662} = -3.19.$$

The probability that a standard normal variable will exceed −3.19 is shown by Table 365 to be more than 0.999. It is virtually certain, therefore, that a random sample would lead to acceptance of the 170 figure if the true mean were 177. Remember that this is the situation in which the proper decision when $M$ exceeds 170 is the same as when $M$ equals 170, namely to continue to use 170. Similar computations were made for $M = 154, 156, 158, \ldots, 174$, and 176, and the OC curve for the lower-tail test is also plotted in Fig. 417.

The computations of the OC curve for the two-tail test are a little more complicated. We accept when $163.98 < \bar{x} < 176.02$. If $M$ is 177, this means, in terms of standard normal variables, that we accept when

$$\frac{163.98 - 177}{3.662} < K < \frac{176.02 - 177}{3.662}$$

that is, when

$$-3.555 < K < -0.268.$$

The probability that $K$ will be between these two limits is 1 minus the sum of two probabilities, (1) the probability that $K$ will be below the lower limit, and (2) the probability that $K$ will be above the upper limit. From Table 365, we find that the first of these probabilities is 0 (to three decimals) and the second is 0.606, so the probability of acceptance is 0.394—about two chances out of five. Similar computations for $M = 154, 156, \ldots, 184, 186$ were also made, and the OC curve is charted in Fig. 417.

Fig. 417 brings out the differences among the three tests. The two-tail test has some chance of detecting excesses of the true mean above 170, though not as good a chance as the upper-tail test. On the other hand, unlike the upper-tail test, the two-tail test has a fairly good chance of detecting shortages in the true mean below 170. Which test to use depends on which OC curve comes nearest to giving the proper decisions for the various values of $M$ that might prevail.

A shortcut method for the problem of this section will be found in Sec. 19.2.

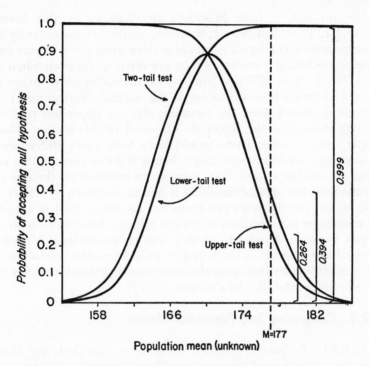

FIG. 417. Operating-characteristic curves of three tests of a mean. Null hypothesis M = 170; standard error of mean 3.662, significance levels 0.1.

**13.2.1.1  *A Technical Refinement: Student's Distribution.*** In more technical treatments of statistics it is usual to make proper allowance for the fact that when the standard deviation is obtained from a sample it is subject to sampling error itself. The method of making this allowance was introduced in 1908 by William Sealy Gosset (1876–1937), a statistician for Guinness, a famous Irish brewery. The firm did not then allow publication of research done by its staff. Gosset was able to obtain relaxation of this rule for the statistical methods he developed, but only on condition that he use a pen name. Today the name Gosset is scarcely known, but the name "Student" is one of the most celebrated in the history of statistics.

Student's distribution, also called the *t*-distribution, requires, in effect, a table like Table 365 or 391 for each sample size—or, rather, for each value of a quantity called the "number of degrees of freedom," which, in the case of the standard deviation when the sample mean is used in the calculations, is $n - 1$, that is, one less than the sample size.

We do not include Student's distribution in this book. In Sec. 13.4.1, Technical Note 1, however, we do give formulas by which to compute more accurate values of $K$ than those given in this section. These formulas, as a matter of fact, are often useful even when tables are at hand, since tables of Student's distribution cover many fewer probability levels than do tables of the normal distribution. For the example discussed here, the formulas give an upper-tail probability of 0.058 instead of the 0.053 we obtained in this section. Had the sample been smaller, there would have been more difference. But for most applications, especially when $n$ is 20 or more, these refinements are not important. Our purpose in mentioning them is partly to warn you that special handling is needed for work with very small samples, but mostly to let you know the nature of Student's *t*, in case you encounter it in reading statistical reports. Student's *t* is simply a normal variable standardized in terms of a standard deviation or standard error which is itself subject to independent sampling error, and Student's distribution is the sampling distribution which allows for sampling variability in *s* as well as in $\bar{x}$.

## 13.2.2 Comparing Two Population Means

13.2.2.1 *Independent Samples.* A set of weight data like that just discussed was collected from another class. This time, 40 observations were obtained. Their sum was 6,716 and the sum of their squares was 1,142,241.

To distinguish the two samples we will use the subscripts 1 and 2:

| First Sample | Second Sample |
|---|---|
| $n_1 = 32$ | $n_2 = 40$ |
| $\sum x_1 = 5,629$ | $\sum x_2 = 6,716$ |
| $\sum x_1^2 = 1,003,483$ | $\sum x_2^2 = 1,142,241$ |
| $\bar{x}_1 = 175.91$ | $\bar{x}_2 = 167.90$ |
| $s_1 = 20.718$ | $s_2 = 19.365$ |
| $s_{\bar{x}_1} = 3.662$ | $s_{\bar{x}_2} = 3.062$ |

*13.2 Tests of Means*

(The value of $s_2$ is obtained from

$$s_2{}^2 = \frac{\sum x_2{}^2 - \dfrac{(\sum x_2)^2}{n_2}}{n_2 - 1} = \frac{14{,}624.6}{39} = 374.9897,$$

and similarly for $s_1$—see Sec. 8.5.2.)

The question arises as to whether the difference of 8.01 between the sample means signifies a real difference between the population means, or is within the range of variation to be expected between random samples from a population in which the individuals vary as much as they do in this case. To put the question differently, if both samples had come from the same population, what is the probability that their means would have differed by as much as 8 or more?

Let $D$ stand for the difference between the population means, $M_2 - M_1$, and let $d$ stand for the difference between the sample means, $\bar{x}_2 - \bar{x}_1$. Is our sample difference, $d$, of $-8.01$, consistent with the idea that the population difference, $D$, is zero?

Values of $d$ in repeated sampling will be nearly normally distributed. The mean of their sampling distribution will be zero if $D$ is zero. To find the standard deviation of the distribution we use a fact not mentioned before: *The standard error of the difference of two independent quantities is the square root of the sum of the squares of the standard errors of the separate quantities.*[1]

In our case, since

$$d = \bar{x}_2 - \bar{x}_1,$$

we have

$$\sigma_d = \sqrt{\sigma_{\bar{x}_1}{}^2 + \sigma_{\bar{x}_2}{}^2} = \sqrt{\frac{\sigma_1{}^2}{n_1} + \frac{\sigma_2{}^2}{n_2}}.$$

Since we do not know $\sigma_1$ and $\sigma_2$, we replace them by $s_1$ and $s_2$, with a slight imprecision:

$$s_d = \sqrt{\frac{(20.718)^2}{32} + \frac{(19.365)^2}{40}} = \sqrt{13.4136 + 9.3751}$$

$$= \sqrt{22.7887} = 4.774.$$

The number 4.774 is the unit in which to measure the departure of the observed difference from that assumed:

$$K = \frac{d - D}{s_d} = \frac{-8.01 - 0}{4.774} = -1.678.$$

---

1. This holds true for the sum of independent quantities as well as for the difference. Also, "standard error" could be replaced by "standard deviation."

Suppose we are testing against a two-sided alternative hypothesis, and at the five percent significance level. Then we would reject the null hypothesis (that $D$ is zero) only if $K$ were outside the range $-1.960$ to $+1.960$ (these values come from Table 391). In this case, therefore, we would accept the null hypothesis. The actual two-tail probability is 0.093 (Table 365).

We might, of course, have in mind a one-sided alternative on either side. The OC curves could be computed essentially as in Sec. 13.2.1. In fact, once we found $\sigma_d$, the standard error of the difference, this problem really became a special case of the problem treated in Sec. 13.2.1, the null hypothesis here being that the population mean is zero.

A shortcut method for the problem of this section will be found in Sec. 19.3.1. The use of Student's $t$ in this case is explained in Sec. 13.4.2, Technical Note 2.

13.2.2.2 *Matched Samples.* The situation treated in the preceding section, where two independent samples are compared, is very different from one in which the individuals of a single sample are measured twice. Before-and-after data are a common case of measuring the same individuals twice, but husbands and wives, pairs of siblings, or pairs of measurements for individuals (for example, head length and head width) would also be examples—any bivariate observations.

Suppose that a group of men is weighed on June 1 and then again on December 1. To determine whether there has been a change in the mean weight of the population of which the group is a sample, it would be a serious error to regard these data as two independent samples. The variation from sample to sample at a given time is so great that only large differences in mean weight would be outside the range of sampling variability. Indeed, it is this fact that would lead us to collect data for such a problem by measuring the same individuals at two dates, rather than by drawing separate samples at the two dates. Most of the variation between separate samples would represent the variation among individuals, and this would swamp the variation between the two dates unless the samples or the variation between dates were large.

In this situation, we must recognize that we have only one sample. For each individual we calculate a weight gain, his weight at the later date minus his weight at the earlier date. Thus we have a single sample of $n$ measurements of change. We analyze this sample exactly as in Sec. 13.2.1, with the null hypothesis that the population mean is 0.

### 13.2 Tests of Means

The data of Table 421 represent artificially generated before-and-after weights of 25 men. The mean gain, $\bar{d}$, is 2.28 lbs., with a standard error, based on $s$ since $\sigma$ is unknown, of 0.639. To find the probability of a gain as large as this, if the population mean gain is 0, compute

$$K = \frac{2.28 - 0}{0.639} = 3.568.$$

The upper-tail probability of so large a gain in the sample, if the population mean gain is 0, is shown by Table 365 only as less than 0.001. Even the two-tail probability is actually less than 0.001. Thus, at any significance level likely to be employed, the data are not consistent with the notion that there has been no change, and we conclude that there has been a real gain in the population mean weight, though apparently a small one.

TABLE 421

WEIGHTS OF 25 MEN, BEFORE AND AFTER A LAPSE OF TIME
(POUNDS)

| Before $x_1$ | After $x_2$ | Gain $d$ | Before $x_1$ | After $x_2$ | Gain $d$ |
|---|---|---|---|---|---|
| 162 | 166 | +4 | 190 | 189 | −1 |
| 191 | 196 | +5 | 184 | 183 | −1 |
| 138 | 136 | −2 | 134 | 136 | +2 |
| 182 | 190 | +8 | 150 | 153 | +3 |
| 159 | 160 | +1 | 145 | 147 | +2 |
| | | | | | |
| 138 | 141 | +3 | 150 | 150 | 0 |
| 136 | 139 | +3 | 176 | 177 | +1 |
| 185 | 187 | +2 | 152 | 151 | −1 |
| 162 | 169 | +7 | 200 | 198 | −2 |
| 195 | 200 | +5 | 166 | 174 | +8 |
| | | | | | |
| 142 | 139 | −3 | 173 | 176 | +3 |
| 168 | 167 | −1 | 200 | 205 | +5 |
| 145 | 151 | +6 | | | |

$$\sum d = 57 \qquad \sum d^2 = 375 \qquad \bar{d} = \frac{\sum d}{n} = \frac{57}{25} = 2.28$$

$$s^2 = \frac{\sum d^2 - \frac{(\sum d)^2}{n}}{n-1} = \frac{375 - 129.96}{24} = 10.21$$

$$s = 3.195 \qquad s_{\bar{d}} = \frac{3.195}{\sqrt{25}} = 0.639.$$

Since this case has been reduced to a special case of the problem of Sec. 13.2.1, everything said there about alternative tests and their

OC curves, and about the refinement of Student's $t$, applies here without change.

Had two independent samples been used to study this problem, each sample would have had to contain 2,222 observations (a total of 4,444) to give as reliable conclusions as are given by this sample of 25 measured before and after (50 measurements). This illustrates the potential importance of proper statistical planning before collecting data. In calculating this, we have assumed that the population standard deviations are 20 for individual weights and 3 for individual gains. (The problem, which is based on artificial data taken from a random number table, was set up with these numbers as population values.) Then the standard error of the difference between the means of the two samples of 2,222 would be $\sqrt{\dfrac{400}{2,222} + \dfrac{400}{2,222}}$ or 0.6, and the standard error of the mean gain for the sample of 25 measured twice would be $\dfrac{3}{\sqrt{25}}$, also 0.6.

A shortcut method for dealing with the problem of this section is given in Sec. 19.3.2.

### 13.2.3 Comparing Several Population Means: Analysis of Variance

Suppose that the 25 men whose weights are recorded in Table 421 are divided into 4 age groups: the first 4 men in the youngest group, the next 9 in the second group, the next 7 in the third group, and the last 5 in the oldest group. The data, using the "before" weights from Table 421, are shown this way in Table 423. The means of the four groups vary from 159 to 178. If we had simply divided the 25 men at random into four groups of these sizes, there would have been some variation among the sample means. Before we conclude that the means of the four populations differ, we must see if the amount of variation in the sample means is greater than could be expected in a random grouping.

The method of doing this is based on the fact that $\sigma_{\bar{x}} = \sigma/\sqrt{n}$. We will compute the standard deviation of the four means directly, and compare it with the standard deviation within groups. Actually, it is simpler to deal with variances, the squares of the standard deviations, and the corresponding relation $\sigma_{\bar{x}}^2 = \sigma^2/n$.

To explain the idea of the analysis, we shall suppose initially that all four groups had been of the same size, say 6 (with a total of only

## 13.2 Tests of Means

TABLE 423

WEIGHTS OF 25 MEN, BY AGE GROUPS

|   | Youngest | Next to Youngest | Next to Oldest | Oldest | |
|---|---|---|---|---|---|
|   | 162 | 159 | 190 | 152 | |
|   | 191 | 138 | 184 | 200 | |
|   | 138 | 136 | 134 | 166 | |
|   | 182 | 185 | 150 | 173 | |
|   |   | 162 | 145 | 200 | |
|   |   | 195 | 150 |   | |
|   |   | 142 | 176 |   | |
|   |   | 168 |   |   | |
|   |   | 145 |   |   | All Ages |
| $n$ | 4 | 9 | 7 | 5 | 25 |
| $\Sigma x$ | 673 | 1,430 | 1,129 | 891 | 4,123 |
| $\Sigma x^2$ | 114,893 | 230,728 | 184,913 | 160,589 | 691,123 |
| $\bar{x}$ | 168.25 | 158.89 | 161.29 | 178.20 | 164.92 |
| $s$ | 23.528 | 20.967 | 21.685 | 21.288 | 21.562 |

24). After this explanation, we shall give a computational procedure which both simplifies the arithmetic and allows for the fact that in most actual examples, as in this one, the groups are not of the same size.

Consider, then, four samples each of size 6. Under the null hypothesis that the population means are the same, these four samples can be regarded as four independent samples from the same normal distribution. Compute the sample means, $\bar{x}_1$, $\bar{x}_2$, $\bar{x}_3$, and $\bar{x}_4$.

Now these four means can be regarded as a sample from the population of means of samples of size 6. The population of means of 6 will have a standard deviation equal to the standard deviation of the observations, divided by the square root of 6 (Sec. 11.4.2). That is,

$$\sigma_{\bar{x}}^2 = \frac{\sigma^2}{6}$$

or

$$6\sigma_{\bar{x}}^2 = \sigma^2,$$

where, for simplicity in what follows, we have used variances, the squares of the standard deviations. The analysis involves three stages: (1) Compute a sample estimate of the left hand side of this equation, treating the means as a sample. (2) Compute a sample estimate of the right hand side of the equation, using the variation within the individual samples. (3) Finally, compare these two sample estimates, allowing for some discrepancy due to sampling error. If the number obtained directly from the means is significantly bigger than the number obtained from the variation of the observations within samples, we conclude that something has been added to the ordinary sampling variation among the means—in other words, that the samples come from populations with different means.

To estimate the left hand side of the equation, compute $\bar{x}$, the mean of the sample means, and then compute

$$s_A{}^2 = 6 \times \frac{\sum (\bar{x}_i - \bar{x})^2}{k - 1},$$

where $k$ is the number of samples, 4 in this case, and the subscript $A$, for "among," indicates that this estimate is based on the variation among groups.

To estimate the right hand side of the equation, compute the variances, $s_i{}^2$, of the $k$ individual samples, and find their mean:

$$s_W{}^2 = \frac{\sum s_i{}^2}{k},$$

where the subscript $W$, for "within," indicates that this estimate is based on the variation within groups.

Thus two independent estimates of $\sigma^2$, $s_A{}^2$ and $s_W{}^2$, are computed from these four samples. Under the null hypothesis that the four samples are drawn from the same population, we would expect $s_A{}^2$ and $s_W{}^2$ to be the same, within the limits of the appropriate pattern of chance variation.

If, however, the four samples are from populations with different means, the value of $s_A{}^2$ will be inflated. That is, the differences among $\bar{x}_1$, $\bar{x}_2$, $\bar{x}_3$, and $\bar{x}_4$ will reflect not only the variation among observations from the same population, which is measured by $s_W{}^2$, but also the differences among the means of the populations from which the samples came. If $s_A{}^2$ exceeds $s_W{}^2$ by an amount larger than we are willing to attribute to chance, we decide that the four sample means are not all drawn from the same normal population.

The following is an easy way of computing $s_A{}^2$ and $s_W{}^2$ for the case of unequal groups, illustrated by the data of Table 423.

*To calculate $s_A{}^2$:*

*Step* 1: For each group, calculate $(\sum x)^2/n$, and add these quantities for all groups. (The result may be regarded as the sum of the squares of the 25 observations "adjusted" by replacing them by their group means.)

*13.2 Tests of Means*

*Step* 2: For all groups combined compute $(\sum x)^2/n$ and subtract it from the result of Step 1. (The result is the sum of the squared deviations of the 25 "adjusted" observations from the general mean.)

*Step* 3: Divide the result of Step 2 by $k - 1$, that is, one less than the number of groups. This gives $s_A{}^2$.

In our case, we have:

*Step* 1:

$$\frac{(673)^2}{4} + \frac{(1430)^2}{9} + \frac{(1129)^2}{7} + \frac{(891)^2}{5}$$

$$= 113,232.25 + 227,211.11 + 182,091.57 + 158,776.20$$

$$= 681,311.13.$$

*Step* 2:

$$\frac{(4123)^2}{25} = 679,965.16.$$

Difference $\qquad$ 1,345.97.

*Step* 3: $\qquad s_A{}^2 = \dfrac{1,345.97}{3} = 448.6567.$

*To calculate* $s_W{}^2$:

*Step* 1: Calculate the sum of the squares of all 25 original observations, and subtract from it the result of Step 1 in the computation of $s_A{}^2$. (This gives the sum of the squared deviations of all 25 original observations from their respective group means.)

*Step* 2: Subtract the number of groups from the total number of observations, and divide the difference into the result of Step 1. This gives $s_W{}^2$.

For our data:

*Step* 1: $\qquad 691,123 - 681,311.13 = 9,811.87.$

*Step* 2: $\qquad s_W{}^2 = \dfrac{9,811.87}{25 - 4} = 467.2319.$

It turns out, therefore, that the variance among individuals within the same groups, 467, is more than enough to account for the variance implied by the differences among the group means, 449. There is, therefore, no suggestion in these data that mean weight is related to age.

This method of testing the equality of a group of means is an elementary case of what is known as the *analysis of variance*. This elementary version involves an assumption that the variability is the same in all the populations being compared, whether or not the means are the same. Any further development of the method would be out of place in this volume. A new idea, however, has been introduced, namely that of separating the variation within groups from that between groups, deducing from the variation between groups what the magnitude of the variation within groups must be if it is sufficient to explain the variation between groups, then observing the variation within groups directly and determining whether, within an allowance for sampling variability, it is as large as inferred.

In Sec. 19.4.1 we give a simpler method of comparing several means. Should you want to carry through an actual test yourself, we suggest that you use the method of Sec. 19.4.1.

**13.2.3.1** *The F Distribution.* Had $s_A^2$ been larger than $s_W^2$, we would have faced the question whether the excess was more than could be attributed to chance. This question is answered by computing the variance ratio, or $F$ ratio, so named for Sir Ronald A. Fisher (see Sec. 1.4.2), who introduced the method being discussed, and a wide range of related methods, in 1924:

$$F = \frac{s_A^2}{s_W^2}.$$

Statisticians are equipped with rather elaborate tables of the $F$ distribution, in which they select a page corresponding with the significance level, a column corresponding with the number $(k - 1)$ used in the denominator of $s_A^2$, and a row corresponding with the number $(\sum n - k)$ used in the denominator of $s_W^2$. There they find the rejection level of $F$—the value which, if exceeded, leads to rejection of the null hypothesis that the population means are equal.

In this book we do not give tables of the $F$ distribution. We do give two approximate methods of getting the desired probabilities. Like the method for getting $t$ probabilities (referred to in Sec. 13.2.1.1 and given in Sec. 13.4.1, Technical Note 1), both of these methods yield specific probabilities rather than merely ranges, as do the usual tables. One of these methods is given in Sec. 13.4.3, Technical Note 3. The other, a graphical method, is described in Sec. 19.6.3. For the present example, these methods show a probability of 0.43 that $F$ would be as large as, or larger than, its value of 0.96. The variation among the means is, therefore, not significant.

**13.2.3.2** *Selected Comparisons.* There is a strong temptation to select the two sample means that differ most and compare them by the methods of Sec. 13.2.2.1. To apply those methods to samples selected because their means differ would, however, be misleading. The question answered by the methods of Sec. 13.2.2.1 is, If two samples are selected at random from the same population, what is the probability that their means will differ by a given amount or more? The question pertinent here, however, is, If several samples are selected at random from the same population, what is the probability that the greatest difference between any two means will be a given amount or more? Questions of this kind can be answered by techniques developed recently. While they are not complicated, they do require special tables and entail a good deal of explanation and interpretation beyond the scope of this book.[2]

## 13.3
## TESTS OF PROPORTIONS

### 13.3.1 Testing an Assumption about a Population Proportion

This subject was covered rather fully in Chap. 12 to illustrate the principles of statistical decision procedures. Here we will simply give a concise summary; Chap. 12 should be referred to for more detail and for illustrative calculations.

If the proportion in a population having a certain characteristic is $P$, the proportion $p$ having the characteristic in samples of $n$ (if $n$ is not too small and $P$ not too near 0 or 1) will be approximately normally distributed with mean $P$ and standard error $\sqrt{P(1-P)/n}$.

When we use the normal distribution, which is continuous, to approximate this discontinuous distribution, it is advantageous to make a continuity adjustment. For an upper-tail probability, the adjustment involves replacing $p$ by $p - \dfrac{1}{2n}$, and for a lower-tail probability, $p$ is replaced by $p + \dfrac{1}{2n}$.

For an upper-tail probability, then,

$$K = \frac{p - \dfrac{1}{2n} - P}{\sqrt{\dfrac{P(1-P)}{n}}},$$

---

2. For a clear exposition of one such method, and references to other literature on the subject, see Frederick Mosteller and Robert R. Bush, "Selected Quantitative Techniques," Chap. 8 in Gardner Lindzey (editor), *Handbook of Social Psychology* (Cambridge, Mass.: Addison-Wesley Publishing Company, Inc., 1954), especially pp. 304–307.

and for a lower-tail probability

$$K = \frac{p + \dfrac{1}{2n} - P}{\sqrt{\dfrac{P(1 - P)}{n}}}.$$

For a two-tail probability, the smaller of these two one-tail probabilities is doubled. (The upper-tail probability is smaller if $p$ exceeds $P$, the lower-tail probability if $p$ is less than $P$.)

These probabilities are compared with the significance level of the test. If the probability is below the significance level, the null hypothesis is rejected, otherwise it is accepted.

An alternative procedure is to find the rejection level of $p$, by taking the rejection level of $K$ from Table 391 and solving the appropriate equation above for $p$. For example, in testing against a two-sided alternative at the 5 percent significance level, the rejection levels of $K$ are $\pm 1.960$, so values of $p$ such that

$$1.960 = \frac{p - \dfrac{1}{2n} - P}{\sqrt{\dfrac{P(1 - P)}{n}}}$$

and

$$-1.960 = \frac{p + \dfrac{1}{2n} - P}{\sqrt{\dfrac{P(1 - P)}{n}}}$$

will give the rejection levels of $p$. These are

$$p = P + \frac{1}{2n} + 1.960 \sqrt{\frac{P(1 - P)}{n}}$$

and

$$p = P - \frac{1}{2n} - 1.960 \sqrt{\frac{P(1 - P)}{n}}.$$

That is, if $p$ exceeds the first value, or if it falls below the second value, the null hypothesis is rejected.

Given the rejection levels of $p$, the OC curve is found by assuming various values of $P$ and for each calculating the probability that a random sample would produce a value of $p$ leading to acceptance.

A graphical approximation to the test of this section is given in Sec. 19.6.4.2.

## 13.3.2 Comparing Two Sample Proportions

13.3.2.1 *Independent Samples.* Suppose we have two independent samples of sizes $n_1$ and $n_2$, with sample proportions $p_1$ and $p_2$. Let $P_1$ and $P_2$ represent the population proportions, $D$ represent $P_2 - P_1$, the difference between the population proportions, and $d$ represent $p_2 - p_1$, the difference between the sample proportions.

Then the sampling distribution of $d$ is approximately normal, with mean $D$ and standard error

$$\sigma_d = \sqrt{\sigma_{p_1}^2 + \sigma_{p_2}^2} = \sqrt{\frac{P_1(1 - P_1)}{n_1} + \frac{P_2(1 - P_2)}{n_2}}.$$

Again, we do not know $P_1$ and $P_2$. If we are testing the null hypothesis that $P_1 = P_2$, we may make an estimate, $p$, of their common value, $P$, from both samples combined:

$$p = \frac{n_1 p_1 + n_2 p_2}{n_1 + n_2}.$$

Then we will use

$$s_d = \sqrt{\frac{p(1 - p)}{n_1} + \frac{p(1 - p)}{n_2}}.$$

The standardized normal variable requires a continuity adjustment, since the possible values of $d$ are discontinuous. The amount of the adjustment in this case[3] is

$$\frac{n_1 + n_2}{2n_1 n_2},$$

and again the sign is minus for an upper-tail probability and plus for a lower-tail probability. The standardized normal variable is, for an upper-tail probability,

$$K = \frac{p_2 - p_1 - \dfrac{n_1 + n_2}{2n_1 n_2}}{\sqrt{\dfrac{p(1 - p)}{n_1} + \dfrac{p(1 - p)}{n_2}}}$$

---

3. This adjustment, generally known as *Yates' correction*, is based on special considerations besides those of the kind involved in our earlier continuity adjustments, and you cannot expect to verify it for yourself.

and, for a lower-tail probability,

$$K = \frac{p_2 - p_1 + \dfrac{n_1 + n_2}{2n_1n_2}}{\sqrt{\dfrac{p(1-p)}{n_1} + \dfrac{p(1-p)}{n_2}}}.$$

While this presentation brings out the idea of the test procedure, an easy way to make the calculations is to set up the original data in a 2 × 2 table, or double dichotomy, like Table 430. Designate the samples as "first" and "second" in such a way that the required probability is an upper-tail probability. That is, if the alternative hypothesis is one-sided, let $P_2$ be the population proportion that will be larger under the alternative hypothesis, and call the corresponding sample the "second sample." If a two-sided alternative is involved, so that a two-tail probability is required, let $p_2$ be the larger of the two sample proportions; that is, call the sample with the larger sample proportion the second sample.

TABLE 430

A 2 × 2 TABLE, OR DOUBLE DICHOTOMY,
FOR COMPARING TWO PROPORTIONS

| Sample | Occurrences | Nonoccurrences | Total |
|--------|-------------|----------------|-------|
| First  | $a$         | $b$            | $n_1$ |
| Second | $c$         | $d$            | $n_2$ |
| Total  | $a + c$     | $b + d$        | $n_1 + n_2$ |

In Table 430,

$a$ represents the number of occurrences in the first sample,
$b$ represents the number of nonoccurrences in the first sample,
$c$ represents the number of occurrences in the second sample,
$d$ represents the number of nonoccurrences in the second sample.

Then

$$K = \left(bc - ad - \frac{n_1 + n_2}{2}\right)\sqrt{\frac{n_1 + n_2}{n_1 n_2 (a + c)(b + d)}}.$$

To illustrate, suppose that a question pertaining to the restriction of parking in a certain area is asked of a sample of nine car owners and also of a sample of six non-owners. One of the owners and four of the non-owners favor the restriction. We wish to test the null hypothesis that there is no difference between owners and non-owners in the proportion favoring the restriction, against the one-sided alternative that the proportion favoring the restriction is higher among non-owners. Suppose the test is at the five percent level.

The non-owners will be called the second sample, since their population proportion is higher under the null hypothesis. Table 430 then has the specific numbers shown in Table 431, where $a = 1$, $b = 8$, $c = 4$, $d = 2$. From these,

$$ad = 2, \qquad bc = 32, \qquad n_1 + n_2 = 15, \qquad a + c = 5, \qquad b + d = 10,$$

$$K = \left(32 - 2 - \frac{15}{2}\right)\sqrt{\frac{15}{9 \times 6 \times 5 \times 10}} = \frac{22.5}{\sqrt{180}} = 1.677.$$

TABLE 431

NUMBERS FAVORING A CERTAIN RESTRICTION ON PARKING, BY CAR OWNERSHIP

| Sample | In Favor | Opposed | Total |
|---|---|---|---|
| Owners | 1 | 8 | 9 |
| Non-owners | 4 | 2 | 6 |
| Total | 5 | 10 | 15 |

Since the upper-tail five percent rejection level for $K$ is 1.645 (see Table 391), the null hypothesis is rejected and the alternative, that a larger proportion of non-owners favors the restriction, is accepted. The actual one-tail probability is shown by Table 365 as 0.047.[4]

A graphical approximation for the test of this section is given in Sec. 19.6.4.4.

13.3.2.2 *Matched Samples.* A statistics examination included the following statement, to be marked true or false:

A sample that includes 5 percent of its population is more reliable as an indicator of the characteristics of its population than is a sample which includes only 0.1 percent of its population.

The statement is false, of course, for the reliability of a sample depends on the method of sampling, the variability of the population,

---

4. Computation of an OC curve for this test would introduce complications. When $D$ is not 0, so that $P_1$ and $P_2$ differ, $\sigma_d$ depends not on $D$, the difference between $P_1$ and $P_2$, but on the actual values of $P_1$ and $P_2$. A test which would have a good chance of detecting a difference of 5 percentage points ($D = 0.05$) when $P_1$ and $P_2$ are 0.01 and 0.06, might have a poor chance of detecting the difference between 0.47 and 0.52. In fact, $D$ is not a very good descriptive measure of the difference between proportions, for the difference between 0.01 and 0.06 might, depending on the purpose, be a "large" and important difference, but the difference between 0.47 and 0.52 a "small" and unimportant one. We would like to plot the OC curve not for values of $D$ but for values of some quantity which corresponds better with the importance of the difference between two proportions, and whose meaning is the same whatever the two proportions it compares. Such measures exist, but are not included in this book.

and the actual number of observations in the sample, but ordinarily not on the fraction of the population included in the sample (see Sec. 11.4.3). At the time the examination was given, this subject had been covered only by assigned readings, but not by class discussion. Of 34 students, 22 marked the statement true and only 12 marked it correctly as false.

After the subject had been discussed in class, the same question was included in another examination. This time 15 marked the statement true, but 19 marked it correctly as false. Does this represent a statistically significant improvement? In other words, is the result evidence that in the population of which the 34 students are a sample, some improvement in the proportion answering the question correctly is achieved by the processes to which these 34 students were subjected—an examination question, class discussion, and passage of time?

It certainly will not do to analyze these data as if they were two independent samples. Some of the 34 students may have known the answer from other sources; they would have answered correctly both times regardless of what happened. Some may have had the wrong idea indelibly impressed in their minds, and given the wrong answer both times. In short, the variability between these two sets of data if the class discussion had no effect would be much less than that between two independent samples of 34.

The problem cannot be analyzed fully from the data given so far. We need data of the kind shown in Table 432, where for each stu-

TABLE 432

NUMBER OF CORRECT ANSWERS (0 OR 1) GIVEN A TRUE-FALSE
QUESTION ON EACH OF TWO EXAMINATIONS

| Student | Examination | | Gain |
| | First | Second | |
|---|---|---|---|
| 1 | 0 | 0 | 0 |
| 2 | 0 | 0 | 0 |
| 3 | 1 | 1 | 0 |
| 4 | 0 | 0 | 0 |
| 5 | 0 | 1 | +1 |
| 6 | 0 | 0 | 0 |
| 7 | 0 | 0 | 0 |
| 8 | 1 | 0 | −1 |
| . | . | . | . |
| . | . | . | . |
| . | . | . | . |
| 34 | 0 | 0 | 0 |

### 13.3 Tests of Proportions

dent his scores (number of correct answers to the question—that is, either 1 or 0) on both examinations are shown. From these bivariate data, we compute a gain for each student, $-1$, 0, or $+1$. The problem is analogous to the weight-gain problem of Table 421. Our null hypothesis is that the mean gain is zero; here, the alternative is that it is positive.

Whether the mean gain is positive is simply a question of whether the gains of $+1$ significantly outnumber the gains of $-1$. Thus, the data of Table 432 can be summarized in a table like Table 433.

TABLE 433

STUDENTS CLASSIFIED BY RESPONSES TO THE SAME
TRUE-FALSE QUESTION ON TWO EXAMINATIONS

| Second Examination | First Examination | | Total |
|---|---|---|---|
| | Right | Wrong | |
| Right | 9 | 10 | 19 |
| Wrong | 3 | 12 | 15 |
| Total | 12 | 22 | 34 |

We have to test whether the null hypothesis, that changes from right to wrong are as frequent as changes from wrong to right, is consistent with the observations which show that of 13 changes, 10 were from wrong to right. This reduces the problem to that of Sec. 13.3.1, with $P = 0.5$, $p = 0.769$, and $n = 13$. Since we are testing against the one-sided alternative that $P$ exceeds 0.5, an upper-tail probability is required, and the standard normal variable is

$$K = \frac{p - \frac{1}{2n} - \frac{1}{2}}{\frac{1}{2}\sqrt{\frac{1}{n}}} = \frac{0.769 - 0.038 - 0.5}{0.5 \times \sqrt{\frac{1}{13}}} = \frac{0.231}{0.139} = 1.662.$$

At the 5 percent level, this would be just significant, since the 5 percent value of $K$ is 1.645 (Table 391).

A graphical approximation for the test of this section is given in Sec. 19.6.4.2.

### 13.3.3  Comparing Several Population Proportions

The comparison of several population proportions is based on much the same principle as the comparison of several means. The actual standard deviation among the proportions is compared with that to be expected from the relation

$$\sigma_p = \sqrt{\frac{P(1-P)}{n}}.$$

To illustrate the method, five samples of beads have been formed by combining the groups of samples of 20 shown in Table 107A, samples 1–7 there becoming sample I here, etc., as shown in Table 434. (Table 434 also shows the calculations to be discussed below.) The percentages red for the five large samples range from 11 to 17. We could, of course, have used the data of Table 107A directly, as 50 samples of 20—this would, in fact, provide a better test of our sampling process—but this would have made the arithmetic needlessly cumbersome for an illustration.

TABLE 434

NUMBER AND PROPORTION OF RED BEADS IN FIVE SAMPLES

| Sample Number | Samples from Table 107A | Number in Sample $n_i$ | Number Red $X_i$ | Proportion Red $p_i$ | $X_i^2$ | $\dfrac{X_i^2}{n_i}$ |
|---|---|---|---|---|---|---|
| I | 1–7 | 140 | 16 | 0.114 | 256 | 1.8286 |
| II | 8–16 | 180 | 26 | 0.144 | 676 | 3.7556 |
| III | 17–26 | 200 | 34 | 0.170 | 1,156 | 5.7800 |
| IV | 27–37 | 220 | 32 | 0.145 | 1,024 | 4.6545 |
| V | 38–50 | 260 | 44 | 0.169 | 1,936 | 7.4462 |
| Total | | 1,000 | 152 | 0.152 | | 23.4649 |

*Source:* Table 107A.

$$(152)^2 \div 1,000 = 23.1040$$
$$\text{Difference} \quad \overline{0.3609}$$

$$s_A^2 = \frac{0.3609}{4} = 0.0902$$

$$s^2 = 0.152 \times 0.848 = 0.1289$$

$$x^2 = \frac{0.3609}{0.1289} = 2.7998$$

The basic analysis is similar to the comparison of several means (Sec. 13.2.3). An estimate $s_A^2$ of $\sigma^2$ is based essentially on the variance of the $p$'s. A second estimate $s^2$ represents the variance of the individual observations. If the null hypothesis is true, $s_A^2$ and $s^2$ will differ only within a pattern of chance variation which can be calculated. If the null hypothesis is not true, $s_A^2$ will tend to be inflated by the fact that the samples come from different populations.

Easy computations of $s_A^2$ and $s^2$ are as follows:

### 13.3 Tests of Proportions

*To calculate $s_A{}^2$:*

The method shown in Table 434 is used. Let $X_i$ represent the number of occurrences in the $i$th sample.

*Step* 1: For each sample, compute $\dfrac{X_i{}^2}{n_i}$ and add these for all samples. The sum is 23.4649 for our example.

*Step* 2: From the sum, subtract a similar quantity computed from all samples combined. The amount subtracted in our example is 23.1040, leaving 0.3609.

*Step* 3: Divide by one less than the number of groups, thus obtaining $s_A{}^2$. In the example, we divide by 4 and find $s_A{}^2 = 0.0902$.

*To calculate $s^2$:*

The value of $s^2$, the variance of the individual observations, is

$$s^2 = p(1 - p).$$

In this case, $p = 152/1000$, the proportion of red beads in all 1000 observations, so

$$s^2 = 0.1289.$$

Since 0.1289 is more than the 0.0902 required to explain the variation among the sample proportions, we accept the null hypothesis that the population means are all equal.

13.3.3.1 *The Chi-Square $(\chi^2)$ Distribution.* To make an actual test, in a case like this, statisticians compute a quantity designated by the square of the lower-case Greek letter chi (pronounced "ki," as in "kite"):

$$\chi^2 = \frac{(k - 1)s_A{}^2}{s^2}.$$

Here $\chi^2 = \dfrac{(5 - 1) \times 0.0902}{0.1289} = 2.7991$. Special tables are available for the chi-square distribution, which was discovered in 1876 by the German physicist F. R. Helmert (1843–1917) and again in 1900 by the English statistician Karl Pearson (1857–1936). These tables show rejection levels for various levels of significance. They are usually shown on separate lines for each value of a quantity called the "number of degrees of freedom," which in this kind of problem is $k - 1$, one less than the number of samples.

The normal distribution will give reasonably satisfactory approximations, however, if we calculate

$$K = \sqrt{2\chi^2} - \sqrt{2k - 3},$$

which may be regarded as a standard normal variable, the upper-tail probability being the one required.

For the example of Table 434,

$$K = \sqrt{5.5982} - \sqrt{7} = 2.3661 - 2.6458 = -0.2797.$$

From Table 365 we find the upper-tail probability to be 0.610.

Thus, about three times in five, on the average, five random samples from the same population would differ as much as, or more than, the five samples of Table 434. We therefore accept the null hypothesis that the samples come from populations having the same proportion red.

A method of finding chi-square probabilities more accurately is shown in Sec. 13.4.4, Technical Note 4. A method of approximating the probabilities graphically is given in Sec. 19.6.2.

## 13.4
## TECHNICAL NOTES

### 13.4.1 Technical Note 1: Student's t Distribution (Secs. 13.2.1 and 13.2.2.2)

Denote by $t$ the quantity denoted by $K$ in Secs. 13.2.1 and 13.2.2.2; that is,

$$t = \frac{\bar{x} - M}{s_{\bar{x}}} \quad \text{(Sec. 13.2.1)}$$

or

$$t = \frac{\bar{d}}{s_{\bar{d}}} \quad \text{(Sec. 13.2.2.2)}.$$

Having found a value of $t$, compute $K$ from the following formula:

$$K = t\left(1 - \frac{t^2 + 1}{4f}\right),$$

where $f = n - 1$ in the present cases.

In the first example of Sec. 13.2.1,

$$t = \frac{175.91 - 170}{3.662} = 1.614,$$

so $t^2 = 2.605$. Also, $f = 32 - 1 = 31$. Hence

$$\frac{t^2 + 1}{4f} = \frac{3.605}{124} = 0.029$$

and

$$K = 1.614(1 - 0.029) = 1.614 \times 0.971 = 1.567.$$

**13.4 Technical Notes**

Table 365 shows the probability now as 0.059, instead of 0.053—not enough difference to matter for most practical purposes.

For even greater accuracy, especially when $n$ is quite small, the more elaborate formula

$$K = t\left(1 - \frac{t^2 + 1}{4f} + \frac{13t^4 + 8t^2 + 3}{96f^2}\right)$$

may be used. This always gives a value between the two already obtained, so unless it matters where within that range the correct probability lies, it will not be worthwhile to use the more complicated formula. For illustrative purposes, however, we apply it here. Since $t^4$ equals 6.786,

$$\frac{13t^4 + 8t^2 + 3}{96f^2} = \frac{88.218 + 20.840 + 3}{92256} = \frac{112.058}{92256}$$

$$= 0.001.$$

The factor 0.971 found from the simpler formula is therefore increased to 0.972, and now $K = 0.972 \times 1.614 = 1.569$, for which the probability shown in Table 365 is 0.058, instead of 0.059 as from the simpler formula. Had $n$ been smaller, say 5 or 6, there would have been more difference between the two results.

**13.4.2   Technical Note 2: Student's t for Two Independent Samples (Sec. 13.2.2.1)**

Again denote by $t$ what was denoted by $K$ in Sec. 13.2.2.1:

$$t = \frac{d - D}{s_d}.$$

Then let

$$f = \frac{(n_1 - 1)(n_2 - 1)}{(n_2 - 1)c^2 + (n_1 - 1)(1 - c)^2},$$

where

$$c = \frac{s_{\bar{x}_1}^2}{s_{\bar{x}_1}^2 + s_{\bar{x}_2}^2}.$$

These values of $t$ and $f$ can now be used to compute $K$ from either formula in Technical Note 1.

For the example of Sec. 13.2.2.1,

$$n_1 = 32 \qquad\qquad n_2 = 40$$

$$s_{\bar{x}_1}^2 = 13.414 \qquad\qquad s_{\bar{x}_2}^2 = 9.375$$

$$t = -1.678.$$

Hence,

$$c = \frac{13.414}{22.789} = 0.5886,$$

$$f = \frac{31 \times 39}{39 \times 0.3464 + 31 \times 0.1692} = 64.46.$$

*Further Test Procedures*

Using the simpler formula of Technical Note 1,

$$\frac{t^2 + 1}{4f} = \frac{3.8157}{257.84} = 0.015.$$

Then

$$K = -1.678 \times 0.985 = -1.653,$$

for which the two-tail probability is 0.098, instead of 0.093 as shown by the simpler method of Sec. 13.2.2.1. The more elaborate refinement would lower this probability slightly.

### 13.4.3   Technical Note 3: The F Distribution (Sec. 13.2.3)

The normal distribution can be used instead of the $F$ distribution by going through the following arithmetic, which unfortunately involves a cube root. Let

$$G = \sqrt[3]{F}, \qquad a = \frac{2}{9(k - 1)}, \qquad b = \frac{2}{9(\sum n - k)}.$$

Then

$$K = \frac{(1 - b)G + a - 1}{\sqrt{bG^2 + a}}$$

is a standard normal variable, and the upper-tail probability from Table 365 is appropriate.

In the example of Sec. 13.2.3,

$$F = \frac{448.6567}{467.2319} = 0.9602;$$

$$G = \sqrt[3]{0.9602} = 0.9866;$$

$$a = \frac{2}{9(4 - 1)} = 0.0741;$$

$$b = \frac{2}{9(25 - 4)} = 0.0106;$$

$$K = \frac{0.9894 \times 0.9866 + 0.0741 - 1}{\sqrt{0.0106 \times (0.9866)^2 + 0.0741}} = \frac{0.0502}{0.2905} = 0.173.$$

Then the upper-tail probability is 0.431. That is, if individuals varying as much as those of Table 423 are divided at random into four groups, about three times in seven, on the average, the group means will differ as much as, or more than, do the group means of Table 423.

### 13.4.4   Technical Note 4: The $\chi^2$ Distribution (Sec. 13.3.3)

If there are fewer than four degrees of freedom, that is, five groups for problems of the kind discussed in Sec. 13.3.3, it may be worthwhile to make

### 13.5 Conclusion

the following calculation, which gives more precise results, but requires a cube root: Let

$$g = \sqrt[3]{\frac{x^2}{k-1}} \qquad \text{and} \qquad a = \frac{2}{9(k-1)}.$$

Then

$$K = \frac{g + a - 1}{\sqrt{a}}$$

is a standard normal variable, and the upper-tail probability from Table 365 is the required probability.

For the example of Sec. 13.3.3 (Table 434),

$$g = \sqrt[3]{\frac{2.7998}{4}} = \sqrt[3]{0.7000} = 0.8879;$$

$$a = \frac{2}{9 \times 4} = 0.0556;$$

$$\sqrt{a} = 0.2357;$$

$$K = \frac{0.8879 + 0.0556 - 1}{0.2357} = -0.240.$$

The upper-tail probability is shown by Table 365 to be 0.595, instead of the 0.610 obtained in Sec. 13.3.3.1.

## 13.5
## CONCLUSION

The tests presented in this chapter deal with two common statistical measures, means and proportions. With each type of measure it may be desirable:

(1) To test the parameter of a single population.
(2) To compare the parameters of two populations.
(3) To compare the parameters of several populations.

In testing a single population or in comparing two populations, there are really three distinct tests of any null hypothesis, according to whether the alternative hypothesis is that the parameter, or the difference between the two parameters, is above, below, or on either side of the value specified by the null hypothesis.

For tests of several populations, the null hypothesis considered here is that the parameters tested are the same for all the populations. The only alternative considered is that they are not the same. Such alternatives as that they differ in specified systematic ways, or that the sample values are too close together to be independent, have not been considered.

A test is made by computing a statistic from the sample or samples and comparing it with a rejection number (or numbers, if the test is two-sided). The rejection number is determined so that if the null hypothesis is true the probability of a sample for which the statistic passes the rejection number will equal some predetermined level of significance (risk of error of the first kind). Frequently, instead of computing a rejection number, the probability that the statistic would be as far as it actually is in the direction of the alternative hypothesis is computed on the assumption that the null hypothesis is true. If this probability is less than the level of significance, the null hypothesis is rejected. The two methods are equivalent, but computing the probability has the advantage of showing what significance levels would lead to rejection, namely all those higher than the probability computed from the sample. To put it differently, the computed probability shows what risk of error of the first kind would be involved in rejecting the null hypothesis on the basis of the sample at hand.

Several new ideas have been introduced in discussing the tests of this chapter. One of these is that the standard deviation (error) of the result of adding or subtracting several independent statistical variables is the square root of the sum of the squares of their standard deviations (errors).

Another is that when two or more populations are compared it makes a fundamental difference whether the samples are independent or are made up of matched observations, such as observations on the same individual at different times. In the case of matched, or multivariate, observations, the sampling variability may be considerably less than if different individuals (independent samples) were used; so matched samples may give considerable gains in statistical efficiency, in the sense that as good an OC curve from independent samples would require more observations.

Another idea introduced in this chapter is that of allowing for sampling variability in a measure of sampling variability itself, when it is computed from a sample. Ordinarily, however, the magnitude of the correction resulting from this allowance is not great enough to require replacing the normal distribution by Student's $t$ distribution, which makes the allowance.

The analysis of variance principle was also introduced. The principle here is to compute the standard deviation of several sample statistics and compare it with what would be inferred on the basis of the standard deviations within the several samples. The usual test statistic is the $F$ ratio or variance ratio; we have not introduced the special tables needed to use it, but have shown how to use the normal distribution instead.

When several sample proportions are compared, the appropriate sampling distribution is the chi-square distribution. Again, we have avoided the special tables needed for this by showing how to use the normal distribution instead.

In Chap. 19, on shortcut methods, some of these same problems will be discussed again, and somewhat simpler methods—sometimes nearly as efficient—will be shown.

Problems of statistical inference fall into two general categories, test or decision procedures, and methods of estimation. Methods of estimation seek to determine, with reasonable allowances for sampling error, the parameters of a population, rather than to test preconceived notions about them. Estimation is rather closely connected with testing, however, by the principle that any value of the parameter that would not have been rejected in a test is a reasonable one to include in the range within which the parameter is estimated to lie. We turn to estimation in the next chapter.

## DO IT YOURSELF

In the following examples, you may use the approximate procedures given in the main text of Chap. 13, rather than the more accurate ones given in Sec. 13.4.

EXAMPLE 441A

For the Rockwell hardness data of Table 206, find the upper-tail probability for the sample mean if $M$ were really 58. Find the two-tail probability.

EXAMPLE 441B

For Example 441A, compute the OC curve for a test which rejects the null hypothesis if $\bar{x} \geq 62$ and accepts otherwise.

EXAMPLE 441C

For the data of Example 594, compute one- and two-tail probabilities for the observed difference between mean costs for Ford and Chevrolet, under the null hypothesis that the means are equal.

## Further Test Procedures

### EXAMPLE 442A

For the data of Example 596, compute one- and two-tail probabilities for the observed mean differences in loss of tensile strength, under the null hypothesis that the true mean difference is 0.

### EXAMPLE 442B

For the data of Example 615, compute one- and two-tail probabilities for the difference between observed proportions under the null hypothesis that the true proportions are equal.

### EXAMPLE 442C

In a well known text on marketing research, the author assumes that 2,016 questionnaires are shuffled "so that they are in a chance order" and then divided into consecutive groups of 200 each, omitting the last 16 questionnaires. He gives the following frequencies of "yes" answers to a certain question for each successive group of 200:

138,     163,     189,     150,     165,     149,     158,     185,     141,     159.

Do you believe that random shuffling was really achieved? Explain.

# Chapter 14

# *Estimation*

## 14.1
### INTRODUCTION

Now we come to a second main area of statistical inference, the estimation of population parameters on the basis of the information in a sample. To illustrate this application, suppose the problem is to estimate mean annual income of families in a certain city during a certain year. On the basis of sample information a statistician might estimate this mean income to be $4,500, and add that he is "confident" that the true mean, or parameter, is between $4,100 and $4,900. The first statement is an example of a *point estimate*, the second, of a *confidence interval estimate*.

We shall see how such estimates are made and how they are interpreted. As in the preceding chapter, we shall consider only means and proportions. Also, we shall consider only one method of sampling, simple random sampling. These cases will serve to elucidate the principles, and they are among the most important applications. Other statistics, or more elaborate probability sampling methods, introduce no new principles, but are sometimes more complicated in detail.

At the outset, we must re-emphasize an idea that pervades statistical inference. This is the idea of evaluating sampling *processes*, rather than specific samples. Almost always in practice an investigator must be content with a *single* sample, which may be "good" or "bad," but he can draw inferences from this single sample only through knowledge of the process by which the single sample was produced. This is reflected in the sampling distribution of the statistic being used to estimate a population parameter.

## 14.2
## PRINCIPLES OF ESTIMATION

### 14.2.1 Point Estimation

14.2.1.1 *The Problem.* Even though any single number based on a sample almost surely will not coincide with the parameter, there are many purposes for which a single number is useful. But what single number should be used as a point estimate? This question probably has not occurred to you, for the answer is so obvious in most cases that the question is not even noticed. If in a sample of 20 beads, for example, 13 red beads are observed, it seems "just common sense" to take 13/20 as the estimate of the proportion of red beads in the population. The sample proportion is indeed the best estimate of the population proportion from a simple random sample. Consider, however, the following example:

EXAMPLE 444A  ESTIMATING FROM SERIAL NUMBERS

A certain kind of equipment has been numbered serially by the manufacturer (compare Example 20B). In a random sample of ten, the highest serial number is 929,261. If the numbering system begins with 1 and is really consecutive, clearly the maximum number in the population is almost certainly greater than this. But how much greater should the point estimate be? We have frequently posed this question in discussion, and have received a wide variety of ingenious suggestions. Before reading on, you might see what you can think of yourself; and, more important, try to decide what criteria you would use to choose among alternative suggestions.

Actually, a good estimator in cases like this is not $g$, the greatest observation in the sample of $n$, but $\left(\dfrac{n+1}{n}g\right) - 1$. For our example, $g = 929{,}261$, so the point estimate is 1,022,186. Since the figure 929,261 was in reality obtained by taking the largest number in a random sample of 10 from a population consisting of the integers from 1 to 1,000,000, this estimate is 22,186 (2.2 percent) too large; the maximum observation, in contrast, is 70,739 (7.1 percent) too small.

EXAMPLE 444B  PREDETERMINED NUMBER OF OCCURRENCES

As another example, suppose the digits in a table of random digits are counted until ten odd ones have been obtained. Let $n$ be the number of digits counted to reach ten odd digits. Then a little reflection will show that the obvious estimator, $10/n$, the proportion of odd digits in the sample, will, on the average, tend to give too high an estimate of the proportion of odd digits in the population. We drew ten such samples from the Rand table of random digits. The values of $n$ were 18, 18, 23, 17, 14, 20, 15, 23, 22, and 17, giving a mean of the ten sample proportions of 0.549.

The point is that in each sample every digit is equally likely to be odd or even *except* the last digit; the last digit is bound to be odd, and this raises the proportion odd for all samples above 0.5. An estimator free of this bias is $(c - 1)/(n - 1)$, where $c$ is the predetermined number of occurrences. The mean of such unbiased estimates for our ten samples is 0.524. (Ten samples, of course, do not prove the point; they do happen to illustrate a point that can be proved mathematically.)

EXAMPLE 445   ESTIMATING QUANTILES OF A NORMAL DISTRIBUTION

To estimate the ninth decile of a normal population, the ninth decile of the sample could be computed (Sec. 8.4). The average of the ninth deciles of a large number of samples will tend to equal the ninth decile of the population. But the dispersion of the sampling distribution, as measured by the standard error of the ninth decile, will be considerably larger than if $\bar{x}$ and $s$ had been computed from the sample and $\bar{x} + 1.28s$ taken as the estimate of the ninth decile. (The value 1.28 is from Table 391.)

It is not our purpose here to explain why the point estimators given for these three examples are good ones, though this does follow from general ideas to be developed in this chapter. Rather, our purpose here is to show that the choice of point estimates is not always obvious.

The term *estimator* is used to refer to a formula into which the observations of a sample are to be substituted to compute an *estimate* of a population parameter. An estimator is, then, a general formula, and an estimate is a number obtained by applying the formula to a particular sample.

14.2.1.2   *Maximum Likelihood.*   The principle by which point estimators are generally chosen is called the principle of *maximum likelihood*. The underlying idea, which was introduced in 1921 by Sir Ronald Fisher (see Sec. 1.4.2), is to consider every possible value that the parameter might have, and for each value compute the probability that the particular sample at hand would have occurred if that were the true value of the parameter. Of all possible values of the parameter, the one to be chosen as the estimate is the one for which the probability of the actual observations is greatest. Formulas that will give such estimates are called maximum likelihood estimators.

*Estimation*

To put the principle of maximum likelihood in concrete terms, consider again the sample of 20 beads containing 13 red ones, first discussed in Sec. 4.3.2. Letting $P$ denote the probability that a bead will be red, that is, the unknown parameter we seek to estimate, the probability of any specific sequence of 20 beads among which 13 are red and 7 not red is

$$P^{13}(1 - P)^7,$$

by the multiplication rule (Sec. 10.6.2). It can be computed that there are 77,520 different sequences of 13 red and 7 green beads, so

TABLE 446

PROBABILITY OF 13 RED BEADS IN A SAMPLE OF 20,
UNDER VARIOUS ASSUMPTIONS ABOUT THE POPULATION
PROPORTION RED

| Population Proportion Red | Probability of 13 Red in Sample of 20 |
|---|---|
| 0.0 | 0.0 |
| 0.1 | 0.000 000 004 |
| 0.2 | 0.000 013 |
| 0.3 | 0.001 018 |
| 0.4 | 0.014 563 |
| 0.5 | 0.073 929 |
| 0.6 | 0.165 882 |
| *0.65* | *0.184 401* |
| 0.7 | 0.164 262 |
| 0.8 | 0.054 550 |
| 0.9 | 0.001 970 |
| 1.0 | 0.0 |

*Source:* **National Bureau of Standards,** *Tables of the Binomial Probability Distribution* (Washington: Government Printing Office, 1949), pp. 34–35; except for $P = 0.1$, which was computed by us.

the probability of a sample with 13 red beads is $77,520\, P^{13}(1 - P)^7$, by the addition rule (Sec. 10.6.1).

Values of the probability of the sample under various assumptions about $P$ are shown in Table 446 and Fig. 447. The greatest probability of the actual sample is obtained by assuming that $P = 0.65$, the sample proportion. It can be shown mathematically that the value of $P$ that maximizes the likelihood of any sample is $p$, the sample proportion; Fig. 353A illustrates this. Thus, $p$ is the maximum likelihood estimator of $P$.

One warning: The reason for estimating $P$ to be 0.65 is *not* the fact that if $P$ were 0.65 the most probable sample result would be

### 14.2 Principles of Estimation

$p = 0.65$; as a matter of fact, $p = 0.65$ is the most probable result in a sample of 20 for any $P$ from 0.619 to 0.667. The reason for estimating $P$ to be 0.65 is that a sample with $p = 0.65$ is more probable if $P = 0.65$ than if $P$ has any other value.

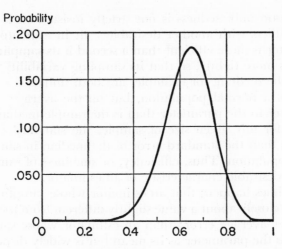

FIG. 447. Probability of 13 red beads in a sample of 20, under various assumptions about the population proportion red.

Source: Table 446.

**14.2.1.3** *Unbiasedness and Efficiency.* Maximum likelihood estimators are not necessarily *unbiased*. That is, the average of the estimates for a large number of samples may tend to be some other value than the population parameter. When this is the case, and the amount of the bias does not depend on the unknown parameter, an adjustment is often made in the estimator to eliminate the bias. This is the case, for example, when $n - 1$ is used in the denominator of the estimator $s^2$ for the variance $\sigma^2$.

The fact that an estimator is unbiased does not mean that the estimate for any particular sample is equal to the population parameter. The correct interpretation is that the mean of the sampling distribution of the estimate is equal to the parameter being estimated. Even though this does not tell us whether the *particular* sample estimate is close or far from the population parameter, there is comfort in the knowledge that there is no tendency to err systematically.

It is not always essential or even important, however, that estimators be strictly unbiased. A bias known to be slight is often unimportant, and frequently in more advanced applications of statistics, slightly biased estimators may be used by preference. What *is* objectionable is a sampling and estimation procedure for which the magnitude of the bias is unknown, as it is for non-probability sampling methods.

One reason unbiasedness is not strictly insisted on is that sometimes it conflicts with another desirable trait in estimators, *efficiency*. One estimator is more efficient than a second if its sampling distribution clusters more tightly, so that its sampling variability is less. The mean and the median, for example, are both unbiased estimators of the mean of a normal population, but on the average, the sample mean is closer to the parameter than is the sample median. The reason is that for any given size of sample, the standard error of the mean is less than the standard error of the median in sampling from a normal population. Thus, efficiency, or smallness of sampling variability, is a desirable characteristic of an estimator.

It sometimes happens that an estimator whose sampling distribution clusters closely about a value slightly different from the parameter has a smaller average error[1] than an estimator whose sampling distribution has the parameter as its mean but is widely dispersed. Such situations arise especially when complicated sampling plans are used, as in some surveys of human populations. On the whole it is best to have as small a mean square error as possible, even if this entails using a biased estimator—it is better, that is, to accept bias if by so doing the mean square error can be reduced. In the simple situations discussed in this book, however, no conflict arises; estimators can be, and most of those we discuss are, both efficient and unbiased at the same time.

14.2.1.4 *Minimax Estimation.* Other criteria than unbiasedness and efficiency, and other principles than maximum likelihood, are used in selecting point estimates, but they are beyond the scope of this book. One recent development that should be mentioned, however, is the *minimax principle*, the purpose of which is to minimize the maximum possible average cost of errors in estimates. Minimax estimation was introduced in 1939 by the American statistician Abraham Wald (1902–1950).

The sampling distribution of a proposed estimator makes it possible to calculate the average cost of errors for the estimator, assuming the cost of a given discrepancy between an estimate and the parameter

---

1. Such average errors are measured by the *mean square error*, that is, the square root of the arithmetic mean of the squared deviations between the estimates for different samples and the population parameter being estimated (see Sec. 7.4.4.1).

is known. These average costs depend on what assumption is made about the population parameter, since sampling distributions are always deduced on the basis of assumptions about the population. One possible estimator may have a low average cost of errors if the population parameter is assumed to be 10, but a high average cost if the parameter is assumed to be 20; while another estimator may have a high average cost if the parameter is 10, but a low average cost for 20. For each possible estimator the maximum average cost is found—that is, the average cost for that value of the parameter at which the particular estimator has the poorest average. Then, of all the estimators, that one is selected whose maximum average cost is least—whose worst performance is least bad, in other words. This estimator minimizes the maximum average cost of errors that could occur for any value of the parameter, and is called a minimax estimator.

If the minimax principle is applied to the sample of 20 beads of which 13 are red, the resulting estimate of the population proportion red is not 0.650 but 0.623. The formula for the minimax estimator, $p^*$, is

$$p^* = \frac{\sqrt{n}\, p + \frac{1}{2}}{\sqrt{n} + 1}.$$

This formula is based on the assumption that the cost of an error is proportional to the square of the difference between the estimate and the parameter.

Although minimax estimation procedures are the subject of much current research in theoretical statistics, they cannot yet be recommended for practical applications.

## 14.2.2 Interval Estimates: Confidence Intervals

Since we know that a point estimate is unlikely to be precisely equal to the parameter, we need to indicate the margin of error to which it is subject. A way to do this is to specify an interval within which we may be confident that the parameter does lie. Such an interval is called a *confidence interval*. The confidence attached to it is measured by a *confidence coefficient*, which is an objective probability, not a subjective evaluation of degree of belief. A 95 percent confidence interval is an interval which, in a sense that we shall explain shortly, has a probability of 0.95 of being "correct." Confidence intervals were introduced in 1937 by the Polish-English-American statistician Jerzy Neyman (born 1894).

The two *confidence limits* that bound a confidence interval are statistics: they are numbers computed from samples. They are therefore subject to sampling fluctuations. If repeated samples are drawn from the same population, the confidence intervals will vary from sample to sample. The parameter to be estimated by them, however, remains fixed. Some of the confidence intervals are "correct," in the sense that the limits include the parameter; others are incorrect.

Since we do not know the parameter, we cannot, of course, identify the correct and incorrect confidence intervals. But the confidence coefficient tells us an important fact about the sampling distribution of confidence intervals: it tells us what proportion of samples will lead to confidence intervals that are correct. The confidence coefficient, therefore, describes a property of the *process* by which the confidence interval was obtained; it does not describe the individual confidence interval. To repeat an analogy that we used in another connection (Sec. 4.6.1), the concept of confidence is like the concept of a fair deal in cards, which applies to the process of dealing, not to any specific hand that has been dealt.

The point has been expressed nicely by a former student of one of the authors:

> ... the meaning of confidence is a subtle point. A very homely example may help. The game of horseshoe consists of throwing a horseshoe at a peg. A ringer occurs when the shoe rings or encloses the peg. What is the interpretation of the statement that the probability is .95 that a ringer will be thrown? Does it mean that 95 percent of the shoes thrown will ring the peg? Or does it mean that 95 percent of the time the peg will come up through the horseshoe? The correct statement is fairly obvious. It is a particularly good example, because the wider the spread of the shoe, the more chance there is of making a ringer, other things being equal.[2]

When a confidence interval and its confidence coefficient are stated, they indicate that the interval results from a process—and by process we mean both the method of drawing the sample and the method of computing the confidence limits from it—which, in indefinitely many repetitions, will give correct intervals for the proportion of samples specified by the confidence coefficient.

The problem, then, is to find formulas for computing the confidence limits so that their sampling distribution will have the property specified by the confidence coefficient. The basic principle for doing this is quite simple—once you see it.

As usual in statistical inference, we start by making assumptions about the population and from these assumptions and the laws of

---

2. R. Clay Sprowls, *Elementary Statistics for Students of Social Science and Business* (New York: McGraw-Hill Book Company, Inc., 1955), p. 104.

probability we deduce the behavior of samples. Suppose the confidence coefficient desired is 95 percent. We consider in turn each possible value that the parameter might have. For each possible value, we compute the sampling distribution of our statistic, and determine a limit above which the statistic will go only 2.5 percent of the time if that value of the parameter is correct, and a limit below which the statistic will go only 2.5 percent of the time. Then when a sample is observed, we estimate that the parameter is one of those for which the sampling distribution includes within its central 95 percent zone the observed value of the statistic.[3]

To illustrate, suppose that the standard deviation of weights is 20.718 pounds in the population from which the 32 observations of Table 173 are drawn.[4] The standard error of the mean for a sample of 32 from this population is then

$$\sigma_{\bar{x}} = \sigma/\sqrt{n} = 20.718/\sqrt{32} = 3.662.$$

If the population mean $M$ is 160 pounds, the probability is 0.95 that the mean of a sample will be within the range

$$M \pm 1.960\sigma_{\bar{x}} = 160 \pm 7.18,$$

or from 152.8 to 167.2. (The value 1.960 is taken from Table 391.) Similarly, if $M$ is 180, the probability is 0.95 that a sample mean will be between 172.8 and 187.2. Computations of this kind could be made, of course, for any value of $M$. The results are summarized in Fig. 452A, where the horizontal axis represents possible values of $M$ and the vertical axis, possible values of $\bar{x}$. The vertical lines above $M = 160$ and $M = 180$ represent the intervals just computed. The two sloping lines bound the tops and bottoms of all such intervals.[5]

Now turn to Fig. 452B, in which the band has been reproduced without showing the vertical lines which are its underlying fabric. Our particular sample of 32, as we saw in Sec. 7.4.1, has $\bar{x} = 175.91$. The confidence interval is constructed as shown in Fig. 452B, by drawing a horizontal line between the boundaries at the height

3. The 95 percent zone is "central" on a probability scale, but not necessarily on a scale of the statistic. For example, if the statistic can range from 0 to 1, as in the case of a sample proportion, the central zone probability-wise may be at an extreme proportion-wise; it might run from 0.060 to 0.153, as is actually the case when $n = 150$ and $P = 0.1$.

4. It is merely for simplicity of exposition that we assume the population standard deviation to be known; this assumption will be removed in Sec. 14.3.3. The value 20.718 is chosen so that the numerical details of this illustration will articulate with Sec. 14.3.3.

5. For this particular statistic, the two boundaries are straight lines, so that only two intervals need be computed explicitly. For other statistics, however, the boundaries may curve.

175.91, and noting that the horizontal range covered by this line extends from 168.7 to 183.1. This range is a 95 percent confidence interval for $M$.

To see why this method works, let us suppose for the moment that, contrary to fact, $M$ is actually known and is 180. Then, as we have seen, the probability is 0.95 that a sample mean will lie between

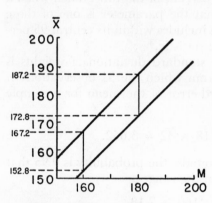
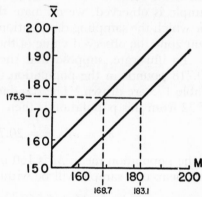

FIG. 452A. Method of determining confidence interval estimators.

FIG. 452B. Method of making confidence interval estimates.

172.8 and 187.2 (see Fig. 452A). For any such sample mean, the method of making a confidence interval estimate shown in Fig. 452B leads to a horizontal line which intersects the vertical line above $M = 180$. If this happens, $M = 180$ is included—correctly—in the confidence interval.

For $M = 180$, there is also a probability of 0.05 that a sample mean will lie above 187.2 or below 172.8. If this happens, the horizontal line will be either entirely to the right or entirely to the left of the vertical line above $M = 180$, and $M = 180$ will be *outside* the confidence interval.

We have generated artificially ten random samples of 32 from a normal population with $M = 180$, and $\sigma_{\bar{x}} = 3.662$. The means of these samples were:

| | | | | |
|---|---|---|---|---|
| 179.82, | 183.73, | 189.66, | 179.02, | 185.80, |
| 178.53, | 182.20, | 182.26, | 177.45, | 178.16. |

The confidence intervals corresponding with these means are shown graphically in Fig. 453. The horizontals for nine of the samples intersect the vertical at $M = 180$, and thus include $M = 180$ as a possible value for $M$. The horizontal for $\bar{x} = 189.66$, the third sample, does not include 180; its interval, as shown, is from 182.5 to 196.9. Thus, nine of ten intervals are correct, the other is incorrect, which (in a sample of ten) is not inconsistent with a population proportion of 0.95 correct.

Everything we have said about $M = 180$ applies to *any* value of $M$. We see, therefore, that the probability that a confidence interval

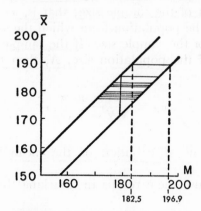

FIG. 453. Confidence intervals from ten samples from a normal population with $M = 180$, $\sigma_{\bar{x}} = 3.662$.

will be correct—0.95, in this example—is the same whatever may be the value of the unknown parameter.

To calculate 95 percent confidence intervals, therefore, one way is to take as the lower confidence limit the value of the parameter for which the observed statistic lies at the 97.5th centile, often called the upper $2\frac{1}{2}$ percent point, of its sampling distribution; and take as the upper confidence limit the value of the parameter for which the statistic lies at the 2.5th centile, or lower $2\frac{1}{2}$ percent point, of its distribution. Put in terms of significance tests, we include in the 95 percent confidence interval each value of the parameter which would not be rejected if it were the null hypothesis in a two-sided significance test at the 5 percent level.

## 14.3
## ESTIMATION OF MEANS

### 14.3.1 Review of Sampling Distribution of Means

In Sec. 11.4 we studied the sampling distribution of the sample mean, $\bar{x}$. There were three main conclusions:

(1) The mean of the sampling distribution of means is equal to the mean of the population of individual measurements; that is, $M_{\bar{x}} = M_x$. Another way of saying this is that $\bar{x}$ is an unbiased estimator of $M$.

(2) The standard deviation of the sampling distribution of the mean is equal to the standard deviation of the population divided by the square root of the sample size; that is, $\sigma_{\bar{x}} = \sigma/\sqrt{n}$. This is true regardless of the population from which the samples come and almost regardless of the sample size. If the sample size, $n$, is a substantial fraction of the population size, $N$, then the slightly more complicated formula,

$$\sigma_{\bar{x}} = \sqrt{\frac{N - n}{N - 1}} \frac{\sigma}{\sqrt{n}},$$

is preferable.

(3) The sampling distribution of the mean is approximately normal.

We shall now use these results in attacking the problem of estimation.

## 14.3.2   Point Estimates of Means

Suppose we must use some single number as our estimate of the unknown mean of a population from which we have drawn a sample. What number should we use?

From what we now know, it seems plausible that we should use the common sense estimator, $\bar{x}$. If we do so, we will have the assurance of an unbiased estimator of the mean of the population. It turns out, in fact, that $\bar{x}$ is the maximum likelihood estimator of the mean of a normal population; that is, the probability of any particular sample is larger if we assume $M = \bar{x}$ than if we assume any other value for $M$. Furthermore, it can be proved mathematically that for a normal population, $\bar{x}$ has the maximum possible efficiency, that is, the smallest standard error of any possible estimator of $M$.

In short, common sense is confirmed in this case: $\bar{x}$ is the best estimator of a single value for the mean of a normal population based on a simple random sample.

## 14.3.3   Confidence Intervals for the Mean

We saw in Sec. 14.2.2 that the lower limit of a 95 percent confidence interval for the mean is given by that value of $M$ for which

*14.3 Estimation of Means*

$\bar{x}$ lies at the 97.5th centile of the sampling distribution. Similarly, the upper limit of the confidence interval is given by the value of $M$ for which $\bar{x}$ is at the 2.5th centile.

Let us find numerically a 95 percent confidence interval for the weight data of Table 173, a problem for which a graphical solution has already been provided in Sec. 14.2.2. The mean of these 32 observations has already been found to be 175.91, the standard deviation 20.718, and the standard error 3.662 (Sec. 13.2.1). We want to find the value of $M$ for which 175.91 lies at the 97.5 centile, or upper 2.5 percent point, of the sampling distribution of $\bar{x}$. The standard normal variable is

$$K = \frac{175.91 - M}{3.662}$$

and the upper 2.5 percent point of a standard normal variable is 1.960 (Table 391). Therefore, we want to find $M$ so that

$$\frac{175.91 - M}{3.662} = 1.960.$$

We denote this value of $M$ by $m_L$:

$$m_L = 175.91 - (1.960 \times 3.662) = 175.91 - 7.18 = 168.73.$$

The upper confidence limit, which we shall denote by $m_U$, is the value of $M$ for which

$$\frac{175.91 - M}{3.662} = -1.960,$$

since $-1.960$ is the lower 2.5 percent point for a standard normal variable. So

$$m_U = 175.91 + (1.960 \times 3.662) = 175.91 + 7.18 = 183.09.$$

Thus, we are 95 percent confident that the population mean lies between 168.7 and 183.1 lbs.

In general, we may write

$$m_L = \bar{x} - Ks_{\bar{x}}$$

and

$$m_U = \bar{x} + Ks_{\bar{x}},$$

where $K$ is to be determined from the confidence coefficient, $C$, as follows:

(1) For $m_L$, $K$ is the upper $\frac{1}{2}(1 - C)$ point of the standard normal distribution shown in Table 391.

(2) For $m_U$, $K$ is the lower $\frac{1}{2}(1 - C)$ point of the standard normal distribution shown in Table 391.

In explaining confidence intervals in Sec. 14.2.2 we took $\sigma_{\bar{x}}$ as known, while our actual computation, as shown above, is based on $s_{\bar{x}}$. The justification for using $s_{\bar{x}}$ instead of $\sigma_{\bar{x}}$ is the same as the justification for using $s_{\bar{x}}$ instead of $\sigma_{\bar{x}}$ in tests of significance (see Sec. 13.2.1). For small samples, less than 10, say, or perhaps 20, special adjustments may be in order to allow for the fact that $s$ as well as $\bar{x}$ is subject to sampling errors. Such refinements, based on Student's $t$, are given in Sec. 14.7.1.

The fact that we use $s_{\bar{x}}$ instead of $\sigma_{\bar{x}}$ means that in repeated samples, confidence intervals would differ from one sample to another not only in location, as in Fig. 453, but in width as well. It is still true, however, that for any confidence coefficient—say, 95 percent—the probability is 0.95 that a sample will produce a confidence interval that will include the parameter $M$.

As a second example of confidence limits for a mean, consider the data of Sec. 13.2.2.1, where two samples of weights were obtained. The difference between the means, $\bar{x}_2 - \bar{x}_1$, was $-8.01$ and the standard error of the difference was 4.774. Then 95 percent confidence limits for the difference $M_2 - M_1$ in the population means are

$$-8.01 \pm (1.960 \times 4.774) = -8.01 \pm 9.36 \quad \text{or} \quad -17.4 \text{ to } +1.4.$$

The fact that the 95 percent confidence interval includes both positive and negative values corresponds with the fact that a two-sided significance test at the five percent level leads to accepting the null hypothesis that the population means do not differ. Sec. 14.7.1 shows how sampling error in $s_1$ and $s_2$ could have been allowed for in this problem.

Finally, we shall compute a confidence interval for the mean weight gain of the data of Table 421. For variety, let us use a confidence coefficient of 0.99, for which $K = 2.576$ (Table 391). Since the mean gain was 2.28 lbs. and the standard error 0.639 lbs., the 99 percent confidence interval is

$$2.28 \pm (2.576 \times 0.639) = 2.28 \pm 1.65,$$

that is, from 0.63 to 3.93 lbs. Since this 99 percent confidence interval lies entirely above zero, the null hypothesis would be rejected by a two-tail test at the one percent level, or by an upper-tail test at the 0.5 percent level.

Sometimes point and interval estimates are combined by writing a result as 2.28 ± 1.65, as above. In such cases it is necessary to indicate explicitly what is meant by the " ±1.65," for such notation is used with different meanings. Sometimes " ±1.65" means that the standard error is 1.65, sometimes that the standard deviation is 1.65, and sometimes—when (as is common in the physical sciences) the writer is following statistical practice of the pre-World War I era—it means that 1.65 is 0.674 times the standard error or 0.674 times the standard deviation (called "probable error," because a standard normal variable is equally likely to be inside or outside to range from −0.674 to +0.674). Even when it is made clear that " ±1.65" gives a confidence interval, it is necessary to indicate the confidence coefficient. One difficulty with this method of combining point and interval estimates is that the point estimate may not be at the center of the confidence interval; it lies at the center for means, but ordinarily does not (if computed exactly) for proportions or standard deviations.

## 14.4
## ESTIMATION OF PROPORTIONS

### 14.4.1  Review of Sampling Distribution of Proportions

In Sec. 11.5 we saw that:

(1) The mean of the sampling distribution of proportions is equal to the population proportion, $P$. The interpretation of this is subject to the same warnings as given in Sec. 14.3.1 for the corresponding fact about means.

(2) The standard error of a proportion is equal to $\sqrt{P(1 - P)/n}$. If the sample size, $n$, is a substantial fraction of the population size, $N$, the relation

$$\sigma_p = \sqrt{\frac{N - n}{N - 1} \times \frac{P(1 - P)}{n}}$$

is preferable.

(3) The sampling distribution of $p$ is approximately normal, provided that $n$, $P$, and $1 - P$ are not too small. For confidence interval problems, this means that the normal distribution is satisfactory only if the resulting confidence limits, $p_L$ and $p_U$, are within the range of $P$ for which the normal approximation is satisfactory.

## 14.4.2 Point Estimates of Proportions

Common sense is right in suggesting that the proportion $p$ observed in a sample is the best single number to take as an estimate of the population proportion $P$. This estimator has all the virtues that $\bar{x}$ has for estimating $M$ for the normal distribution. In particular it is unbiased and has maximum efficiency.

Thus, from our sample of Sec. 4.3.2, in which 13 red beads appeared in a sample, the point estimate of $P$ is 0.65. This estimate was discussed at some length in Sec. 14.2.1.2, especially Table 446 and Fig. 447, which should be referred to again now.

From the total of 1,000 beads from Population I of Sec. 4.3.3, $p = 0.548$, and from the 1,000 from Population II of Sec. 4.3.4, $p = 0.152$, so these are our point estimates of $P_1$ and $P_2$.

## 14.4.3 Confidence Intervals for Proportions

Consider the 1,000 beads from Population I (Table 105). Here $n = 1,000$ and $p = 0.548$. The lower limit of a 90 percent confidence interval, $p_L$, is that value of $P$ for which $p = 0.548$ lies at the upper 5 percent point of the sampling distribution. This is given by finding $P$ such that

$$\frac{0.548 - 0.0005 - P}{\sqrt{\dfrac{P(1 - P)}{1,000}}} = 1.645,$$

where the 1.645 is from Table 391. Perhaps you once took an algebra course and studied quadratic equations. This is, we dare say, the first time you have confronted one since. If so, the chances are poor of your being able to solve it. Fortunately, the solutions needed here can usually be approximated satisfactorily and easily.

When the sample is large or $p$ is in the range from 0.3 to 0.7, there is little inaccuracy in replacing $P(1 - P)$ in the denominator by $p(1 - p)$. Hence we have, in general,

$$\frac{p - \dfrac{1}{2n} - P}{\sqrt{\dfrac{p(1 - p)}{n}}} = K,$$

or in this case

$$\frac{0.548 - 0.0005 - P}{\sqrt{\dfrac{0.548 \times 0.452}{1,000}}} = 1.645.$$

Then, in general,

$$p_L = p - \frac{1}{2n} - K \sqrt{\frac{p(1-p)}{n}},$$

or for our data,

$$p_L = 0.548 - 0.0005 - (1.645 \times 0.0157) = 0.522.$$

For the upper limit of the confidence interval, $p_U$, the general formula is

$$p_U = p + \frac{1}{2n} + K \sqrt{\frac{p(1-p)}{n}}.$$

In this case,

$$p_U = 0.548 + 0.0005 + (1.645 \times 0.0157)$$

$$= 0.574.$$

Thus the 90 percent confidence interval for $P_1$ (using the subscript 1 to denote the first population) is 0.522 to 0.574.

As a second example, consider the 1,000 beads from Population II, for which $p = 0.152$ (Table 107A). Then, for a 90 percent confidence interval,

$$p_L = 0.152 - 0.0005 - 1.645 \sqrt{\frac{0.152 \times 0.848}{1,000}}$$

$$= 0.152 - 0.0005 - (1.645 \times 0.0114)$$

$$= 0.133.$$

For the upper limit,

$$p_U = 0.152 + 0.0005 + (1.645 \times 0.0114)$$

$$= 0.171.$$

Hence, the 90 percent confidence interval for $P_2$ is 0.133 to 0.171.

This interval is shorter than the one for $P_1$—its length is 0.038 instead of 0.052. For a given sample size and confidence coefficient, confidence intervals are narrower the farther $p$ is from 0.5. For a given value of $p$, the intervals are narrower the larger the sample size, and the smaller the confidence coefficient.

As a final example, we will determine a 90 percent confidence interval for the difference, $D$, between the population proportions,

**Estimation**

$P_1$ and $P_2$, of the sampling demonstrations. The lower limit of the interval is obtained by finding the value of $D$ for which

$$\frac{d - \dfrac{n_1 + n_2}{2n_1n_2} - D}{\sqrt{\dfrac{p_1(1 - p_1)}{n_1} + \dfrac{p_2(1 - p_2)}{n_2}}} = K,$$

where $d = p_2 - p_1$, the difference in the sample proportions, $K$ is the upper $\frac{1}{2}(1 - C)$ probability point of the standard normal distribution (shown in Table 391), and $C$ is the confidence coefficient. Letting $d_L$ represent the lower limit of the interval,

$$d_L = d - \frac{n_1 + n_2}{2n_1n_2} - K\sqrt{\frac{p_1(1 - p_1)}{n_1} + \frac{p_2(1 - p_2)}{n_2}}.$$

Similarly,

$$d_U = d + \frac{n_1 + n_2}{2n_1n_2} + K\sqrt{\frac{p_1(1 - p_1)}{n_1} + \frac{p_2(1 - p_2)}{n_2}}.$$

For our data, $d = -0.396$, $n_1 = 1,000$, $n_2 = 1,000$, $p_1 = 0.548$, $p_2 = 0.152$, and $K = 1.645$; so

$$\frac{n_1 + n_2}{2n_1n_2} = \frac{1,000 + 1,000}{2 \times 1,000 \times 1,000} = 0.001,$$

$$\frac{p_1(1 - p_1)}{n_1} = \frac{0.548 \times 0.452}{1,000} = 0.000248,$$

$$\frac{p_2(1 - p_2)}{n_2} = \frac{0.152 \times 0.848}{1,000} = 0.000129.$$

Then

$$d_L = -0.396 - 0.001 - 1.645\sqrt{.000377}$$

$$= -0.429,$$

and

$$d_U = -0.396 + 0.001 + 0.032$$

$$= -0.363.$$

The 90 percent confidence interval for $P_2 - P_1$ is, therefore, $-0.429$ to $-0.363$.

A confidence coefficient for a proportion, or for a difference between two proportions, is to be interpreted as the minimum probability, rather than the exact probability, that random sampling will lead to an interval that includes the parameter. The discontinuities of the binomial distribution make it impossible to draw vertical bars like those in Fig. 452A so that they cover exactly 95 percent (or some other fixed percentage) of the sampling distributions. Each is therefore drawn so as to include at least 95 percent. For practically all values of $P$ the probability of a sample that leads to a correct interval is greater than the confidence coefficient.

A more accurate method of computing confidence intervals for proportions, based on solving the quadratic equation at the beginning of this section, is explained in Sec. 14.7.2. A graphical shortcut method is explained in Sec. 19.6.4.3.

## 14.5
## CONFIDENCE INTERVALS AND DECISION PROCEDURES

### 14.5.1 Relation of Estimates to Tests

Choosing a confidence coefficient raises much the same questions as choosing a significance level for a test. Confidence intervals, too, involve two kinds of risk:

(1) The interval may fail to include the parameter. This is analogous to an error of the first kind in testing hypotheses. By making the confidence coefficient high enough, this risk can be made as small as may be desired.

(2) The interval may include too many wrong values—that is, be too wide to be useful. This is analogous to an error of the second kind in testing hypotheses. It is of little help to know that a needle you are seeking is in a certain haystack, however precisely you may be told the location of the haystack. To know which cubic inch the needle is in, however, may be all that is needed for the practical purpose of finding it.

In general, the more specific an inference based on a sample, the more information it gives—but the greater the risk that it is wrong. The more reliable the statement, the more likely it is to be so vague as to convey little information. The Delphic oracle operated with a confidence coefficient of 1 by making completely ambiguous predictions.

A confidence interval, instead of leading to a choice between two hypotheses formulated in advance, leads to a division of all possible hypotheses into two groups, those that are consistent with the evidence and those that are not, "consistent" being interpreted in terms of a specific probability. The first are included within the confidence limits; the second are outside the confidence limits.

In the psychiatric study discussed in Chap. 12, a sample of $n = 100$ with $p = 0.56$ was not sufficient evidence for discarding the hypothesis that $P = 0.50$, using a one-sided test at the ten percent level of significance. Suppose, however, that instead of the test of significance described in Chap. 12, the investigators had computed an 80 percent confidence interval, obtaining the result 0.491 to 0.629. From this interval they could see that:

(1) The hypothesis that $P = 0.5$ cannot be rejected at a 20 percent level of significance in a two-tail test or a 10 percent level in a lower-tail test, since $P = 0.5$ is included within the 80 percent confidence interval.

(2) On the other hand, the hypothesis that $P = 0.60$ cannot be rejected either.

These two facts correspond with two practical conclusions: (1) The experiment has failed to show that the therapy was effective. (2) The experiment has also failed to show that the therapy was not effective, if we assume as before that $P = 0.60$ represents an effectiveness of practical importance. In short, the experiment was inconclusive at the stated level of significance. The confidence interval brings out this unhappy fact somewhat more obviously than the test of significance, which simply says that we cannot reject $P = 0.50$.

This again illustrates the importance, in considering tests of significance, of taking into account not only the risk of falsely rejecting the null hypothesis, that is, the risk of Type I error, but also the risk of Type II error, which in Sec. 12.6 was shown to be 0.24 when $P = 0.60$. Indeed, we see again the importance in planning an experiment of considering the entire operating-characteristic curve of the test.

Confidence intervals and tests of significance, when properly interpreted, lead to the same practical conclusions. It is the emphasis that is different. We can distinguish the following situations in practice.

(1) *Clear-cut hypotheses are not formulated in advance:* Confidence intervals are appropriate.

(2) *Clear-cut hypotheses are formulated in advance:* Tests of significance are appropriate, but confidence intervals are valuable for a full interpretation of the result: (a) For a finding of nonsignificance, the confidence interval shows vividly whether the risk of Type II error is high or low, that is, whether the result "nonsignificant" is indicative of positive evidence against the alternative hypothesis, or simply re-

flects insufficient evidence against the null hypothesis. (b) For a finding of significance, the confidence interval narrows the possibilities consistent with acceptance of the alternative hypothesis. For example, suppose that the psychotherapy experiment had yielded a $p$ of 0.60. This result would be significant statistically: we would conclude that the treatment was effective. But then the question arises, How effective? The answer would be given by a confidence interval. The 80 percent confidence interval would extend from 0.532 to 0.668.

## 14.5.2 Sample Size

We have already discussed the problem of sample size in the context of tests of significance (Sec. 12.9). There we saw that an increase in sample size steepens an OC curve. A steeper OC curve means sharper discrimination between the null and alternative hypotheses. But this sharper discrimination is obtained only at a cost, the cost of additional observations. The sample size should be increased only as long as the increase in accuracy is worth more than it costs to increase the sample size.

The same general principles apply in estimation. As before, the sample size appropriate to a given investigation is a compromise between the accuracy needed in the results and the cost of obtaining accuracy.

If the amount of resources—time, money, etc.—to be used for the study is fixed, there is no problem. If, to take an oversimplified example, 1,000 dollars—no more and no less—is available and must be used for interviewing, and an interview costs one dollar on the average, then the sample size will be 1,000. Suppose, however, that the investigator estimates the accuracy of such a sample. Accuracy can be expressed in terms of the standard error of the mean, since the confidence interval is obtained by adding and subtracting a certain multiple of the standard error to the observed statistic. For the mean, the standard error is $\sigma/\sqrt{n}$. By making an estimate of $\sigma$ (on the basis of earlier studies, or simply by a "guesstimate" on the basis of general knowledge), the investigator can thus get an idea of the accuracy attainable with 1,000 interviews. If this accuracy is either more or less than is necessary for the problem at hand, consideration can be given to changing the amount of money to be spent.

Suppose, for illustration, that in a proposed study of family income the standard deviation is guessed as $3,000. A series of calculations like those at the top of the following page might then be made.

| $n$ | 1 | 100 | 400 | 900 | 1,600 | 2,500 | 3,600 ... 10,000 |
|-----|---|-----|-----|-----|-------|-------|-------------------|
| $\dfrac{\sigma}{\sqrt{n}} = \dfrac{3000}{\sqrt{n}}$ | 3,000 | 300 | 150 | 100 | 75 | 60 | 50 ... 30 |

For each sample size, there will be a cost. The investigator then selects the best combination of costs and accuracy.

This brief discussion of the determination of sample size, though much oversimplified, should provide a glimpse of the principles by which such problems are approached.

### 14.5.3 Asymmetrical Confidence Intervals

Suppose that the 32 weights for which a 95 percent confidence interval of 168.7 to 183.1 lbs. was computed in Sec. 14.3.3 were to be used to determine a safe average weight for calculating the total passenger weight of airplanes. From the viewpoint of safety, it would be important not to use too low a figure. The danger of underestimating the mean would be of primary concern. In such a case, we want to be sure that the risk of underestimating the mean is controlled at a prescribed level, but we want to use the lowest figure consistent with this requirement.

In this case, we would want an upper confidence limit, rather than a confidence interval. We would want to say, for example, that we are 95 percent confident that the population mean weight does not exceed a certain figure; and we want to use the highest figure of which this could be asserted.

Such confidence limits can be obtained by computing

$$m_U = \bar{x} + K s_{\bar{x}},$$

where $K$ is now the upper $1 - C$ point of a standard normal distribution (instead of the $\frac{1}{2}(1 - C)$ point as for a confidence interval), $C$ being the confidence coefficient. For the 32 weights, $\bar{x} = 175.91$ and $s_{\bar{x}} = 3.662$, so the 95 percent upper confidence limit is

$$m_U = 175.91 + (1.645 \times 3.662) = 181.9$$

(1.645 being obtained from Table 391). If repeated random samples were drawn, only five percent of the samples would result in values of $m_U$ less than the parameter.

For a safety problem of this sort, a confidence coefficient higher than 95 percent would probably be used. For 99.9 percent confidence, for example, $K = 3.090$ and $m_U = 187.23$. As a practical matter, the airplane problem involves more complications than just determining the population mean. The variability of individual weights must also be allowed for, to control the risk that any particular plane load will exceed the safe limit.

A lower confidence limit could also be computed if circumstances made it appropriate. This would perhaps be the case if you were paying for passenger loads on a weight-per-load basis, and the load weights were based on an average weight per man. In such a case, you might accept a lower confidence coefficient than would be acceptable to safety authorities, say 90 percent. The two limits will then constitute an 89.9 percent confidence interval, in the sense that the probability that a random sample will produce an interval bracketing the parameter will be 0.899; but it will not be a symmetrical confidence interval. A symmetrical confidence interval is symmetrical in the sense that each limit treated separately has the same confidence coefficient, not in the sense of being symmetrical about a point estimate.

## 14.6
## OTHER PROBLEMS OF ESTIMATION

We have considered estimation for simple random sampling only, and only for means and proportions. These cases serve to present the logic of the methods used in making point or interval estimates, and to illustrate the details. In making estimates from data collected by more complicated probability sampling designs, the basic principles are similar but the details are more intricate. The interpretation of the results, however, is identical.

Point estimates and confidence intervals can be computed for other parameters than means or proportions, for example, standard deviations, correlation coefficients, medians, etc. If you encounter point or interval estimates, even of unfamiliar statistics, you should be helped by the basic ideas given here. Suppose you are told that the sample correlation coefficient between two variables is 0.796. Without knowing anything of the technical meaning of correlation, you should recognize that this coefficient is computed from a sample, and that the true correlation may be considerably different from 0.796. Then before worrying about the practical interpretation of the sample correlation coefficient, you should ask:

(1) Was the sampling process ultimately random, that is, was it a probability sample?

(2) If the sample was a probability sample, from what population was it drawn?

(3) What is the confidence interval?

Similar questions would be asked for any other statistic—say, a median.

Let us examine these three questions in more detail:

(1) Often the first question must be answered, "No, a non-probability sample was used." In such cases, it is not possible to attach a numerical probability to confidence in the estimate, or even to make a good estimate.

(2) Estimates based on samples drawn from restricted populations do not justify conclusions about broader populations, at least without additional information or assumptions. It is commonly said that much of the empirical information in certain branches of psychology applies to college sophomores. This is an exaggeration, but it emphasizes an important point. In medicine, generalization from clinical populations to broader populations is notoriously hazardous.

(3) Even with a probability sample from exactly the population we want to investigate, we should always consider the confidence interval as well as the point estimate. The temptation in practical work is to forget the confidence interval and credit the point estimate with more precision than it has. No great reliance should be placed on deductions based on a point estimate unless the deductions would hold also for other values in the confidence interval. And in using a confidence interval, the confidence coefficient must be kept in mind, for it measures the risk that the parameter is not really in the interval.

## 14.7
## TECHNICAL NOTES

### 14.7.1 Technical Note 1: Use of Student's t in Confidence Limits for Means (Sec. 14.3.3)

The formulas of Sec. 14.3.3 are slightly imprecise because they fail to allow for the fact that $s$, and consequently $s_{\bar{x}}$, are subject to sampling error. For samples of ten, or perhaps 20, or less, a refinement (corresponding with that of Sec. 13.4.1) is to use, instead of $K$, a quantity

$$t = K\left(1 + \frac{K^2 + 1}{4f}\right),$$

with $f = n - 1$ in this case. A more elaborate refinement is to replace $K$ by

$$t = K\left[1 + \frac{K^2 + 1}{4f} + \frac{(K^2 + 3)(5K^2 + 1)}{96f^2}\right].$$

### 14.7 Technical Notes

For the first problem of Sec. 14.3.3, where we obtained 95 percent confidence limits 168.7 and 183.1, the simpler formula gives

$$t = 1.960 \left[ 1 + \frac{(1.960)^2 + 1}{4 \times 31} \right] = 1.960 \times 1.039 = 2.036.$$

Since $2.036 \times 3.662 = 7.46$, the confidence limits are 168.4 and 183.4. The more elaborate refinement gives

$$t = 1.960 \left( 1.039 + \frac{[(1.960)^2 + 3][5(1.960)^2 + 1]}{96(31)^2} \right)$$

$$= 1.960 \left( 1.039 + \frac{6.8416 \times 20.2080}{96 \times 961} \right) = 1.960(1.039 + 0.001)$$

$$= 2.038.$$

Then $2.038 \times 3.662 = 7.46$, and the confidence limits are still 168.4 and 183.4. If the sample had been smaller, there would have been more difference between the two values of $t$, and between them and $K$; but in general, the implication of this example—that $K$ is adequate for most practical purposes—is correct.

If $t$ is used for confidence limits for the difference between the means of two independent samples, the value of $f$ must be taken from Sec. 13.4.2, but otherwise the procedure is as just shown.

#### 14.7.2 Technical Note 2: Quadratic Confidence Limits for Proportions (Sec. 14.4.3)

More precise confidence limits for proportions can be obtained by solving the quadratic equation in the first paragraph of Sec. 14.4.3. The solution is

$$\frac{2np' + K[K \pm \sqrt{K^2 + 4np'(1 - p')}]}{2(n + K^2)}$$

where for the lower limit, $p_L$,

$$p' = p - \frac{1}{2n},$$

$\pm$ is taken as minus;

and for the upper limit, $p_U$,

$$p' = p + \frac{1}{2n},$$

$\pm$ is taken as plus.

To illustrate the calculations, we compute the upper limit of the 90 percent confidence interval for $n = 20$, $p = 0.15$. Here (from Table 391), we find $K = 1.645$, $K^2 = 2.706$, $p' = 0.175$, and $np' = 3.5$.

*Step* 1: $\sqrt{K^2 + 4np'(1 - p')} = \sqrt{2.706 + 4 \times 3.5 \times 0.825}$

$$= \sqrt{14.256} = 3.776.$$

*Step* 2: $2np' + K[K + (\text{result of Step 1})] = 7 + 1.645 \, [1.645 + 3.776]$

$$= 15.918.$$

*Step* 3: $2(n + K^2) = 2(20 + 2.706) = 45.412.$

*Step* 4: Step 2 divided by Step 3:

$$\frac{15.918}{45.412} = 0.351.$$

Thus,

$$p_U = 0.351.$$

## 14.8
## CONCLUSION

When a population parameter is estimated from a sample, the estimate may be expressed as a single number, called a point estimate, or as an interval, called a confidence interval. For many purposes it is essential to use point estimates, either for convenience in using the estimate, or because a practical decision depends on whether the parameter is above or below some critical level. Point estimates can not be expected to coincide with the population parameter, however, since they are subject to sampling variability, and confidence intervals indicate the extent of the allowance that must be made for this, taking account of sample size and the degree of confidence required.

The most widely used principle for making a point estimate is to select as the estimate the value of the parameter which maximizes the probability of the observed data—that is, renders them more probable than if any other value is assumed to be the parameter. This principle is known as the principle, or method, of maximum likelihood.

The basic principle in making interval estimates is that the sampling distribution of the intervals, when they are computed from repeated samples, must be such that they cover the true parameter a specified proportion of the time. This proportion is called a confidence coefficient. It is an objective, mathematical probability—the probability that random sampling will produce a sample for which the confidence interval includes the parameter.

Point estimates of a mean or proportion are made by simply using the sample mean or sample proportion. In the case of some other parameters, the selection of a good estimator is not obvious. Even where there is an "obvious" choice, moreover, it may be biased, that is, tend to equal not the parameter but some other value, when many sample values are averaged. More serious, its sampling variability may be greater than that of some other statistic, and its average discrepancy from the parameter (measured by its mean square error) may be greater.

Confidence intervals for a mean or proportion may be based on the normal distribution. This involves some degree of approximation in the case of means if the standard error has to be based on a sample estimate, rather than on the population value, of the standard deviation. In the case of proportions it involves the approximation of replacing the binomial by the normal distribution. In both cases, however, the normal distribution is ordinarily satisfactory enough.

Confidence intervals are wider the smaller the sample size for a given population and confidence coefficient, the more variable the population for a given sample size and confidence coefficient, and the larger the confidence coefficient for a given sample size and population.

Confidence intervals are closely related to significance tests, in that any value of the parameter which would be accepted if it were the null hypotheses of a test with a corresponding significance level is included in the confidence interval, and other values are excluded.

Sometimes one-sided confidence limits are used instead of intervals, where it is important to know how much larger the parameter may be than the point estimate (or how much smaller) but it is far less important how much smaller it may be (or how much larger).

This chapter concludes Part III, on the basic principles of statistical inference. We started from recognition that samples vary, even from a fixed population, so the results of a sample depend on chance. The pattern of sampling variability of a statistic, called its sampling distribution, depends on the parameter, however, and it is this dependence that makes it possible to infer characteristics of a population from a sample. For such inferences, we must be able to deduce from various assumptions about the population what the relevant sampling distributions would be. The deductions about sampling distributions can be made if, but only if, probability sampling is used; then the laws of mathematical probability enable us to deduce sampling distributions.

The sampling distributions of many important statistics are well approximated by the normal distribution, even when the parent population of the sample producing the statistic is not normal. When a sampling distribution is normal, the mean and standard error of the sampling distribution, plus a table of the standard normal distribution, is all that is needed for obtaining the sampling distribution of the statistic.

Once we have the necessary sampling distributions, the basic principle of testing and of interval estimation is to regard the data as consistent with those values of the parameter for which the sampling distributions include the observed data within the range of reasonably probable results, specifying "reasonably probable" in terms of a confidence coefficient (or "not reasonably probable" in terms of a significance level) and in relation to alternative hypotheses. The principle of point estimation by maximum likelihood is to select that value of the parameter for which the sample is more probable than it is for any other value of the parameter.

We turn now to a series of topics of special interest and importance—the planning of experiments and surveys, quality control, correlation, time series, and shortcut methods.

## DO IT YOURSELF

EXAMPLE 470

It will be well to review Secs. 11.2.2, 11.2.2.1, and 11.4 before doing this one. The numbers 8, 2, 6, 2 are obtained from four spins of the ten-sided die described in Sec. 10.3. Considering these numbers as a sample of $n = 4$:

(1) Compute the sample mean, $\bar{x}$.

(2) What is the mean of the sampling distribution of $\bar{x}$?

(3) What is the standard error of the sampling distribution? [Hint: What is $\sigma$ for the parent population?]

(4) What shape does the sampling distribution have?

(5) What is the population mean, $M$?

(6) Compute a 95 percent confidence interval for $M$ from the sample mentioned above. Does your confidence interval include $M$?

(7) Draw 25 samples of $n = 4$ from Table 632. Compute 90 percent confidence intervals for $M$ from each sample. What proportion of your confidence intervals include $M$?

(8) Suppose for the 25 samples required in (7) that you did not know $\sigma$, but estimated $\sigma$ by the standard deviation of each sample. Compute 90 percent confidence intervals from each sample, using the number 2.31 (see Sec. 14.7.1) instead of the value of $K$ used in (7), to allow for the fact

**Do It Yourself**

that the estimate $s$ is based on a very small sample. What proportion of these confidence intervals include $M$?

(9) Comment on the differences between your results in (7) and (8). If you had taken 10,000 samples instead of 25, would you expect the proportion of correct inferences to be higher, lower, or about the same for the method of (8)?

EXAMPLE 471A

Calculate 80, 90, 95, and 99 percent confidence limits for $P$ if
(1) $n = 100, p = 0.56;$
(2) $n = 1,000, p = 0.20;$
(3) $n = 20, p = 0.65.$

EXAMPLE 471B

In a pharmaceutical company a new enzyme was being considered in the hope of increasing the yield in a certain manufacturing process. The yield for a batch of product was expressed as a ratio between the actual yield and the theoretical yield computed from formulas based on past experience. For example, a yield of 101.3 meant that 1.3 percent more product was obtained from the batch than the formula indicated. The new enzyme was tried out on 41 batches, and an average yield of 125.2 was obtained, with a standard deviation of 20.1.

(1) Compute a 99 percent confidence interval for the true yield obtained by the enzyme. State carefully all the assumptions on which your calculation is based.

(2) Assuming that the assumptions referred to in (1) are correct, would you conclude that the true yield obtained by the new enzyme is above 100? Explain.

(3) What is wrong with the following argument, advanced by a member of the company's research department? Comment fully on all relevant aspects of the problem.

... There is no good theoretical reason for believing that the mean yield should be higher than 100 with the enzyme. Moreover, in the 41 batches studied, the standard deviation of individual yields was 20.1. In my opinion, 20 points represents a large fraction of the difference between 125.2 and 100. There is no real evidence that the enzyme increases yield. ...

EXAMPLE 471C

Review the second illustration in Sec. 14.2.1.1, and the reasoning used there to show that $c/n$ will give, on the average, too high an estimate of the proportion of occurrences, if sampling continues until $c$ occurrences are observed. Similar reasoning has sometimes been used, fallaciously, to explain the excess of male over female births:

Attempts have been made to explain the excess of the births of boys over those of girls by the general desire of fathers to have a son who would perpetuate the name. Thus, by imagining an urn filled with an infinity of white and black balls in equal number, and supposing a great number of persons each of whom draws a ball from this urn and continues with the intention of stopping when he shall have extracted a white ball, it has been thought that this intention would render the number of white balls extracted larger than the number of black ones.[6]

What is wrong in this reasoning?

EXAMPLE 472A

(1) Compute the maximum likelihood and minimax estimates of a population proportion when the sample consists of only one observation:
(a) if the observation is an occurrence;
(b) if the observation is a nonoccurrence.
Which estimates seem more "sensible" to you, and why?

(2) For samples of ten from a binomial population, make a table showing, for each possible sample result, the values of $p$ and $p^*$, the maximum likelihood and minimax estimates of $P$. Which estimates seem more "sensible" to you, and why?

(3) Compare the estimators $p$ and $p^*$ when $n = 0$. Which seems more "sensible" to you, and why?

EXAMPLE 472B

(Continuation of Example 381D.)

(3) Compute the sampling distribution of the median, and compute the mean and standard deviation of this distribution.

(4) To estimate the population median, which estimator would you prefer, $\bar{x}$ or the sample median? Explain.

---

6. Pierre Simon, Marquis de Laplace, *A Philosophical Essay on Probabilities*, translated from the Sixth French Edition by Frederick Wilson Truscott and Frederick Lincoln Emory (New York: Dover Publications, Inc., 1951), pp. 167–168. This *Essay* was first published in 1819. The Truscott-Emory translation has been slightly altered in our quotation. Laplace follows the passage quoted with a clear explanation of the fallacy.

# PART IV

# SPECIAL TOPICS

# Chapter 15

# Design of Investigations

## 15.1
## THE PROBLEM OF DESIGN

A statistical design is a plan for the collection and analysis of data. Preparation of a statistical design, like preparation of the design of a house, requires careful thought about what is wanted, visualization of possible ways of getting it, and selection of one way.

The need for well-thought-out blueprints and specifications is less frequently recognized in statistical investigations than in home building. Many investigators proceed under the impression that statistical problems begin only after the data have been collected. By that time, competent analysis is likely to show that the hoped-for dream home is really a nightmare, or that its cost has been exorbitant. (See Secs. 6.1 and 6.5.)

Like misuses of statistics generally, examples of bad statistical design, such as the following, can be instructive.

EXAMPLE 475  A STATISTICAL TRAFFIC JAM

ROUTE 17 AUTO JAM 10 MILES LONG CAUSED BY THRUWAY BOARD'S POLL. STERLINGTON, N.Y., May 21.—A traffic jam that stalled a double line of autos for ten miles north of here developed tonight during a poll being conducted by the New York State Thruway Commission on Route 17.

Because of the unusually heavy traffic brought out by the fine weather, as many as 40,000 vehicles were affected or will be before the poll is completed at 7 A.M. tomorrow, observers estimated.

The poll, decided upon to help determine the southern route of the proposed superhighway between New York City and Buffalo, started at 7 A.M.

**475**

today in this little community two miles north of Sloatsburg in Rockland County and twenty-eight miles north of the George Washington Bridge.

At that time state police started stopping every fourth automobile on Route 17, one of the main arteries to up-state areas, while canvassers for the commission asked the drivers these three questions: "Where did you come from?" "Where are you going?" "How often do you make this trip?"[1]

It seems likely that this study cost more than was necessary to attain its objectives: If 40,000 vehicles were involved, a 25 percent sample was undoubtedly larger than necessary for data of the detail and accuracy needed. Perhaps the 25 percent sample size originated in the widespread, but fallacious, notion that a large percentage of the population must be sampled. Other criticisms, however, are more serious. By creating a traffic jam, the investigators may have altered the very thing they were studying: motorists may have taken alternate routes when they heard about the jam, or changed their plans once they were caught in it. The wording of the questions, as reported in the article, seems open to criticism (see Sec. 5.6.2). The sample, though large in number of motorists, was taken on only one day— and that, incidentally, a Sunday. The same or even a smaller number of interviews spread over a larger number of days would have been better.

Evolving a good design, of course, is much harder than criticizing a bad one. It requires skill and ingenuity to use statistical theory effectively in solving specific problems of design. For solving specific problems, statisticians have evolved an elaborate and continually growing kit of techniques. While these techniques do not reduce statistical design to the selection of a good recipe from a cookbook, they, and the ideas underlying them, are extremely useful in attacking new problems.

Detailed classification or listing of techniques of statistical design is impracticable here, in part because many of the techniques are relatively specialized to particular fields of inquiry, and in part because even a moderately detailed listing would be beyond our scope. All such techniques, however, have in common an attempt to achieve the highest reliability for a given cost (or the smallest cost for a given reliability), that is, to achieve highest *efficiency*. Efficiency is sought by working backwards from a visualization of the final analysis. Various alternative plans are considered and the sampling errors and costs associated with each are appraised as well as possible, using all the

---

1. *New York Times*, May 22, 1950.

available evidence. The alternative appearing to offer the highest efficiency is chosen.[2]

To indicate the kinds of ideas that statisticians contribute to the design of investigations, we shall first present a series of simple, more or less disconnected examples of the ways in which good design can contribute to efficiency. Then we shall discuss some important ideas of design which have evolved in two important types of inquiry, usually called *experiments* and *surveys*.

## 15.2
## SIMPLE ILLUSTRATIONS OF GOOD DESIGN

EXAMPLE 477   WEIGHING TWO SMALL OBJECTS ON A
BALANCE SCALE

Suppose that a laboratory technician has two small objects whose weights he wishes to determine. He uses a balance scale, which consists of two pans at either end of a lever that rests like a teeter-totter on a pivot midway between the pans. The object to be weighed is placed in one pan and known weights are added to the other until the scale balances on the pivot. The error of the scale, measured by $\sigma$, the standard deviation of the population that would be produced by repeated weighings of the same object, is assumed to be the same no matter how heavy the object being weighed, within reason.

The natural method of weighing the two small objects in such a balance scale would be to weigh each in turn, thus obtaining for each an estimated weight with standard error $\sigma$. To reduce the standard error, each object could be weighed twice and the mean of the two readings used. Each mean would have then a standard error of $\sigma/\sqrt{2}$ and a total of four weighings would have been made.

A simple but ingenious modification of this design would achieve the same accuracy from a total of two weighings instead of four. First put both objects on one pan, and get the sum of their weights. Next, put one object on each pan and get the difference of their weights, by finding the additional weight necessary to balance the scale. Then the weight of one object is estimated as one-half the sum of these two results, and the weight of the other is estimated as one-half the difference of the two results. It can be proved by simple mathematics (an extension of the rule given in Sec. 13.2.2.1 for the standard deviation of the sum or difference of two random variables)

---

2. The problem of determining the scale of investigation (sample size), which we have treated briefly in Secs. 12.9 and 14.5.2, involves not only attainment of the highest reliability for a given cost, but also a decision about how much reliability is worth obtaining in view of cost. In this chapter we shall not discuss this latter problem.

that each estimate so made has the same standard error as if it were based on the mean of two direct weighings, namely $\sigma/\sqrt{2}$. Thus, two weighings made this way are as good as four made the "common sense" way.

### EXAMPLE 478A   DIFFERENCE IN MEAN WEIGHTS

In Sec. 13.2.2.2, an example was given in which a random sample of 25 men measured on two occasions gave the same accuracy in testing for a change in mean weight as would have been obtained with 4,444 men divided equally between two independent random samples. The particular numbers follow from certain assumptions made in constructing the example, but the conclusion is realistic, as can be seen by the following analogy. Suppose we wanted to decide whether or not two ordinary weighing scales, A and B, give essentially the same results. One design is to weigh a sample of, say, 25 men on each scale and use the analysis of Sec. 13.2.2.2. A second design is to weigh a sample of 25 men on scale A and a second, independent sample of 25 on scale B, and use the analysis of Sec. 13.2.2.1. With the second design, the same weighing error would be present as for the first, but the difference in mean weights recorded by the two scales would chiefly reflect differences in the true weights of the two samples of men. The first design is clearly preferable.

The idea illustrated here brings out why matched samples are preferable to independent samples in a wide variety of studies for which matching is feasible. A common application is the "panel" technique for measuring such things as shifts in political preference, income changes, and purchases of commercial products.

### EXAMPLE 478B   MEASURING THE THICKNESS OF A SHEET OF PAPER

Suppose you were given an ordinary ruler graduated in sixteenths of an inch and asked to measure the thickness of a page of this book. Obviously, your "reading" would be only a crude guess. You could do much better with the same ruler by measuring the thickness of 600 pages and dividing by 600, or by counting the pages making a thickness of one inch and dividing one by this number.

This design idea has many applications, such as obtaining the average weight of small manufactured parts, measuring a car's average mileage per gallon of gasoline, and measuring the velocity of light.

### EXAMPLE 478C   DRAWING A STRAIGHT LINE

Suppose you wish to draw a straight line starting at a given point and oriented in a specified direction. You need to establish one point lying in the desired direction in order to draw the line. Should you establish the point 1 inch or 10 inches from the initial point? The answer is intuitively obvious: the point 10 inches away would ordinarily be preferred. Use of this point would tend to reduce the discrepancy between the actual and the intended orientation of the line. An important statistical application of this idea, discussed in Sec. 17.4.4, note 6, page 546, occurs in the design of experiments intended to estimate the linear relationship between two quantitative variables, one of which is under control of the investigator.

# 15.3
# DESIGN OF EXPERIMENTS

In 1935, Sir Ronald A. Fisher laid the foundation for the subject which has come to be known by the title of his book, *The Design of Experiments*. Since 1935 the theory of experimental design has been developed much further and its range of applications extended to almost all the laboratory sciences, basic and applied, and to business and the social sciences as well. In Secs. 2.8.3 and 2.8.4 we described in some detail two studies, one of vitamin supplementation and the other of cloud seeding, which illustrate the kinds of problems falling within the scope of experimental design. We shall illustrate some of the important ideas of experimental design by discussing a few elementary designs.

## 15.3.1 Randomized Groups

Shorn of certain complications that need not concern us, the vitamin supplementation experiment (Sec. 2.8.3) involved the formation of two groups of soldiers, one of which was given vitamin supplementation, and the comparison of physical performance between these groups by an analysis like that of Sec. 13.2.2.1. Two important statistical questions arise in such an experiment. The first question relates to the method by which the original group is to be divided into two groups, and the second to the number of men to be assigned to each group.

You should not have to guess at this stage about the method by which the subdivision into two groups should be accomplished (see Sec. 9.6.1): it should be done at random. Randomization is essential not only to this simple experimental design but also to more complicated designs. Without randomization there would be little reason to worry about efficiency because there would be no valid basis on which to analyze the experimental data. The following illustration conveys the point forcibly:

EXAMPLE 479 EFFECTIVENESS OF BCG VACCINATION

Each of a group of physicians was assigned a group of children from tubercular families and told to vaccinate half of them with a vaccine called

"BCG." The following table records the TB deaths which occurred in the subsequent six-year period, December, 1926, through December, 1932:

| | Cases | TB deaths | Percent |
|---|---|---|---|
| Vaccinated | 445 | 3 | 0.67 |
| Controls | 545 | 18 | 3.30 |

Analysis of these data by the method of Sec. 13.3.2.1 suggests that the difference in observed death rates is more than could be expected between two randomly formed groups. We cannot, however, attribute the difference to the effect of BCG because the groups were *not* randomly formed. At this stage of the actual study, it was suggested that the physicians' initial choice of children to vaccinate might have tended to favor children in families where the consent of parents was easy to obtain, and that these children might possibly be less prone than others to tuberculosis. A second experiment was then designed in which the physicians were told to vaccinate every other child—not strict randomization but an improvement over the first study. After eleven more years the following results were available:

| | Cases | TB deaths | Percent |
|---|---|---|---|
| Vaccinated | 556 | 8 | 1.44 |
| Controls | 528 | 8 | 1.52 |

This difference could easily be accounted for by chance alone; hence there is no evidence that BCG did any good.[3]

As to the second issue in the two-group design, the number of observations in each group, an equal number in each group gives the highest efficiency unless the cost of the treatment is an important limitation. For, if the null hypothesis is true, it is easy to show that the standard error of the difference formula of Sec. 13.2.2.1 is less for an equal than for an unequal division of the total number of observations.

## 15.3.2 Randomized Blocks

The vitamin supplementation experiment involved one complication that we can now profitably consider. The actual experiment consisted essentially of four small experiments, one for each of four platoons. In each platoon the design described in Sec. 15.3.1 was used; that is, half the men in each platoon, selected at random, were given vitamin supplementation. This experiment illustrates a *randomized block* design, each of the platoons constituting a "block".

3. Milton I. Levine and Margaret F. Sackett, "Results of BCG Immunization in New York City," *American Review of Tuberculosis*, Vol. 53 (1946), pp. 517–532.

### 15.3 Design of Experiments

The idea underlying a randomized block design is essentially the idea of stratified sampling, which we shall discuss also in Sec. 15.4.2. Each platoon had its own leadership, its own sleeping quarters, its special modifications of the activity schedule, and, as the experiment progressed, its own special history and tradition. The physical performance of men within a platoon, therefore, was likely to be relatively homogeneous and the standard deviation of performance within platoons to be relatively small. The analysis of this design, an extension of the analysis of variance ideas of Sec. 13.2.3, uses a standard error based on the average standard deviation of performance within platoons. Since this standard deviation is relatively small, at least if the blocks are homogeneous, the standard error is also relatively small and the experiment is more efficient than the simpler design of Sec. 15.3.1.

Another way of expressing the same idea is to say that this randomized block design prevents differences among the platoons from being reflected in the experimental error.

### 15.3.3 Factorial Designs

Suppose that the school of education at a large university wants to evaluate three methods of presenting lectures in elementary physics —movie, closed circuit television, and "live." It also wants to evaluate different times of day at which the lectures might be given: morning, midday, and afternoon. These two factors, method and time of presentation, can be evaluated simultaneously by the design shown in Table 481.

TABLE 481

DESIGN TO EVALUATE METHOD AND TIME OF DAY IN
ELEMENTARY PHYSICS COURSE
(Number of Sections)

| Time | Method | | | Total |
|------|--------|------|------|-------|
| | Movie | TV | Live | |
| Morning | 1 | 1 | 1 | 3 |
| Midday | 1 | 1 | 1 | 3 |
| Afternoon | 1 | 1 | 1 | 3 |
| Total | 3 | 3 | 3 | 9 |

Nine hundred students in elementary physics would be randomly assigned to one of nine sections, each of 100 students. All students would take the same examinations, and the performance of each class would be measured by its mean score on the examinations.

To compare methods of presentation, we would compare the average performances of the three sections exposed to each method (see bottom row of Table 481). To compare times of day, we would compare the average performance of the three sections at each time of day (see the right-hand column of Table 481).

Thus, both factors—method and time—are appraised in a single experiment. This simultaneous appraisal is possible because of the *balance* built into the design. For example, the three sections exposed to movies include a morning, midday, and afternoon class, and the same is true of the three sections exposed to TV and of the three sections exposed to live lecturing. Similarly, the three morning sections include one with each method of teaching, as do the three midday and three afternoon sections. The balance of the design avoids one complexity that made difficult the interpretation of the data on beauty and brains (Example 277), in which lack of balance resulted in a confounding of the year-in-school effect with the appearance effect.

This type of design is called a two-factor *factorial* design. The idea extends to factorial experiments with more than two factors.

### 15.3.4 Latin Squares

The idea of balance illustrated in the preceding section may be regarded as an extension of the idea of matching, exemplified in the matched-sample design for measuring change in weight (Example 477). Still further extensions of these ideas are possible, one simple but ingenious example being the *Latin square*. Suppose that in the problem of Sec. 15.3.3 three different textbooks, *A*, *B*, and *C*, are also to be evaluated. Each of the books could be used in three classes according to the arrangement shown in Table 483.

The arrangement of the texts in Table 483 is such that (1) each text appears three times; (2) each text appears once and only once in each row; (3) each text appears once and only once in each column. Thus each text is used once at each time of day and once for each method of instruction. The texts can be compared by comparing average performance of the three sections using each text.

The methods can be compared as before, since the new factor—textbook—is introduced so that each method uses each text once. Similarly, each time of day can be compared as before, since texts *A*, *B*, and *C* are used at each time of day. Thus, although no two ob-

TABLE 483

Design to Evaluate Method, Time of Day, and Textbook, in Elementary Physics Course

(Letters represent textbook used)

| Time | Method | | |
|---|---|---|---|
| | Movie | TV | Live |
| Morning | B | C | A |
| Midday | C | A | B |
| Afternoon | A | B | C |

servations are comparable, the groups that we want to compare are all comparable.

In practice, the scores of individual students as well as the mean scores of all students in each section would be used in analyzing the designs of this section and the last. Also, individual scores on aptitude tests or grades in earlier courses might increase the effectiveness of the experiment, just as initial performance ratings were used in the analysis of the vitamin supplementation experiment (Sec. 2.8.3).

## 15.4
## SURVEYS

Surveys are distinguished from experiments mainly by the fact that the investigator is trying to measure what would have happened even if his study had not taken place, rather than to measure the effect of responses to stimuli he deliberately introduces. This distinction blurs under close logical scrutiny, but it serves well enough to distinguish, for example, the vitamin supplementation experiment from a pre-election opinion poll.

The theory of survey design is an even more recent development than the theory of experimental design, most of the major developments having occurred since 1935. As with experimental design, we shall describe a series of simple designs which illustrate some of the major ideas.

### 15.4.1 Simple Random Sampling

Simple random samples can sometimes be used in surveys without modification, as in studies of college students at a university or the customers of a public utility. A list of the people in the population is then usually available and it is easy to select a simple random sample with the aid of a table of random numbers (Sec. 10.9.2).

## 15.4.2 Proportional Stratified Sampling

Suppose that a large company contemplating a group life insurance program wants to know, among other things, the mean amount of life insurance already carried by its employees. A simple random sample could easily be selected, using, say, the company's payroll as a list of the population. It is likely, however, that the average amount of life insurance is quite different for different categories of employees; this fact could be exploited to attain a more efficient sampling design.

For simplicity, suppose that there are just two categories of employees, office and plant, in relative frequencies 0.2 and 0.8 (which can be ascertained from the payroll). The true means for each group may be represented by $M_1$ and $M_2$, respectively, and these are, of course, unknown. By using the formula for weighted means (Sec. 7.4.2), the over-all mean, $M$, can be written as

$$M = 0.2M_1 + 0.8M_2.$$

A simple random sample of $n$ observations would yield an estimate $\bar{x}$ of $M$ which can also be written as a weighted mean,

$$\bar{x} = \frac{n_1}{n} \bar{x}_1 + \left(1 - \frac{n_1}{n}\right) \bar{x}_2,$$

where $n_1$ is the number of office employees who happen to be included in the sample, and $\bar{x}_1$ and $\bar{x}_2$ are the sample means observed, respectively, for the office and plant employees who happened to be chosen in the sample. All three of these quantities—$n_1$, $\bar{x}_1$, and $\bar{x}_2$—are subject to sampling error.

A *stratified sampling* design would eliminate one of these sources of sampling error, that attributable to sampling fluctuation of $n_1$. Each of the two groups could be regarded as a *stratum*, and an independent random sample drawn from each stratum. The two sample means thereby obtained, $\bar{x}_1$ and $\bar{x}_2$, could then be weighted by 0.2 and 0.8 to form the estimate

$$\bar{x} = 0.2\bar{x}_1 + 0.8\bar{x}_2.$$

This estimate is subject to only two sources of sampling error, those in $\bar{x}_1$ and $\bar{x}_2$. It uses the population value of $n_1/n$, namely 0.2, instead of a value based on a sample, and therefore subject to sampling error.

In stratified sampling, it is common, though neither necessary nor always desirable, to make the individual samples proportional to the numbers of observations in the strata. In our illustration, for example, the sample from the first stratum would comprise 20 percent of the total sample size and that from the second stratum would comprise 80 percent. Unless observations from different strata are more alike than observations from the same strata, a situation rarely encountered, this design assures that the estimate $\bar{x}$ from the stratified design is more efficient than the estimate $\bar{x}$ from the simple random design, total sample sizes being the same. A subsidiary but important practical advantage is that, for estimating $M$, the individual samples can be combined and the over-all sample mean computed by the usual formula, thus saving the explicit use of a weighted average.

This example is one of the few illustrations in which a lapse in design could be largely compensated in analysis. Suppose that a simple random sample had been used. Instead of using the sample mean $\bar{x}$ to estimate $M$, we could weight the means $\bar{x}_1$ and $\bar{x}_2$ by the *true* weights 0.2 and 0.8. That is, we could use the estimator $0.2\bar{x}_1 + 0.8\bar{x}_2$ instead of $(n_1/n)\bar{x}_1 + (n_2/n)\bar{x}_2$. This estimator would be nearly as efficient as the estimator from the stratified random sample just described.

EXAMPLE 485 RE-ANALYSIS OF *Literary Digest* POLL

During the 1936 presidential campaign, when the *Literary Digest* poll was showing a substantial lead for the Republican over the Democratic candidate, a statistician made a re-analysis of the data, based on the fact that the *Digest* tabulated preferences separately for those who had voted Republican and Democratic in the 1932 election. The proportion voting Republican in 1932 was substantially higher in the *Digest's* 1936 poll than in the election returns of 1932. For each state the *Digest's* proportions Republican in 1936 for those voting Republican in 1932 and for those voting Democratic in 1932 were, therefore, weighted according to the election returns of 1932, instead of according to the frequencies in the *Digest* data. This showed the two candidates about even. Thus no prediction was justified by the *Digest's* data. As it turned out, the Democratic candidate carried every state except Maine and Vermont.[4]

## 15.4.3 Nonproportional Stratified Sampling

Suppose we wanted to estimate the proportion of automobiles whose license plates bear descriptive legends ("America's Dairyland," "Land of Opportunity," etc.). This problem is suitable for a stratified design, using states as strata. Here, however, it would obviously be inefficient to allocate sample sizes among the states in proportion to

4. William L. Crum, *Wall Street Journal*, October 30, 1936, p. 1.

the total number of license plates. In fact, it would be inefficient to take more than a single observation in any one state. A total of 48 observations, one from each state, would yield an errorless estimate, assuming, of course, that the total number of license plates in each state was known. This sampling design illustrates *nonproportional stratified sampling*.

This example brings out the idea that any observations beyond the first are wasteful if the stratum is completely homogeneous. This idea can be easily extended to show that for strata of the same size, the sample sizes should be proportional to the standard deviations of the characteristic being studied. In the example of the preceding section, suppose that most plant employees held little or no insurance, but that office employees held amounts varying from little or none for the office boys, to amounts in the hundreds of thousands of dollars for the top executives. Under these conditions, more than 20 percent of the total sample should be allocated to the office stratum and less than 80 percent to the plant stratum. The reasoning is that, up to a point, the precision of the weighted mean is decreased less by reducing the number of observations in the more homogeneous stratum below a proportional share, than it is increased by adding the same number of observations to the less homogeneous stratum.

In general, the maximum precision will be attained by computing $N\sigma$ for each stratum, where $N$ is the number of observations in the population in the stratum and $\sigma$ is the standard deviation within the stratum, and allocating the sample among the strata in proportion to these quantities. That is, if $N\sigma$ for a particular stratum is 10 percent of the sum of similar quantities for all strata, allocate 10 percent of the sample to that stratum. The reasoning is this: Suppose the number of observations in the sample from one stratum is reduced. This will lower the precision (that is, increase the standard error) of the mean for this stratum, hence lower the precision of the weighted mean. But if a corresponding number of observations is added to the sample from another stratum, this will raise the precision for that stratum, hence raise the precision of the weighted mean. Whenever collecting an observation from one stratum rather than another will result in a net increase in the precision of the weighted mean, the transfer should be made.

The standard error of the mean of a stratified sample is easy to compute, the formula being

$$s_{\bar{x}} = \sqrt{\left(\frac{N_1}{N}\right)^2 \frac{s_1{}^2}{n_1} + \left(\frac{N_2}{N}\right)^2 \frac{s_2{}^2}{n_2} + \cdots + \left(\frac{N_k}{N}\right)^2 \frac{s_k{}^2}{n_k}} = \sqrt{\sum_{i=1}^{k} \left(\frac{N_i}{N}\right)^2 \frac{s_i{}^2}{n_i}},$$

where

$$\bar{x} = \frac{N_1}{N}\,\bar{x}_1 + \frac{N_2}{N}\,\bar{x}_2 + \cdots + \frac{N_k}{N}\,\bar{x}_k = \sum_{i=1}^{k}\frac{N_i}{N}\,\bar{x}_i,$$

$N_i/N$ is the proportion of the observations in the population that are in the $i$th stratum, $n_i$ is the sample size in the $i$th stratum, $s_i$ is the estimated standard deviation for the $i$th stratum. (As a matter of fact, $s_{\bar{x}}$ above is the standard error of any weighted mean whose weights are not subject to sampling errors, with $N_i/N$ representing the weight of the $i$th sample.)

## 15.4.4 Cluster Sampling

Suppose a study were being made of the incomes of families in a large city. A random sample of 100 city blocks might be drawn and 10 families selected at random from each block. These 1,000 families would not provide as accurate a mean as if 1,000 families were drawn at random from the entire city. To see why this is, imagine an exag-/ gerated situation in which all families in a block have the same income. Then means of samples of 1,000, drawn 10 each from 100 blocks, would have practically the same variability as means of samples of 100 drawn purely at random (assuming that there are so many blocks that a simple random sample of 1,000 would include few families from the same blocks). To put it differently, after one family is selected from a block, under this exaggerated assumption, the other nine are redundant—they simply repeat the information provided by the first family.

In reality, the families in a block are not identical in income, but they are similar. After one family's income is obtained, the others are partially redundant. Nine more families from the same block do not reduce the standard error of the mean as much as nine more families each chosen at random from the whole city.

But nine more families from the same block do not add as much to the cost of the sample as nine more families chosen at random from the whole city. The realistic choice, therefore, may be between perhaps 25 families from each of 100 blocks, and 1,000 families at random from the whole city. If the similarity within blocks is not too great, 2,500 families chosen 25 each from 100 blocks may give a smaller standard error than 1,000 chosen purely at random.

Such a sample is called a *cluster sample*. Clusters of observations are formed on some basis that reduces costs, usually geographical compactness. Geographical compactness almost always entails rela-

tive homogeneity of observations within the clusters, as in our illustration of incomes of families in the same block. Sometimes, however,
clusters can be formed in such a way as to gain both cheapness and
heterogeneity. If families are used as clusters in estimating the sex
ratio, cluster sampling is not only cheaper but also more efficient
than simple random sampling with the same number of individuals.

### 15.4.5 Systematic Sampling

A common survey design can be illustrated by a survey of student
opinion.

To select a sample of, say, 250 students from a student body of
10,000, the first step, with this design, would be to divide 10,000 by
250 to obtain a "sampling interval," here 40. Next, select a random
number from 01 to 40. Suppose that this number turns out to be 17;
then start at the 17th student in the student directory and select every
40th one thereafter, that is, the 57th, 97th, and so on. This is a *systematic sample*.

A systematic sample bears some resemblance to a stratified sample
with a single observation in each stratum, the strata in our illustration
being the consecutive groups of 40 students. This resemblance gives
intuitive insight into the reason, other than convenience, for using a
systematic design. The implicit division of the population into strata
may produce more efficient estimators of means or proportions for
the same reason as in proportional stratified sampling generally: there
may be similarity within strata. In the example of the student directory,
this stratification effect might be important if the listing of students
was according to year in school, field of study, or some other characteristic that might be related to attitude. If the listing were alphabetical, then the sampling would presumably be effectively random, or
nearly so—though presumptions of this kind are often the downfall
of sampling investigations, as when initial letters of last names are
related (in an unforeseen way) to nationality and nationality is related
(in an unforeseen way) to attitude on the particular issue. If the listing were effectively random, the systematic design would not benefit
from stratification effects, but it would provide a convenient way of
getting a simple random sample.

A danger in systematic sampling is that the characteristics being
studied may have a certain pattern or periodicity in the list. On census
record sheets, for example, the first names on the sheets tend to be
predominantly male, gainfully employed, and above average in income. The reason is that the enumerators are instructed to start in a

certain block at the corner house (which tends to have a higher rental value than houses in the middle of the block), and in the household to start with the head (usually male and the breadwinner). Similarly, in inspecting every tenth item on a production line, it may be that at a previous stage every tenth item has been inspected and made perfect.

A simple modification has almost all the potential advantages and almost none of the potential disadvantages of systematic sampling. In the student opinion study, this alternative design would involve selecting a random number from 01 to 40 for *every* interval of 40, not just the first. If the random numbers 17, 02, 14, 35, . . . were obtained, the following students would be selected:

$$00 + 17 = 17\text{th}$$
$$40 + 02 = 42\text{nd}$$
$$80 + 14 = 94\text{th}$$
$$120 + 35 = 155\text{th, etc.}$$

This is a proportional stratified sample with a single observation per stratum.

## 15.4.6 Area Sampling

Area sampling is an application of cluster sampling—with other design features interwoven—to obtain a probability sample of widely-dispersed populations, such as the families of the United States, the deer in Michigan, or the jute in Bengal. It is based on a simple idea: the items in the population can be associated with geographical areas. By drawing a probability sample of these areas, and sampling appropriately within them, it is possible to obtain a probability sample of the population. The parts of the population in the areas can be thought of as clusters, just as in the earlier example of families in blocks (Sec. 15.4.4).

To illustrate how this idea can be used along with other design ideas in obtaining a sample of families in the United States, we shall describe a design which, although simplified, incorporates many of the features used in actual designs.

There are about 3,070 counties in the United States, and these may be used as areas. To select 50 counties, the 3,070 counties are divided into 50 strata, each stratum consisting of a relatively homogeneous group of counties. One county from each stratum is chosen at random. The counties may be given probabilities of selection proportional to the numbers of families in them.

At this stage, we have a stratified random sample made up of 50 counties. The next step is to draw a sample of the families in each of these 50 counties. Since this procedure is similar in each county, we illustrate by one county.

As a first step, a map of the county is subdivided into small land areas, such as city blocks or small country areas bounded on all sides by roads. Suppose there are 6,000 of these small areas. A random, or stratified random, sample of these areas, say 20, is chosen. In each of the 20 small areas, one-fifth, say, of the families are chosen at random.

Every family in the county has the same conditional probability of inclusion once the county is chosen:

(1) The probability of any small land area's being chosen is 20/6,000 or 1/300.

(2) In any area, the probability of any family's being selected, if the area is chosen, is 1/5.

(3) Therefore, the probability that any family in the county will be chosen is $(1/300) \times (1/5) = 1/1,500$. This probability is the same for every family in the county.

By extension of this approach, it is possible to assure that every family in the entire United States has an equal probability of inclusion in the sample. In practice, we might not want to assign every family an *equal* probability of inclusion, but it is always necessary that each family have a *known* probability other than zero.

Usually, the standard error of the mean for an area sample is larger than the standard error of the mean for a simple random sample with the same number of observations, despite the fact that both stratification and clustering are incorporated in these designs. The reduction in the standard error due to stratification is typically more than offset by the increase due to clustering. It is not appropriate to use simple standard error formulas, such as $\sqrt{p(1-p)/n}$, in estimating the standard errors of estimates from area samples.

## 15.5
## CONCLUSION

A design is a plan for obtaining and analyzing data. A design should be selected by visualizing the analysis of data obtainable under alternative plans, appraising their standard errors and costs, and choosing what then appears to be the best alternative. For any given cost, this procedure implies selection of the design with the smallest sampling error—that is, the most efficient design. The selection of a design requires careful planning *in advance* of data collection and analysis. Thoughtlessness in the selection of a design can at best be only partially compensated by energetic and intelligent analysis of the actual data. In fact, analysis of data from a poorly designed study is likely to resemble an autopsy: whatever is learned will be of value for the future, but not for the patient.

Every statistical investigation presents unique design problems, and from efforts to meet these problems an elaborate science of statistical design has arisen. Many of the important ideas underlying this science can be illustrated by simple examples, such as measuring the thickness of a sheet of paper with a ruler or drawing a straight line.

The science of experimental design concerns arranging stimuli or treatments in such a way that inferences about the effects of these treatments can be drawn and their reliability measured. By the random selection of experimental units, it is possible to remove ambiguity about the causal interpretation of observed associations. While randomization is the essential ingredient of all experimental design, there are many devices for increasing the precision of the inferences that can be made with given confidence. For example, a large experiment may be subdivided into several smaller ones, in each of which conditions are relatively homogeneous, and several factors may be evaluated simultaneously by application of the idea of balance.

The sample survey is a second major area of development of statistical design. A whole body of techniques has grown up to meet the objectives of valid inferences about widely dispersed human populations, and many of these techniques can be applied to other problems as well. One major idea is *stratification*, in which the population is subdivided into relatively homogeneous subpopulations, or strata, and a random sample drawn from each. A second major idea is clustering, by which the population is subdivided into relatively small groups or *clusters* which can be studied together economically, and a random sample of clusters drawn. In practical survey designs, clustering, stratification, and other techniques are usually interwoven in order to obtain the highest precision consistent with cost considerations.

Both experiments and surveys involve a host of practical problems which can only be combated by careful design. The nonresponse, or missing observation, problem is the most serious of these. In experiments, missing observations may seriously complicate analysis, while nonresponse in surveys may inject serious bias into inferences. These problems, while they must always be kept in mind, and have been discussed in earlier chapters, are outside the range of this chapter.

## DO IT YOURSELF

EXAMPLE 492A

The data of the experiment described in Example 479 contain internal evidence that the original instructions were not carried out by the physicians. What is this evidence?

EXAMPLE 492B

In the first half of Example 479, at what level of significance (one-tail test) is the difference between 0.67 percent and 3.30 percent?

EXAMPLE 492C

A "panel" is a sample from which measurements are made at different time periods. For example, a sample of voters might be interviewed about voting intentions several times during the course of a political campaign. One important advantage of the use of a panel is that brought out in Example 478A. What potential disadvantages can you think of, using the political opinion panel as an example?

EXAMPLE 492D

In the vitamin supplementation experiment (Secs. 2.8.3, 15.3.1, 15.3.2), suppose that two of the four platoons were selected at random and that all soldiers in these platoons received vitamin supplementation, while all soldiers in the remaining two platoons were controls. What advantages and disadvantages do you see for this design as compared with the design actually used?

EXAMPLE 492E

Comment on the following quotation:

Control in experimentation requires constancy or uniformity in all elements affecting the results, except the *one* variable which is being tested.

EXAMPLE 492F

Comment on the following quotation:

A city might be made up of racial and economic groups of such diverse opinions on the matter being studied that it would not do to leave their representation in the sample to pure chance, as in . . . random sampling. . . . It would be necessary to sample each group in accordance with its proportionate part of the city's total registration of voters.

EXAMPLE 493A

*Quota sampling* superficially resembles stratified sampling, in that the population is divided into subgroups. Interviewers are assigned quotas for the number of people they are to interview from each of these subgroups, for example, the number of men and women, young and old, rich and poor. What is the major difference between quota sampling and stratified random sampling? Why is this difference important?

EXAMPLE 493B

Suppose that in an area sample of a certain city, blocks are stratified according to the median rent at the time of the last census into three strata: high rent, medium rent, low rent. Within each stratum a simple random sample of blocks is chosen, and all families within each block are asked to give their current incomes. Someone objects to this approach on the ground that use of out-of-date rental figures will result in biased estimates of income from this sample. How would you answer this objection?

EXAMPLE 493C

Give an example of a problem for which you would almost certainly not want to use systematic sampling.

# Chapter 16

# Statistical Quality Control

## 16.1
## NATURE AND PURPOSES OF STATISTICAL QUALITY CONTROL

The statistical surveillance of repetitive processes—called statistical quality control, after the field in which it has been widely applied —is one of the most useful and economically important applications of the principles of statistical decision presented in Chap. 12. Simple, standardized, and graphic methods have been devised which are easily and quickly mastered. Literally thousands of decisions are made from them as an hour by hour, day by day routine, mostly in business—manufacturing, administration, accounting, purchasing— but also in such fields as public health, safety, and laboratory analysis.

Statistical quality control serves excellently to illustrate the principles of statistical decision procedures, not only because it is important in its own right, but because it is simple—as its widespread, routine application requires—yet involves all of the main ideas: acceptance and rejection rules, risks of error, operating-characteristic curves, etc. It thus serves excellently to make more concrete some of the ideas of Chap. 12. It also provides an opportunity to introduce an important idea, not included in Chap. 12, that of sequential sampling, in which not only the decisions to accept or reject are provided for, but also the decision to withhold judgment and collect more data.

The kind of repetitive process to which statistical quality control might be applied is typified by a machine turning out a large number of presumably "identical" pieces—rivets, plastic bottle caps, card-

**494**

board boxes, filled containers, etc. But statistical quality control is applied to many other kinds of repetitive process—printers' errors, accountants' mistakes, stitches skipped in garment manufacturing, new cases of communicable disease, school absences, cashiers' shortages, complaints from customers, orders received, library books missing, airline reservations, yards gained by a football play—in short, almost anything in which, under more or less constant conditions, there is a large volume of activity made up of distinct units that are repeated many times.

Systematic statistical quality control originated in the work of the American industrial statistician, Walter A. Shewhart (born 1891). Its greatest development came during World War II, when a short, intensive training course developed at Stanford University was given repeatedly in industrial centers throughout the country under the auspices of the War Production Board. The American Society for Quality Control (whose membership badge is shaped like a $\sigma$ and contains a picture of a control chart), although organized only at the end of the war, now has many more members than the combined memberships of the American Statistical Association and the Institute of Mathematical Statistics, plus the Biometric, Econometric, Psychometric, Sociometric, and similar Societies.

The purposes for which statistical quality control is used are of two types, typified in industry by (1) *process control*, which aims at evaluating future performance, and (2) *acceptance inspection*, which aims at evaluating past performance. In both cases, samples are drawn and from them decisions are made about the population, which for process control is the infinite number of possible results of further repetitions of the same process, and for acceptance inspection is the qualities of a finite group of items, called a *lot*, already in existence.

## 16.1.1 Process Control

Process control, the principal tool of which is the control chart (already discussed in Sec. 4.8), determines whether a process is in a state of statistical control. A process is said to be in a state of statistical control if the variation is such as would occur in random sampling from some stable population. If this is the case, the variation among the items is attributable to chance and there is no point to seeking special causes for individual cases. When the process is in control, if its performance is unsatisfactory the only remedy is some change in the process. But when the process is out of control, it should be possible to locate specific causes for the variation, and by removing them to improve the future performance of the process.

In a sense, the object of process control is to evaluate items not yet produced; for when the process is in control, it is relatively safe to predict the future items on the basis of past output. Note that process control does not aim at judging whether the process is satisfactory, only whether it is in a state of statistical control and hence predictable.

### 16.1.2 Acceptance Inspection

The object of acceptance inspection is to evaluate a definite lot of material that is already in existence and about whose quality a decision must be made. This is done by inspecting a sample of the material, using definite statistical standards to infer from the quality of the sample whether the whole lot is acceptable. The standards in acceptance inspection are set according to what is required of the product, rather than by the inherent capabilities of the process, as in process control.

### 16.1.3 Producer's or Seller's Risk vs. Consumer's or Buyer's Risk

Suppose that a sample is taken from a lot or process. If the sample contains a specified number of defective items or more, the lot is "rejected" or the process is declared "out of control"; otherwise the lot is "accepted" or the process declared "in control." Such a sampling plan entails two risks. First, there is the risk that a good lot will yield a bad sample; second, there is the risk that a bad lot will yield a good sample. Often the first is called the producer's risk or the seller's risk, and the second the consumer's risk or the buyer's risk.

As we have seen in Sec. 12.4, these two errors are referred to in general in statistics as errors of the first and second kinds. Though this terminology has the advantage of generality, it does not bring out the practical differences in the two types of errors in commercial quality control applications as well as do the terms seller's and buyer's risks, or producer's and consumer's risks.

A sampling plan can be evaluated by computing the probabilities of these two kinds of errors. A certain quality is defined as acceptable, and the probability is calculated that a lot or process of this quality would produce a sample leading to rejection or an out-of-control alarm. Another quality is defined as rejectable—there is a zone of comparative indifference between the acceptable and rejectable quality levels—and the probability is calculated that a lot or process of this quality will produce a sample leading to acceptance or an in-control decision. These two probabilities measure the producer's and consumer's risks, respectively.

### 16.1.4  Operating Characteristics in Quality Control

Actually, there is no reason to evaluate a sampling procedure only at the two qualities labeled acceptable and rejectable, especially since difficult and arbitrary decisions may be involved in designating these two quality levels. Instead, the probability of getting a sample leading to acceptance or to an in-control judgment, can be calculated for any quality. The result will be an OC curve like that in Fig. 497.

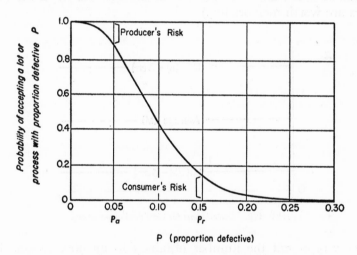

**FIG. 497.**  Operating-characteristic curve for a typical sampling plan.

This curve is completely analogous to the curves shown in Figs. 389B, 391, and 417. The points on the horizontal scale represent possible lot or process qualities, and the height of the curve shows the probability that a lot (or process) of this quality will be accepted (or said to be in control), assuming the specified sampling plan is in use. In Fig. 497 it has been assumed that the acceptable and rejectable qualities are measured as proportions of the items that are defective, and are $P_a = 0.05$ and $P_r = 0.15$; from the OC curve, the producer's and consumer's risks are seen to be both a little more than 0.10 in this example. (The sampling plan of Fig. 497 calls for accepting the lot if three or fewer defectives are found in a sample of 40.)

## 16.2
## PROCESS CONTROL

### 16.2.1   Basic Principles of Process Control

Suppose a sample is drawn from a normally distributed population with mean $M$ and standard deviation $\sigma$. Then tables of the normal distribution, such as Table 365, show what proportions of the observations will lie within given distances of $M$, measuring the distances as multiples of $\sigma$. In particular, about two-thirds of the observations will be in the interval $M \pm \sigma$, about 95 percent in the interval $M \pm 2\sigma$, and 99.7 percent in the interval $M \pm 3\sigma$. (These figures are worth memorizing.)

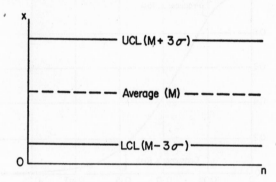

FIG. 498.   Control chart for individual observations.

Let $x$ represent the number obtained in an observation. Then, the idea behind control chart analysis is this: draw a chart with possible values of $x$ on the vertical axis and on the horizontal axis a series of integers, beginning with 1, to represent the sequence of observations—the first observation to be obtained, the second, and so on. Draw horizontal lines at heights that correspond to $M + 3\sigma$ and to $M - 3\sigma$. The line at $M + 3\sigma$ is called the *upper control limit* or UCL and that at $M - 3\sigma$ the *lower control limit* or LCL. Then observations are plotted successively, and, if they are normally distributed, only three in a thousand, on the average, will fall outside of the control limits as long as the population does not change. Such a chart might look like Fig. 498 (see also Fig. 123). The dashed line at height $M$ is convenient in interpreting the data. In few practical cases do we know $M$ and $\sigma$, so we use point estimates, $\bar{x}$ and $s$, obtained from past data, making periodic revisions as more data accumulate.

While the preceding paragraph explains the idea, in practice control charts are seldom used for single observations. Instead, means of small samples are used. There are three reasons for this:

(1) Individual observations are more variable than means. Hence the control limits for observations have to be set quite far apart or the risk of Type I error will be large. But when the limits are far apart the risk of Type II error is large. That is, the power of single observations to discriminate between in-control and out-of-control processes is low; the process has to go far out of control before trouble is likely to be detected in a single observation.

(2) Individual items often are not normally distributed, so the risks may not be at all what they seem from the normal curve. The observations are ordinarily normal enough, however, so that means of very small samples—means of samples of four are widely used for control charts—are nearly enough normally distributed. (See Sec. 11.3.1, on the central limit theorem.) If samples of size $n$ are used, the control limits are at $M \pm 3\sigma_{\bar{x}} = M \pm 3\sigma/\sqrt{n}$.

(3) If observations are grouped on a "rational" basis—for example, items produced by one machine or from one batch of raw material, or records made by one clerk—the average of the variabilities within groups provides an appropriate measure of the variability to be expected between groups if the process is in control. The over-all variability, which would have to be used if the observations were not grouped, has the disadvantage of reflecting any out-of-control variability among groups, and thereby incorporating it into the allowance for in-control variability.

Sometimes $2\sigma_{\bar{x}}$ is used instead of $3\sigma_{\bar{x}}$ ("two-sigma limits" instead of "three-sigma limits," in the jargon of statistical quality control); this reduces the consumer's risk but raises the producer's risk from 0.003 to 0.046. An LCL does not always exist. In sampling for proportion defective, for example, the true proportion may be so low that zero defectives in the sample would not be unusual.

Fig. 500 exhibits the relationship between the control chart and the normal distribution. With fixed control limits, a change in $M$ will be equivalent to a vertical shift of the normal distribution; this will increase the probability of a point falling outside the limits. (Why?) An increase in $\sigma$ will also increase this probability, and a decrease in $\sigma$ will decrease it. Thus the control chart for averages tends to detect process changes that involve changes in the mean or increases in the variability, and—perhaps more important—tends to

Statistical Quality Control

detect occasional "bloopers"; but it will not detect decreases in variability.

Sometimes separate control charts are used for variability, the range, $R$, being the usual measure of variability. To distinguish them, control charts based on means are called $\bar{x}$ Control Charts or $\bar{x}$ Charts, and those based on the range are called $R$ Control Charts or $R$ Charts. The usual objections to the range as a measure of dispersion (see Sec. 8.2) are unimportant here. In particular, all samples are of the same size, so variations in the range correspond with

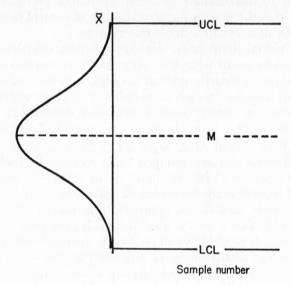

FIG. 500. Relation between control chart and normal distribution.

variations in the variability; and the range is subject to very little more sampling variability than the standard deviation for samples not larger than, say, eight. As a matter of fact, the mean range within groups is usually used for setting control limits on $\bar{x}$, rather than the mean standard deviation.

The control chart for means has some similarities in principle to the analysis of variance method of testing the null hypothesis that a group of population means are equal (see Sec. 13.2.3). In both methods, the variability *within* samples is used to deduce the variability *among* samples. Then, if the actual variability among samples agrees, except for an allowance for sampling discrepancy, with the amount consistent with the within-sample variability, the null hypothesis (that the process is in control, or that all samples are from populations having the same mean) is accepted; otherwise, the null hypothesis is rejected. The control chart differs from the analysis of variance procedure for testing a set of means in that (1) the variability within samples usually is measured by ranges instead of standard deviations; (2) the test criterion is an extreme value of any individual mean, rather than a swollen standard deviation among the set of means. This second point means that the control chart procedure is more likely than the standard analysis of variance to detect those departures from control (the null hypothesis) in which all population means are the same but one; but the control chart procedure is less likely to detect those departures in which there is a relatively small, but general, swelling of the variability of the means.

Sometimes it is suggested that control limits should be one-sided. If, for example, the higher $\bar{x}$ the better the quality, why regard a high $\bar{x}$ as evidence that the process is out of control? This misses the point that control limits describe the variability inherent in the process, as shown by past experience, not what is hoped or feared about the process. And as for out-of-control points on the too-good side, "thar's gold in them thar hills." Determining why such a high quality occurred may point the way to changes in the process that will make this high quality usual, just as determining why low quality occurred may point the way to changes in the process that make low quality rarer. Serendipity is often the source of much bigger gains than are envisaged when a statistical quality control program is initiated.

## 16.2.2 Illustrative Control Charts

EXAMPLE 501 TRAVEL EXPENSES

The data of Table 502A represent the expenses for 50 consecutive trips by sales engineers of a large organization during a given time period.[1] The figures show the total expense of each trip for meals, hotels, cabs, etc., but do not include transportation between cities. The arithmetic mean for these 50 expenses is $26.49, the standard deviation $10.10.

A first approach to a control chart for the expenses of individual trips is illustrated by Fig. 503. The upper control limit is set at

$$\$26.49 + (3 \times \$10.10) = \$26.49 + \$30.30 = \$56.79.$$

No lower limit is shown because it would fall below zero. Trips 1 and 33 were out of control, and should be investigated.

---

1. Although this example is based on a real case, these are not the real data.

TABLE 502A

EXPENSES FOR FIRST SET OF 50 TRIPS

| Trip Number | Expense (dollars) | Trip Number | Expense (dollars) | Trip Number | Expense (dollars) | Trip Number | Expense (dollars) |
|---|---|---|---|---|---|---|---|
| 1 | 64.92 | 13 | 29.15 | 26 | 21.11 | 39 | 28.30 |
| 2 | 30.70 | 14 | 19.58 | 27 | 27.82 | 40 | 28.92 |
| 3 | 27.45 | 15 | 31.96 | 28 | 17.85 | 41 | 22.26 |
| 4 | 33.44 | 16 | 26.41 | 29 | 15.91 | 42 | 20.20 |
| 5 | 16.38 | 17 | 27.09 | 30 | 27.21 | 43 | 12.45 |
| 6 | 35.05 | 18 | 25.22 | 31 | 32.49 | 44 | 28.96 |
| 7 | 19.26 | 19 | 33.68 | 32 | 13.09 | 45 | 25.96 |
| 8 | 21.74 | 20 | 23.16 | 33 | 61.50 | 46 | 23.68 |
| 9 | 8.72 | 21 | 27.30 | 34 | 24.53 | 47 | 17.57 |
| 10 | 22.24 | 22 | 18.17 | 35 | 16.37 | 48 | 29.96 |
| 11 | 42.83 | 23 | 28.81 | 36 | 28.91 | 49 | 27.21 |
| 12 | 30.77 | 24 | 31.94 | 37 | 35.26 | 50 | 21.64 |
| | | 25 | 18.82 | 38 | 20.64 | | |

Fifty more trips were then recorded as they occurred. The data are given in Table 502B and are plotted as Trip Numbers 51–100 in Fig. 503. Of these last 50, only one (trip 67) falls above the UCL. This one was investigated. It turned out to be an unusually prolonged trip, 14 days, and the expense is quite reasonable in view of that. We shall see in Chap. 17 that a more elaborate control chart could have allowed for the length of trip; it will show us that trips 1 and 33 are in control when duration is allowed for, but that trip 99, a short one, is out of control.

TABLE 502B

EXPENSES FOR SECOND SET OF 50 TRIPS

| Trip Number | Expense (dollars) | Trip Number | Expense (dollars) | Trip Number | Expense (dollars) | Trip Number | Expense (dollars) |
|---|---|---|---|---|---|---|---|
| 51 | 21.60 | 63 | 32.41 | 76 | 32.07 | 89 | 15.35 |
| 52 | 22.21 | 64 | 22.48 | 77 | 20.50 | 90 | 29.15 |
| 53 | 45.20 | 65 | 29.01 | 78 | 25.48 | 91 | 17.45 |
| 54 | 17.98 | 66 | 12.64 | 79 | 39.29 | 92 | 34.22 |
| 55 | 18.00 | 67 | 76.40 | 80 | 43.83 | 93 | 31.24 |
| 56 | 35.17 | 68 | 18.40 | 81 | 34.69 | 94 | 29.33 |
| 57 | 28.20 | 69 | 24.08 | 82 | 25.37 | 95 | 46.61 |
| 58 | 17.19 | 70 | 28.01 | 83 | 37.30 | 96 | 24.08 |
| 59 | 29.63 | 71 | 21.93 | 84 | 37.61 | 97 | 37.21 |
| 60 | 55.20 | 72 | 20.73 | 85 | 16.80 | 98 | 22.12 |
| 61 | 44.69 | 73 | 29.13 | 86 | 24.65 | 99 | 43.83 |
| 62 | 23.23 | 74 | 19.84 | 87 | 16.01 | 100 | 39.06 |
| | | 75 | 31.51 | 88 | 46.08 | | |

This preliminary approach to a control chart can be improved in two important respects. The use of means of samples of four, or some other small

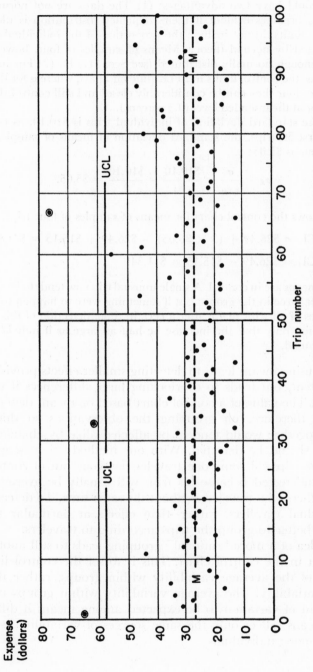

FIG. 503. Travel expense control chart, individual trips. Three-sigma control limits computed from first 50 trips.

number, would have two advantages: (1) The data are not normally distributed, as becomes evident if their frequency distribution is plotted, so three-sigma limits do not include the proportion of the individual observations expected by normal theory. Means of samples of four, however, will be nearly enough normally distributed (see Sec. 11.3.1). (2) For individual observations, the control limits must be quite far apart, whereas for the means of samples of four they can be considerably closer and still control the risk of Type I error at the intended level (0.3 percent).

Since the standard deviation of individual trips is $10.10, as computed from the first 50 trips, the standard deviation of means of samples of four will be taken as $5.05:

$$\sigma_{\bar{x}} = \frac{\sigma}{\sqrt{n}} = \frac{\$10.10}{\sqrt{4}} = \frac{\$10.10}{2} = \$5.05.$$

Fig. 505 shows the control chart for means of samples of four trips.

UCL = $26.49 + (3 \times $5.05) = $26.49 + $15.15 = $41.64.

LCL = $26.49 − $15.15 = $11.34.

All 25 means are in control. A single unusual trip now tends to be swamped by the other three in the group; but if something were to happen to increase the expenses of all trips made after a certain time, detection of this increase would require only that the increase be half as large as if individual trips were being used.

Thus, using means leads to detecting smaller effects provided they are persistent, but leads to overlooking large differences if they are transitory. The value of a control chart based on means depends considerably, therefore, on grouping the observations so that those grouped together would tend to be affected alike by changes of the kind that should be detected. With our method—that is, grouping consecutive trips—if some one traveler has gone out of control, this may be undetected because his trips will usually be averaged with those of others who traveled at about the same time. To detect effects on individual travelers, rather than effects at particular times, it would be better to group the trips according to travelers.

The idea of a more "rational" grouping leads to still another improvement in the control chart. This is to set the control limits on the basis of the average variability within groups, rather than the over-all variability. The average variability within groups indicates the amount of variation to be expected among means if differences among means reflect only the same sources of variability as do differences among individuals.

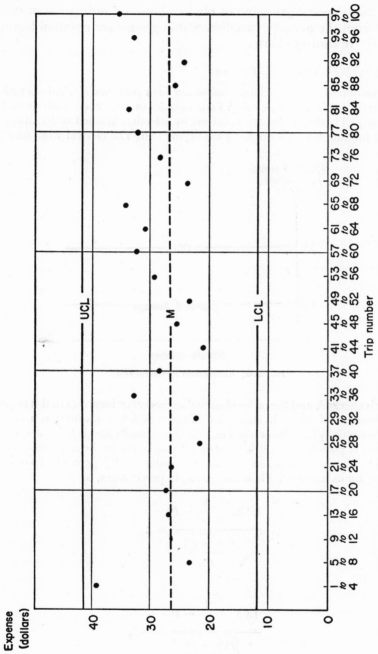

FIG. 505. Travel expense control chart, means of four trips. Three-sigma control limits computed from first 50 trips.

*Statistical Quality Control*

Thus, rational grouping of observations and estimation of variability from the average variation within groups are essential features of effective control charts.

EXAMPLE 506 CLERICAL ERRORS

One department of a large mail-order firm performs clerical operations on about 25,000 orders per day. Four samples of 100 orders each are taken each day. These samples are obtained by selecting finished orders more or less at random from an outgoing conveying belt. The clerical work on each

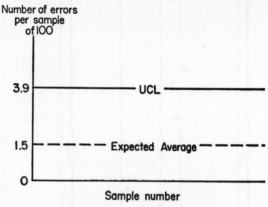

FIG. 506. Control chart for clerical errors.

order is checked, and the order classified as correct or incorrect in this respect. The number of errors in successive samples of 100 is plotted on a control chart essentially like the one in Fig. 506. The actual chart is large and colorful, and is posted where it is seen by all employees in the department concerned. The mean number of incorrect orders is 1.5 percent. Two-sigma control limits are used. These are given, in percentages, as

$$\frac{\text{UCL} - \dfrac{1}{2n} - P}{\sqrt{\dfrac{P(1-P)}{n}}} = 2$$

and

$$\frac{\text{LCL} + \dfrac{1}{2n} - P}{\sqrt{\dfrac{P(1-P)}{n}}} = -2,$$

or

$$\text{UCL} = P + \left(\frac{1}{2n} + 2\sqrt{\frac{P(1-P)}{n}}\right)$$

and

$$\text{LCL} = P - \left(\frac{1}{2n} + 2\sqrt{\frac{P(1-P)}{n}}\right).$$

Here

$$\text{UCL} = 0.015 + 0.005 + 0.024 = 0.044,$$

$$\text{LCL} = 0.015 - 0.029 < 0.$$

Thus, the upper control limit would be 4.4 errors per hundred; that is, four or fewer errors in a sample of 100 would be regarded as in control, five or more as out of control. No lower control limit exists.

Actually, $P = 0.015$ is too small for the normal distribution to approximate the binomial adequately, even for samples as large as 100. It is customary in statistical quality control work, however, to use the normal approximation under almost any circumstances, in the interest of ultra-simplicity. It is also customary to omit the continuity adjustment, and in this case the upper control limit actually used was 0.039—as it happens, a better figure in this instance because of the non-normality. In some ways, the most important thing in routine quality control is to establish and enforce a definite uniform control limit. If the limit is calculated a little inaccurately, this may be unimportant for practical purposes, especially since the appropriate risks usually cannot be decided with great precision.

A question that comes to mind about this application is this: Even though the clerical work is fairly standard, is it not necessary to allow for the fact that some errors are more important or more costly than others? It is true that the nature of the errors is important information; in fact, records are kept of all errors discovered, whether the sample is out of control or not. But the cost of a particular error is not important for the purposes the control chart serves. The control chart is not a remedial device, to catch and correct outgoing errors, but a measuring device to show how many errors are being made. The cost of perfect accuracy would be prohibitive, and an average of 1.5 errors in 100 orders is a satisfactory compromise between the costs of errors and the costs of accuracy. The control chart is helpful in maintaining this standard. Many of the errors do not affect the customer, and may not affect the company seriously. Should there be important differences in types of errors, separate control charts could be established for each type. If the cost of each error were clearly measurable, a control chart could be based on the average cost of errors per order, instead of on the percentage of orders involving errors.

EXAMPLE 508  FILLING CONTAINERS

A common application, when control charts are first introduced in a plant, involves the filling of containers. A large oil company, for example, in filling quart cans of motor oil, allows enough extra so that the actual amount poured into the crankcase will rarely be less than a full quart. The filling machine has some variation in the amount it puts in the cans, and there is some variation in the amount that clings to a can when it is emptied, so the average content of cans labeled "one quart" must exceed a quart. A study preparatory to establishing control limits found that the process was badly out of control. Tracing down the causes of this greatly reduced the variability in the filling of the cans. The average was then set so that the lower three-sigma control limit for the volume poured out would be one quart. This new average was below the old because of the reduced variability and the more exact allowance for variation, and annual savings in excess of $50,000 were realized. One reason this kind of application is often made initially is that the savings are clearly measurable, which is not always true of improvements in quality, customer good will, employee morale, and other gains.

## 16.2.3  Selection of Control Limits

It is important to distinguish between the unsystematic inspection and supervision which often goes under the name of "quality control," and statistical quality control. The former does not say when or how samples should be taken or how large they should be, ordinarily does not have the advantages that go with graphic presentation, and does not enforce a clear, objective standard for "take action" or "skip it." The statistical quality control chart makes use of well-thought-out, tested rules, and avoids the indecision, inconsistency, and arbitrariness of haphazard quality control. Statistical control is based on the fact that repeated random samples from a fixed population will vary, but in a predictable pattern.

Control limits at $M \pm 3\sigma$ are commonly used in American quality control work. The reason that $3\sigma$ is used, instead of 1 or 2 or some other multiple of $\sigma$, is that 3 is both a "conservative" figure and a round number. The statistician desires a "conservative" control limit that will result in few false alarms; generally he personally gets into more trouble by sounding false alarms than by occasionally overlooking real trouble. Of course, if he drew the control limits at $10\sigma$, he would never sound any false alarm; on the other hand, he would almost never detect trouble of even a gross sort. The determination of control limits should properly be based on the costs of the two kinds of errors in the specific problem. On the other hand, in mass production the advantages of simplicity for routine application by people with relatively little training—the advantages of standardization, in short—ordinarily are thought to outweigh the advantages of special study of each case.

There is another kind of reason for preferring conservative control limits. If the distribution is not normal, the departure from normality need not be very great to change the three false alarms in a thousand to thirty or even more false alarms in a thousand. For instance, consider the distribution of travel expenses, which is somewhat skewed to the right (that is, values above the mean are fewer but more extreme than those below the mean). In such a distribution more than 0.135 percent of the observations (the value for the normal distribution) exceed the upper $3\sigma$ control limit. On the travel expense control chart for individual trips, three (or more than 20 times the number predicted by the normal distribution) of the expenses exceed the upper control limit of $M + 3\sigma$. When means of groups are used, however, the non-normality danger is unimportant.

In this connection, it may be remarked that even if a distribution is not normal, not over $\frac{1}{9}$ (about 11 percent) of the observations in *any* population can be more than three standard deviations away from the mean. This is a special case of a general principle that not more than a fraction $1/K^2$ of the observations in a population can lie more than $K$ times the standard deviation away from the mean. This mathematical relation bears the impressive name, *Tchebycheff's inequality*, after the Russian mathematician Pafnuti Lvovitch Tchebycheff (1821–1894), although it was apparently discovered first by the French mathematician, J. Bienaymé (1796–1878). It follows that not over $\frac{1}{4}$ of the observations can lie more than $2\sigma$ from the mean, and so on. This rule, however, depends on $\sigma$, the true standard deviation of the population, not $s$, the standard deviation of a sample. The rule applies to the proportion of all the observations in the *population* falling outside the limits $M \pm K\sigma$, not to the proportion in any given sample.

# 16.3
# ACCEPTANCE INSPECTION

## 16.3.1 Principles of Acceptance Inspection

A typical application of acceptance inspection is to determine whether a batch of items, called an *inspection lot* or simply a *lot*, that has been delivered by a supplier is of acceptable quality. Another application is to a lot that is complete and ready for shipment to customers, to make sure that it is of adequate quality. Still another application is to a lot of partly completed material, to determine whether it is of high enough quality to justify further processing. A specific instance of the last is the sampling of invoices received from a supplier, to see if their accuracy justifies payment in advance of detailed auditing.

In acceptance inspection, decisions are made separately for each lot. Each lot is inspected and, on the basis solely of information taken from that lot, is either accepted or rejected. The decision is influenced by the past performance of the seller or supplier only insofar as that may have led to accepting large risks, or to insisting on small ones. Standards are set primarily according to the quality requirements in the use to which the product is to be put.

Process inspection is a different technique. In process inspection some point in the process is selected, and by sampling at that point, a judgment is reached as to whether the process is in control. Information is collected about the process, and the product is accepted or rejected on the basis of this information. If it were known that a lot had been produced by a process that was in control while the lot was being produced, and the quality level at which the process was in control were also known, there would be no point to acceptance inspection. If the quality level of the process were not high enough, it would be necessary to examine each item in the lot to find the acceptable ones, and if it were high enough it would be purely a matter of chance if too many defectives were found in a sample. As a matter of fact, there is a growing movement among large firms to insist that a certified control chart be submitted with each lot of material they purchase. Their own inspection can then be quite limited, serving merely to audit the supplier's control charts. Aside from the direct saving, an important indirect effect is that the supplier is thereby induced to learn about statistical quality control, and the result usually is to improve his quality or lower his costs. Some large firms have actually provided quality control training programs for their small suppliers, since the simplicity of statistical quality control procedures makes them useful even in quite small plants.

The terms "buyer" and "seller" are used in connection with acceptance sampling merely to distinguish between the maker and the user of the product. Often the buyer and seller are in the same plant or organization; no financial transaction is necessarily implied by the use of the terms. Sometimes the application is not to a business matter at all. Perhaps "supplier" and "receiver" would be more appropriate. The terms "accept" and "reject" simply describe the two alternative decisions. When a lot is rejected, it does not necessarily mean that it is destroyed or scrapped, or even that its purchase is refused. The lot may be examined completely and the defective items removed, it may be used for some alternative purpose, it may be purchased at a reduced price, etc.

### 16.3.2 Importance of the Operating-characteristic Curve

Suppose that a seller is supplying material that is acceptable if not more than two percent of the items are defective, according to a certain definition of "defective." He will want a sampling plan which will accept nearly all lots that have two percent defectives or fewer. He may not be much interested in the number of bad lots which, if produced, would pass the inspection plan. The buyer, however, will have a different viewpoint. He will be especially interested in a plan that will reject most of the bad lots that may be submitted; he may not be much interested in whether all good lots are passed or not. (Actually, of course, price and cost considerations give both buyer and seller an interest in both aspects of a sampling plan.)

With any given sampling plan, the acceptance or rejection of a lot is a matter of chance, since it depends on a random sample; the *probability* of acceptance, however, depends on the true quality of the lot. It is important to understand that the probability of acceptance is the probability that *if* a lot of a certain quality is offered it will be accepted. It is not the probability that if a lot is accepted it will be of a certain quality. The latter would be proportional to the product of two probabilities: (1) the probability that a lot of the stated quality will be offered, and (2) the probability that if a lot of the stated quality is offered it will be accepted. Only the second of these two probabilities can be controlled by an inspection plan. For example, if all lots submitted are of the same quality, that will be the quality of the lots accepted. This fact is expressed by a saying among acceptance inspection men that "quality cannot be inspected into the product, it must be built in." The fact that a particular sampling plan would accept 20 percent (say) of submitted lots having 50 percent defectives, does not mean that in practice some (much less 20 percent) of the accepted lots will be 50 percent defective. If no 50-percent-defective lots are offered, none can be accepted; and correspondingly, if nothing but 50-percent-defective lots are offered, any lots that are accepted will necessarily be 50 percent defective. (A review of Sec. 12.5 will help to clarify the points involved here. The material in the next few paragraphs is essentially the same, but in a different context.)

The relation between the probability of acceptance and the quality of the lot can be represented by an operating-characteristic curve, usually referred to as an OC curve. A typical OC curve resembles Fig. 497. The percent of lots accepted decreases as poorer and poorer quality is considered. If the lot has zero percent defectives, it is certain to be accepted and hence the OC curve always starts from 100. If the lot has 100 percent defectives, it is certain to be rejected and the OC curve must be at zero for that lot-quality. The OC curve is the essential basis for evaluating what a sampling plan will accomplish. That is, it is the appropriate criterion for judging the *statistical characteristics* of sampling plans, and for comparison with their costs.

An ideal sampling plan would accept all good lots and reject all bad lots, without error. To achieve this, however, it would be necessary to inspect all, or practically all, of the lot. Suppose that two percent defectives, or fewer, is acceptable quality and over two percent unacceptable, and suppose that a lot is of size 400,000. Then if a sample of 399,999 showed 8,000 defectives it would not be certain (and 100 percent probability means certainty) that the lot was not over two percent defective, for the final item might be defective and 8,001 is more than two percent of 400,000. Thus, unless some uncertainty is tolerated in the immediate vicinity of two percent defective, 100 percent inspection will be necessary.

Even with 100 percent inspection (or 200 percent—inspecting each item twice) there would not be certainty in practice of accepting all good lots and rejecting all bad lots. Inspection involves human and mechanical factors that are fallible. There is a tendency for inspectors to be less careful if a sample is large than if it is clear that much depends on how each item in a small sample is classified. Furthermore, after an inspector has looked at a few thousand items, they all begin to look alike to him. As a matter of fact, a thorough acceptance sampling program should include control charts on the inspection process, controlling the accuracy and output of the inspectors.

The steepness of the OC curve depends on the sample size. The larger the sample, the steeper the curve, and the smaller the zone between the qualities that are almost always accepted and the qualities that are almost always rejected. Hence, the degree of discrimination or sharpness of a test is largely dependent upon the size of the sample. Where one sample can differentiate between two percent and ten percent defectives for given risks, a larger sample may differentiate with the same risks between three percent and nine percent, or between two percent and seven percent, depending on the acceptance criterion.

The location of the OC curve is determined by the maximum number of defective items allowable for acceptance, called the *acceptance number*. If the acceptance number is made larger, the curve is shifted to the right. If the acceptance number is made smaller, the curve is shifted to the left. The quality discrimination is essentially as sharp, but the acceptance number determines which qualities are discriminated between—whether, say, between two percent and seven percent defectives or between 20 percent and 30 percent.

## 16.3.3 Illustrative Sampling Inspection Plans

EXAMPLE 513A  SINGLE SAMPLING

The following sampling plan is often used when acceptable quality is defined as 0.5 percent or fewer defective items in a lot: Draw a random sample of 75 items and classify each as defective or nondefective. If one or fewer are defective, accept the lot; if two or more are defective, reject the lot. Such a plan may be summarized as:

$$n = 75, \qquad A = 1, \qquad R = 2,$$

where $n$ is the sample size, $A$ the acceptance number, and $R$ the rejection number. (The context distinguishes $R$ for rejection number from $R$ for range.) The OC curve of this plan is shown in Fig. 515 and will be discussed in comparing the plan with the three following plans.

EXAMPLE 513B  DOUBLE SAMPLING

Another plan, which also is widely used when the acceptable quality level is 0.5 percent, is as follows: Draw a random sample of 50. Accept if none are defective, reject if three or more are defective. If one or two are defective draw a second sample, this time of size 100. Accept the lot if both samples together (150 items) contain two or fewer defectives, and reject it if they contain three or more defectives. These instructions may be summarized as follows:

$$n_1 = 50, \qquad A_1 = 0, \qquad R_1 = 3,$$
$$n_2 = 100, \qquad A_2 = 2, \qquad R_2 = 3.$$

The OC curve of this plan is included in Fig. 515.

EXAMPLE 513C  MULTIPLE SAMPLING

The following plan, also widely used when 0.5 percent defective represents acceptable quality, is an elaboration of the double sampling plan to seven possible samples:

**Statistical Quality Control**

| | | |
|---|---|---|
| $n_1 = 20$ | $A_1 = *$ | $R_1 = 2$ |
| $n_2 = 20$ | $A_2 = *$ | $R_2 = 2$ |
| $n_3 = 20$ | $A_3 = 0$ | $R_3 = 2$ |
| $n_4 = 20$ | $A_4 = 1$ | $R_4 = 3$ |
| $n_5 = 20$ | $A_5 = 1$ | $R_5 = 3$ |
| $n_6 = 20$ | $A_6 = 1$ | $R_6 = 3$ |
| $n_7 = 20$ | $A_7 = 2$ | $R_7 = 3$ |

* means that acceptance is not possible at this sample size.

Note that the acceptance and rejection numbers, $A$ and $R$, apply to the combined number of defectives found up to that point. The OC curve of this plan is shown in Fig. 515.

### EXAMPLE 514   SEQUENTIAL SAMPLING

This plan, again for cases where 0.5 percent defective represents acceptable quality, carries multiple sampling to the limit by using successive samples of one and comparing the accumulated number of defectives with acceptance and rejection numbers given by the following formulas:

$$A_n = -0.9585 + 0.0197n,$$

$$R_n = +1.2305 + 0.0197n.$$

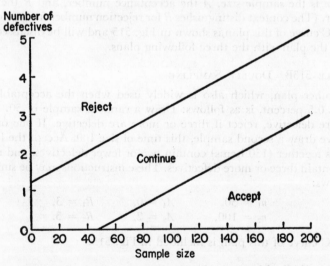

FIG. 514.  **Graphic representation of a sequential sampling plan.**
Slope: 0.0197.
Intercepts: −0.9585,
  +1.2305.

### 16.3 Acceptance Inspection

TABLE 515

THREE ACCEPTANCE SAMPLING PLANS FOR ACCEPTABLE
QUALITY LEVEL 0.5 PERCENT

| Type of Sampling | Sample | Sample Size | Combined Samples | | |
|---|---|---|---|---|---|
| | | | Size | Acceptance number | Rejection number |
| Single | First | 75 | 75 | 1 | 2 |
| Double | First | 50 | 50 | 0 | 3 |
| | Second | 100 | 150 | 2 | 3 |
| Multiple | First | 20 | 20 | * | 2 |
| | Second | 20 | 40 | * | 2 |
| | Third | 20 | 60 | 0 | 2 |
| | Fourth | 20 | 80 | 1 | 3 |
| | Fifth | 20 | 100 | 1 | 3 |
| | Sixth | 20 | 120 | 1 | 3 |
| | Seventh | 20 | 140 | 2 | 3 |

*Source:* Statistical Research Group, Columbia University, *Sampling Inspection* (New York: McGraw-Hill Book Company, 1948), p. 288.

\* Acceptance not permitted until three samples have been inspected.

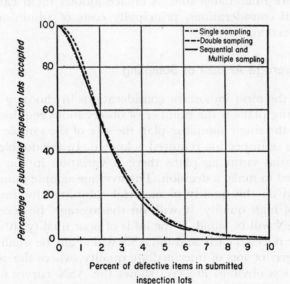

FIG. 515. Operating-characteristic curves for the three sampling plans of Table 515.
Source: Same as Table 515.

An acceptance number given by the formula for $A_n$ is rounded down to the next lower integer; for example, the acceptance number given by the formula for $n = 75$ is 0.52, but to require 0.52 or fewer defective items is equivalent to requiring zero. Similarly, rejection numbers given by the formula for $R_n$ are rounded up to the next integer.

Such a sequential plan is often presented graphically, as in Fig. 514. After each observation, a point is plotted above the value of $n$ representing the accumulated sample size, at a height representing the accumulated number of defects. If the point falls above the upper line, rejection is indicated. If it falls below the lower line, acceptance is indicated. If it falls between the two lines, another observation is to be taken.

The plans of Examples 513A, 513B, and 513C are summarized in Table 515.

The OC curves of the four sampling plans—single, double, multiple, and sequential—are practically alike. They are shown in Fig. 515. (The OC curve for the sequential sampling plan is so close to that for the multiple sampling plan that it is not shown separately.) The probability of acceptance is 95 percent when the quality is 0.4 to 0.6 percent defective, 50 percent when the quality is 2.2 to 2.3 percent defective, and 10 percent when the quality is 4.8 to 5.2 percent defective. Thus, as far as the decisions based on them are concerned, the plans are interchangeable. A choice among them can be based on practical considerations, principally costs of administration and actual inspection.

## 16.3.4 Average Amount of Sampling

One of the most important considerations in choosing among the four sampling plans is the number of observations required for a decision. For the single sampling plan the size of the sample is fixed, so the amount of inspection required is known. In the double, multiple, and sequential sampling plans there is variation in the number of items needed to make a decision. The average sample number (ASN) will depend on the quality of material submitted for inspection. If the lot is of high quality, it will, on the average, be accepted early and the ASN will be small. If the lot is of poor quality, it will, on the average, be rejected early and the ASN will again be small. The ASN will be larger for lots of intermediate quality, where the appropriate decision is less obvious. Fig. 517 shows the ASN curves for the four plans.[2] The pattern here is fairly typical: sequential sampling requires

2. The values of $\bar{n}$ were computed on the assumption that the first samples in double and multiple sampling will be inspected completely, that inspection of later samples

the smallest average sample number, multiple sampling next, and single or double (usually single) most. Sequential sampling commonly requires a third to a half fewer observations, on the average, than single sampling.

Sometimes it is important to minimize the maximum sample size that might be required to reach a decision. Single sampling does this best, double sampling next, multiple next, and sequential least.

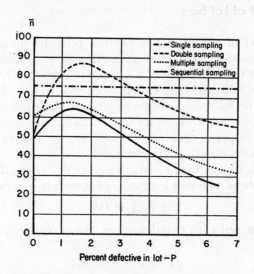

FIG. 517. Average sample number, n̄, for four sampling plans having the equivalent operating-characteristic curves of Fig. 515.

*Source: Statistical Research Group, Columbia University, Techniques of Statistical Analysis* (New York: McGraw-Hill Book Company, 1947), p. 240.

Sometimes it is desirable to minimize the variability in the number of observations required. Single sampling does this most effectively, but no consistent order prevails among the other types.

Sequential methods were developed in 1943 by the American mathematical statistician Abraham Wald (1902–1950). Wald was a refugee from Austria who, when he came to Columbia University in 1938, had had no exposure to modern statistical theory. Between 1940 and his death in 1950 in an airplane crash in India, however, he virtually revolutionized the mathematical theory of statistics—decision theory, for example, was originated by him. Sequential analysis was

---

will be curtailed as soon as the rejection number is reached, but that curtailing will not be applied to accepted lots.

developed for use in scientific research and development, but it spread so widely and so fast in acceptance inspection—it was in use in over 6,000 plants within two years of Wald's initial work—that many people now think of it exclusively in that connection. Actually, almost any problem in significance testing is amenable to sequential analysis, as are many problems in estimation.[3]

## 16.3.5 Effect of Lot Size

The subject of acceptance inspection provides an opportunity to bring out again that the size of a lot (that is, population) ordinarily does not affect the reliability of a sample of a given size. The path to this conclusion followed here differs from that in Sec. 11.4.3.

Consider the following simple single-sampling plan: Take a sample of two; reject the lot if either item is defective, accept if both are nondefective. Suppose this plan is applied to lots of various sizes which all contain exactly 20 percent defectives.

If the lot is of size five, of which one is defective, the probability that the sample of two will lead to acceptance is (see Sec. 10.7):

$$\tfrac{4}{5} \times \tfrac{3}{4} = 0.6.$$

For a lot of size ten with two defectives, the probability of acceptance is

$$\tfrac{8}{10} \times \tfrac{7}{9} = 0.6222;$$

for lot size 20 with four defectives,

$$\tfrac{16}{20} \times \tfrac{15}{19} = 0.6316;$$

for lot size 100 with 20 defectives,

$$\tfrac{80}{100} \times \tfrac{79}{99} = 0.6384;$$

and for lot size infinity with 20 percent defectives,

$$0.8 \times 0.8 = 0.6400.$$

These and similar results are summarized in Table 519A.

Table 519A shows that the sample of two gives almost exactly the same protection for all lot sizes, certainly for all lot sizes that are at

---

3. A general treatment of sequential testing is beyond the scope of this book. For an elementary but thorough account, see Statistical Research Group, Columbia University, *Sequential Analysis of Statistical Data: Applications* (New York: Columbia University Press, 1945), in which problems like the acceptance sampling problem are treated in Sec. 2. One frequent misapprehension should be dispelled: there is no danger that sampling will fail to terminate, even if the population proportion equals the slope of the two parallel lines of Fig. 514.

*16.3 Acceptance Inspection*

TABLE 519A

PROBABILITY THAT TWO ITEMS WILL BOTH BE NONDEFECTIVE,
IF SAMPLING IS FROM A LOT THAT CONTAINS 20 PERCENT DEFECTIVES,
FOR VARIOUS LOT SIZES

| Lot Size | Probability of Acceptance |
|---|---|
| 5 | 0.6000 |
| 10 | 0.6222 |
| 20 | 0.6316 |
| 50 | 0.6367 |
| 100 | 0.6384 |
| 1,000 | 0.6398 |
| Infinity | 0.6400 |

least ten times the sample size. Thus, as was shown also in Sec. 11.4.3, *the reliability of a sample depends on the number of observations in the sample, almost regardless of what proportion of the population these constitute.* The reason is that the only effect of population size comes through the change in the percentage of good items remaining in the population as the sample is withdrawn. When the sample contains only a small fraction of the lot, this change is negligible. For practical purposes, therefore, the reliability of samples is computed on the assumption that the population is infinitely large; if the population is really only five times the sample size or smaller, the sample will be a little more reliable than is indicated on this basis.

In contrast, consider the following set of single-sampling plans, one for each lot size: Take a sample of ten percent of the lot; accept the lot if ten percent or less of the items in the sample are defective; reject if more than ten percent are defective. The probabilities of acceptance, still assuming the lots to be 20 percent defective, are a little complicated to compute, but for lot sizes 100, 200, and 300 they are shown in Table 519B. Thus, the protection given by this

TABLE 519B

PROBABILITY THAT A TEN PERCENT SAMPLE WILL HAVE TEN PERCENT OR
FEWER DEFECTIVES, IF SAMPLING IS FROM A LOT THAT CONTAINS
20 PERCENT DEFECTIVES, FOR VARIOUS LOT SIZES

| Lot Size $N$ | Sample Size $n$ | Acceptance Number $A$ | Probability of Acceptance |
|---|---|---|---|
| 100 | 10 | 1 | 0.36 |
| 200 | 20 | 2 | 0.19 |
| 300 | 30 | 3 | 0.11 |

plan varies considerably with the lot size, in contrast with the plan involving a constant sample size, which gives practically the same protection for any lot size.

As a general rule, if a sample is not large enough to judge a lot of two million items, it is not large enough to judge two thousand items. If it is large enough to judge two thousand items, it is large enough to judge two million. In other words, population or lot size is irrelevant to the adequacy of a sample. For practical purposes, this holds unless the sample contains more than 20 percent of the population. That is, while lot size theoretically makes a difference in all cases, the difference is not large enough to be of practical importance unless the sample contains about 20 percent or more of the population. This principle has wide implications. For example, in a sample of the people in the United States, if a given degree of accuracy for the whole country requires a certain sample size, the same accuracy for five regions requires five samples of this size.

It must never be forgotten that many things other than size influence the reliability of a sample, especially the method of sampling.

There *is* a sound reason for varying sample size with lot size in acceptance inspection. Although the discrimination attained by a sample of a given size is essentially the same regardless of lot size, the cost of inspection, per unit produced, depends on the lot size. If, for example, it costs 10 cents to inspect an item, and 100 items are inspected from each lot, the inspection cost is $10 per lot. If the lots are of size 1,000, this is 1 cent per unit produced; if the lots are of size 5,000, the cost is 0.2 cent per unit produced. If the discrimination achieved by a sample of 100 is just worth the cost of 1 cent when the lot size is 1,000, it will be more than worth its cost of 0.2 cent when the lot size is 5,000; so for lots of 5,000 it will pay to buy more discrimination by increasing the sample size to some point (less than 500) where there is both greater discrimination and lower cost. Similarly, for lots of 500, the discrimination achieved would be worth less than its cost of 2 cents, so it will pay to reduce the sample size to some point (greater than 50) where there is both less discrimination and higher cost. These economic considerations, of balancing constant protection from samples of a given size against decreasing cost as lot size increases, mean that the larger the lot, the larger, other things the same, should be the sample.

## 16.4
## CONCLUSION

Statistical quality control is interesting because it represents two groups of simple, widely used techniques that illustrate many of the fundamental ideas of statistical decision-making.

The first of these, process control, is essentially a method of judging whether the future performance of a process can be expected to be like the past, and the criterion is whether the latest results are within the normal range of variation in the past. The second, acceptance inspection, is a method of basing a decision about an already existing group of things on measurements of a sample of them.

There are usually alternative sampling procedures that are statistically equivalent, that is, have essentially the same operating characteristics, among which a choice can be based on cost. Single, double, multiple, and sequential sampling—all of which can be used for tests of significance, whatever the field of application—illustrate one group of alternative sampling plans. With types of sampling that allow for withholding a decision and collecting more data, the sample size depends on chance, but its probability distribution depends on the characteristics of the population being sampled. Sequential sampling typically permits the smallest samples on the average; single sampling permits the smallest maximum samples.

We have touched on only a few highlights of statistical quality control, those that are of most general statistical interest, or that best illuminate basic statistical principles. Those interested in the subject for its own sake should refer to one of the textbooks in the field— several of which are larger than this book.[4]

## DO IT YOURSELF

Example 521A

For the first 50 samples of Fig. 109, calculate upper and lower three-sigma control limits for $p$. Which samples, if any, are out of control?

Example 521B

Group the data of Table 206 into consecutive subsamples of 5 and calculate upper and lower three-sigma control limits for $\bar{x}$, using the average of the within-group standard deviations. Which samples, if any, are out of control?

Example 521C

(1) Make a histogram combining the 100 observations of Tables 502A and 502B to demonstrate the lack of normality.

(2) If you can use a table of logarithms, replace each observation in Tables 502A and 502B by its logarithm, and make a histogram showing the

---

4. At least as good an entry point as any to the literature of statistical quality control is Eugene L. Grant, *Statistical Quality Control* (revised ed.; New York: McGraw-Hill Book Company, 1952). This includes many references to other works in the field.

distribution of the 100 logarithms. Compare this with the histogram of the actual observations, and comment on the differences.

### EXAMPLE 522A

A large lot has ten percent defective items. An experiment simulating sampling from such a lot can be devised by letting successive digits from Table 632 represent items inspected. For example, zero can represent a defective item, the other digits a nondefective one.

(1) Draw ten samples of 75 items each, and use the single-sampling plan of Table 515 to determine acceptance or rejection. How many of these samples would lead you to reject the lot?

(2) Draw ten double samples (do not use any of the digits in the single samples of (1)), and use the double-sampling plan of Table 515 to determine acceptance or rejection.[5] How often do you reject? What is the average sample size for your ten samples?

(3) Draw ten multiple samples (do not duplicate digits used in (1) and (2)), using the plan of Table 515. How often do you reject? What is your average sample size?

(4) Draw ten sequential samples (do not duplicate digits used earlier), using the plan of Fig. 514. How often do you reject? What is your average sample size?

(5) How do the average sample sizes observed in (1), (2), (3), and (4) compare with the theoretical results of Fig. 517?

### EXAMPLE 522B

Same as Example 522A, but let $P = 0.01$.

### EXAMPLE 522C

Same as Example 522A, but let $P = 0.02$.

### EXAMPLE 522D

Consider the following single-sampling plan: $n = 5$, $A = 1$, $R = 2$.

(1) Compute the height of its OC curve at $P = 0$, .1, .2, .3, .4, .5, .6, .7, .8, .9, and 1.0. [Hint: See Sec. 11.2.1.]

(2) Using Table 632 to simulate random sampling (see, for example, Example 522A), draw 20 independent samples of five for each of these values of $P$: .1, .2, .3, .4, .5, .6, .7, .8, and .9, and apply the plan. Record the number of times the lot is accepted at each value of $P$.

(3) Plot your results in (1) and (2) on the same graph, connecting the points of (1) by a smooth curve. Do the results appear to be consistent?

---

5. If you exhaust the supply of random digits going through them in a certain order, you can use a new order. For example, if first you read across the rows, now read down the columns. You can also read along diagonals, or every tenth digit, etc. These devices are not as satisfactory as a fresh supply, of course, for while they give different orders, any chance deviations in the over-all frequencies will be the same for all orders.

*Do It Yourself*

EXAMPLE 523

Set up and operate a control chart for some repetitive activity in which you are engaged or which you observe regularly. For example, you might use the number of people late or absent from a regular meeting (perhaps a class), the cost of your lunches, the time to travel a certain route, the number of runs scored by your favorite baseball team (perhaps using games in each series against the same opponent as a basis for grouping and averaging).

# Chapter 17

# *Relationships between Variables*

## 17.1
### INTRODUCTORY SURVEY

### 17.1.1 Introduction

So far in our study of statistical inference, we have been concerned mainly with methods of analyzing data consisting of one observation (one measurement or one count) on each item in the sample.

Problems dealing with the relationship between two variables are usually subsumed under the general heading *correlation*. Technically, "correlation" is concerned only with expressing the degree of a certain special type of relationship between two variables. In practical problems, it is often more important to find out what the relation actually is, in order to estimate or predict one variable (the *dependent* variable) from knowledge of another variable (the *independent* variable); and the statistical technique appropriate to such a case is called *regression analysis*, or often *least squares*.

### 17.1.2 A Homely Example

Suppose you want to determine the electricity consumption per minute of an electric clothes dryer in your home. One way to go about it would be to keep track of the minutes of operation of the dryer and of the kilowatt hours on the electric meter. Assume, for a moment, that all other electric consumption in the house is absolutely the same every day. Then at the end of each day you would record the minutes of operation of the dryer and the kilowatt hours of elec-

**524**

## 17.1 Introductory Survey

tricity consumed in the house that day. After a number of weeks you could plot a chart like Fig. 525, with a point for each day, the *abscissa* (horizontal distance) representing the minutes of operation and the *ordinate* (vertical distance) representing the increment in kilowatt hours. If the time-rate of electricity consumption of the dryer were

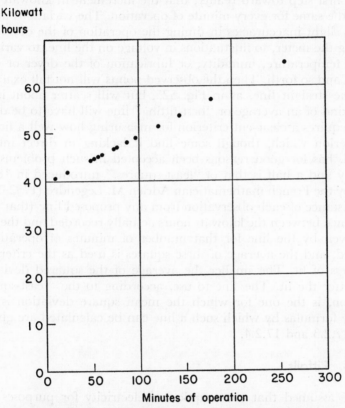

FIG. 525. Minutes of operation of an electric clothes dryer and increment in electric meter reading. (Hypothetical.)

always the same, the points for the various days would all lie along a straight line. The *slope* of this line (kilowatt hours rise per minute of operating time) would represent the rate of electricity consumption for the dryer. The *intercept* (height of the line at zero minutes dryer operating time) would represent the fixed consumption of electricity in the household for other purposes.

This example, though greatly oversimplified, brings out the principle of *correlation* analysis, namely to compare the values of the dependent variable for different values of the independent variable.

### 17.1.3 Least Squares

No skill in analysis or interpretation would be required in an example as oversimplified as that just stated. Suppose, however, to take a first step toward reality, that the increment in kilowatt hours is not the same for every minute of operation. The variations may be due to slight inaccuracies in timing the operation of the dryer or in reading the meter, to fluctuations in voltage on the line, to variations in the temperature, humidity, or lubrication of the dryer or of the meter, and so forth. Then the observed points will not fall exactly on any one straight line, as in Fig. 525, but will scatter about it, and some kind of an average or "best-fitting" line will have to be drawn. This requires a clear-cut criterion for measuring how well a line fits. A criterion which, though some find it lacking in direct intuitive appeal, has for good reasons been accepted in such problems for a century and a half is that of "least squares," introduced in 1805 or 1806 by the French mathematician Adrien M. Legendre (1752–1833). The distance of each observation from any proposed line (that is, the difference between the kilowatt hours actually recorded and the number given by the line for that number of minutes of operation) is squared, and the average of these squares is used as the criterion of goodness of fit. The smaller the average of the squared deviations, the better the fit. The line to use, according to the "least-squares" criterion, is the one for which the mean square deviation is least. Simple formulas by which such a line can be calculated are given in Secs. 17.2.3 and 17.2.4.

### 17.1.4 Pitfalls

We assumed that consumption of electricity for purposes other than the dryer remained always the same. Actually, of course, the other consumption will vary. This variation will simply cause more scatter about the line than there would be if only such factors as we mentioned in the preceding section were operating, and the line will still represent the average consumption for various periods of dryer operation. If we jump to cause-and-effect conclusions, however, we may find ourselves in the deep end of a pitfall.

Suppose, for example, that the habits of the household are such that on days when the dryer operates, a washing machine operates for about the same number of minutes and an electric ironer for about twice as long. Then the electricity consumption on days when the dryer has operated an hour will, on the average, exceed the consumption on days when the dryer has not operated at all; but the excess will represent not only the hourly consumption of the dryer but also the hourly consumption of the washer plus twice the hourly consumption of the ironer. The fact that in the household studied electricity consumption tends to go up this much for each hour's use of the dryer is, of course, true. To say that the dryer *causes* this much electricity consumption would, however, be very misleading to one considering, say, whether to replace an electric by a gas dryer.

Similar factors might cause an underestimate of the consumption caused by the dryer. Suppose that the washer and dryer are operated on alternate days, and consume about the same amount of electricity. Then electricity consumption will be about the same, on the average, on those days when the dryer is operated as on those days when it is not, leading to the correct conclusion that operation of the dryer is not associated with increased consumption of electricity in the house, but possibly to the incorrect interpretation that the dryer consumes no electricity.

Indeed, operation of the dryer might even be associated with *reduced* consumption of electricity. To see this possibility, suppose that on days when the housewife operates the dryer, she does not operate (or at least reduces her operation of) the washer, ironer, electric oven, electric hot water heater, television set, radio, vacuum cleaner, electric radiator, reading lamps, etc.

## 17.1.5   Multiple Correlation

These difficulties can be attacked by allowing for the effects on consumption of all the major electrical equipment. The idea of *multiple correlation* can be seen by considering two pieces of electrical equipment, say a dryer and a washer. Record for each day the number of minutes of operation of each, and the kilowatt hours consumed. Plot the number of minutes of dryer operation as abscissa and the number of minutes of washer operation as ordinate, and through this point imagine erecting a line perpendicular to the paper, its height representing the kilowatt hours of electricity consumed. Such a perpendicular line is erected for each day for which we have data. Under assumptions of constancy in consumption like those of Sec. 17.1.2, the tops of these lines will all lie in a plane. The rise in this plane for a minute's operation of the dryer will represent the consumption of the dryer, and the rise for a minute's operation of the washer will represent the consumption of the washer.

*Relationships between Variables*

When variation is present, the tops of the perpendiculars will not all fall in one plane, and again the least-squares criterion is used to fit a plane: that plane is used which has the smallest possible average for the squared differences between kilowatt hours actually recorded and kilowatt hours given by the plane for the observed washer and dryer operating times.

Any number of "causal" factors can be handled by multiple correlation, as long as there are more observations than factors. While the procedure cannot be described graphically for more than two causal factors, or independent variables, the formulas and ideas are essentially the same.

### 17.1.6 Pitfalls in Multiple Correlation

No matter how many factors are introduced into a multiple correlation analysis, there is danger that factors not introduced explicitly may be correlated with those introduced explicitly. Thus, if we take account of the washer, dryer, and ironer, it may be that from a causal viewpoint the ironer's effect is overstated because the television set is always operated during ironing, or that the effects of all three are understated because the cooking stove and space heaters are used less when the laundry equipment is used than on other days.

Example 528   Empty Freight Cars

A study of railway costs found, by a multiple correlation analysis of costs of seventy-six large railroads in 1948, that operating costs per ton-mile of freight hauled, tend to go down as the ratio of empty to loaded freight-car miles goes up. This result is patently absurd, so the author investigated it and reported that it occurred because mineral-hauling railroads have high ratios of empty to loaded freight-car miles and also achieve operating economies through long trains. Since mineral hauling had not been taken into account as an explicit causal factor, it showed up implicitly in the ratio of empty to loaded car miles, a variable correlated with mineral hauling. Had the effect not been so strong as to produce a patently absurd result, it might have led simply to an understatement of the effect of empty-car miles on costs, without the relation between empty-car miles and mineral hauling being noted.[1]

### 17.1.7 Curvilinear Correlation

So far we have talked as if all relations are linear, that is, as if the effect of an increment in one of the independent variables is the same

---

1. George H. Borts, *Cost and Production Relations in the Railway Industry*, unpublished doctoral dissertation, University of Chicago, 1953.

whether the absolute level of the variable is high or low. We assumed, for example, that the electricity consumed by a dryer is the same during the first, seventeenth, sixtieth, or any other minute. Actually, a dryer may use more current the first minute, to start the motor and bring the coils up to temperature, or while the clothes are very wet, etc. If there is wide variation of the actual observations about the fitted line it may be difficult to detect such nonlinearity, yet the location of the fitted line may depend appreciably upon an essentially arbitrary assumption about the nature of the curve. A guiding principle in scientific investigations is to use the simplest shape that is consistent with the available information ("Occam's razor"). When it is appropriate, curvilinear relations can also be fitted by least squares.

EXAMPLE 529   DEMAND FOR STEEL

In an attempt to measure the relation of the quantity of steel sold to steel prices, and more specifically to measure the elasticity of demand, defined as the ratio of a small percentage change in quantity to the corresponding small percentage change in price, five different relationships were fitted to the data. These involved different variables and different shapes, but all seemed reasonable. The resulting elasticities ranged from +0.52 (indicating that a given percentage increase in price will be accompanied by a percentage *increase* in quantity sold that is half as great) to −0.88 (indicating that a given percentage increase in price will be accompanied by a percentage *decrease* in quantity sold that is seven-eighths as great). Thus, the conclusions depended in large part on the variables chosen and the type of relation among them that seemed reasonable a priori.[2]

## 17.1.8   Correlation and Causation

Despite its pitfalls, correlation analysis can be one of the most useful devices of statistics. The thing it accomplishes is to show whether variables, as a matter of actual experience, have varied together, and if so what the relationship has been. It is thus particularly valuable in the early, exploratory stages of an investigation, in revealing relationships to be investigated, or indicating the magnitude of a relationship. On the other hand, a relationship that cannot be explained in terms of causal relations from the relevant field of knowledge, but rests solely on empirical association, leaves much unanswered. As an eminent medical statistician has said,

---

2. Theodore O. Yntema and others, *United States Steel Corporation Studies, Prices and Costs,* in Hearings before the Temporary National Economic Committee, Seventy-sixth Congress of the United States, *Investigation of Concentration of Economic Power, Part 26: Iron and Steel Industry* (Washington: Government Printing Office, 1940).

. . . if an essential biologic association is to be established as a definitive scientific conclusion, that is to say, if it is to be considered *"proved,"* the population must not be anything else except an *experimental population.* An association found in a purely statistical investigation made on an existent population, by which I mean an investigation which is retrospective as regards either of the variables concerned, however strongly it may suggest association as a *presumptive* conclusion, is tentative until it is corroborated fully by means of experiment. I am not here referring to "association" in a purely statistical, descriptive sense. If proper study of a given population shows that there is positive correlation between stature and weight, then it is a descriptive fact that tall individuals in that population are on the average heavier than short individuals. But there is no concluding even here that there is a necessary biologic relation between stature and weight; we do not know for instance that the correlation would exist if the population were placed on a different diet.[3]

The discussion of Sec. 9.6 is pertinent here. Errors in correlation analysis are, for the most part, misinterpretations of correlations that are real enough. So-called "spurious correlation" usually turns out on investigation to be spurious interpretation of a valid correlation. Thus, in Example 528, the positive correlation was real enough; costs really are lower for those railroads operating many empty-car miles. What would have been spurious, as the author himself warned, would have been a conclusion that hauling cars empty causes costs to go down. Similarly, in Examples 78D to 79D (relating to feet and handwriting, storks' nests, propaganda leaflets, business school alumni, and the Kenny treatment), the associations were real enough; the dangers lay in misinterpretation.

The term "spurious correlation" should be reserved for correlations that are illusory rather than real, for example those that have no counterpart in the real phenomena they purport to describe, but are introduced by arithmetic operations or by methods of selecting data (as in Sec. 17.4.5). A spurious interpretation may, however, be given to a perfectly valid correlation.

## 17.2
## FITTING A REGRESSION LINE

### 17.2.1 Regression as a Problem of Estimation

Each member of a class of 70 men was instructed to write down two numbers: (1) his height to the nearest inch, and (2) his weight to the nearest pound. It was then explained that the data would be

---

3. Joseph Berkson, "The Statistical Study of Association between Smoking and Lung Cancer," *Proceedings of the Staff Meetings of the Mayo Clinic,* Vol. 30 (1955), p. 323.

collected only from a sample of ten, since the data for all members of the class would require too much arithmetic for classroom computations. The sample of ten was selected by the use of a table of random numbers, and the results summarized as in Table 531.

TABLE 531

HEIGHTS AND WEIGHTS FOR SAMPLE OF TEN MEN

| Man | Height (inches) | Weight (pounds) |
|---|---|---|
| 1 | 74 | 195 |
| 2 | 70 | 191 |
| 3 | 72 | 225 |
| 4 | 72 | 205 |
| 5 | 67 | 180 |
| 6 | 73 | 184 |
| 7 | 69 | 182 |
| 8 | 69 | 152 |
| 9 | 72 | 193 |
| 10 | 68 | 175 |

Before calculating relationships from data like these, it is important to plot the data and see the picture as a whole. The two-dimensional chart used for this purpose is called a "scatter diagram" (see

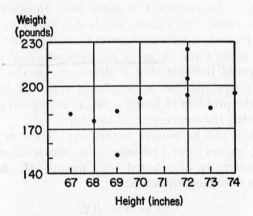

FIG. 531. Scatter diagram for heights and weights of ten men.

Example 201). As a general rule, the horizontal axis represents the independent variable, the vertical axis the dependent variable. In this case, we assume that weight is the dependent variable, that is, the variable to be predicted. Fig. 531 is a scatter diagram of the data.

*Relationships between Variables*

The scatter diagram suggests that, on the average, weight increases with height, though at any given height there is a good deal of variability in weight.

It is common to denote the independent variable by $X$, the dependent variable by $Y$. In regression analysis, an algebraic relationship between $Y$ and $X$ is sought, telling what $Y$ will be, on the average, for any specific value of $X$. For linear regression, this can be represented geometrically by a straight line, as in Fig. 532.

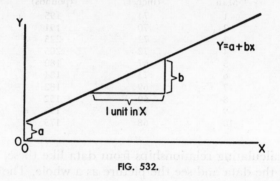

**FIG. 532.**

At any value of $X$, the height of the line tells the average value of $Y$. When $X$ is 0, the average $Y$ is equal to $a$. In other words, $a$ tells where the line crosses the $Y$ axis, and is therefore called the *intercept*. The other coefficient, $b$, tells how much $Y$ changes with a unit change in $X$, and is termed the slope or the *regression coefficient*. If $b$ were negative, it would indicate that $Y$ decreases, on the average, as $X$ increases. Any straight line is completely specified by its intercept and slope, so the problem of fitting a linear regression line reduces to that of estimating the intercept and the slope.

The problem can be looked at this way: There is a sample of $n$ pairs of $(X, Y)$ values from a population in which values of $Y$ are independently normally distributed with means, $M$, that depend on $X$ through the following equation:

$$M = A + BX.$$

The standard deviation of $Y$ for a given $X$, denoted by $\sigma_{Y \cdot X}$, is assumed to be the same for all values of $X$, but is usually unknown. In effect, then, we have a series of separate populations for $Y$, one population for each value of $X$. These populations are assumed to be normally distributed. For any particular value of $X$ the mean of the population of $Y$'s is $A + BX$. The problem is to make estimates, $a$

and $b$, of the parameters $A$ and $B$ of the population, on the basis of a sample.

The principle of maximum likelihood (Sec. 14.2.1.2) leads in this case to selecting $a$ and $b$ by what is called the method of least squares. This method, as has been said, selects $a$ and $b$ so that the mean of the squared vertical deviations between the data and the line will be as small as possible. A "vertical deviation" is the vertical distance from

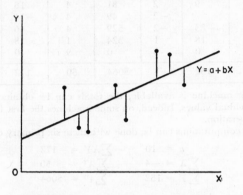

FIG. 533. **Deviations from a regression line.**

an observed point to the line, as depicted in Fig. 533 by the vertical segments. For each individual in the sample, its deviation is squared and the line is so placed as to make the sum of these squared deviations a minimum. This could be done by trial and error, but that is not necessary; the least squares line can be found from convenient formulas for $a$ and $b$.

## 17.2.2  Six Primary Computations from the Data

The labor of computation is often greatly diminished by "coding" the data, much as in the shortcut calculation of means and standard deviations in Chaps. 7 and 8. Each weight was, therefore, reduced by 175 and each height by 71; but no divisor (change of the unit) was used. Then

$$Y = \text{weight} - 175;$$

$$X = \text{height} - 71.$$

Eventually we will need to know $\Sigma Y$, $\Sigma X$, $\Sigma Y^2$, $\Sigma X^2$, and $\Sigma XY$. Table 534 shows the computation of these quantities. The last two columns, $(Y + X)$ and $(Y + X)^2$, are used in checks.

TABLE 534

DATA NEEDED IN REGRESSION COMPUTATIONS

| Weight | Height | $Y$ | $X$ | $Y^2$ | $X^2$ | $XY$ | $(Y+X)$ | $(Y+X)^2$ |
|--------|--------|-----|-----|-------|-------|------|---------|-----------|
| 195 | 74 | 20 | 3 | 400 | 9 | 60 | 23 | 529 |
| 191 | 70 | 16 | −1 | 256 | 1 | −16 | 15 | 225 |
| 225 | 72 | 50 | 1 | 2500 | 1 | 50 | 51 | 2601 |
| 205 | 72 | 30 | 1 | 900 | 1 | 30 | 31 | 961 |
| 180 | 67 | 5 | −4 | 25 | 16 | −20 | 1 | 1 |
| 184 | 73 | 9 | 2 | 81 | 4 | 18 | 11 | 121 |
| 182 | 69 | 7 | −2 | 49 | 4 | −14 | 5 | 25 |
| 152 | 69 | −23 | −2 | 529 | 4 | 46 | −25 | 625 |
| 193 | 72 | 18 | 1 | 324 | 1 | 18 | 19 | 361 |
| 175 | 68 | 0 | −3 | 0 | 9 | 0 | −3 | 9 |
| Total | | 132 | −4 | 5064 | 50 | 172 | 128 | 5458 |

If a calculating machine is available, the totals can be obtained directly without writing down individual values. Indeed, on some machines the first five sums can all be obtained in one operation.

All subsequent computations can be done with these six primary quantities:

$$n = 10 \qquad \sum XY = 172$$
$$\sum X = -4 \qquad \sum X^2 = 50$$
$$\sum Y = 132 \qquad \sum Y^2 = 5064$$

As a check, we note that

$$128 = 132 - 4,$$
$$5458 = 5064 + 50 + (2 \times 172).$$

## 17.2.3 The Slope

The formula for the maximum likelihood or least-squares estimator of the slope is

$$b = \frac{\sum (X - \bar{X})(Y - \bar{Y})}{\sum (X - \bar{X})^2}.$$

This may be written in the following form, more convenient for computations:

$$b = \frac{\sum XY - \dfrac{(\sum X)(\sum Y)}{n}}{\sum X^2 - \dfrac{(\sum X)^2}{n}}.$$

Substituting the data into the second form gives

$$b = \frac{172 - \dfrac{(132)(-4)}{10}}{50 - \dfrac{(-4)^2}{10}} = \frac{172 + 52.8}{50 - 1.6} = \frac{224.8}{48.4}$$

$$= 4.645 \text{ pounds per inch.}$$

Thus, on the average, an inch increase in height is accompanied by 4.64 pounds increase in weight.

### 17.2.4 The Intercept

Once the estimated slope, $b$, is known, the estimated intercept, $a$, is easy to compute:

$$a = \bar{Y} - b\bar{X} = \frac{\sum Y}{n} - b\frac{\sum X}{n}.$$

For our data,

$$a = \frac{132}{10} - 4.645\frac{(-4)}{10} = 13.2 - (4.645)(-0.4)$$

$$= 15.058 \text{ pounds.}$$

The height of the line, then, is 15.06 at $X = 0$. Another way to put it, is that the height is 13.2 at $X = -0.4$; that is, at $\bar{X}$, the height of the line is $\bar{Y}$. Least-squares straight lines always pass through the point with coordinates $\bar{X}$ and $\bar{Y}$.

### 17.2.5 Equation of the Line

We now have the estimated regression line

$$Y = 15.058 + 4.645X.$$

To put this in terms of weight and height, remember that

$$Y = \text{weight} - 175; \qquad X = \text{height} - 71.$$

Substituting these for $Y$ and $X$ in the regression line, we have

$$\text{Weight} - 175 = 15.058 + 4.645 \text{ (height} - 71)$$
$$\text{Weight} = 175 + 15.058 + 4.645 \text{ height} - 329.795$$
$$\text{Weight} = -139.737 + 4.645 \text{ height.}$$

Since "weight" here represents the calculated or estimated average value for a given value of height, not a weight actually observed, it will be referred to as "calculated weight." The final regression line is, then,

Calculated weight (lbs.) $= -139.737 + 4.645$ height (in.).

### 17.2.6 Check on Accuracy

The algebraic sum of the deviations about a regression line (as about an arithmetic mean) should total zero. In Table 536 we find the predicted value, "calculated weight," for each value of height, subtract it from the actual weight, and add the differences, $d$. The differences add up to zero in this case, though in other cases rounding errors might result in a slight departure from zero. This does not

TABLE 536

ACTUAL AND PREDICTED VALUES OF WEIGHT

| Actual Height | Actual Weight | Calculated Weight | Deviation $d =$ Actual − Calculated |
|:---:|:---:|:---:|:---:|
| 74 | 195 | 204.0 | −9.0 |
| 70 | 191 | 185.4 | 5.6 |
| 72 | 225 | 194.7 | 30.3 |
| 72 | 205 | 194.7 | 10.3 |
| 67 | 180 | 171.5 | 8.5 |
| 73 | 184 | 199.3 | −15.3 |
| 69 | 182 | 180.8 | 1.2 |
| 69 | 152 | 180.8 | −28.8 |
| 72 | 193 | 194.7 | −1.7 |
| 68 | 175 | 176.1 | −1.1 |
| Total | 1882 | 1882.0 | 0 |

necessarily mean that all the calculations are right, but if this check had failed, they would definitely be wrong. Actually, one of the best checks, which should always be made, is to plot the data and the line, and study the fit. Table 536 shows the individual deviations for later reference; otherwise it would have been enough simply to notice that the sum of the calculated weights equals the sum of the observed weights.

## 17.3
## SAMPLING VARIABILITY

### 17.3.1 Standard Error of Estimate

The standard deviation of the values of $Y$ for a given $X$ is called the *standard error of estimate* of $Y$ from $X$, and is, as was said, denoted by $\sigma_{Y \cdot X}$. Its estimate from a sample, $s_{Y \cdot X}$, is the standard deviation of the differences, $d$, between actual and estimated weights. In this estimate, $n - 2$ is used in the denominator, instead of $n - 1$, because the deviations whose squares we are averaging are based on two quantities ($a$ and $b$) computed from the sample, not just one ($\bar{x}$) as in the simple standard deviation. The standard error of estimate

*17.3 Sampling Variability*

of weight from height, as estimated from a sample of $n$, is

$$s_{\text{weight}\cdot\text{height}} = \sqrt{\frac{\sum d^2}{n-2}}.$$

Since our coding involved only subtracting constants from the observations, this is exactly the same as

$$s_{Y \cdot X} = \sqrt{\frac{\sum(Y - Y_c)^2}{n-2}},$$

where $Y_c$ stands for the value computed, for a given $X$, from $15.058 + 4.645X$, the regression equation before decoding. Had our coding also involved dividing by some constant, $C$, $s_{Y \cdot X}$ would have had to be multiplied by $C$ to give $s_{\text{weight}\cdot\text{height}}$. For computational purposes, it is easier to use

$$s_{Y \cdot X} = \sqrt{\frac{\sum(Y - \bar{Y})^2 - b\sum(X - \bar{X})(Y - \bar{Y})}{n-2}}.$$

From the computation of $b$ we already know (see Sec. 17.2.3) that

$$\sum(X - \bar{X})(Y - \bar{Y}) = 224.8,$$

since this is the numerator used in calculating $b$. Also

$$\sum(Y - \bar{Y})^2 = \sum Y^2 - \frac{(\sum Y)^2}{n} = 5064 - \frac{(132)^2}{10} = 3321.6.$$

So, substituting for $b$, $\sum(Y - \bar{Y})^2$, $\sum(X - \bar{X})(Y - \bar{Y})$, and $n$ in the formula for $s_{Y \cdot X}$, gives

$$s_{Y \cdot X} = \sqrt{\frac{3321.6 - (4.645)(224.8)}{10 - 2}} = \sqrt{284.68} = 16.87,$$

which is also the standard error of estimate of weight from height.

This value of $s_{Y \cdot X}$ is, of course, a sample estimate of $\sigma_{Y \cdot X}$, the true population standard error of estimate. If $\sigma_{Y \cdot X}$ were known to be 20, say, this would be interpreted as meaning that about 95 percent of the weights are within 40 pounds of the (true) line. This may seem a wide margin of uncertainty. It means that weight cannot be predicted accurately from height alone, because of the inherent variability of weight for a given height. Larger samples will show the location of the line more reliably, but probably will not reduce $s_{Y \cdot X}$ much. If the line is used for predicting weights from heights, there are two sources of error: (a) a sample does not show exactly where the true (population) line is; and (b) the inherent variability causes a large margin of uncertainty in predicting individual weights—in other words, people of the same height vary quite a lot in weight, and this puts an inherent limitation on the accuracy of predicting weight from height. Taking larger samples will reduce errors due to (a), but not those due to (b).

It is interesting to compare $s_{Y \cdot X}$ with $s_Y$, the standard deviation of weight when height is disregarded. For this sample of ten,

$$s_Y = \sqrt{\frac{\sum(Y - \bar{Y})^2}{n - 1}} = \sqrt{\frac{3321.6}{9}} = \sqrt{369.07} = 19.21.$$

The standard deviation is, therefore, reduced from 19.2 for men of all heights to 16.9 for men of the same height. This is not a large reduction, and suggests that height does not help a great deal—though, it does apparently help a little—in estimating weight. It suggests, in other words, that the relation between individual weights and heights is not very close, even though average weights may vary some with height. We will pursue this point in Sec. 17.4.4.

### 17.3.2 Confidence Interval for the Slope

Repeated random samples from the same population would give varying values of $b$, and the pattern of variability would be a normal distribution with mean equal to $B$, the population slope. The standard deviation of the sampling distribution of $b$, called the standard error of $b$, is given by:

$$s_b = \frac{s_{Y \cdot X}}{\sqrt{\sum(X - \bar{X})^2}}.$$

In the present example,

$$s_b = \frac{16.87}{\sqrt{48.4}} = 2.42.$$

A 95 percent confidence interval for $b$ is[4]

$$b \pm 2.30 s_b = 4.645 \pm 5.566 = -0.92 \text{ to } 10.21.$$

---

4. The value 2.30 is used instead of 1.96 because for samples this small (which, however, are not usual in practical work) it is more accurate to use Student's $t$-distribution than the normal distribution. Student's distribution allows for the fact that $s$ may differ appreciably from $\sigma$ in small samples. See Sec. 14.7.1 for an explanation of Student's $t$, and for the formula leading approximately to the value 2.30. Using the more elaborate formula, we find, since in this case

$$f = n - 2 = 8,$$

$$t = 1.9600 \left(1 + \frac{3.8416 + 1}{4 \times 8} + \frac{6.8416 \times 20.2080}{96 \times 64}\right) = 1.9600(1 + 0.1513 + 0.0225)$$

$$= 2.301$$

*17.3 Sampling Variability*

It is clear that the estimate 4.645 of the slope is not a very precise one; in fact, this sample does not establish, at the 95 percent confidence level, that the true slope is positive. With data as variable as the weight data, and with no greater spread of the heights, a sample of ten is too small to give a precise estimate of the slope with much reliability.

As the last statement implies, the standard error of the slope of a regression line is smaller (1) the smaller the variability of the dependent variable for fixed values of the independent variable—that is, the smaller $\sigma_{Y \cdot X}$; (2) the larger the sample size (for this increases the number of terms $(X - \bar{X})^2$, all positive, to be summed for the denominator), and (3) the larger the dispersion of the independent variable (for this increases the size of the terms $(X - \bar{X})^2$ to be summed for the denominator).

## 17.3.3 Confidence Band for the Line

In repeated samples from the same population the values of $a$ and $b$ would, as has been pointed out, be subject to sampling variation. This means that the estimate of $Y$, $Y_c$, for any value $X$ would be subject to sampling variation. The values of $Y_c$ would be normally distributed around the population mean value of $Y$ for that $X$, with a standard deviation, as usual called a standard error, of

$$s_{Y_c} = s_{Y \cdot X} \sqrt{\frac{1}{n} + \frac{(X - \bar{X})^2}{\sum (X - \bar{X})^2}} .$$

The value of $X$ to be used in $(X - \bar{X})^2$ is the particular value at which the standard error of $Y_c$ is wanted. The other quantities in the formula, $n$, $\bar{X}$, and $\sum (X - \bar{X})^2$, pertain to the whole sample, so are the same in all computations of $s_Y$ from a given sample. Computations of $s_{Y_c}$ for various values of $X$ are shown in Table 540. By choosing values of $X$ that are symmetrical with respect to $\bar{X}$, the computations needed for the lower half of the table duplicate those of the upper half. Values of $Y_c$ and of $2.30 s_{Y_c}$ for use in the 95 percent confidence intervals, $Y_c \pm 2.30 s_{Y_c}$, are also shown. The fitted line, the 95 percent confidence interval for the line (in terms of $W_c = Y_c + 175$), and the original data are plotted in Fig. 540. This confidence interval is for the true line, not for the observations; these are scattered about the true line. The confidence interval for the line is

---

This refinement is not necessary if the sample size is, say, 20 or more (that is, $f = 18$ or more); in that case, $K$ may be used as shown in Table 391.

*Relationships between Variables*

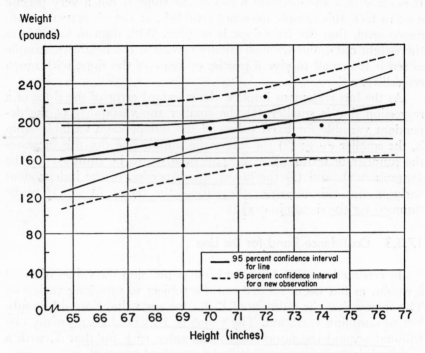

FIG. 540. Regression of weight on height, ten men; 95 percent confidence interval for the line; and 95 percent confidence interval for a new observation.

$$W_c = -139.737 + 4.645H. \quad s_{W \cdot H} = s_{Y \cdot X} = 16.87.$$

TABLE 540

COMPUTATION OF STANDARD ERROR OF ESTIMATE OF LINE, AND 95 PERCENT CONFIDENCE INTERVAL FOR THE LINE

| $H$ | $X$ | $X-\bar{X}$ | $(X-\bar{X})^2$ | $\dfrac{1}{n}+\dfrac{(X-\bar{X})^2}{\Sigma(X-\bar{X})^2}$ | $s_{Y_c}$ | $2.30s_{Y_c}$ | $Y_c$ | $W_c$ | $W_c$ $+2.30s_{Y_c}$ | $W_c$ $-2.30s_{Y_c}$ |
|---|---|---|---|---|---|---|---|---|---|---|
| 64.6 | −6.4 | −6 | 36 | 0.8438 | 15.50 | 35.65 | −14.67 | 160.33 | 195.98 | 124.68 |
| 66.6 | −4.4 | −4 | 16 | 0.4306 | 11.07 | 25.46 | −5.38 | 169.62 | 195.08 | 144.16 |
| 68.6 | −2.4 | −2 | 4 | 0.1826 | 7.21 | 16.58 | +3.91 | 178.91 | 195.49 | 162.33 |
| 70.6 | −0.4 | 0 | 0 | 0.1000 | 5.33 | 12.26 | +13.20 | 188.20 | 200.46 | 175.94 |
| 72.6 | +1.6 | +2 | 4 | 0.1826 | 7.21 | 16.58 | +22.49 | 197.49 | 214.07 | 180.91 |
| 74.6 | +3.6 | +4 | 16 | 0.4306 | 11.07 | 25.46 | +31.78 | 206.78 | 232.24 | 181.32 |
| 76.6 | +5.6 | +6 | 36 | 0.8438 | 15.50 | 35.65 | +41.07 | 216.07 | 251.72 | 180.42 |

$H$ represents height, that is, $X + 71$. $\bar{X} = -0.4$.

$W_c$ represents calculated weight, that is, $Y_c + 175$.

$s_{Y_c}$ is computed by multiplying the square root of the number in the preceding column by $s_{Y \cdot X} = 16.87$.

broad enough to include a line with a small negative slope. The reason the interval widens as heights are further and further removed from the mean, where it is narrowest, is that errors in the slope are magnified when projected a considerable distance. The larger the sample, the narrower, on the average, will be the confidence interval for the line at a given $X$.

## 17.3.4 Confidence Interval for a New Observation

The standard error of the difference between a new observation, $Y$, and the computed value, $Y_c$, for the corresponding value of $X$ is (see Sec. 13.2.2.1)

$$s_{Y-Y_c} = \sqrt{s_Y^2 + s_{Y_c}^2}$$

or

$$s_{Y-Y_c} = s_{Y \cdot X} \sqrt{\frac{n+1}{n} + \frac{(X - \bar{X})^2}{\sum(X - \bar{X})^2}}.$$

These values can be computed from Table 540 by simply adding one to each value in the fifth column, then carrying through the rest of the computations. The details of such a calculation are shown later (Sec. 17.5), but the results in this case are shown in Fig. 540 by the dashed lines outside the confidence interval for the line itself. These dashed lines could be used as control limits for individual observations.

The process to which the 95 percent confidence coefficient applies is the whole process of drawing a sample of $n$, computing the limits, and drawing another observation. Ninety-five percent of the times that this process is repeated, on the average, the new observation will lie within the limits. If the same limits are used for a group of observations, the random factors affecting the limits will tend to affect all the observations alike, so the proportion of them included in the long run will be either more or less than 95 percent. Limits, called tolerance limits, which may be expected, with a given confidence coefficient, to include a specified proportion of future observations can be computed, but are somewhat more complicated.

Increasing the sample size will tend to narrow the confidence interval for a new observation only insofar as it reduces the confidence interval for the line. Even for an infinitely large sample, the width would not be zero; instead it would be $2 \times 1.96\sigma_{Y \cdot X}$. Thus, a lower limit on the width is set by the fact that individuals vary in weight for a given height, and even complete knowledge of the population distribution of weight at a given height will not make possible exact predictions of individual weights from height.

## 17.4
## SOME SPECIAL TOPICS

### 17.4.1 Interchanging the Dependent and Independent Variables

With the sample showing weight and height, the line for estimating weight from height was

$$W_c = -139.737 + 4.645H.$$

This is not the appropriate equation for estimating height from weight. The predicted height should be obtained from a different line, one that minimizes the sum of squared *horizontal* deviations in Fig. 531. To look at it another way, if we had planned to estimate height from weight, we would have plotted height on the vertical axis, weight on the horizontal axis, and then proceeded with the same general method that we actually used. This would have resulted in the appropriate line for estimating mean height from weight, and would differ from what we get by solving the equation above for height in terms of weight.

### 17.4.2 Several Independent Variables

The regression method we have discussed can be extended in many ways that go beyond the scope of this book. More than two variables can be included; for instance, other factors besides height, for example age, might be useful in predicting weight.

Consider the case in which $Y$ is to be estimated from two independent variables, say $X_1$ and $X_2$. Predicting weight from height and age would be an example. The fitted regression plane may be written

$$Y_c = a + b_1X_1 + b_2X_2.$$

To show the formulas for the estimates $b_1$ and $b_2$, it is convenient to express each variable as a deviation from its sample mean. Let

$$y = Y - \bar{Y}, \qquad x_1 = X_1 - \bar{X}_1, \qquad x_2 = X_2 - \bar{X}_2.$$

Then

$$b_1 = \frac{\sum x_2{}^2 \sum x_1 y - \sum x_1 x_2 \sum x_2 y}{\sum x_1{}^2 \sum x_2{}^2 - (\sum x_1 x_2)^2}$$

**17.4 Special Topics**

and

$$b_2 = \frac{\sum x_1^2 \sum x_2 y - \sum x_1 x_2 \sum x_1 y}{\sum x_1^2 \sum x_2^2 - (\sum x_1 x_2)^2}.$$

The formula for $a$ is

$$a = \bar{Y} - b_1 \bar{X}_1 - b_2 \bar{X}_2 = \frac{\sum Y - b_1 \sum X_1 - b_2 \sum X_2}{n}.$$

The computations are not tedious. From the original or coded data the following primary quantities are computed: $n$, $\sum Y$, $\sum X_1$, $\sum X_2$, $\sum YX_1$, $\sum YX_2$, $\sum X_1 X_2$, $\sum Y^2$, $\sum X_1^2$, $\sum X_2^2$, much as in the simple computation in Table 534. All of the quantities used in the formulas for $b_1$ and $b_2$ can be computed according to the following examples:

$$\sum x_2^2 = \sum X_2^2 - \frac{(\sum X_2)^2}{n},$$

$$\sum x_1 y = \sum X_1 Y - \frac{(\sum X_1)(\sum Y)}{n}.$$

### 17.4.3  Curvilinear Regression

It may be that weight is related nonlinearly to height, and that better results could be obtained using a curved line such as

$$W_c = a + b_1 H + b_2 H^2,$$

where $W_c$ = calculated weight, $H$ = height. The computations would be carried through exactly as before, letting $X_1 = H - 71$ and $X_2 = (H - 71)^2$ for each observation. Indeed, $(H - 71)^2$ is merely another example of an additional variable, like age.

The simple linear regression is, however, widely useful. A straight line will fit any curve fairly well over a limited interval. This brings up an important point: Be very wary of extrapolating a regression line much beyond the range of the data on which it is based. The true relationship may be curved, even though a straight line fits well over the observed range (see Fig. 547). Furthermore, even if the true line is straight, the value of $b$ found from a sample is subject to sampling error, and a small error in $b$ makes a big difference in the estimated values far from the sample mean of the independent variable.

There are statistical tests for checking on the adequacy of a straight line. They are somewhat technical but afford good insurance against using a straight line when unwarranted. Another good protection is the common-sense rule mentioned twice already: Always plot and inspect the line and the data carefully.

## 17.4.4 Correlation Coefficients

The correlation coefficient, denoted by $r$, may be regarded as the ratio of two standard deviations. The numerator, $s_{Y_c}$, is the standard deviation of the *calculated* values, $Y_c$, corresponding with the values of $X$ in the sample. The numerator, therefore, is what the standard deviation of $Y$ would be *if* all the observations were exactly on the line. In Table 536, this is the standard deviation of the values in the third column, which is 10.77 ($n - 1$ is used in computing this numerator standard deviation, to agree with the denominator). The denominator is simply $s_Y$, the standard deviation of the *observed* values of $Y$ in the sample. In Table 536, this is the standard deviation of the values in the second column, which was calculated in Sec. 17.3.1 to be 19.21. Thus, the correlation coefficient for the height-weight data is

$$r = \frac{s_{Y_c}}{s_Y} = \frac{10.77}{19.21} = 0.56.$$

In other words, the variation (standard deviation) of the weights calculated from the heights is 56 percent as large as the variation (standard deviation) of the weights actually observed. To put it differently, the standard deviation would have been only 56 percent as large as it is if the observations had all been on the fitted line at their respective values of $X$. The correlation coefficient is considered positive or negative according to whether $b$, the slope of the regression line, is positive or negative.

Ordinarily, $r$ is computed from the direct formula

$$r = \frac{\sum (X - \bar{X})(Y - \bar{Y})}{\sqrt{\sum (X - \bar{X})^2 \sum (Y - \bar{Y})^2}} = \frac{\sum xy}{\sqrt{\sum x^2 \sum y^2}} = \frac{224.8}{\sqrt{48.4 \times 3321.6}}$$
$$= \frac{224.8}{401.0} = 0.56.$$

Note that all the quantities needed for this computation are provided by the regression analysis.

The correlation coefficient cannot be less than $-1$ or greater than $+1$. If it is near $+1$ or $-1$, the variability of the calculated values is nearly as great as that of the original data; in other words, one variable is capable of "explaining"[5] nearly all of the variation in

---

5. So that you will not be tempted to read "causing" into the word "explaining," we put quotation marks around it. "Correlation does not necessarily mean causation," as the saying goes.

the other variable. When $r$ is near zero, neither variable "explains" much of the variation in the other. The correlation coefficient, incidentally, is the same whichever variable is used to predict the other; this is clear from the direct formula.

The deviation of any observation $Y$ from $\bar{Y}$ could be written as the sum of two deviations, the deviation of $Y$ from the regression value $Y_c$ for its value of $X$, and the deviation of $Y_c$ from $\bar{Y}$:

$$Y - \bar{Y} = (Y - Y_c) + (Y_c - \bar{Y}).$$

It is also true—it will test your algebra to prove it—that

$$\Sigma(Y - \bar{Y})^2 = \Sigma(Y - Y_c)^2 + \Sigma(Y_c - \bar{Y})^2.$$

For the height-weight data, this last equation becomes

$$3321.6 = 2277.4 + 1044.2.$$

The left side represents the total variation in the data as measured by the sum of the squared deviations between the observations and their mean. The first term on the right measures the variation not "explained" by the regression line, as measured by the sum of the squared deviations of the observations from the regression line. The second term on the right measures the variation "explained" by the regression line as measured by the sum of the squared deviations of the regression values from the mean. In these terms, the correlation coefficient can be thought of as the square root (with appropriate sign) of the fraction of the total variation that is "explained" by the regression line,

$$r = \sqrt{\frac{1044.2}{3321.6}} = \sqrt{0.3144} = 0.56.$$

Correlation coefficients are often used when standard errors of estimate would be more appropriate. The standard error of estimate is a measure of the average discrepancy between the observations and the fitted line. It measures the goodness of fit, in units in which the dependent variable is measured. The correlation coefficient, however, is a relative measure. It depends not only on how well the line fits the observations, but also on how much dispersion there is in the observations. For example, a study of the relation of weight to height in which heights were all within a narrow range, say 69 to 71 inches, would show a low correlation coefficient, while a study which made a point of including many men under 65 inches and over 75 inches[6] would show a high correlation coefficient. The low correlation coefficient in the sample restricted to a narrow range of heights reflects the fact that knowing the exact height within that range scarcely improves an estimate of weight over an estimate based on the average weight of all men in the range. On the other hand, the margin of error with which weight can be estimated from height is shown by the standard error of estimate, and this will be about the same whether the data cover a narrow or a wide range of heights.

### 17.4.5  Hazards of Ratios

One can imagine hearing this argument: It stands to reason that a person with zero height must have zero weight; therefore the regression line for weight and height should go through the origin—have an intercept of zero. This argument is completely irrelevant, however, because the data relate only to a range of values for $X$ and $Y$ that is far removed from the origin. A regression line based on data for a certain group of adult males cannot be applied to boys or infants, much less to embryos or dimensionless creatures from Mars. The line will be useful only for the population from which the sample is drawn and within the range covered by the sample. Forcing the line through the origin may ruin the fit within the range of the data.

Suppose, for illustration, that the over-all relation between weight and height is something like that shown in Fig. 547. The data, however, are confined to the area of the small rectangle—about 62 to 77 inches. Even though the total relation is nonlinear, the fit of a straight line within the rectangle is almost perfect. If, however, a straight line were forced to pass through the origin, it would take the path of the dashed line in the rectangle—a very poor fit.

People often fall into this trap in a slightly more subtle way. In studying two related variables, say $X$ and $Y$, they attempt to allow for the effect of $X$ by forming the ratio $Y/X$. From the height-weight data, for example, they might calculate the ratio of weight to height for each man, average the ratios, and use this average ratio to esti-

---

6. Random sampling for regression analysis requires only that random samples be obtained of the values of the dependent variable for those values of the independent variable included in the study. The values of the independent variable can be selected arbitrarily. A good way to select them is so as to make $\sum (X - \bar{X})^2$ large, thereby reducing the standard error of the slope. The largest value of $\sum (X - \bar{X})^2$ is attained by taking half the observations at the lowest possible value and half at the highest possible value of the independent variable. It is wise to use several, well-spread-out values of the independent variable, however, in order to be able to check on the linearity of the relation.

**17.4 Special Topics**

mate weight from height. In general, this method is inferior to the regression line method; it is roughly equivalent to forcing a regression line to pass through the origin. It will, therefore, tend to over-

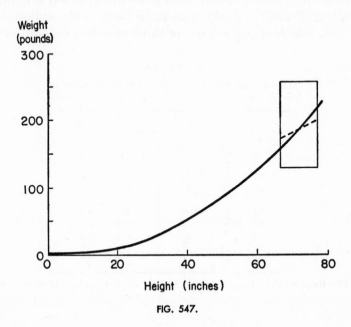

FIG. 547.

estimate the weight of short men and underestimate the weight of tall men, or vice versa. The only time this method should be considered is when the population regression line is straight and passes through the origin.[7]

Another way to get into trouble with ratios, and see correlations that are of no real interest, occurs when more than two variables are involved. Call the variables $X$, $Y$, and $Z$. The relation of $Y$ to $X$ is wanted, but $Z$ is thought to complicate the relationship. Suppose, for example, that the relation of weight to height is wanted. Age may affect this relationship. So the ratio of weight to age and of height to age might be computed, and the relation of the ratios taken as showing whether people who are tall for their ages are also heavy for their ages.

---

7. If the line is to be forced through the origin, the maximum likelihood or least-squares estimator of the slope is

$$b' = \frac{\sum XY}{\sum X^2}.$$

*Relationships between Variables*

The difficulty is that relating $Y/Z$ to $X/Z$ is, in part, relating the reciprocal of $Z$ to itself. Suppose that all values of $X$, $Y$, and $Z$ are positive, and that $X$ and $Y$ are unrelated. Then whenever $Z$ happens to be large, $X/Z$ and $Y/Z$ will tend to be small. When $Z$ happens to be small, $X/Z$ and $Y/Z$ will tend to be large. This is illustrated in Table 548, which shows ten sets of three two-digit random numbers

TABLE 548

TEN SETS OF THREE TWO-DIGIT RANDOM NUMBERS
AND THE RATIOS OF THE FIRST TWO TO THE THIRD

| $X$ | $Y$ | $Z$ | $X/Z$ | $Y/Z$ |
|-----|-----|-----|-------|-------|
| 77 | 82 | 99 | 0.8 | 0.8 |
| 96 | 67 | 9 | 10.7 | 7.4 |
| 22 | 87 | 67 | 0.3 | 1.3 |
| 34 | 24 | 84 | 0.4 | 0.3 |
| 45 | 56 | 92 | 0.5 | 0.6 |
| 92 | 33 | 100 | 0.9 | 0.3 |
| 57 | 30 | 5 | 11.4 | 6.0 |
| 2 | 2 | 26 | 0.1 | 0.1 |
| 82 | 65 | 21 | 3.9 | 3.1 |
| 100 | 85 | 11 | 9.1 | 7.7 |

*Source:* The Rand Corporation, *A Million Random Digits* (Glencoe, Illinois: Free Press, 1955).

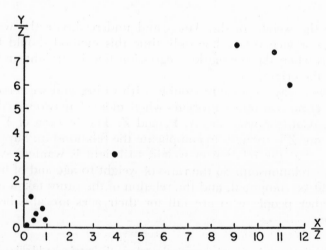

FIG. 548.  Scatter diagram of $\frac{X}{Z}$ and $\frac{Y}{Z}$, where $X$, $Y$, and $Z$ are two-digit random numbers. Sample size ten.
Source: Table 548.

(reading 00 as 100), and the ratios of the first two to the third. Fig. 548 is a scatter diagram of the ratios, and shows clearly the tendency of the two ratios to be correlated even though $X$ and $Y$ are not. (You should plot $X$ and $Y$ yourself to see that they look unrelated. What little relationship appears is due entirely to the eighth observation (2,2).)

If $X$ and $Y$ actually are related, the real relationship will be distorted by the spurious relation resulting from introducing $Z$ into both sides of the relation. A proper way to study a relation involving three or more variables is that indicated in Secs. 17.1.5 and 17.4.2, multiple regression.

## 17.5
## AN ILLUSTRATION

EXAMPLE 549   EXPENSE AND DURATION OF TRIPS

The analysis of the travel expenses in Sec. 16.2.2 disregarded the durations of the trips. There is nothing actually wrong with neglecting duration of trip, but it is an inefficient way to carry out the analysis because chance variation enters in two ways, in duration of trip, and in expense for a given duration. The first source of variation can be eliminated by allowing for duration of trip. The allowance can be made by regression techniques.

The data needed are in Table 550. In Fig. 551, individual expenses are plotted against trip duration measured in quarters of a day. As anyone would guess, travel expense tends to be greater, the greater the duration of a trip.

The center line has been fitted to the data by the method of least squares. The line obtained is

$$Y_e = 11.97 + 4.59X,$$

where $Y$ denotes the vertical height of the line corresponding to the horizontal distance $X$; that is,

Estimated expense $= 11.97$ plus $4.59$ per day.

The computations are shown in Table 552.

The dashed lines drawn above and below the fitted line are analogous to the three-sigma control limits of Fig. 503, though here they are based on all 100 observations instead of just the first 50. The difference is that these limits allow for the duration of a trip. They are 99.7 percent confidence intervals for a new observation (see Sec. 17.3.4). Computation of the limits is shown in Table 553. The curvature is much less noticeable here than in Fig. 540 because of the smaller standard error of estimate and larger sample size.

TABLE 550

Travel Expense, Length of Trip, and Expense per Day, 100 Trips

| Trip Number | Length (days) | Expenses (dollars) | Expense per Day | Trip Number | Length (days) | Expenses (dollars) | Expense per Day |
|---|---|---|---|---|---|---|---|
| 1 | 7.75 | 64.92 | 8.38 | 51 | 3.00 | 21.60 | 7.20 |
| 2 | 3.50 | 30.70 | 8.77 | 52 | 3.00 | 22.21 | 7.40 |
| 3 | 6.50 | 27.45 | 4.22 | 53 | 7.25 | 45.20 | 6.23 |
| 4 | 4.00 | 33.44 | 8.36 | 54 | 4.75 | 17.98 | 3.79 |
| 5 | 1.75 | 16.38 | 9.36 | 55 | 1.50 | 18.00 | 12.00 |
| 6 | 4.75 | 35.05 | 7.38 | 56 | 3.75 | 35.17 | 9.38 |
| 7 | 2.75 | 19.26 | 7.00 | 57 | 5.00 | 28.20 | 5.64 |
| 8 | 3.50 | 21.74 | 6.21 | 58 | 1.50 | 17.19 | 11.46 |
| 9 | 1.00 | 8.72 | 8.72 | 59 | 2.50 | 29.63 | 11.85 |
| 10 | 3.50 | 22.24 | 6.35 | 60 | 9.75 | 55.20 | 5.66 |
| 11 | 4.25 | 42.83 | 10.08 | 61 | 5.00 | 44.69 | 8.94 |
| 12 | 4.75 | 30.77 | 6.48 | 62 | 3.75 | 23.23 | 6.19 |
| 13 | 4.75 | 29.15 | 6.14 | 63 | 6.25 | 32.41 | 5.19 |
| 14 | 2.50 | 19.58 | 7.83 | 64 | 2.00 | 22.48 | 11.24 |
| 15 | 4.75 | 31.96 | 6.73 | 65 | 2.00 | 29.01 | 14.50 |
| 16 | 2.50 | 26.41 | 10.56 | 66 | 2.50 | 12.64 | 5.06 |
| 17 | 3.75 | 27.09 | 7.22 | 67 | 14.00 | 76.40 | 5.46 |
| 18 | 2.50 | 25.22 | 10.09 | 68 | 2.75 | 18.40 | 6.69 |
| 19 | 3.25 | 33.68 | 10.36 | 69 | 2.00 | 24.08 | 12.04 |
| 20 | 2.75 | 23.16 | 8.42 | 70 | 3.25 | 28.01 | 8.62 |
| 21 | 3.75 | 27.30 | 7.28 | 71 | 3.00 | 21.93 | 7.31 |
| 22 | 2.00 | 18.17 | 9.08 | 72 | 3.00 | 20.73 | 6.91 |
| 23 | 4.25 | 28.81 | 6.78 | 73 | 3.75 | 29.13 | 7.77 |
| 24 | 4.00 | 31.94 | 7.98 | 74 | 2.25 | 19.84 | 8.82 |
| 25 | 1.75 | 18.82 | 10.75 | 75 | 4.50 | 31.51 | 7.00 |
| 26 | 1.75 | 21.11 | 12.06 | 76 | 3.75 | 32.07 | 8.55 |
| 27 | 6.00 | 27.82 | 4.64 | 77 | 2.00 | 20.50 | 10.25 |
| 28 | 2.00 | 17.85 | 8.92 | 78 | 3.75 | 25.48 | 6.79 |
| 29 | 2.00 | 15.91 | 7.96 | 79 | 4.00 | 39.29 | 9.82 |
| 30 | 2.75 | 27.21 | 9.89 | 80 | 4.25 | 43.83 | 10.31 |
| 31 | 4.25 | 32.49 | 7.64 | 81 | 5.00 | 34.69 | 6.94 |
| 32 | .50 | 13.09 | 26.18 | 82 | 4.00 | 25.37 | 6.34 |
| 33 | 9.00 | 61.50 | 6.83 | 83 | 2.00 | 37.30 | 18.65 |
| 34 | 5.50 | 24.53 | 4.46 | 84 | 5.25 | 37.61 | 7.16 |
| 35 | 1.25 | 16.37 | 13.10 | 85 | 1.50 | 16.80 | 11.20 |
| 36 | 2.00 | 28.91 | 14.46 | 86 | 2.50 | 24.65 | 9.86 |
| 37 | 3.25 | 35.26 | 10.85 | 87 | 1.75 | 16.01 | 9.15 |
| 38 | 2.75 | 20.64 | 7.51 | 88 | 6.50 | 46.08 | 7.09 |
| 39 | 2.00 | 28.30 | 14.15 | 89 | 1.00 | 15.35 | 15.35 |
| 40 | 3.25 | 28.92 | 8.90 | 90 | 4.75 | 29.15 | 6.14 |
| 41 | 2.25 | 22.26 | 9.89 | 91 | 2.50 | 17.45 | 6.98 |
| 42 | 1.50 | 20.20 | 13.47 | 92 | 4.00 | 34.22 | 8.56 |
| 43 | 1.50 | 12.45 | 8.30 | 93 | 2.75 | 31.24 | 11.36 |
| 44 | 2.00 | 28.96 | 14.48 | 94 | 3.25 | 29.33 | 9.02 |
| 45 | 4.00 | 25.96 | 6.49 | 95 | 4.50 | 46.61 | 10.36 |
| 46 | 4.00 | 23.68 | 5.92 | 96 | 2.50 | 24.08 | 9.63 |
| 47 | 2.75 | 17.57 | 6.39 | 97 | 6.25 | 37.21 | 5.95 |
| 48 | 3.00 | 29.96 | 9.99 | 98 | 2.75 | 22.12 | 8.04 |
| 49 | 2.75 | 27.21 | 9.89 | 99 | 2.50 | 43.83 | 17.53 |
| 50 | 1.50 | 21.64 | 14.43 | 100 | 4.25 | 39.06 | 9.19 |
| | | | | Total | 353.25 | 2,818.79 | 899.90 |

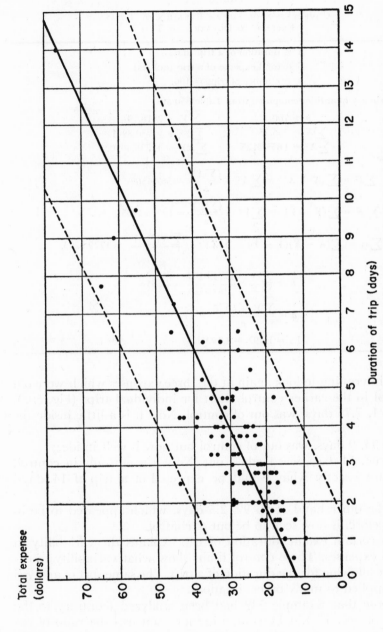

Total expense
(dollars)

Duration of trip (days)

**FIG. 551.** Travel expense control chart allowing for duration of trip. Three-sigma limits for individual trips based on 100 observations.
Relation: Expense = $11.97 plus $4.59 per day.
Standard error of estimate: $6.47.
Source: Tables 550, 552, and 553.

Relationships between Variables

### TABLE 552

COMPUTATION OF LINEAR RELATION OF TOTAL
EXPENSE TO DURATION OF TRIP

$X$ = duration of a trip (days).

$Y$ = total expense of a trip (dollars).

$n$ = number of trips = 100.

The six primary quantities computed from Table 550 are

$$n = 100 \qquad \sum Y = 2818.79$$
$$\sum X = 353.25 \qquad \sum Y^2 = 91730.8941$$
$$\sum X^2 = 1635.3125 \qquad \sum XY = 11736.4050.$$

From these

$$\sum x^2 = \sum (X - \bar{X})^2 = \sum X^2 - \frac{(\sum X)^2}{n} = 387.4569$$

$$\sum y^2 = \sum (Y - \bar{Y})^2 = \sum Y^2 - \frac{(\sum Y)^2}{n} = 12275.1235$$

$$\sum xy = \sum (X - \bar{X})(Y - \bar{Y}) = \sum XY - \frac{(\sum X)(\sum Y)}{n} = 1779.0293.$$

Then

$$b = \frac{\sum xy}{\sum x^2} = \frac{1779.0293}{387.4569} = 4.5916$$

$$a = \frac{\sum Y - b \sum X}{n} = 11.97$$

$$Y_c = 11.97 + 4.59X.$$

It is interesting to look again at the three expenses which were out of control in the earlier control chart for individual trips (Fig. 503).

Trip 1, 7.75 days, was out of control. Now it is a little inside the limits.

Trip 33, 9 days, was out of control but now is well inside.

Trip 67, 14 days, was farthest out. Now it is not only in control, but almost exactly what would be expected of a trip of 14 days' duration.

On the other hand, Trip 99, 2.5 days, which appeared to be in control before, is now seen to be out of control.

To a certain extent, duration of trip explains the variability in the total expenses. These control limits show what variability to expect after allowing for the duration of a trip. The same sort of analysis can be applied to many other situations.

Suppose that Example 549 had been analyzed (contrary to the warning of Sec. 17.4.5) by computing for each trip the ratio of expense to duration, and calculating the mean of these 100 ratios. The mean of the ratios is *$9.00 per day.*

### 17.5 An Illustration

TABLE 553

COMPUTATION OF CONTROL LIMITS $Y_c \pm 3s_{Y-Y_c}$ FOR LINEAR RELATION
OF TOTAL EXPENSE TO DURATION OF TRIP

$$s_{Y \cdot x} = \sqrt{\frac{\sum(Y - \bar{Y})^2 - b\sum(X - \bar{X})(Y - \bar{Y})}{n - 2}} \qquad s_{Y-Y_c} = s_{Y \cdot x}\sqrt{\frac{n + 1}{n} + \frac{(X - \bar{X})^2}{\sum(X - \bar{X})^2}}$$

$$= \sqrt{\frac{12275.1235 - (4.5916)(1779.0293)}{98}} \qquad = 6.47\sqrt{\frac{101}{100} + \frac{(X - 3.53)^2}{387.4569}}.$$

$= 6.47 = $ standard error of estimate.

| $X$ | $Y_c$ | $s_{Y-Y_c}$ | $Y_c + 3s_{Y-Y_c}$ | $Y_c - 3s_{Y-Y_c}$ |
|---|---|---|---|---|
| 0 | 11.97 | 6.60 | 31.77 | −7.83 |
| 1 | 16.56 | 6.56 | 36.24 | −3.12 |
| 2 | 21.15 | 6.52 | 40.71 | 1.59 |
| 3 | 25.74 | 6.50 | 45.24 | 6.24 |
| 4 | 30.33 | 6.50 | 49.83 | 10.83 |
| 5 | 34.92 | 6.52 | 54.48 | 15.36 |
| 6 | 39.51 | 6.55 | 59.16 | 19.86 |
| 7 | 44.10 | 6.60 | 63.90 | 24.30 |
| 8 | 48.69 | 6.67 | 68.70 | 28.68 |
| 9 | 53.28 | 6.75 | 73.53 | 33.03 |
| 10 | 57.87 | 6.84 | 78.39 | 37.35 |
| 11 | 62.46 | 6.95 | 83.31 | 41.61 |
| 12 | 67.05 | 7.07 | 88.26 | 45.84 |
| 13 | 71.64 | 7.21 | 93.27 | 50.01 |
| 14 | 76.23 | 7.36 | 98.31 | 54.15 |
| 15 | 80.82 | 7.52 | 103.38 | 58.26 |

On the other hand, the total expense for the 100 trips is $2818.79, and the total duration is 353.25 days. The ratio of these gives the expense as *$7.98 per day.*

The regression analysis, however, has shown the daily expense as *$4.59 per day.*

Why do these figures differ? What is the meaning of each? Which one best represents the average expense per day?

First, consider the two results of the ratio method, $9.00 and $7.98. Whenever an average is computed, it is an average over some kind of units. The $9.00 figure (which may be called the average of the ratios) is an average based on one number for each *trip.* The $7.98 figure (which may be called the ratio of the averages, referring to average expense per trip and average duration per trip) is an average based on one number for each *day.* Thus, the mean daily expense *per trip* is $9.00; but the mean daily expense *per day* is $7.98.

Either of these ratios could be expected to prevail (with allowance for sampling error) in the future only if the distribution of trips by duration were to be the same in the future; for there is a definite relation between the ratios and the duration: the shorter the duration, the higher, on the average, the ratio of expense to duration. In fact, the mean ratio of expense to duration for trips of duration $D$ is $\$4.59 + (\$11.97/D)$. The $\$4.59$ represents the incremental or marginal cost, that is, the amount by which the total expense increases, on the average, for each day's increase in duration. The $\$11.97$ represents a fixed or overhead cost per trip, and in the ratio of expense to duration it is spread more thinly over the days of a long trip than over the days of a short trip.

The five trips with the highest average expense per day are shown in Table 554A.

TABLE 554A

FIVE TRIPS WITH HIGHEST AVERAGE EXPENSE PER DAY

| Trip Number | Expense per Day |
|---|---|
| 32 | $26.18 |
| 65 | 14.50 |
| 83 | 18.65 |
| 89 | 15.35 |
| 99 | 17.53 |

Now, Table 550 shows that all of these trips were of shorter duration than average. The durations were, respectively, 0.50, 2.00, 2.00, 1.00, and 2.50 days. Table 554B gives a comparison between actual total expense and expense predicted by the regression equation $Y_c = 11.97 + 4.59X$. Thus, after allowing for duration, Trips 32 and 89 are actually less expensive than average. Trip 65 is above average

TABLE 554B

ACTUAL TOTAL EXPENSE AND TOTAL EXPENSE PREDICTED BY REGRESSION EQUATION, FIVE TRIPS OF TABLE 554A

| Trip Number | Actual Total Expense | Calculated Total Expense ($Y_c$) |
|---|---|---|
| 32 | $13.09 | $14.26 |
| 65 | 29.01 | 21.15 |
| 83 | 37.30 | 21.15 |
| 89 | 15.35 | 16.56 |
| 99 | 43.83 | 23.44 |

even after allowing for duration of trip, but Fig. 551 shows that it is well within the upper control limit. In fact, it is closer to the regression line than quite a few other trips. Trip 83 does not reach the upper control limit, but it is still unusually high, even after allowing for its relatively short duration. Fig. 551 shows that Trip 95 is nearly as much on the high side as Trip 83, but the duration of Trip 95 was 4.50 days, and its expense per day, $10.36, was not particularly unusual. Finally, both the regression approach and the expense per day show Trip 99 to be unusually high.

The fundamental reason for not using either the ratio of the averages or the average of the ratios to calculate the cost per day is that total expense is not proportional to duration of trip. The average total expense is related to duration, it varies with duration, it is a linear function of duration—but it is *not* proportional to duration. If each total expense were strictly proportional to duration of trip, then the average of the ratios, the ratio of the averages, and the slope of the regression line would all be equal. But even if total expense is proportional to duration on the average, but not in each trip separately, the three measures will differ.

## 17.6
## CONCLUSION

Relationships between quantitative variables may be studied by regression lines, which show how the mean value of one variable is associated with the values of other variables, and by correlation coefficients, which show to what extent the variation in one variable can be "explained" by variation in another variable on which the mean value of the first variable depends. Thus, correlation coefficients are measures of the strength of association, and regression lines express the actual relation of a variable to others.

The principle of maximum likelihood leads to estimating the parameters of a regression equation in such a way that the line minimizes the sum of the squared vertical deviations of the observations from the line. The mean of these squared deviations, using 2 less than their number as the divisor, is an estimate of the variance of the observations about the line; its square root is called the standard error of estimate. From the standard error of estimate can be computed standard errors of the slope of the regression line, of the height of the line at any value of the independent variable (including the intercept when the independent variable is zero), and of the difference between a new observation and the line. From these standard errors, corresponding confidence intervals can be computed.

The use of ratios instead of, or as part of, regression analysis is fraught with hazards. To allow for the effect of one variable on another by dividing the first into the second implies that the two vary proportionally, whereas regression analysis allows for variation that is not proportional. Also, the relation between two ratios, each to the same third variable, may be in some degree spurious, a result of using the third variable on both sides of the relation.

Many misuses of correlation and regression involve fallacious interpretations of valid relations. In particular, causal connections are often imputed to correlated variables. A valid relation may be explained in various ways besides causation, one of the most common being the relation of each variable to some other variables.

Like other tools, correlation and regression methods can be extremely valuable when properly handled, but misleading when improperly handled.

## DO IT YOURSELF

### Example 556A

For Example 201, let $Y$ = score on Test 2, and $X$ = score on Test 1. (If you wish, you may first code your observations to simplify computations, for example, by subtracting 100 from each score.)

(1) Compute $a$ and $b$ in $Y = a + bX$. Check both by plotting your line and by the method of Table 536.

(2) Compute $s_b$. Is $b$ significantly different from zero?

(3) Compute $s_{Y \cdot X}$.

(4) Compute $s_{Y_c}$ for $X$ = 25, 50, 75, 100, 125, 150, 175, 200.

(5) Plot 95 percent confidence limits for the regression line. (Multiply $s_{Y_c}$ by the factor given in Table 540.)

(6) Compute and plot 95 percent confidence limits for a new observation.

(7) Compute $r$.

### Example 556B

Same requirements as Example 556A, for the regression of *height* on *weight*, that is, height = $a + b$ (weight), using the data of Table 531.

On the same chart, plot your regression line and the one obtained in Sec. 17.2.5. Why is your line different? Under what circumstances would each line be useful?

### Example 556C

In Example 556B, show that $r$ is the same whether computed from $s_{Y_c}/s_Y$ or $s_{X_c}/s_X$.

### Example 556D

Do Example 243A, or if you already have, reconsider your answer in the light of this chapter.

*Do It Yourself*

EXAMPLE 557

This example illustrates a simple descriptive technique useful in studying the relationship between a dependent variable and two independent variables. The following data are coded data derived from actual laboratory tests of a certain steel product. $Y$ represents tensile strength of the product. $X_1$ and $X_2$ represent concentrations of two elements thought to be related to tensile strength.

TABLE 557

TENSILE STRENGTH AND CONCENTRATIONS OF TWO ELEMENTS,
30 SAMPLES OF A STEEL PRODUCT

| Test Number | $Y$ | $X_1$ | $X_2$ | Test Number | $Y$ | $X_1$ | $X_2$ |
|---|---|---|---|---|---|---|---|
| 1 | 184 | 21 | 26 | 16 | 88 | 16 | 15 |
| 2 | 70 | 8 | 16 | 17 | 170 | 24 | 18 |
| 3 | 94 | 22 | 16 | 18 | 74 | 16 | 8 |
| 4 | 120 | 18 | 29 | 19 | 107 | 16 | 18 |
| 5 | 116 | 20 | 21 | 20 | 112 | 16 | 23 |
| 6 | 174 | 20 | 27 | 21 | 152 | 20 | 26 |
| 7 | 104 | 18 | 19 | 22 | 123 | 16 | 17 |
| 8 | 117 | 15 | 28 | 23 | 114 | 20 | 15 |
| 9 | 26 | 10 | 17 | 24 | 114 | 16 | 23 |
| 10 | 110 | 19 | 13 | 25 | 87 | 18 | 19 |
| 11 | 104 | 19 | 14 | 26 | 100 | 16 | 24 |
| 12 | 140 | 20 | 23 | 27 | 110 | 20 | 17 |
| 13 | 160 | 22 | 26 | 28 | 99 | 18 | 13 |
| 14 | 47 | 9 | 18 | 29 | 124 | 15 | 16 |
| 15 | 110 | 20 | 11 | 30 | 104 | 16 | 20 |

(1) Plot a scatter diagram with $Y$ as vertical axis and $X_1$ as horizontal. What is the visual impression you get as to the relationship between $Y$ and $X_1$?

(2) Pick out the tests with the ten highest values of $X_2$ and mark their dots with the letter "$H$" (for "high"). Similarly, mark the dots corresponding with the middle ten values of $X_2$ with "$M$," and the lowest ten, with "$L$." How, if at all, does $X_2$ appear to affect the relationship between $Y_1$ and $X_1$?

(3) Parallel to Step (1), using $Y$ and $X_2$.

(4) Parallel to Step (2), ranking by values of $X_1$. (Because of ties, you will not be able to group the data into three exactly equal groups, but this is not essential.)

(5) Summarize what you have learned about the relationship between $Y$ and $X_1$ and $X_2$.

In practice, often Steps (1) and (2) would suffice. While this procedure is really a descriptive rather than analytical one, it is useful even in analytical studies as a preliminary to mathematical computations.

*Relationships between Variables*

EXAMPLE 558A

(1) Fit the regression relationship $Y_c = a + b_1 X_1 + b_2 X_2$ to the data of Example 557.

(2) For each test compute $Y_c$ and $Y - Y_c$. The sum of the deviations $Y - Y_c$ should equal zero except for rounding discrepancies.

(3) State in words the interpretation of your result. Compare with your answer in Step (5) of Example 557.

EXAMPLE 558B

What internal evidence, if any, of inaccurate measurement do you find in the data of Example 557?

# Chapter 18

# *Time Series*

## 18.1
### TIME SERIES PROBLEMS

A time series is a set of observations made at different times. Each observation represents both a quantity and the time when this quantity occurred. Typical time series are the population of the United States at the successive decennial censuses beginning with 1790; the number of games played in successive World Series beginning with 1903; the level of Lake Michigan by months beginning with 1860; the velocity of a missile by milliseconds (thousandths of a second) from firing; the number of shares sold on the New York Stock Exchange daily beginning with, say, the end of World War II; the number of cars on Manhattan Island each minute from midnight to midnight on, say, February 29, 1956.

Time series are sometimes studied simply because of an interest in history. Sometimes, as in the Goldhamer-Marshall study of mental disease (Sec. 2.8.2) the interest is in correlation and analysis of relations between variables. Often, however, the ultimate interest is in the future: prognostication.

Those who attempt to forecast time series often turn to statisticians. Indeed, a common stereotype of the statistician is a sort of astrologer who studies the movements and concatenations of time series, then forecasts business conditions; consequently, "statistician," "seer," and "crystal ball" are often juxtaposed. If you have read—or even skimmed—this far in this book, it certainly will not surprise you to learn that statisticians deal with other problems and by methods other than clairvoyance, but it may shock you to learn that when it comes to time series, there is no Santa Claus.

In the field of time series, there is no well-developed, widely-applicable, tried-and-proven body of techniques for inference com-

**559**

parable to those presented in Part III. Time series is a subject to which much attention is being given by statistical theorists currently, and in which, indeed, there have been some promising developments, especially in the past decade—enough so that a few theorists may feel affronted by the first sentence of this paragraph. But the obstacles are great. For one thing, the techniques that have been developed tend to require formidable computations. It has been pointed out that methods such as those presented in Part III, especially more complicated methods of the same type not included in this book, could not have become so widely understood and used throughout all branches of science without the ready accessibility of modern desk calculators, which are well suited to the calculations involved. Equally effective analyses of time series may be similarly dependent upon the widespread accessibility of high-speed electronic calculators, whose development and distribution has as yet scarcely begun. There are, however, many conceptual as well as computational difficulties.

At any rate, statistics can offer help in handling time series like that it can offer in descriptive statistics. Appropriate methods depend especially closely on the subject matter and the problem, but statisticians have acquired a certain amount of skill, lore, and wariness by which they can save the subject-matter specialist from having to learn everything the hard way—that is, by painful blunders.

## 18.2
## SERIAL CORRELATION

A simple control chart, such as that of Fig. 503, is a kind of time series, for the observations are plotted in the order of their occurrence, though not with reference to their exact time. No special problems arose in our discussion of control charts to forewarn of the difficulties of analyzing time series. On a control chart, the observations are independent, at least under the null hypothesis that the process is in control. Independence implies (see Sec. 10.4) that the observations are as likely to occur in any particular sequence as in any other. It is exactly as likely (and no more likely) that the smallest and largest observations will be consecutive as that the two largest or two smallest—or, indeed, any two observations—will be consecutive. An additional observation taken between two of those in the sample would have brought as much new information as an additional observation taken at any other time, for with independent observations no light is thrown on the value of an observation by knowing the value of adjacent observations.

## 18.2 Serial Correlation

With most time series, however, observations that are consecutive or near together are correlated. If a measurement of the velocity of a missile is 3,000 feet per second at a certain time and 2,950 feet per second a second later, measurements at intervening times will add little information that could not be deduced from these two. Not only the population mean velocity, but the velocity of the particular missile and the errors of the measuring devices are restricted in their possibilities for change within short times. Thus, measurements each millisecond between the two original measurements would bring little, if any, increase in information about the population of velocities. Certainly the standard error of the mean would not be reduced by a factor of $\sqrt{1,000/2}$, or about 22, as would be the case for independent observations if a sample were increased from 2 to 1,000.

The stock market is another example. It is almost as "impossible" that consecutive days will see the all-time high and all-time low level of prices—or even the year's high and low—as it is that consecutive seconds will see the high and low velocities of a missile.

The dependence, or serial correlation, of observations that are close in time is, however, a little more subtle and difficult to understand than has been made clear yet. Insofar as the relation between observations is accounted for by changes in the population mean from one time to another, there is no new problem. The problem arises from relations among the *deviations* of the observations from the means of their respective populations.

Though the height-weight data of Sec. 17.2 are not a time series, they can serve to illustrate the point about *serial correlation*, that is, correlation of values that are adjacent or near in a series. The weights at successive heights are *not* serially correlated. Knowing that observations at, say, 69 and 71 inches were above (or below) the population means for those heights would have no bearing on whether an observation at 70 inches would be above or below its population mean. Serial correlation refers to correlation among values of dependent variables *other* than the correlation accounted for by the independent variable (time, in time series). With the missile, for example, if for some reason its velocity at a certain instant exceeds the population mean velocity for that instant in the trajectories of similar missiles under similar conditions, then the velocity a second earlier or later—and certainly a microsecond (millionth of a second) earlier or later—will also be above the corresponding population mean, and probably by about the same amount. Furthermore, if the error in measuring the actual velocity is positive at a given time, it will probably be positive an instant later or earlier, because of lags, inertia, or persistence in the measuring device, though that depends on the device. Thus, the successive observations do not give independent determinations of their respective population means. It is not the fact that the population means are changing that is referred to as serial correlation, but the fact that successive deviations from the population means are correlated.

Serial correlation ordinarily does not prevent the method of least squares from providing reasonably good point estimates of the parameters of a line describing a time series. It does, however, invalidate the usual estimates of standard errors, and consequently interferes with interval estimation or testing of hypotheses. Furthermore, many time series (especially of social phenomena) move in ways sufficiently complicated to make it impractical to fit lines to them mathematically, so that even point estimates may not actually be available.

Serial correlation can create an illusion of cycles where data are merely nonindependent. When some random effect moves a series above its mean, for example, it tends, because of serial correlation, to move back only after several observations. If, as is often the case, the *changes* themselves are serially correlated, this illusion of cycles is even more pronounced. Such movements are, of course, real oscillations; but they are not cycles in the sense of having a regular duration (*period*) or amplitude of rise and fall. So-called "business cycles," for example, are not cycles with the rigid periodicity implied by the term "cycles," but oscillations of variable (and unpredictable) duration and amplitude. Almost any series, if stared at long and hopefully enough, begins to shape up into patterns and cycles. An enterprising new Rorschach[1] may some day develop a test of statistical personality based on a standard set of random, serially correlated time series.

Serial correlation also complicates affairs for those who want to use time series to study the relationship between variables. If each of two series being compared is serially correlated, close agreement at one point is accompanied by close agreement at adjacent points, so what looks like several instances of agreement may simply be one instance seen several times. Similarly, what appears as a number of consecutive instances of nonagreement, or of counter movement, may really be one instance seen repeatedly. To the extent that data are serially correlated, they are redundant—repetitions of a single piece of information, like a phonograph needle stuck in the groove, or election votes under a dictatorship.

---

1. The Rorschach personality test consists of a series of ink blots. The subject describes what each suggests to him, and from these descriptions the tester draws inferences about the personality of the subject.

### 18.2 Serial Correlation

The fact that methods of tackling these difficulties are beyond the scope of this book should not be allowed to create the impression that there are no such methods. The standard error of a mean of correlated observations can be computed if the amount of correlation can be computed. The presence of serial correlation can be tested, and its amount estimated, by comparing the standard deviation of the differences between consecutive observations with the value this standard deviation would be expected to have if the same observations were arranged independently at random. If the observations are represented by $X_1, X_2, \ldots, X_n$ and the differences by $d_1 = X_2 - X_1, d_2 = X_3 - X_2, \ldots, d_{n-1} = X_n - X_{n-1}$, the differences will be smaller in absolute value, on the average, and hence will have a smaller standard deviation, if the observations are serially correlated positively (and larger if the serial correlation is negative) than if they are independent. Test procedures have been developed making use of this principle.

EXAMPLE 563   SERIAL CORRELATION

A simple, artificial example of serial correlation will illustrate some of the basic points just discussed. Suppose that the observations in a time series arise as follows: each value $K$ is an independent drawing from a normal distribution with a mean of zero and standard deviation of one. We have drawn thirty observations in this way from a table of random normal deviates.[2] The results are shown in Table 563 and graphed as a time series in Fig. 564.

TABLE 563

THIRTY INDEPENDENT STANDARD NORMAL VARIABLES
ARRANGED AS A "TIME SERIES"

| Year | $K$ | Year | $K$ | Year | $K$ |
|------|--------|------|--------|------|--------|
| 1 | .551 | 11 | −.298 | 21 | .036 |
| 2 | −.506 | 12 | −.241 | 22 | .932 |
| 3 | −1.077 | 13 | −1.959 | 23 | −2.513 |
| 4 | 2.834 | 14 | .489 | 24 | −2.090 |
| 5 | 1.318 | 15 | 1.086 | 25 | −.342 |
| 6 | .660 | 16 | −.409 | 26 | −.044 |
| 7 | .034 | 17 | −.078 | 27 | .268 |
| 8 | −.222 | 18 | −.614 | 28 | −.885 |
| 9 | −1.566 | 19 | −.782 | 29 | −.456 |
| 10 | .488 | 20 | −.115 | 30 | .069 |

2. The Rand Corporation, *A Million Random Digits with 100,000 Normal Deviates* (Glencoe, Illinois: Free Press, 1955).

*Time Series*

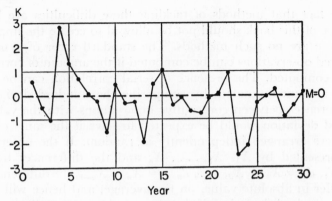

**FIG. 564. Thirty independent standard normal variables arranged as a "time series."**

As you examine Fig. 564, you may be tempted to see a downward trend, or a bias in favor of negative observations, or too many observations outside the interval −2 to +2. There are, indeed, hints of all three of these things, but none of the hints is strong enough to break through the barrier of statistical significance. In fact, Fig. 564 could be regarded as another example of a control chart for an in-control process (see Chap. 16).

Now let us construct another time series in which the population mean $M$ is still zero, but in which the successive observations are no longer independent. In particular, assume that the value of the series in any year is exactly the same as in the preceding year *except* for a random change. Such a series may be generated by cumulating the values of $K$ in Table 563. The series starts at .551 in the first year. In the second year the random deviation is −.506, so the second observation is .551 + (−.506) = .045. Similarly, the third year is .045 + (−1.077) = −1.032. The resulting time series is plotted in Fig. 565.

It requires no special imaginativeness to see patterns of systematic variation in Fig. 565. Actually, however, the population mean has never wavered from zero. The appearance of trend and cycles is due to the serial correlation.

While this particular mechanism should not be taken too seriously as a description of real time series, it does illustrate the problem of dealing with serially correlated data, and suggests how it might be overcome. If we were confronted with a series like that of Fig. 565, we would know how to deal with it statistically *if* we knew how it had been generated. We would then look at the *differences* between successive observations, and this would bring us back to Fig. 564, to which standard methods would apply, since the differences are random, independent drawings from a stable population. In this case we see that it is impossible to find a better method of forecasting the series of Fig. 565 than to say that next year will be identical with this year (point estimate) with a confidence interval based on the standard error of the random deviation $K$, or 1.

In practical statistics, we do not ordinarily know the mechanism, so have to evolve and test hypotheses about it. We need to be able

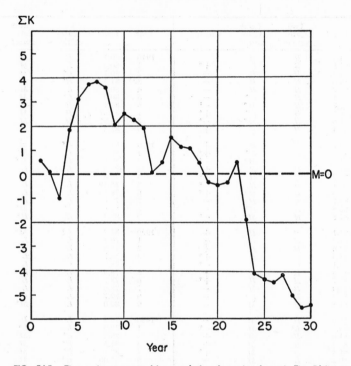

FIG. 565. Time series generated by cumulating the series shown in Fig. 564.

to test whether there is really serial correlation in data like those of Fig. 565 and also, if possible, whether the mean of the population is stable through time. Some simple methods of approaching these problems will be presented in Secs. 18.4 and 18.5.

## 18.3
## AN ILLUSTRATIVE TIME SERIES

To illustrate the problems of analyzing statistical time series, we shall use primarily a single series. This series, shown in Table 566 and Fig. 567, gives the highest monthly level of Lake Michigan-Huron for each year from 1860 to 1955. We see a rather erratic pattern of movement, with an apparent tendency to drift downward.

## TABLE 566

### LAKE MICHIGAN-HURON, HIGHEST MONTHLY MEAN LEVEL
### FOR EACH CALENDAR YEAR, 1860–1955

(Height in Feet above 500)

| Year | Level | Two Cate-gories[a] | Three Cate-gories[b] | Change | | Year | Level | Two Cate-gories | Three Cate-gories | Change |
|---|---|---|---|---|---|---|---|---|---|---|
| 1860 | 83.3 | H | H |    | | 1910 | 80.5 | L | M | − |
| 1 | 83.5 | H | H | +* | | 1 | 80.0 | L | M | −* |
| 2 | 83.2 | H | H | − | | 2 | 80.7 | L* | M | +* |
| 3 | 82.6 | H | H | − | | 3 | 81.3 | H* | M | +* |
| 4 | 82.2 | H | H | − | | 4 | 80.7 | L | M | − |
| 5 | 82.1 | H | H* | − | | 5 | 80.0 | L* | M | −* |
| 6 | 81.7 | H | M* | −* | | 6 | 81.1 | H | M | + |
| 7 | 82.2 | H | H* | +* | | 7 | 81.87 | H | M | + |
| 8 | 81.6 | H | M* | −* | | 8 | 81.91 | H | M | +* |
| 9 | 82.1 | H | H | + | | 9 | 81.3 | H | M | − |
| 1870 | 82.7 | H | H | + | | 1920 | 81.0 | H* | M | − |
| 1 | 82.8 | H | H* | +* | | 1 | 80.5 | L | M | −* |
| 2 | 81.5 | H | M* | −* | | 2 | 80.6 | L | M* | +* |
| 3 | 82.2 | H | H | + | | 3 | 79.8 | L | L | − |
| 4 | 82.3 | H | H | +* | | 4 | 79.6 | L | L | − |
| 5 | 82.1 | H | H | −* | | 5 | 78.49 | L | L | −* |
| 6 | 83.6 | H | H | +* | | 6 | 78.49 | L | L | 0 |
| 7 | 82.7 | H | H | − | | 7 | 79.6 | L | L* | + |
| 8 | 82.5 | H | H* | − | | 8 | 80.6 | L* | M* | + |
| 9 | 81.5 | H | M* | −* | | 9 | 82.3 | H | H* | +* |
| 1880 | 82.1 | H | H | + | | 1930 | 81.2 | H* | M* | − |
| 1 | 82.2 | H | H | + | | 1 | 79.1 | L | L | −* |
| 2 | 82.6 | H | H | + | | 2 | 78.6 | L | L | +* |
| 3 | 83.3 | H | H | +* | | 3 | 78.7 | L | L | −* |
| 4 | 83.1 | H | H | −* | | 4 | 78.0 | L | L | + |
| 5 | 83.3 | H | H | + | | 5 | 78.6 | L | L | +* |
| 6 | 83.7 | H | H | +* | | 6 | 78.7 | L | L | −* |
| 7 | 82.9 | H | H | − | | 7 | 78.6 | L | L | +* |
| 8 | 82.3 | H | H* | − | | 8 | 79.7 | L | L* | − |
| 9 | 81.8 | H | M | − | | 9 | 80.0 | L | M* | +* |
| 1890 | 81.6 | H* | M | − | | 1940 | 79.3 | L | L | − |
| 1 | 80.9 | L* | M | −* | | 1 | 79.0 | L | L* | −* |
| 2 | 81.0 | H | M | + | | 2 | 80.2 | L* | M | + |
| 3 | 81.3 | H | M | +* | | 3 | 81.5 | H* | M | +* |
| 4 | 81.4 | H* | M | − | | 4 | 80.8 | L* | M | −* |
| 5 | 80.2 | L | M | − | | 5 | 81.00 | H* | M | +* |
| 6 | 80.0 | L | M | −* | | 6 | 80.96 | L* | M | −* |
| 7 | 80.85 | L | M | +* | | 7 | 81.1 | H* | M | +* |
| 8 | 80.83 | L* | M | −* | | 8 | 80.8 | L | M* | − |
| 9 | 81.1 | H* | M | +* | | 9 | 79.7 | L | L* | −* |
| 1900 | 80.7 | L* | M | −* | | 1950 | 80.0 | L* | M | + |
| 1 | 81.1 | H* | M | +* | | 1 | 81.6 | H | M* | + |
| 2 | 80.83 | L | M | − | | 2 | 82.7 | H | H | +ᵈ |
| 3 | 80.82 | L* | M | −* | | 3 | 82.1 | H | H* | − |
| 4 | 81.5 | H | M | + | | 4 | 81.7 | H | M | − |
| 5 | 81.6 | H | M | +* | | 5 | 81.5 | H* | M* | −ˢ |
| 6 | 81.5 | H | M | −* | | | | | | |
| 7 | 81.6 | H | M | +* | | | | | | |
| 8 | 81.8 | H | M | +* | | | | | | |
| 9 | 81.1 | H* | M | − | | | | | | |

*Source:* Unpublished data provided by U.S. Lake Survey, Corps of Engineers, U.S. Army. A published hydrograph of the Lake Survey gives these data graphically. Data for certain years are shown to two decimals to avoid ties.

    [a] *H*: 581 ft. or more        [b] *H*: 582 ft. or more
      *L*: Under 581 ft.           *M*: 580 ft. or more but under 582 ft.
    *represents end of a run.          *L*: Under 580 ft.

## 18.3 An Illustrative Time Series

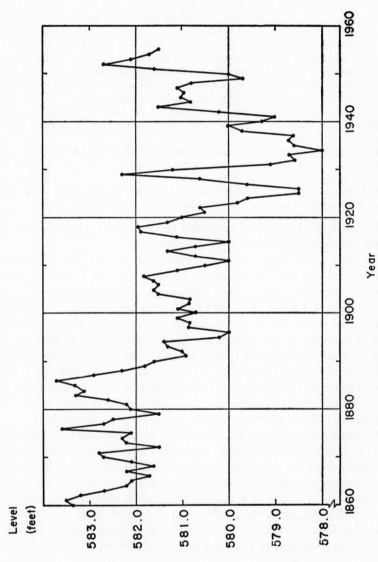

FIG. 567. Lake Michigan-Huron, highest monthly mean level, 1860–1955.

Changes in the level of the lakes are of importance for many reasons:

> Levels of the water surfaces of the Great Lakes have varying effects on three major economic interests—shore property, lake shipping, and hydroelectric power. In general, high levels benefit shipping and power. Increased depths in harbors and channels, which permit vessels to load only an inch or two deeper, permit sizable increases in cargoes, particularly in the huge modern lake freighters. Production of hydroelectric power is obviously facilitated by an abundance of water. But high lake levels are extremely injurious to shore properties, particularly during storms. . . . Periods of low lake levels likewise present problems. For example, maintenance of high flows for power would further decrease the drafts to which vessels on the Great Lakes could be loaded.
>
> There is no way to solve these problems with the lakes in their present unregulated state, since the recurring highs and lows are natural and not manmade. . . .
>
> It has long been recognized that accurate forecasting of lake levels would enable each interest to gain some measure of protection against oncoming highs and lows which might be damaging. A number of studies to this end have been made in the past by leading hydraulic engineers, but until recently it was believed inadvisable to forecast lake levels more than one month in advance.[3]

Prediction of the lake levels may also be of general scientific interest, beyond any one particular practical problem.

If a time series is in a state of statistical control (see Chap. 16), the historical record tells us all we need or can know for predictive purposes, unless we can identify additional variables, as in the travel expense data of Chaps. 16 and 17. As long as a series is in control, the mean of the series is our point estimate for the future, and the upper and lower control limits are the bounds of our confidence interval. A glance at Fig. 567 suggests, and our analysis will confirm, that the lake data are *not* in a state of statistical control. Hence we are interested in exploring ways in which the data depart from statistical control. Do they show a downward trend? Is there some regular cycle? Are there patterns which would help us to predict?

We shall explore questions like these in the balance of this chapter. The explorations we shall make will illustrate, of course, only the statistical facets of a real study of the lake levels. A real study would involve the whole highly developed science of hydrology, and the concomitant study of related series: flows, involutions, temperatures, winds, etc., with their effects on the water level. Such a study would also take account of knowledge gained from similar studies of other bodies of water; of relevant historical events, such as deforesta-

---

3. Louis D. Kirshner and M. Asce, "Forecasting Great Lakes Levels Aids Power and Navigation," *Civil Engineering*, Vol. 24 (1954), pp. 98–99. Lakes Michigan and Huron have a common level and so are regarded as one lake for this purpose. Each of the other Great Lakes, Superior, Erie, and Ontario, has a separate level.

tion and reforestation, canals, dams, and power projects; of relations between depth, volume, and area of a lake; and of many other factors.

The first statistical issue that we shall consider is the question of serial correlation. We will investigate this through runs, or clustering, of like observations.

# 18.4
# RUNS

## 18.4.1  Runs of Several Kinds of Observations

18.4.1.1  *Two Kinds of Observation.*  One way to decide whether there is a real clustering of like observations is to classify the observations into a few broad categories, such as high and low, or high, low, and medium, or perhaps high, high-medium, low-medium, and low, etc. Then a new time series is made up of symbols such as $H$ and $L$, or $H$, $L$, and $M$, or $A$, $B$, $C$, and $D$, etc. In this new series, the number of *runs*, $r$, is counted, a run being a group of consecutive symbols that are the same. (A run may consist of only one symbol.)

EXAMPLE 569   HIGH AND LOW LEVELS OF
LAKE MICHIGAN-HURON

Table 566 illustrates this. In the column headed "Two categories" the letters $H$ and $L$ have been entered according to whether the highest monthly mean level of Lake Michigan-Huron for a given year was 581 feet or more ($H$), or under 581 feet ($L$). The division was made at 581 feet simply because a quick inspection revealed it to be approximately at the midpoint of the range, which is from 578.0 feet (1934) to 583.7 feet (1886).[4] Had the median been readily available, it would have been used as the dividing point. Of the 96 years, 58 are marked $H$ and 38 are marked $L$. A star has been placed at the end of each run; there are 23 runs, that is, $r = 23$.

To interpret the figure 23, it is necessary to know the sampling distribution of $r$. Suppose 58 cards are marked $H$ and 38 marked $L$. All 96 are thoroughly shuffled, and the number of runs counted; and suppose this is repeated a great number of times. How would the values of $r$ vary? More specifically, would 23 be an improbably low value? If so, it would indicate clustering or bunching in the particular sample of 96 that has been observed.

The sampling distribution of $r$ is sufficiently well approximated by a normal distribution. The population mean, under the null hy-

---

4. These are elevations referred to mean tide at New York as of 1935.

pothesis of independence, is

$$M_r = \frac{2n_1n_2}{n} + 1,$$

where $n_1$ and $n_2$ are the numbers of the two kinds of observation and $n = n_1 + n_2$ is the total number of observations. (It is obviously immaterial which frequency is denoted by $n_1$.) The population standard deviation is

$$\sigma_r = \sqrt{\frac{2n_1n_2(2n_1n_2 - n)}{n^2(n - 1)}}.$$

The alternative hypothesis is that there is clustering, and this would result in too few runs; so a lower-tail probability is wanted. The exact probability of $r$ or fewer runs is the same as that of $r + 0.1$ or fewer, $r + 0.9$ or fewer, $r + 0.999$ or fewer, or of any other number not less than $r$ but less than $r + 1$; for since $r$ must be an integer, the only way a value less than $r + 1$ can occur is if it is $r$ or less. The normal approximation gives different probabilities throughout the range from $r$ to $r + 1$ in which the exact probability is constant, so we will take the normal probability at the middle of the interval—at $r + \frac{1}{2}$, that is—as the approximation to the exact probability. Thus, the standard normal variable is

$$K = \frac{r + \frac{1}{2} - M_r}{\sigma_r},$$

and a little algebra shows that this is

$$K = \frac{n(r - \frac{1}{2}) - 2n_1n_2}{\sqrt{\dfrac{2n_1n_2(2n_1n_2 - n)}{n - 1}}}.$$

For the two-category classification of Table 566

$$n = 96, \qquad n_1 = 38, \qquad n_2 = 58, \qquad r = 23,$$

whence

$$2n_1n_2 = 2 \times 38 \times 58 = 4{,}408$$

and

$$K = \frac{(96 \times 22.5) - 4{,}408}{\sqrt{\dfrac{4{,}408(4{,}408 - 96)}{95}}} = \frac{-2{,}248}{447.3} = -5.0.$$

The value 23, then, is five standard deviations below the value to be expected (46.9) if the observations are independent. In other words, the lake has tended to be high for several years at a time and then low for several years. To estimate next year's highest level, this year's level is likely to be closer than the 96-year average.

### 18.4.1.2 *More Than Two Kinds of Observation.*

EXAMPLE 571   HIGH, LOW, AND MEDIUM LEVELS OF LAKE MICHIGAN-HURON

With as many as 96 observations, more than two categories might well be used. In the column of Table 566 headed "Three categories" the years are marked $H$ if the level is 582.0 feet or above, $L$ if it is under 580.0 feet, and $M$ if it is 580 feet or more but under 582.0 feet. Now the number of runs is 22.

The sampling distribution in this case is also approximately normal. If $n_i$ represents the frequency of the $i$th kind of observation (any numbering of the kinds is satisfactory) and $n = \sum n_i$ is the total number of observations, the mean and standard deviation of $r$ are

$$M_r = \frac{n(n+1) - \sum n_i^2}{n},$$

$$\sigma_r = \sqrt{\frac{\sum n_i^2[\sum n_i^2 + n(n+1)] - 2n\sum n_i^3 - n^3}{n^2(n-1)}}.$$

Again, a lower-tail probability is required, so the continuity adjustment is made by adding $\frac{1}{2}$ to $r$. Then the standard normal variable is

$$K = \frac{\sum n_i^2 - n(n - r + \frac{1}{2})}{\sqrt{\dfrac{\sum n_i^2[\sum n_i^2 + n(n+1)] - 2n\sum n_i^3 - n^3}{n-1}}}.$$

These formulas are valid for any number of kinds of observations. For two kinds, they can be simplified to those given in Sec. 18.4.1.1.

For the three-category classification of Table 566, we have:

| Category | $n_i$ | $n_i^2$ | $n_i^3$ |
|----------|-------|---------|---------|
| H | 28 | 784 | 21,952 |
| M | 52 | 2,704 | 140,608 |
| L | 16 | 256 | 4,096 |
| Total | 96 | 3,744 | 166,656 |

$$n(n+1) = 96 \times 97 = 9{,}312, \qquad n^3 = (96)^3 = 884{,}736, \qquad r = 22.$$

Then

$$K = \frac{3{,}744 - (96 \times 74.5)}{\sqrt{\dfrac{3{,}744(3{,}744 + 9{,}312) - 192 \times 166{,}656 - 884{,}736}{95}}}$$

$$= \frac{-3{,}408}{\sqrt{168{,}410.27}} = \frac{-3{,}408}{410.4} = -8.3.$$

Again, the observed number of runs is far less than the number to be expected—58—if the observations were independent.

This test and the preceding one would not, of course, both be applied to one set of data. This has been done here simply to illustrate the alternative tests.

### 18.4.2 Movements Up and Down

18.4.2.1 *Runs Up and Down.* Another relatively simple test shows whether the directions of movement tend to cluster, that is, whether directions of movement tend to persist. It would be possible for high and low values to cluster together simply as a result of a few large changes, with directions of changes varying as for independent observations.

EXAMPLE 572 RUNS UP AND DOWN IN THE LEVEL OF LAKE MICHIGAN-HURON

In Table 566, the fifth column shows whether the maximum level each year was higher ($+$) or lower ($-$) than that of the preceding year. The longest movements in one direction were the two five-year declines from 1861 to 1866 and 1886 to 1891. There have been two four-year movements, the rise from 1879 to 1883 and either a decline from 1922 to 1926, or a rise from 1925 to 1929. In 1925 and 1926, the levels were the same, as far as can be determined from these data. It happens that the total number of runs is the same, 48, whether this change is counted $+$ or $-$. If this were not the case, the number of runs would be counted twice, first treating the change as $+$, then as $-$, and the two numbers of runs would be averaged (even if this gives a number not an integer).

Again, the number of runs up and down, which may be designated $R$, is approximately normally distributed. The mean value, under the null hypothesis that the $n$ numbers of the sample have been arranged at random and independently, is

$$M_R = \frac{2n - 1}{3},$$

and the standard deviation is

$$\sigma_R = \sqrt{\frac{16n - 29}{90}},$$

where $n$ is the total number of *observations*, hence is one more than the number of plus and minus signs. A lower-tail probability is required, since the alternative hypothesis is one-sided—that there will be fewer runs than the null hypothesis indicates—so the continuity adjustment consists of adding $\frac{1}{2}$ to $R$. The standard normal variable, after some algebraic rearrangements, is

$$K = \frac{3R - 2n + 2.5}{\sqrt{\dfrac{16n - 29}{10}}}.$$

For the data of Table 566, $n = 96$ and $R = 48$, so

$$K = \frac{144 - 192 + 2.5}{\sqrt{\dfrac{16 \times 96 - 29}{10}}} = \frac{-45.5}{\sqrt{150.7}}$$

$$= -\frac{45.5}{12.3} = -3.7.$$

Table 365 indicates that the probability of so few runs up and down, if there were no real persistence of movement in the same direction, would be less than 0.001. Thus, the lake evidently has a tendency to move consecutively in the same direction more often than would be the case with independent observations. The clustering of high and low values, therefore, is not due (at least not exclusively) to a few large changes, but in some part to cumulative movements up and down.

18.4.2.2 *Predominance of Upward or Downward Changes.* It might be, of course, that the general level of a series is changing. If the trend were large relative to the oscillations about the trend, this could result in few runs of like kinds of observations and also few runs up and down.

A simple test for trend is to count the number of plus and minus signs. If neither direction of movement predominates, the two signs should be equally numerous, except for chance variations.

The number, $S$, of signs of either kind will be normally distributed, in sequences of independent observations, with mean and standard deviation

$$M_S = \frac{n - 1}{2},$$

$$\sigma_S = \sqrt{\frac{n + 1}{12}},$$

where again $n$ is the number of *observations*, or one more than the number of signs.

Note that successive changes are not independent under the null hypothesis of a random, independent arrangement of the observations. To see this, consider the six possible sequences of the numbers 1, 2, 3. With their corresponding sequences of plus and minus signs they are

| 1 | 1 | 2 | 2 | 3 | 3 |
|---|---|---|---|---|---|
| + | + | − | + | − | − |
| 2 | 3 | 1 | 3 | 1 | 2 |
| + | − | + | − | + | − |
| 3 | 2 | 3 | 1 | 2 | 1 |

Since each sequence is equally probable under the null hypothesis, the probability of a plus or a minus is $\frac{6}{12}$ or $\frac{1}{2}$. But the conditional probability that the second change will be +, given that the first is +, is only $\frac{1}{3}$; and the conditional probability that the second will be −, given that the first is +, is $\frac{2}{3}$. Similar results hold if the first sign is −, namely that the next sign is twice as likely to be the opposite as to be the same. For this reason, $\sigma_S$ is less than (only about 58 percent as great as) if the successive signs were independent.[5]

A two-tail probability is ordinarily appropriate here, since if there is a trend it may be either upward or downward. Letting $S$ represent the number of plus or of minus signs, whichever is less numerous, the standard normal variable, incorporating a continuity adjustment, is

$$K = \frac{n - 2(S + 1)}{\sqrt{\dfrac{n + 1}{3}}}$$

The probability shown by Table 365 is to be doubled.

EXAMPLE 574  PREDOMINANT DIRECTION OF CHANGE IN THE LEVEL OF LAKE MICHIGAN-HURON

The data of Table 566 show $n = 96$, $S = 45\frac{1}{2}$ (plus signs), where the case of no change (1925 to 1926) has been counted as $\frac{1}{2}$ plus and $\frac{1}{2}$ minus. Hence

$$K = \frac{96 - 93}{\sqrt{\dfrac{97}{3}}} = \frac{3}{5.69} = 0.53.$$

---

5. If the signs were independent, the standard deviation would be $\sqrt{\dfrac{n-1}{4}}$. (See Sec. 19.3.2; the $n$ there is the number of signs, so corresponds with $n - 1$ here.)

Clearly no evidence of trend is provided by this test, since the discrepancy is less than one standard deviation. From Table 365 a two-tail probability of 0.60 is found. The upward movements might, of course, be on the average larger, or smaller, than the downward movements. A trend of that kind would not be detected by this test.

# 18.5
# MOVING AVERAGES

The data on the maximum monthly mean level of Lake Michigan, Fig. 567, are again charted in Fig. 576. "Erratic" variations from year to year tend to obscure such matters as the underlying directions of movement of the level, or the exact year in which highs and lows were reached. For some purposes it is desirable to smooth out such irregularities, in the hope of getting a truer picture of the basic movements of the series.

Sometimes this is done by fitting a curved line to the data by least squares, letting time be the independent variable. For a series with as complicated a movement as this one, however, least-squares fitting is likely to be impractical. Simple formulas will probably not follow the data closely, and more complex curves may be excessively laborious computationally—especially if the observations are not evenly spaced in time, as fortunately they are in this example.

Another device for giving a smoother description of a time series is the moving average. The point plotted for each date represents not the one measurement made for that date, but an average of that measurement and several neighboring ones, possibly with unequal weights.

Before illustrating the use of a moving average, let us attempt to clarify its purposes. In Sec. 18.4 we discussed tests which detect various types of departure from control in time series. In the Lake Michigan-Huron data, we found serial correlation and possibly some time-shifts in the population mean, but no shifting of the population mean in any consistent direction. When we speak of "smoothing" the time series, then, we have in mind estimating the true population mean for various years. A number of schemes for "decomposing" time series —especially economic series—into systematic shifts of the population mean on the one hand and random components on the other have been widely expounded and used. The systematic shifts are further subdivided into "trend," "cyclical," and "seasonal" components. These methods involve assumptions about the generating mechanism of time series which may be far from realistic, and may, therefore,

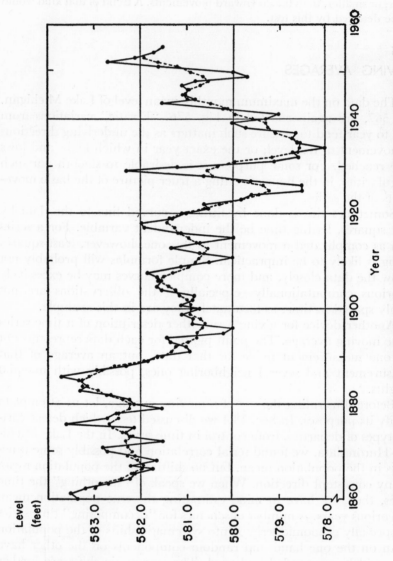

FIG. 576. Lake Michigan-Huron, highest monthly mean level and five-year moving average, 1860-1955.

lead to bad inferences by the time series analyst. The following quotation is a warning to any naive manipulator of time series:

> ... the isolation of cyclical fluctuations is a highly uncertain operation. Edwin Frickey once diligently assembled 23 trend lines fitted by various investigators to pig iron production in the United States, and found that some of the trend lines yield cycles averaging 3 or 4 years in duration while others yield cycles more than ten times as long. This range of results illustrates vividly the uncertainty that attaches to separations of trends and cycles, though it perhaps exaggerates the difficulties. If an investigator fits a trend line in a mechanical manner, without specifying in advance his conception of the secular trend or of cyclical fluctuations, he may get 'cycles' of almost any duration. . . .
>
> It is fairly common for statisticians to assume that the elimination of the secular trend from a time series indicates what the course of the series would have been in the absence of secular movements, and that the graduation of a time series, whether in original or trend-adjusted form, indicates what the course of the series would have been in the absence of random movements. There is no warrant for such simple interpretations. A 'least squares' trend line fitted, for example, to grocery chain store sales in the United States may move majestically on a chart, but the analytic significance of the trend line is obscure. At least some of the 'growth factors' impinging on this branch of business—the addition of meats and vegetables to the grocery line, the rise of supermarkets, special taxes on chain stores—have made their influence felt spasmodically. When a continuous 'trend factor' is eliminated from the data, it is therefore difficult to say what influences impinging on the activity have been removed and what influences have been left in the series. Cyclical graduations are no easier to interpret than trend adjustments. Systematic smoothing of a time series will, indeed, eliminate short-run oscillations produced by random factors; but can it eliminate the influence of powerful random factors—such as a protracted strike, or a succession of bad harvests, or a great war?
>
> There is always danger that the statistical operations performed on the original data may lead an investigator to bury real problems and worry about false ones. . . .[6]

In presenting the method of moving averages, we set forth a relatively simple method of smoothing, and attempt to point out its potential dangers as we explain it. This technique should be regarded primarily as a descriptive rather than an inferential technique. We deliberately omit the full "decomposition" procedure often given in statistics books, as too risky except in the hands of experts.

EXAMPLE 577   FIVE-YEAR MOVING AVERAGE OF ANNUAL HIGHS OF LAKE MICHIGAN-HURON

A five-year moving average has been computed for the lake level series of Table 566, and added to Fig. 576. The method of computing the five-year moving average is illustrated in Table 578.

---

6. Arthur F. Burns and Wesley C. Mitchell, *Measuring Business Cycles* (New York: National Bureau of Economic Research, 1946), pp. 37–38.

*Time Series*

TABLE 578

ILLUSTRATION OF COMPUTATIONS FOR FIVE-YEAR MOVING AVERAGE

| Year | Level[a] | Five-Year Moving | |
|------|----------|------------------|------|
|      |          | Total[b] | Average[a] |
| 1860 | 83.3 |       |       |
| 1    | 83.5 |       |       |
| 2    | 83.2 | 414.8 | 82.96 |
| 3    | 82.6 | 413.6 | 82.72 |
| 4    | 82.2 | 411.8 | 82.36 |
| 5    | 82.1 | 410.8 | 82.16 |
| 6    | 81.7 | 409.8 | 81.96 |
| 7    | 82.2 | 409.7 | 81.94 |
| 8    | 81.6 | 410.3 | 82.06 |
| 9    | 82.1 | 411.4 | 82.28 |
| 1870 | 82.7 | 410.7 | 82.14 |

*Source:* Table 566.

  [a] In feet above 500.
  [b] In feet above 2,500.

The moving total for a given year consists of the observation for that year plus those for the two preceding and the two following years. Each successive moving total is computed from the preceding total by subtracting the earliest observation from the preceding total and adding the first observation after the preceding total. For example, the total for 1863 (413.6) is the 1862 total (414.8) minus the 1860 value (83.3) plus the 1865 value (82.1). The totals are then divided by 5 (that is, multiplied by 0.2) to obtain moving averages.

In interpreting a moving average it is important to remember that the change from one year to the next is in no way affected by the difference between the actual measurements for those two years. From 1867 to 1868, for example, the moving average of Table 578 rises slightly, by 0.12 ft. This reflects the fact that the 1870 value (82.7), which enters the 1868 but not the 1867 average, is 0.6 higher than the 1865 value (82.1), which enters the 1867 but not the 1868 average. The intervening years, 1866 to 1869, are all included in both the 1867 and the 1868 averages.

Moving averages can introduce an appearance of cycles, for moving averages are serially correlated even if the original observations are not. Suppose, for example, that a series is essentially constant except for a sharp, short, upward spasm, as in Fig. 579A. If a moving average is computed, this rise is introduced into the moving average earlier and retained later, but it is damped. In a five-year moving average, for example, it would appear two years earlier and last two years longer—a total of five years instead of one—but it would be only one-fifth as high. A sharp peak is thus converted into a broad plateau, as in Fig. 579B.

This brings out a third characteristic of moving averages, that they may shift the timing of rises and falls. In the example, a rise has been shifted two years earlier. Someone studying the relation of this series to another series which actually moved concurrently, might get

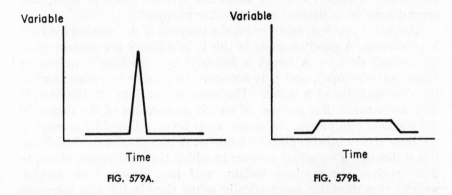

FIG. 579A.                    FIG. 579B.

the notion from the moving average that this series precedes, or leads, the other, and might thereby be put onto false scents in attempting to develop an explanation of the behavior of the series.

Finally, a moving average, at least of this type, cannot be computed for the earliest or for the latest years, since both of these depend on data not available.

There are other types of moving average that are essentially weighted averages in which, unlike the moving averages discussed so far, the weights are not uniform. One of the most common of these occurs when an even number of terms is used in the moving average. Suppose a four-year moving average had been used with the data of Table 566. The average of the first four measurements, 1860 through 1863, would not apply to either 1861 or 1862. The average of the first two four-year moving averages, however, would apply to 1862. But this is really a five-year moving average, 1860 through 1864, with the first and last years weighted only half as much as the three intermediate years. Algebraically,

$$\bar{x}_{1861-62} = \frac{x_{1860} + x_{1861} + x_{1862} + x_{1863}}{4},$$

$$\bar{x}_{1862-63} = \frac{x_{1861} + x_{1862} + x_{1863} + x_{1864}}{4},$$

$$\bar{x}_{1862} = \frac{\bar{x}_{1861-62} + \bar{x}_{1862-63}}{2}$$

$$= \frac{x_{1860} + 2(x_{1861} + x_{1862} + x_{1863}) + x_{1864}}{8}.$$

Obviously, a variety of other weighting systems could be used, and several have been devised for particular purposes.

A special problem arises when the purpose of the moving average is prediction. A good example of this is in military fire-control (that is, aiming) devices. A target is followed by "tracking" equipment (radar or telescope), and it is necessary to predict its course during the time of flight of a missile. The most recent data are the best, if they are correct. But because of erratic movements of the target or the tracking equipment, the course must be established by averaging the most recent observations with others. A common method of doing this is through a weighted average in which the most recent observation receives the greatest weight, and past observations receive weights that diminish geometrically as the time in the past increases arithmetically. For example, if an observation one second old is weighted half as much as the current observation, then an observation two seconds old is weighted one-fourth as much, an observation three seconds old is weighted one-eighth as much, and so on. The sequence of weights might also be 1, 0.9, 0.81, 0.729, etc., or any other sequence in a declining geometric progression, depending on the nature of the application. The averaging (and often the aiming) is done automatically and practically instantaneously by various electro-mechanical devices.

## 18.6
## SEASONAL VARIATION

### 18.6.1 Purpose of Seasonal Adjustment

One source of systematic variation in time series that is often worth analyzing and allowing for is seasonal variation. While seasonal adjustments potentially present many of the same pitfalls a⸱ does removal of secular "trend" or the isolation of "cycles," the within-year patterns are often quite pronounced and regular. Hence adjustments of time series for seasonal variation are, with justification, widely used. If a time series has a marked pattern of variation, within a year, this may hide or exaggerate the basic movement of the series. The level of Lake Michigan, for example, though it has risen or declined in general for periods as long as eight years, has had both substantial rises and substantial falls within every year except one (1951, and even then there was a barely perceptible decline one month). A similar thing is true of many social and economic series, for example, department store sales, marriages, or college graduations. Anyone watching one of these series closely for changes in the basic movement (and the only practically important or successful "forecasting" of economic conditions is that which endeavors to detect changes as soon after they occur as possible) must, therefore, make an allowance for seasonal variation. Many important economic series are published as "seasonally adjusted," meaning that they have been changed to offset purely seasonal factors.

## 18.6.2   Ratio to Moving Average Method

The idea of this method of seasonal adjustment is to determine what fraction of the yearly total is, on the average, represented by each month, each quarter, each week, or other subdivision of the year. The ratio between the actual fraction of the total and a proportionate fraction is the seasonal index for that month. Department stores, on the average, make about one-seventh of their year's sales in December. A proportionate fraction of the year's sales would be one-twelfth.[7] The seasonal index for December is then one-seventh divided by one-twelfth, or 171 percent. Actual sales for any December would, therefore, be "seasonally adjusted" by dividing them by 1.71. July sales, on the other hand, on the average account for only about one-fifteenth of the year's sales. Then the July index is about 1/15 divided by 1/12, or 80 percent. Actual sales in any July would be divided by 0.80 (that is, multiplied by 1.25) to get a seasonally adjusted figure. A seasonally adjusted figure states, therefore, what monthly average is implied for the whole year by the figure observed in a particular month, on the tentative fiction that each year is a scale model of each other year.

EXAMPLE 581   SEASONAL PATTERN OF COLLEGE ENROLMENT

The following example, to bring out the main features of seasonal adjustment, is taken from an analysis of the effect of tuition rates on enrolment at the University of Chicago during the period 1931–42.

---

7. This might be refined slightly by computing the fraction of the year's selling time that comes in December, allowing for Sundays and holidays.

*Time Series*

Enrolment in the Summer, Autumn, Winter, and Spring terms is subject to systematic seasonal variation. This seasonal variation must be allowed for before changes in successive quarterly figures can be interpreted as showing any real change in enrolment. For instance, if one-eighth of all enrolments typically occur in the Winter term, a Winter enrolment of 1,000 students implies, so to speak, a yearly enrolment of 8,000, in units of one full-time student for one term. One-fourth of the implied annual rate gives the *seasonally adjusted* figure of 2,000 for the Winter term.

Enough data are given in Table 583A to show all the steps involved in this method of making seasonal adjustments. First, the yearly moving total that corresponds to each term is the enrolment of that term, plus the enrolments of the two preceding and one succeeding terms. That is, the moving total for the Winter term, 1932, is obtained by adding the enrolments of Summer and Autumn, 1931 and Winter and Spring, 1932. (See the figures which are printed in boldface type in Table 583A.) The moving average is one-fourth of the corresponding moving total.[8]

Next, the actual enrolment for each term is divided by the corresponding moving average; this expresses the term's enrolment as a percentage of the average for the year corresponding to it. To get the seasonal index, the ratios-to-moving average for all Summers, all Autumns, etc., are brought together as in Table 583B and averaged for each season. In this example, the seasonal indexes add to exactly 400.00. If they had added to 399, say, it would have been necessary to multiply each by 400/399. Usually such an adjustment is necessary.

Finally, going back to Table 583A, each actual enrolment figure is divided by the appropriate index, the ratio being the seasonally adjusted enrolment.

The seasonally adjusted data show a decline of 10 percent between the Summer and the Autumn of 1931, where the original data had shown a rise of 18 percent. Thus, "enrolment rises" or "enrolment falls," depending on whether seasonal changes are allowed for or not (see Examples 74B and 74C). Similarly, the original data show an impressive rise of 38 percent between the Summer and Autumn of 1932, but the seasonal adjustment deflates the rise to 5 percent. The seasonal indexes show that the rise from Summer to Autumn averages 32 percent.

---

8. Actually, the third quarter is a little past the center of the year, but the second quarter is a little ahead of the center. It would be better to take as the moving total corresponding with any term an average between (1) the four terms of which the given term is the second and (2) the four terms of which it is the third. Similarly, with monthly data, moving totals are usually centered at the seventh month. That is, the total recorded opposite any given month is the sum of those twelve months for which the given month is the seventh. Here again it would be better to use an average between the two totals in which the given month is the sixth and the seventh, respectively. However, these refinements in centering the moving totals rarely produce a perceptible improvement for monthly data, and not much for quarterly data, so they are scarcely worth the extra trouble, slight though it is.

## 18.6 Seasonal Variation

TABLE 583A

SEASONAL ADJUSTMENT OF ENROLMENT DATA,
UNIVERSITY OF CHICAGO, 1931–1942

| Academic Year | Term | Enrol-ment | Moving Total | Moving Average | Enrolment ÷ Average (percent) | Seasonal Index | Adjusted Enrolment |
|---|---|---|---|---|---|---|---|
| 1931–1932 | Sum | 4,531 | 20,891 | 5,223 | 86.75 | 82.19 | 5,513 |
| | Aut | 5,354 | 20,314 | 5,078 | 105.44 | 108.39 | 4,940 |
| | Win | 5,061 | 19,810 | 4,952 | 102.20 | 105.48 | 4,798 |
| | Spr | 4,864 | 19,110 | 4,778 | 101.80 | 103.94 | 4,680 |
| 1932–1933 | Sum | 3,831 | 19,058 | 4,765 | 80.40 | 82.19 | 4,661 |
| | Aut | 5,302 | 19,104 | 4,776 | 111.01 | 108.39 | 4,892 |
| | Win | 5,107 | 19,043 | 4,761 | 107.27 | 105.48 | 4,842 |
| | Spr | 4,803 | 19,555 | 4,889 | 98.24 | 103.94 | 4,621 |
| . | . | . | . | . | . | . | . |
| . | . | . | . | . | . | . | . |
| . | . | . | . | . | . | . | . |
| 1941–1942 | | | | | | | |

TABLE 583B

COMPUTATION OF SEASONAL INDEX FOR UNIVERSITY
OF CHICAGO ENROLMENT, 1931–1942

| Year | Ratio of Enrolment to Moving Average (percent) | | | |
|---|---|---|---|---|
| | Summer | Autumn | Winter | Spring |
| 1931–1932 | 86.75 | 105.44 | 102.20 | 101.80 |
| 1932–1933 | 80.40 | 111.01 | 107.27 | 98.24 |
| . | . | . | . | . |
| . | . | . | . | . |
| . | . | . | . | . |
| 1941–1942 | | | | |
| Average (Seasonal Index) | 82.19 | 108.39 | 105.48 | 103.94 |

Comparison with the same period a year earlier is not as satisfactory as using seasonally adjusted data. For example, the seasonally adjusted data show a rise between the Summer and Autumn of 1932, indicating that enrolment was increasing, whereas the Autumn of 1932 was slightly below the Autumn of 1931, which would have suggested that enrolment was decreasing. In general, comparison with the corresponding period a year earlier has the drawback of comparison with a single observation which may itself have had a sizable random component. It also has the drawback that the current figure may be below a year ago, even though currently the direction of movement is upward, simply because of a large decline in the intervening year.

## 18.6.3   Difference from Moving Average Method

The method of Sec. 18.6.2 is perhaps the commonest method of seasonal adjustment, at least of economic time series, but other methods might be used in other applications. The method we now describe is identical to the previous one up to the point at which the observation for a given month is compared with the moving average. Then, instead of computing the ratio of the actual figure to the moving average, the *difference* is computed. For each of the twelve months, the differences for each of the years covered by the monthly data are averaged. This average difference is then subtracted from the current observation for that month in order to obtain the seasonally adjusted observation. We shall illustrate the mechanics by an example, and then discuss the conditions under which this method might be preferred to that of Sec. 18.6.2.

EXAMPLE 584   SEASONAL PATTERN OF LAKE LEVEL,
LAKE MICHIGAN-HURON

Table 584 gives the monthly mean elevations of Lake Michigan-Huron for 1860–1862.

TABLE 584

MONTHLY MEAN ELEVATIONS, LAKE MICHIGAN-HURON,
1860–1862

| Year | Jan | Feb | Mar | Apr | May | Jun |
|------|-----|-----|-----|-----|-----|-----|
| 1860 | 582.68 | 582.86 | 582.89 | 583.02 | 583.14 | 583.26 |
| 1861 | 582.00 | 582.09 | 582.48 | 582.58 | 583.00 | 583.16 |
| 1862 | 582.49 | 582.34 | 582.64 | 582.80 | 583.05 | 583.18 |

| Year | Jul | Aug | Sep | Oct | Nov | Dec |
|------|-----|-----|-----|-----|-----|-----|
| 1860 | 583.30 | 583.11 | 582.91 | 582.60 | 582.27 | 582.11 |
| 1861 | 583.29 | 583.53 | 583.22 | 583.10 | 582.87 | 582.70 |
| 1862 | 583.08 | 583.07 | 583.00 | 582.89 | 582.50 | 582.36 |

*Source:* U.S. Army, Corps of Engineers, U.S. Lake Survey.

We first compute a 12-month centered moving average.[9] For July 1860, for example, the computation is as follows:

(1) Compute the total for the twelve-month period in which July 1860 is the 7th month, that is, the twelve months of 1860:

$$582.68 + 582.86 + \cdots + 582.11 = 6{,}994.15.$$

___

9. The refinement of centering is not really necessary, but it is, if anything, an improvement, and was used by the Corps of Engineers in their computations, the results of which we present here.

(2) Compute the total for the twelve-month period in which July 1860 is the 6th month, that is, February 1860 through January 1861. This can be done easily by subtracting the January 1860, and adding the January 1861 figure, to the total of (1):

$$6{,}994.15 - 582.68 + 582.00 = 6{,}993.47.$$

(3) Add the sums obtained in (1) and (2), then divide by 24:

$$\frac{6{,}994.15 + 6{,}993.47}{24} = 582.82.$$

The actual value for July 1860, is 583.30, and this is higher than the moving average by $583.30 - 582.82 = 0.48$. Similarly, the moving average for July 1861 is 582.86, and the actual value is 583.29, so the difference is $583.29 - 582.86 = 0.43$. This process is continued for all the Julies for the years 1860–1951.[10] The average of the 92 July differences was 0.53. That is, on the average, the actual July value was 0.53 feet higher than the July moving average.

In July, 1955, the lake level was 581.37. Since July has been typically high by 0.53, the seasonally adjusted figure is

$$581.37 - 0.53 = 580.84.$$

The deviations for all months are shown in Table 585, which shows that the high month, July, averages 1.00 foot above the low month, February.

TABLE 585

Average Monthly Deviations from 12-Month Moving Average,
Lake Michigan-Huron, 1860–1951

| Month | Deviation in Feet |
|---|---|
| Jan. | −0.45 |
| Feb. | −0.47 |
| Mar. | −0.39 |
| Apr. | −0.16 |
| May | +0.15 |
| Jun. | +0.39 |
| Jul. | +0.53 |
| Aug. | +0.48 |
| Sep. | +0.29 |
| Oct. | +0.08 |
| Nov. | −0.13 |
| Dec. | −0.33 |

*Source:* Louis D. Kirshner and M. Asce, "Forecasting Great Lakes Levels Aids Power and Navigation," *Civil Engineering*, Vol. 24 (1954), p. 100.

These seasonal adjustments are actually used by the Corps of Engineers as an aid in forecasting levels of Lake Michigan-Huron, and similar adjustments are used for the other three Great Lakes.

---

10. At the time the computation was made, more recent data were not available.

### 18.6.4 The Choice of Methods of Seasonal Adjustment

In Secs. 18.6.2 and 18.6.3, we have described two methods of seasonal adjustment which, though similar in general approach, make quite different assumptions about the phenomenon being studied. The ratio method assumes that the effect of seasonal variation is proportional to the level of the series, as estimated by the moving average. Thus, in Sec. 18.6.2, the seasonal enrolment was estimated by averaging percentage deviations from the moving average. For the years 1931 to 1942, the actual Autumn figures averaged about 8 percent higher than the moving average. It is assumed that the yearly percentage deviations for Autumn terms will tend to be about the same for low levels of enrolment as for high. This assumption can, of course, be tested by the data themselves. If it is correct, the percentage deviations for successive Autumn terms should behave like a time series in a state of statistical control.

The difference method, by contrast, assumes that the seasonal effects are independent of the level of the series. For example, if the July lake level averages one-half foot above the moving average during 1860 to 1951, it should tend to be about one-half foot above in years of high and low lake levels alike. The mechanism of lake level change makes it seem plausible that the seasonal forces should be largely independent of the general lake level:

> Mean water-surface evaluations of the lakes are the result of all the factors which either add or subtract water. Water is added by precipitation on the lake surface, tributary stream runoff, diversions into the lakes, condensation on the surface, inseepage, and inflow from the lakes above. Water is subtracted by outflow to the lakes below, evaporation, diversions from the lakes, and outseepage.[11]

Again, however, the accuracy of the assumption can be tested by the data themselves. If the assumption is valid, the yearly sequence of differences for any given month should vary from year to year like a series in a state of statistical control, and in particular should not tend to increase with the level of the lake.

Seasonal adjustment of data is by no means always as straightforward and objective as the foregoing discussion may suggest. Seasonal patterns may themselves be subject to trends and to "cycles." The seasonal pattern of steel production, for example, is a weighted average of the seasonal patterns of all the uses of steel. In times of prosperity, the seasonal pattern of steel for automobiles is weighted

---

11. Kirshner and Asce, *op. cit.*, p. 99.

much more heavily relative to that of tinplate for canning food than is the case in times of depression, for purchases of canned food fluctuate less during a business cycle than do purchases of new automobiles. Thus, the seasonal pattern shifts with the business cycle because of shifts in the proportions of steel going into uses with different seasonal patterns. Since ordinarily one of the first steps in studying business cycles is to make a seasonal adjustment of the data, there is danger of either eliminating part of the cyclical fluctuation or of interpreting as cyclical some fluctuation that is really seasonal.

## 18.7
## CONCLUSION

Time series offer many special difficulties for statistical analysis. Chief among these are (1) that their movements are often complicated, and any meaningful analysis depends primarily upon careful formulation of problems and assumptions by experts in the subject matter to which the series relate, and (2) that the presence of serial correlation among the observations—that is, correlation among the deviations from their respective population means—invalidates most of the common statistical techniques and necessitates special techniques for time series.

A simple test for bunching of similar observations may be made by classifying the observations into a few categories and counting the number of runs, that is, sequences in which all observations are in the same category. A test for persistence of movement in a given direction may be made by marking each observation (except the first) plus or minus, according to its difference from the preceding observation, and counting the number of runs up and down. A similar test, one for trend, may be made from the number of positive and negative changes.

Smoothing of time series is sometimes done by least squares when changes in slope are few. More complicated series are often smoothed by moving averages, that is by averages that include points near the one to which the average applies. Moving averages, like most useful things, must be handled with discretion, for they can introduce the appearance of "cycles" and can shift the timing of changes in direction. Various specially-weighted averages can improve the usefulness of moving averages in particular circumstances.

Allowance can be made for recurrent seasonal movements in a series by determining seasonal indexes. These represent the average ratio between the actual fraction of a year's total which occurs in each month (or other period), and a proportionate fraction (one-twelfth for monthly data). Dividing the actual data by these seasonal indexes provides seasonally adjusted data, that is, data showing the average monthly rate for the year implied by the observed amount for a given month. Complications arise when seasonal patterns themselves change, either with time or with the level of the series. An alternative seasonal adjustment, which is sometimes to be preferred, is based on averages of deviations rather than ratios.

## DO IT YOURSELF

EXAMPLE 588A

Apply the tests of Secs. 18.4.1.1 and 18.4.2 to the data of Table 563, and summarize your conclusions. Are your conclusions consistent with what you would have expected in view of the method by which Table 563 was made?

EXAMPLE 588B

Same requirements as Example 588A, for Fig. 565.

EXAMPLE 588C

Apply the tests of Secs. 18.4.1.2 and 18.4.2 to the travel expense data of Tables 502A and 502B, and summarize your conclusions.

EXAMPLE 588D

The following are ratios of total Republican to total Democratic votes for candidates for the House of Representatives, 1920–1954.

TABLE 588

RATIO OF TOTAL REPUBLICAN TO TOTAL DEMOCRATIC VOTES,
CANDIDATES FOR HOUSE OF REPRESENTATIVES, 1920–1954

| Year | Ratio Republican to Democratic Vote | Year | Ratio Republican to Democratic Vote |
|------|------|------|------|
| 1920 | 1.65 | 1938 | 0.97 |
| 1922 | 1.16 | 1940 | 0.89 |
| 1924 | 1.38 | 1942 | 1.10 |
| 1926 | 1.41 | 1944 | 0.93 |
| 1928 | 1.33 | 1946 | 1.21 |
| 1930 | 1.18 | 1948 | 0.88 |
| 1932 | 0.76 | 1950 | 0.998 |
| 1934 | 0.78 | 1952 | 1.002 |
| 1936 | 0.71 | 1954 | 0.90 |

*Source: Statistical Abstract: 1955*, Table 390, p. 330.

(1) Plot the data.
(2) Analyze the series by the tests of Sec. 18.4 and summarize your conclusions.

*Do It Yourself*

EXAMPLE 589A

Find a time series of interest to you, plot it, and analyze it as in Example 588D.

EXAMPLE 589B

Obtain monthly weather data—for example, monthly mean temperature—for your area, covering at least ten years. Compute seasonal adjustments by the method of Sec. 18.6.3, and use these adjustments to get seasonally-adjusted data. Plot both unadjusted and adjusted data.

EXAMPLE 589C

The following data show total live births by months in the United States, 1948–1953.

TABLE 589

LIVE BIRTHS BY MONTHS IN THE UNITED STATES, 1948–1953

| Month | 1948[a] | 1949[a] | 1950[a] | 1951[b] | 1952[b] | 1953[b] |
|---|---|---|---|---|---|---|
| Total | 3,535,068 | 3,559,529 | 3,554,149 | 3,750,850 | 3,846,986 | 3,902,120 |
| January | 295,494 | 299,255 | 297,276 | 303,538 | 311,626 | 322,488 |
| February | 285,694 | 273,195 | 272,604 | 282,118 | 300,218 | 296,312 |
| March | 300,463 | 300,117 | 294,038 | 312,820 | 317,178 | 315,132 |
| April | 277,636 | 270,770 | 258,868 | 295,924 | 292,028 | 286,962 |
| May | 272,277 | 281,595 | 275,786 | 312,970 | 300,366 | 307,382 |
| June | 267,712 | 285,442 | 293,879 | 306,788 | 311,340 | 321,246 |
| July | 308,010 | 318,218 | 315,538 | 328,208 | 345,452 | 354,464 |
| August | 320,968 | 322,774 | 325,094 | 334,264 | 350,476 | 356,450 |
| September | 312,511 | 312,623 | 315,375 | 328,708 | 343,682 | 347,740 |
| October | 307,720 | 311,480 | 311,905 | 329,166 | 336,136 | 334,202 |
| November | 290,552 | 290,956 | 292,497 | 304,302 | 315,148 | 319,966 |
| December | 296,031 | 293,104 | 301,289 | 312,044 | 323,336 | 339,776 |

[a] National Office of Vital Statistics, *Vital Statistics–Special Reports, National Summaries,* Vol. 37 (1950), p. 152.

[b] *Ibid.,* Vol. 42 (1955), p. 254. Based on a 50 percent sample.

(1) Plot the data.
(2) Construct a seasonal index, using the method of Sec. 18.6.2.
(3) Compute and plot seasonally adjusted data.

EXAMPLE 589D

Here is a record of the winner of the All Star Baseball Game between the American (A) and National (N) League in each of the years it has been played. (No game was played in 1945.)

TABLE 590

WINNERS OF ALL-STAR BASEBALL GAMES, 1933–1955

| Year | Winner | Year | Winner |
|------|--------|------|--------|
| 1933 | A | 1944 | N |
| 1934 | A | 1946 | A |
| 1935 | A | 1947 | A |
| 1936 | N | 1948 | A |
| 1937 | A | 1949 | A |
| 1938 | N | 1950 | N |
| 1939 | A | 1951 | N |
| 1940 | N | 1952 | N |
| 1941 | A | 1953 | N |
| 1942 | A | 1954 | A |
| 1943 | A | 1955 | N |

Analyze this series by the method of Sec. 18.4.1.1. State carefully what your conclusion means.

# Chapter 19

# *Shortcuts*

## 19.1
### THE PLACE OF SHORTCUTS

The preceding eighteen chapters are intended primarily for readers or hearers of material that involves, or should involve, statistical reasoning. Instructions for computing are given so that you can follow the details of our illustrations and try out a few examples of your own, to get the "feel." Often, the computing methods that we have presented are clumsier than they might have been had we not avoided technical material and special apparatus such as tables and nomograms, which require study themselves but do not contribute to the essential statistical ideas.

An example is the problem of finding the probability corresponding with a given value of $F$, the variance ratio, which arose in testing whether several population means are equal (Sec. 13.2.3.1). Practicing statisticians have tables, several pages long, which they use instead of computing the standard normal variable, $K$, which gives only an approximate probability. (Actually, most statisticians have easy access to such tables for only a few probability levels—significance levels—and proper interpolation requires formidable computations.) Study of Sec. 13.2.3 has, we hope, given you an understanding of the elementary principles of the analysis of variance. For serious computations employing it extensively, you will want to obtain and learn to use, the proper tables.[1] Reasonably good approximate probabilities, however, can be obtained by the graphical device described

---

1. Three good collections of tables for statisticians are:

E. S. Pearson and H. O. Hartley (editors), *Biometrika Tables for Statisticians* (Cambridge, England: Cambridge University Press, 1954).

Ronald A. Fisher and Frank Yates, *Statistical Tables for Biological, Agricultural, and Medical Research* (4th ed.; London and Edinburgh: Oliver and Boyd, Ltd., 1953).

A. Hald, *Statistical Tables and Formulas* (New York: John Wiley and Sons, Inc., 1952).

in Sec. 19.6 below. For occasional, informal tests, probably you will do better not to use the method of Sec. 13.2.3.1 at all, but the shortcut substitute for it in this chapter (Sec. 19.4.1).

In other words, this chapter is oriented not toward your reading, but toward your doing it yourself. For this, we recommend quick and easy methods. They may not be as good as some of the methods described earlier *if* those methods are executed perfectly; but they may do better in your hands—just as a box camera takes better pictures in our hands than one complete with power focusing and electronic shutter control.

Moreover, the statistical analogs of the box camera are often used even by experts in circumstances in which the fancier gadgets are not applicable. Shortcut methods, it happens, tend to be more foolproof in interpretation than standard methods, for most shortcut methods are *non-parametric*. This means that they are valid regardless of the population from which the samples come, provided only that the observations are random and independent. In other words, they do not depend on such assumptions as that the population is normal. (Not all non-parametric tests are quick and easy, by any means, even though most quick and easy methods are non-parametric.)

As a matter of fact, we have given three examples of non-parametric tests in the preceding chapter. Tests of runs of several kinds of observations (Sec. 18.4.1) do not depend on the population frequencies of the various kinds of observations. Tests of movements up and down, either of the number of movements in a given direction (Sec. 18.4.2.2) or of runs up and down (Sec. 18.4.2.1) involve exactly the same sampling distribution whatever the distribution of the observations (provided only that "ties"—consecutive observations that are equal —are rare).

Shortcut methods commonly are based on such devices as replacing measurements by a few general classifications (for example,

---

Four special but useful tables for which almost any practicing statistician will find much use are:

*Barlow's Tables of Squares, Cubes, Square Roots, Cube Roots and Reciprocals of all Integer Numbers up to 12,500* (New York: Chemical Publishing Company, Inc., 1944).

*Tables of Normal Probability Functions* (National Bureau of Standards, Applied Mathematics Series 23) (Washington: Government Printing Office, 1953).

*Tables of the Cumulative Binomial Probability Distribution*, Harvard University Computation Laboratory (Cambridge, Mass.: Harvard University Press, 1955).

*A Million Random Digits, with 100,000 Normal Deviates*, The Rand Corporation (Glencoe, Illinois: Free Press, 1955).

Published in diverse places, there are literally scores—indeed hundreds—of useful tables and nomograms which can make a practicing statistician's life easier—if he can find them when he wants them, and can remember how to use them correctly and efficiently.

above or below some value) or by ranks (the smallest observation becomes 1, the next smallest 2, and so forth, the largest becoming $n$). Often the question asked is shifted somewhat in the interest of short-cuts, for example, from a question about the mean to a question about the median.

In addition to shortcut methods, there are various shortcut devices for use with standard methods, such as tables, charts, and nomograms, one of which is presented in Sec. 19.6.

## 19.2
## CONFIDENCE INTERVALS AND SIGNIFICANCE TESTS
## FOR AN AVERAGE

A typical shortcut method is the following: A 95 percent confidence interval for the median can be obtained from a sample of $n$ by finding the integer closest to

$$\frac{n+1}{2} - \sqrt{n},$$

then counting up this number of observations from the smallest and down this number from the largest observation in the sample.

EXAMPLE 593   WEIGHTS

For the 32 weights arrayed in Table 173,

$$\frac{32+1}{2} - \sqrt{32} = 16.5 - 5.7 = 10.8,$$

and the nearest integer is 11. Observation number 11 in the array is 165 pounds, and number 22 ($=32 - 11 + 1$) is 185 pounds. Therefore, a 95 percent confidence interval for the median is 165 to 185. The distribution is sufficiently symmetrical so that the median is probably a reasonable substitute for the mean.

This interval is about one-third wider than the interval computed from the same data in Sec. 14.3.3. Such greater width is typical. The reason is that in Sec. 14.3.3 we took account of the actual values of the observations. Here we have taken account only of the facts that 11 of the observations are 165 or less and 11 are 185 or more.

We would replace $\sqrt{n}$ by $1.3\sqrt{n}$ for a 99 percent confidence interval or by $0.8\sqrt{n}$ for a 90 percent confidence interval.

Had the problem been to make a two-sided test at the 5 percent level of the null hypothesis that the population median has some specified value, we would have computed the same confidence interval, and then accepted the null hypothesis if it specified a value in the interval or rejected it if it specified a value outside.

## 19.3
## COMPARING TWO AVERAGES

### 19.3.1 Independent Samples

A quick and easy test for the problem discussed in Sec. 13.2.2.1, where two populations are to be compared on the basis of two independent samples, can be made by ranking all observations of both samples combined from 1 to $N$, where $N = n_1 + n_2$. (If there are ties, give each tied observation the mean of the ranks for which it is tied; but if more than one-fourth of the observations are involved in ties, the method is not suitable without a special adjustment.) Then compute $R$, the sum of the ranks of either sample—the smaller sample is easier. Next, compute the approximate standard normal variable

$$K = \frac{2R \pm 1 - n(N+1)}{\sqrt{\dfrac{n(N+1)(N-n)}{3}}},$$

where $n$ is the size of the sample from which $R$ is taken, and $\pm 1$ is taken as $-1$ if $2R$ exceeds $n(N+1)$ or as $+1$ if $n(N+1)$ exceeds $2R$. If $2R = n(N+1)$, $K$ is taken as 0. The probability of this value of $K$ is then taken from Table 365 and doubled for a two-tail test. (For a lower-tail test, $\pm 1$ is always taken as $+1$, and for an upper-tail test as $-1$, regardless of the value of $R$.)

EXAMPLE 594  OPERATING COSTS PER MILE FOR FORDS
AND CHEVROLETS

In 1953 a large firm analyzed records for a random sample of cars purchased in 1951 and operated during 1952 by company salesmen. For each car, operating expenses—gas, oil, repairs and preventive maintenance—during the months of February, May, July, and December, 1952 were ascertained from the salesmen's reports. The total expense was then divided by the total mileage during these months. The resulting expenses per mile for 17 Fords and 18 Chevrolets are shown in Table 595A. Observations tied to two decimals have been carried to an additional decimal to break the tie.

*First*, we rank the observations from 1 to $N$, that is, 1 to 35, as in Table 595B.

## 19.3 Comparing Two Averages

TABLE 595A

OPERATING COSTS PER MILE, IN 1952, CHEVROLETS AND FORDS
PURCHASED IN 1951 (Cents per Mile)

| Chevrolet | | Ford | |
|---|---|---|---|
| 3.926 | 4.08 | 4.70 | 1.56 |
| 3.45 | 3.67 | 4.15 | 4.29 |
| 2.00 | 2.94 | 4.55 | 1.74 |
| 2.28 | 5.90 | 3.31 | 2.17 |
| 3.494 | 2.18 | 2.13 | 1.97 |
| 4.25 | 5.39 | 4.686 | 4.689 |
| 2.38 | 2.74 | 2.68 | 2.87 |
| 3.02 | 3.492 | 2.36 | 3.17 |
| 3.26 | 2.70 | 3.934 | |

TABLE 595B

RANKING OF OBSERVATIONS OF TABLE 595A

| Rank | Obser-vation | Make | Rank | Obser-vation | Make | Rank | Obser-vation | Make |
|---|---|---|---|---|---|---|---|---|
| 1 | 1.56 | F | 13 | 2.74 | C | 25 | 3.934 | F |
| 2 | 1.74 | F | 14 | 2.87 | F | 26 | 4.08 | C |
| 3 | 1.97 | F | 15 | 2.94 | C | 27 | 4.15 | F |
| 4 | 2.00 | C | 16 | 3.02 | C | 28 | 4.25 | C |
| 5 | 2.13 | F | 17 | 3.17 | F | 29 | 4.29 | F |
| 6 | 2.17 | F | 18 | 3.26 | C | 30 | 4.55 | F |
| 7 | 2.18 | C | 19 | 3.31 | F | 31 | 4.686 | F |
| 8 | 2.28 | C | 20 | 3.45 | C | 32 | 4.689 | F |
| 9 | 2.36 | F | 21 | 3.492 | C | 33 | 4.70 | F |
| 10 | 2.38 | C | 22 | 3.494 | C | 34 | 5.39 | C |
| 11 | 2.68 | F | 23 | 3.67 | C | 35 | 5.90 | C |
| 12 | 2.70 | C | 24 | 3.926 | C | | | |

*Second*, the sum of the ranks for, say, Ford is computed from Table 595B as

$$R = 1 + 2 + 3 + 5 + 6 + 9 + \cdots + 33$$
$$= 294.$$

*Third*, since $n = 17$ and $N = 35$, and since $2R < n(N + 1)$, we compute the lower-tail probability from

$$K = \frac{2R + 1 - n(N + 1)}{\sqrt{\dfrac{n(N + 1)(N - n)}{3}}} = \frac{(2 \times 294) + 1 - (17 \times 36)}{\sqrt{\dfrac{17 \times 36 \times 18}{3}}}$$

$$= \frac{-23}{\sqrt{3672}} = -\frac{23}{60.60} = -0.38,$$

which corresponds with a two-tail probability of 0.70 (Table 365).

These data do not indicate, therefore, a significant difference between the two makes of car in operating cost per mile. The differences among cars of the same make are sufficient to account for the apparent difference between the makes.

This test is called the *Wilcoxon test,* or—to distinguish it from the test of the next section—the *Wilcoxon two-sample test.*

### 19.3.2 Matched Samples

To compare two population means when the two samples are matched, the differences between corresponding observations are computed, as in Sec. 13.2.2.2. Any differences that are precisely zero are ignored, and $n$, the number of pairs, is reduced accordingly. The absolute values of the remaining differences (that is, the differences without regard to sign) are ranked. (If some differences are tied, each is assigned the mean of the ranks tied for; but if more than one-fourth of the differences are involved in ties, the method is not applicable without a special adjustment.) The sum of the ranks is computed for all those differences that were negative, and another sum of ranks for all those differences that were positive. (As a check, those two sums of ranks should total $n(n + 1)/2$, where $n$ is the number of non-zero differences.) Let $T$ be the *smaller* of the two sums of ranks. Then

$$K = \frac{2T + 1 - \dfrac{n(n + 1)}{2}}{\sqrt{\dfrac{n(n + 1)(2n + 1)}{6}}}$$

is approximately a standard normal variable. The two-tail probability obtained from Table 365 is the probability of as much difference between the two samples as observed, if the null hypothesis (that the two population averages are the same) is true.[2]

EXAMPLE 596    STRESS AND LOSS OF TENSILE STRENGTH

In a study of the corrosive effects of a salt-hydrogen peroxide solution on a certain alloy, one problem was to find out if there was a significant difference in response to stress. Two samples of the alloy, one subjected to stress and one not, were immersed simultaneously in the solution and the loss in tensile strength was afterward measured for each sample. This was

---

2. For an upper-tail test (to test against the one-sided alternative hypothesis that the average difference is positive), let $T$ be the sum of the *positive* ranks (whether or not this is the smaller sum of ranks), and replace $2T + 1$ by $2T - 1$. For a lower-tail test, proceed as described in the text, but use the one-tail probability shown in Table 365 without doubling it.

repeated for 11 other pairs of samples. The results are shown in Table 597A below.

TABLE 597A

PERCENT LOSS IN TENSILE STRENGTH,
IMMERSION IN CORROSIVE SOLUTION OF
PAIRED SAMPLES, STRESSED AND UNSTRESSED

| Test Number | Unstressed | Stressed | Difference |
|:-----------:|:----------:|:--------:|:----------:|
| 1  | 6.4 | 9.2 | 2.8  |
| 2  | 4.6 | 7.9 | 3.3  |
| 3  | 4.6 | 7.3 | 2.7  |
| 4  | 6.4 | 8.0 | 1.6  |
| 5  | 3.2 | 5.7 | 2.5  |
| 6  | 5.2 | 7.6 | 2.4  |
| 7  | 6.5 | 5.7 | −0.8 |
| 8  | 4.9 | 4.1 | −0.8 |
| 9  | 4.3 | 8.1 | 3.8  |
| 10 | 5.6 | 6.5 | 0.9  |
| 11 | 3.7 | 6.9 | 3.2  |
| 12 | 4.6 | 6.0 | 1.4  |

To facilitate the analysis, we have, in Table 597B, ranked the differences of Table 597A according to their absolute values.

TABLE 597B

RANKING OF DIFFERENCES OF TABLE 597A
ACCORDING TO ABSOLUTE VALUE

| Rank | Absolute Value of Difference | Sign of Difference |
|:----:|:----------------------------:|:------------------:|
| 1.5 | 0.8 | − |
| 1.5 | 0.8 | − |
| 3   | 0.9 | + |
| 4   | 1.4 | + |
| 5   | 1.6 | + |
| 6   | 2.4 | + |
| 7   | 2.5 | + |
| 8   | 2.7 | + |
| 9   | 2.8 | + |
| 10  | 3.2 | + |
| 11  | 3.3 | + |
| 12  | 3.8 | + |

First, the sum of ranks for negative differences is $1.5 + 1.5 = 3$. Hence $T = 3$. While a check is not really needed in this example, we note that the sum of the ranks for positive differences is $3 + 4 + \cdots + 11 + 12 = 75$, and $3 + 75 = 78$, which in turn equals $\frac{1}{2}n(n + 1)$.

Second, noting that $n = 12$, we obtain

$$K = \frac{2T + 1 - \frac{n(n+1)}{2}}{\sqrt{\frac{n(n+1)(2n+1)}{6}}} = \frac{(2 \times 3) + 1 - 78}{\sqrt{\frac{12 \times 13 \times 25}{6}}}$$

$$= \frac{-71}{25.50} = -2.78,$$

and from Table 365 we obtain a lower-tail probability of less than 0.003. Clearly, stress is associated with a greater loss of tensile strength.

A confidence interval for the median difference can be found by treating the $n$ differences as in Sec. 19.2.

This test is called the *Wilcoxon signed-rank test*. It is generally superior to (that is, has a steeper operating-characteristic curve than) another shortcut test often used for the same problem, namely the *sign test*. The sign test may be useful, however, if many of the differences are ties. In the sign test, the numbers of positive and negative differences are counted. Let $s$ be the number of occurrences of the less numerous of the two signs, and $n$ be the total number of signs—that is, the number of pairs of observations *less* the number in which the difference was zero, so far as the available data show. Then

$$K = \frac{2s + 1}{\sqrt{n}} - \sqrt{n}$$

is approximately a standard normal variable, and again the two-tail probability is obtained from Table 365.

To illustrate the calculations, we apply the sign test to the data of Table 597A.

EXAMPLE 598 ALTERNATIVE ANALYSIS OF EXAMPLE 596

In this example, $n = 12$ and $s = 2$. Hence

$$K = \frac{2s + 1}{\sqrt{n}} - \sqrt{n} = \frac{5}{\sqrt{12}} - \sqrt{12} = 1.443 - 3.464$$

$$= -2.02,$$

and from Table 365 we obtain a lower-tail probability of about 0.02. The greater loss of tensile strength associated with stress again appears to be real.

## 19.4
## COMPARING SEVERAL AVERAGES

### 19.4.1 Independent Samples

A quick and easy method here is very much like that for two samples presented in Sec. 19.3.1. If there are $k$ samples, of sizes $n_1$, $n_2, \ldots, n_k$, a total of $N$ in all samples, all $N$ observations are ranked (giving ties the mean of the ranks tied for) and the sums of the ranks, $R_1, R_2, \ldots, R_k$, are computed for the separate samples. Then compute

$$H = \frac{12}{N(N+1)} \sum_{i=1}^{k} \frac{R_i^2}{n_i} - 3(N+1).$$

If all groups contain at least 3 observations, this variable has approximately the chi-square distribution, which is not covered in this book (except for a brief explanation in Secs. 13.3.3.1 and 13.4.4); but a method of finding the approximate probability graphically is shown later in this chapter (Sec. 19.6.2). The "degrees of freedom" for this case is $k - 1$. Alternatively, the probability of $H$ can be approximated from the standard normal variable

$$K = \sqrt{2H} - \sqrt{2k - 3}$$

as in Sec. 13.3.3.1, or from the more accurate standard normal variable formula given in Sec. 13.4.4.

EXAMPLE 599  OPERATING COSTS PER MILE FOR FORDS,
CHEVROLETS, AND PLYMOUTHS

The study described in Example 594 included data on a sample of 18 Plymouths as well as the 17 Fords and 18 Chevrolets. Data on all three makes are shown in Table 600A, the figures for Ford and Chevrolet being taken from Table 595A, but with additional decimals where necessary to resolve ties.

*First*, rank the observations from 1 to $N$, that is, 1 to 53, as in Table 600B. An intermediate step which facilitates the ranking is to list the data, with their appropriate symbols indicating make, in class intervals, say 1.00 to 2.00, 2.00 to 3.00, etc., or 1.00 to 1.50, 1.50 to 2.00, etc.; then the rankings within classes are easily made.

*Second*, compute the sum of the ranks, $R_i$, for each make.

Chevrolet: $R_1 = 5 + 8 + 10 + \cdots + 51 + 52 = 530.$

Ford: $\quad\ \ R_2 = 1 + 3 + 4 + \cdots + 48 + 49 = 454.$

Plymouth: $R_3 = 2 + 9 + 13 + \cdots + 50 + 53 = 447.$

TABLE 600A

Operating Costs per Mile in 1952, Chevrolets, Fords,
and Plymouths, Purchased in 1951
(Cents per Mile)

| Chevrolet | | Ford | | Plymouth | |
|---|---|---|---|---|---|
| 3.926 | 4.08 | 4.70 | 1.56 | 2.43 | 2.524 |
| 3.45 | 3.67 | 4.15 | 4.29 | 2.98 | 3.10 |
| 2.00 | 2.94 | 4.55 | 1.74 | 3.04 | 3.53 |
| 2.28 | 5.90 | 3.31 | 2.17 | 4.94 | 3.06 |
| 3.494 | 2.18 | 2.13 | 1.97 | 3.15 | 2.57 |
| 4.25 | 5.39 | 4.686 | 4.689 | 2.46 | 3.48 |
| 2.382 | 2.74 | 2.68 | 2.87 | 3.34 | 5.94 |
| 3.02 | 3.492 | 2.36 | 3.17 | 2.384 | 2.516 |
| 3.26 | 2.70 | 3.934 | | 2.27 | 1.61 |

TABLE 600B

Ranking of Observations of Table 600A

| Rank | Obser-vation | Make | Rank | Obser-vation | Make | Rank | Obser-vation | Make |
|---|---|---|---|---|---|---|---|---|
| 1 | 1.56 | F | 19 | 2.68 | F | 37 | 3.494 | C |
| 2 | 1.61 | P | 20 | 2.70 | C | 38 | 3.53 | P |
| 3 | 1.74 | F | 21 | 2.74 | C | 39 | 3.67 | C |
| 4 | 1.97 | F | 22 | 2.87 | F | 40 | 3.926 | C |
| 5 | 2.00 | C | 23 | 2.94 | C | 41 | 3.934 | F |
| 6 | 2.13 | F | 24 | 2.98 | P | 42 | 4.08 | C |
| 7 | 2.17 | F | 25 | 3.02 | C | 43 | 4.15 | F |
| 8 | 2.18 | C | 26 | 3.04 | P | 44 | 4.25 | C |
| 9 | 2.27 | P | 27 | 3.06 | P | 45 | 4.29 | F |
| 10 | 2.28 | C | 28 | 3.10 | P | 46 | 4.55 | F |
| 11 | 2.36 | F | 29 | 3.15 | P | 47 | 4.686 | F |
| 12 | 2.382 | C | 30 | 3.17 | F | 48 | 4.689 | F |
| 13 | 2.384 | P | 31 | 3.26 | C | 49 | 4.70 | F |
| 14 | 2.43 | P | 32 | 3.31 | F | 50 | 4.94 | P |
| 15 | 2.46 | P | 33 | 3.34 | P | 51 | 5.39 | C |
| 16 | 2.516 | P | 34 | 3.45 | C | 52 | 5.90 | C |
| 17 | 2.524 | P | 35 | 3.48 | P | 53 | 5.94 | P |
| 18 | 2.57 | P | 36 | 3.492 | C | | | |

As a check, note that $R_1 + R_2 + R_3 = 1,431 = \frac{1}{2}N(N+1)$.

*Third*, since $n_1 = 18$, $n_2 = 17$, $n_3 = 18$, calculate $H$ as follows:

$$H = \frac{12}{N(N+1)} \left( \frac{R_1^2}{n_1} + \frac{R_2^2}{n_2} + \frac{R_3^2}{n_3} \right) - 3(N+1)$$

$$= \frac{12}{53 \times 54} (15,605.56 + 12,124.47 + 11,100.50) - (3 \times 54)$$

$$= 162.811 - 162$$

$$= 0.811.$$

*Fourth*, calculate $K$ from $H$:

$$K = \sqrt{2H} - \sqrt{2k - 3}$$
$$= 1.274 - 1.732$$
$$= -0.458.$$

From Table 365 we see that the probability of a larger $K$, and hence (approximately) of a larger $H$, is about 0.7. Thus the data do not demonstrate any real differences in operating cost among the three makes.

This test is called the *Kruskal-Wallis test*.

## 19.4.2 Matched Samples

Sec. 13.2.3 discussed only independent samples because analysis of variance techniques for handling matched samples (such techniques are called two-criterion analyses of variance) would have introduced too many new technical details without a comparable reward in new ideas. Ranking methods of handling the problem, however, require no new technical material.

Suppose there are $n$ matched sets, each set containing $k$ observations. (A more usual way of thinking of this is as $k$ samples of $n$ observations each, in which the observations are matched.) Simply rank the observations of each set from 1 to $k$, giving ties the mean of the ranks for which they are tied. Then for each of the $k$ samples compute the sum of the ranks, calling the sums $R_1, R_2, \ldots, R_k$; actually, the sum of the squares, $\sum R^2$, is required. Now compute

$$W = \frac{12 \sum R^2}{nk(k + 1)} - 3n(k + 1).$$

This again has approximately the chi-square distribution, the "degrees of freedom" being $k - 1$, and approximate probabilities can be obtained graphically by the method described later in this chapter (Sec. 19.6.2), or from the standard normal variable

$$K = \sqrt{2W} - \sqrt{2k - 3},$$

or by the more exact method of Sec. 13.4.4.

Example 601 Judgment of Handwriting

The data of Table 602, showing ranks assigned to twelve samples of handwriting by each of five judges, serve to illustrate the method. This example serves also as a reminder that the original data need not be quantitative for many of the methods of this chapter; so long as the data can be ranked, these methods can be used.

TABLE 602

RANKS ASSIGNED TO 12 SAMPLES OF HANDWRITING
BY FIVE JUDGES

| Judge | Handwriting Sample | | | | | | | | | | | |
|---|---|---|---|---|---|---|---|---|---|---|---|---|
| | A | B | C | D | E | F | G | H | I | J | K | L |
| 1 | 12 | 8 | 11 | 10 | 9 | 6 | 5 | 4 | 7 | 1 | 2 | 3 |
| 2 | 11 | 12 | 10 | 9 | 7 | 8 | 6 | 4 | 1 | 3 | 5 | 2 |
| 3 | 10 | 11 | 12 | 9 | 7 | 8 | 6 | 5 | 3 | 4 | 1 | 2 |
| 4 | 9 | 12 | 11 | 10 | 8 | 7 | 4 | 3 | 5 | 6 | 2 | 1 |
| 5 | 12 | 10 | 11 | 9 | 4 | 7 | 8 | 1 | 6 | 2 | 5 | 3 |
| Total | 54 | 53 | 55 | 47 | 35 | 36 | 29 | 17 | 22 | 16 | 15 | 11 |

*Source:* Frederick Mosteller and Robert R. Bush, "Selected Quantitative Techniques," Chap. 8 in *Handbook of Social Psychology*, edited by Gardner Lindzey (Cambridge, Mass.: Addison-Wesley Publishing Company, 1954), p. 319.

Mosteller and Bush credit the data to J. P. Guilford, *Psychometric Methods* (New York: McGraw-Hill Book Company, 1936), p. 247.

From the data of Table 602,

$$R_1 = 54, \qquad R_2 = 53, \qquad \cdots, \qquad R_{12} = 11$$

$$n = 5 \qquad k = 12 \qquad \sum R_i^2 = 15{,}696.$$

Then

$$W = \frac{12 \times 15{,}696}{5 \times 12 \times 13} - 3 \times 5 \times 13$$

$$= 241.477 - 195 = 46.5.$$

The probability of so large a value of $W$, under the null hypothesis of only chance agreement in the rankings, can be approximated by computing the standard normal variable

$$K = \sqrt{2W} - \sqrt{2k - 3}$$

$$= 9.644 - 4.583 = 5.06.$$

The one-tail probability is shown by Table 365 only as less than 0.001, so at normal significance levels the null hypothesis is rejected. There is more than chance agreement among the judges in ranking the handwriting specimens.

The graphical approximation described in Sec. 19.6.2 is an even quicker and easier way to find the probability for a value of $W$, and is usually accurate enough. The method of Sec. 13.4.4 is more laborious but more accurate.

This test is called the *Friedman test.*

## 19.5
## RELATION BETWEEN TWO VARIABLES

A quick and easy way to test whether two variables are correlated is based on the correlation between their ranks. If there are $n$ pairs of observations, $X$ and $Y$, rank the values of $X$ from 1 to $n$ and also rank the values of $Y$ from 1 to $n$. (Ties are given the mean of the ranks tied for.) Compute the $n$ differences between ranks, square each difference, and add the squares, calling the total $\sum d^2$. Then compute

$$K = \frac{n(n+1)(n-1) - 6(\sum d^2 \pm 1)}{n(n+1)\sqrt{n-1}}$$

where, for a two-tail test, $\pm$ is taken as $+$ if $6\sum d^2$ is less than $n(n+1)$ $(n-1)$ and as $-$ if $6\sum d^2$ is greater than $n(n+1)(n-1)$. The two-tail probability is found from Table 365. Should a one-sided test be wanted, $\pm$ would be taken as $-$ if the alternative hypothesis were that the two variables move in the same direction, and as $+$ if the alternative hypothesis were that the two variables move in opposite directions.

As an illustration of the rank correlation test, we use the data of Table 603.

EXAMPLE 603   CORRELATION BETWEEN TWO LABORATORY
TESTS

Table 603 gives the results of two tests of 10 light-bulb filament wires, together with the rankings needed for this test.

TABLE 603
RESULTS OF TWO TESTS OF 10 LIGHT-BULB FILAMENT WIRES

| Test I | Test II | Rank I | Rank II | $d$ | $d^2$ |
|--------|---------|--------|---------|-----|-------|
| 276 | 14.2 | 1 | 1 | 0 | 0 |
| 293 | 15.6 | 5 | 4 | 1 | 1 |
| 288 | 16.1 | 3 | 5 | −2 | 4 |
| 305 | 15.2 | 7 | 3 | 4 | 16 |
| 315 | 14.6 | 9 | 2 | 7 | 49 |
| 306 | 21.4 | 8 | 10 | −2 | 4 |
| 286 | 19.4 | 2 | 8 | −6 | 36 |
| 289 | 18.9 | 4 | 7 | −3 | 9 |
| 296 | 18.5 | 6 | 6 | 0 | 0 |
| 335 | 20.8 | 10 | 9 | 1 | 1 |

*Source:* Data of Jennett and Dudding, taken from H. A. Freeman, *Industrial Statistics* (New York: John Wiley and Sons, 1942), p. 117.

It is obvious that the two tests differ as to level, but this fact may not be any more important than the corresponding fact for centigrade and fahrenheit thermometers, if the two rise and fall together. The rank correlation test will test whether there is more than a chance relation between the two series.

*First*, calculate $\sum d^2$ by summing the right-hand column of Table 603,

$$\sum d^2 = 0 + 1 + 4 + \cdots + 0 + 1 = 120.$$

*Second*, since $n = 10$, find

$$K = \frac{n(n+1)(n-1) - 6(\sum d^2 + 1)}{n(n+1)\sqrt{n-1}} = \frac{990 - (6 \times 121)}{330}$$

$$= \frac{264}{330} = 0.80.$$

From Table 365 we find a two-tail probability of about 0.42. Hence, there is no evidence here of correlation between the two tests.

Just for practice, however, we show the calculation of the rank correlation coefficient itself:

$$r' = 1 - \frac{6\sum d^2}{n(n+1)(n-1)} = 1 - \frac{6 \times 120}{10 \times 11 \times 9}$$

$$= 0.27.$$

This is a measure of the association analogous to the correlation coefficient (Sec. 17.4.4).

This measure is called the *Spearman rank-correlation coefficient*, and the corresponding test the *Spearman rank-correlation test*.

The runs tests described in Sec. 18.4.2 can also be used as tests of association. The pairs of observations can be arranged in order according to one variable and the number of increases and decreases in the other variable counted and tested as in Sec. 18.4.2.2.

## 19.6
## BINOMIAL PROBABILITY PAPER

### 19.6.1   The Nature of the Graph Paper

The most useful and versatile graphical shortcut available for statistical work is a special graph paper called *Binomial Probability Paper*.[3] This is pictured in Fig. 605.

---

3. Designed by Frederick Mosteller of Harvard University and John W. Tukey of Princeton University, and published by the Codex Book Company, Inc., Norwood, Mass. Regular weight paper is Codex catalog number 32,298; lightweight, 31,298.

## 19.6 Binomial Probability Paper

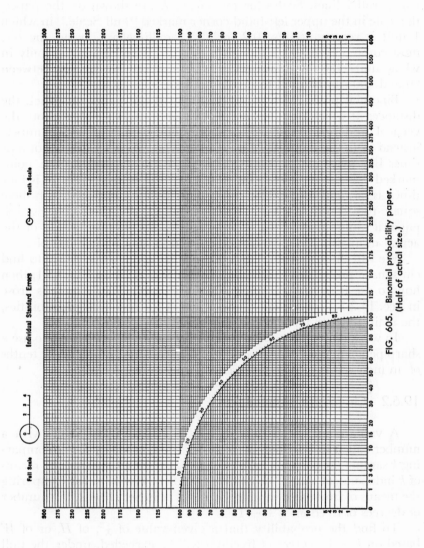

FIG. 605. Binomial probability paper.
(Half of actual size.)

The basic achievement of this paper is enabling the user to obtain the approximate value of a standard normal variable, $K$, simply by measuring the distance between a point and a line, or between two parallel lines. Scales for measuring $K$ are shown on the paper, the scale in the upper left-hand corner marked "Full Scale," in which 1 unit is represented by 0.2 inch (5.1 millimeters), being the one most commonly used. Although the "Full Scale" is marked only in whole units, the divisions on the horizontal or vertical axis between 90 and 100 can be used to divide the units into fifths.

Binomial probability paper is simply graph paper on which the distances at which numbers are plotted are proportional, on both the vertical and horizontal axes, to the square roots of the numbers instead of to the numbers. The point on the axis marked 4 is not four times but twice as far from 0 as is the point marked 1. The point marked 9 is three times as far from 0 as is that marked 1, 16 is four times as far, and so on. The same effects that are achieved with these square-root coordinate scales could be achieved with ordinary graph paper if square roots of numbers were used for plotting instead of the actual numbers; but of course this would be much more trouble.

We will show first how to use binomial probability paper to find chi-square probabilities such as arose in Secs. 19.4 and 13.3.3.1, then how to use it to find variance ratio, or $F$, probabilities such as arose in Sec. 13.2.3.1, and finally how to use it for binomial probabilities, the main purpose for which it was designed.

In working with binomial probability paper it is necessary to use a sharp pencil and a good straight edge (a scale in centimeters or tenths of an inch is an added convenience), and to be precise.

## 19.6.2 Chi-square Probabilities

A value of $\chi^2$ (read "chi square") is always associated with a number called its "degrees of freedom," designated by $f$. For comparing $k$ sample proportions as in Sec. 13.3.3.1, for comparing the means of $k$ independent samples by ranks as in Sec. 19.4.1, or for comparing the means of $k$ matched samples by ranks as in Sec. 19.4.2, the number of degrees of freedom is $k - 1$.

To find the probability that a given value of $\chi^2$, of $H$, or of $W$ based on $k - 1$ degrees of freedom will be exceeded, under the null hypothesis that all samples come from the same population, proceed as follows: First draw a 45° diagonal line starting at the origin, that is, a line through $(0, 0)$ and $(300, 300)$. Then plot a point whose horizontal coordinate is $2k - 3$, and whose vertical coordinate is $2\chi^2$, $2H$,

### 19.6 Binomial Probability Paper

or $2W$. Measure the *vertical* distance of the point from the diagonal line; the sign is $+$ if the point is above the line, $-$ if below. For measuring, mark off the distance on the edge of a slip of paper, and place this against the Full Scale in the upper left-hand corner of the binomial probability paper (using millimeters, tenths of an inch, or the

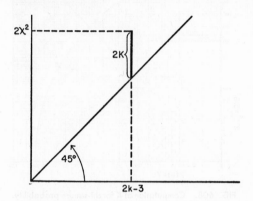

FIG. 607. Computation of $K$ for chi-square probability using binomial probability paper.

subdivisions between 90 and 100 on the axes to get more accuracy). Then *half* the distance so measured is a unit normal deviate. In other words, if the distance is 4 units (20 millimeters, or 0.8 inches) or more, the null hypothesis is rejected at the 5 percent level of significance.

In general, if $f$ is the number of degrees of freedom, the horizontal coordinate is $2f - 1$. In the problems treated here, $f = k - 1$.

Fig. 607 sketches the computation of a chi-square probability on binomial probability paper.

#### Example 607   Comparing Sample Proportions

In Sec. 13.3.3 we computed $\chi^2 = 2.80$ for the five sample proportions of Table 434. Since $k = 5$, we plot a point with horizontal coordinate $2k - 3 = 7$ and vertical coordinate $2\chi^2 = 5.6$; see Fig. 608. The vertical distance from this point to the 45° line is shown by the solid line to the left of the symbol $2K$ in Fig. 608. On the "Full Scale" we mark off this distance and obtain a result of about 0.5, which must be halved, to 0.25. Since the point is below the 45° line, we take this as negative, or $-0.25$. Hence we obtain $K = -0.25$, which may be compared with $-0.24$ by the best approximation given in Sec. 13.4.4. The upper-tail probability corresponding with $K = -0.25$ is shown in Table 365 as 0.599, which compares favorably with our earlier result of 0.595.

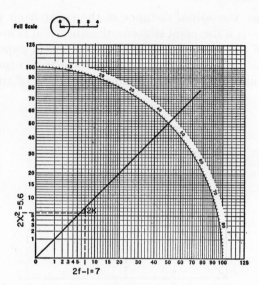

FIG. 608. Computation of K for chi-square probability.

### 19.6.3 Variance Ratio (F) Probabilities

In Sec. 13.2.3.1 we had occasion, in discussing the comparison of several population means, to compare two standard deviations.

To use binomial probability paper to find the probability of a given discrepancy between $s_1$ and $s_2$, if both are estimates of the same value $\sigma$, we work with

$$S_1 = f_1 s_1{}^2, \qquad S_2 = f_2 s_2{}^2,$$

where $f_1$ and $f_2$ are the "numbers of degrees of freedom." For the comparison of the means of $k$ samples, containing altogether $n$ observations,

$$f_1 = k - 1, \qquad f_2 = n - k.$$

The calculation, sketched in Fig. 609, consists of these steps:

*First*, plot a point whose horizontal coordinate is $S_1$ and whose vertical coordinate is $S_2$. If the values of $S_1$ and $S_2$ are too large to plot, divide each by the same constant (usually 10, 100, 1,000, etc.) before plotting.

*Second*, draw a line through the origin and this point.

*Third*, plot a point whose horizontal coordinate is $\frac{1}{2}f_1$ and whose vertical coordinate is $\frac{1}{2}f_2$.

### 19.6 Binomial Probability Paper

*Fourth,* measure the *perpendicular* distance from the point to the line—extending the line if necessary to meet the perpendicular from the point. The measurement is made in terms of the Full Scale in the upper left-hand part of the page. (Alternatively, the distance can be measured in millimeters and divided by 5, or in inches and multiplied by 5.) The sign is − if the point is below (to the right of) the

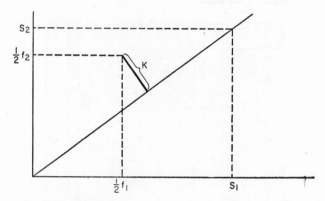

FIG. 609. Computation of K for variance ratio (F) probability using binomial probability paper.

line, + if it is above (to the left). Then the distance so measured is approximately a standard normal variable, whose probability can be obtained from Table 365.

EXAMPLE 609 COMPARING SAMPLE MEANS

In Sec. 13.2.3, in comparing the means of 4 samples, we found

$$f_1 = k - 1 = 4 - 1 = 3,$$
$$f_2 = n - k = 25 - 4 = 21,$$
$$s_1^2 = 448.6567,$$
$$s_2^2 = 467.2319,$$

where the subscript "1" corresponds with the previous subscript "$A$" and "2" with "$W$."

To test the significance of the discrepancy between $s_1$ and $s_2$ using binomial probability paper, proceed as follows:

*First,*

$$\tfrac{1}{2}f_1 = 1.5,$$
$$\tfrac{1}{2}f_2 = 10.5,$$
$$S_1 = f_1 s_1^2 = 1{,}346,$$
$$S_2 = f_2 s_2^2 = 9{,}812.$$

To facilitate plotting, we divide $S_1$ and $S_2$ by 100, obtaining 13.5 and 98.1.

*Second,* plot the point whose horizontal coordinate is 13.5 and whose vertical coordinate is 98.1. Then draw a straight line connecting this point with the origin. See Fig. 610.

*Third,* plot the point whose horizontal coordinate is $\frac{1}{2}f_1 = 1.5$ and whose vertical coordinate is $\frac{1}{2}f_2 = 10.5$.

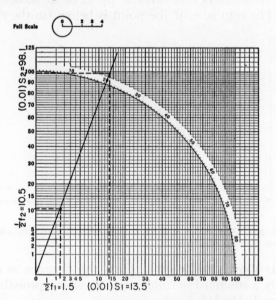

**FIG. 610.** Computation of $K$ for variance ratio ($F$) probability.

*Fourth,* measure the perpendicular distance between the point just plotted in the third step and the line plotted in the second step. In this example, the point is almost precisely on the line, hence the upper-tail probability from Table 365 is about 0.50, as compared with the earlier more precise result of 0.43.

## 19.6.4 Binomial Probabilities

19.6.4.1 *Introduction.* In using double square-root paper for binomial problems, one axis represents the number of occurrences, $X$, and the other the number of nonoccurrences, $n - X$. Usually the number of occurrences is plotted vertically, and the number of nonoccurrences horizontally; since this contradicts the usual custom of letting $X$ denote a horizontal coordinate, sometimes the number of occurrences is denoted by $Y$.

### 19.6 Binomial Probability Paper

A population proportion, $P$, is represented by a straight line through all the combinations $(n - X, X)$ that correspond with the value of $P$ and different values of $n$. Such a line always starts at the origin. Finding a second point on the population line is facilitated by the quarter-circle[4] printed on the paper. This circle corresponds to all samples of 100. Thus, the point on the circle with vertical coordinate $X = 100P$ provides a second point on the population line.

A sample result is represented on binomial probability paper by a point with horizontal coordinate $n - X$, the number of observations in the sample *not* having the trait under consideration, and vertical coordinate $X$, the number of occurrences. As we shall see in a moment, however, points are usually shifted a little in an adjustment analogous to the continuity adjustment in the numerical normal approximation to the binomial (see Sec. 11.4.4.1).

The fact underlying various uses of binomial probability paper in binomial problems is that, for a given population line, the probability that a sample point will be within a given perpendicular distance of the line is given fairly accurately by measuring the distance in the units shown on the Full Scale and regarding this distance as a standard normal variable, whose probability may be found in Table 365. For example, given a population line, lines drawn parallel to it 2 full-scale units away will include 95 percent of sample points (strictly, 95.4 percent). (The sample points for a given sample size, $n$, will fall on a circle centered at the origin.) Correspondingly, if a circle of radius 2 is drawn around a sample point, all population lines for which the sample is within the 95 percent range will pass through the circle, and population lines for which the sample is outside the 95 percent range will not pass through the circle.

Continuity adjustments on binomial probability paper are made by moving a sample point *either* one unit to the right, to a horizontal coordinate of $n - X + 1$, *or* one unit upward, to a vertical coordinate of $X + 1$. If the probability required is an *upper-tail* (the *right* tail) probability—the probability of an observed value $p$ or a higher value—the point is moved to the *right*; that is, it is plotted at $(n - X + 1, X)$. If the probability required is a *lower-tail* probability, the point is moved *up*; that is, it is plotted at $(n - X, X + 1)$. For a *two-tail* probability, if the point lies below the line (that is, if $p < P$), find the lower-tail probability and double it; if the point lies above the line ($p > P$), find the upper-tail probability and double it.

---

4. The numbers printed in the quarter-circle represent angles in degrees from vertical. We will make no use of these numbers.

**Shortcuts**

## 19.6.4.2   Testing a Hypothesis about P.

EXAMPLE 612   ROUNDING WEIGHTS TO 0 AND 5

Consider the example of Sec. 6.2.1, having to do with the tendency to round weights to 0 and 5. Under the null hypothesis, that the weights are rounded to the nearest pound, the population proportion of observations ending in 0 or 5 is $P = 0.2$. The sample size, $n$, is 32, and the number of occurrences of final 0's and 5's, $X$, is 14. The calculation, which is shown in Fig. 612, proceeds as follows:

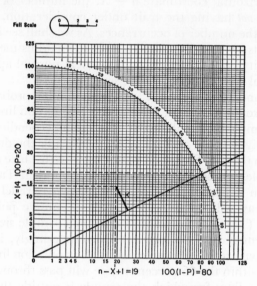

FIG. 612.   Testing a hypothesis about P.

*First*, plot the point on the quarter circle whose horizontal coordinate is $100(1 - P) = 100 \times 0.8 = 80$, and whose vertical coordinate is $100P = 100 \times 0.2 = 20$. Connect this point by a straight line with the origin, as in Fig. 612. This is the population line for $P = 0.2$.

*Second*, plot the point whose horizontal coordinate is $n - X + 1 = 32 - 14 + 1 = 19$, and whose vertical coordinate is $X = 14$. This is the sample point. We use $n - X + 1$ instead of $n - X$ because an upper-tail test is required. Had a lower-tail test been required we would have used $n - X$ but changed the vertical coordinate to $X + 1$.

*Third*, draw the perpendicular line (shown as a solid line on Fig. 612) from the sample point $(n - X + 1, X)$ to the population line drawn in Step 1. Measure this on the Full Scale to obtain $K$. Here $K$ is about 2.8, which corresponds (see Table 365) with an upper-tail probability of about 0.003. A numerical computation gives $K = 3.14$, with an upper-tail probability of about 0.001.

19.6.4.3 *Confidence Interval Estimate for P.* The problem of getting confidence intervals for proportions from binomial probability paper is essentially the reverse of the one just solved. To emphasize this and, more generally, to emphasize the close relationship between confidence intervals and tests, we shall first briefly discuss Fig. 612 from the confidence interval viewpoint. Suppose that we had observed 14 occurrences in 32 trials, and that we wished to calculate the lower limit for a one-sided 99.74 percent confidence interval, for which $K$ is seen from Table 365 to be 2.8. Imagine that in Fig. 612 the plotting sequence was different: that the observation was plotted with coordinates $n - X + 1 = 19$ and $X = 14$, and the lower line drawn whose perpendicular distance from this point is $K = -2.8$. We would do this by drawing a circle of radius 2.8 about the plotted point, and then finding (graphically) the straight line starting at the origin that is below the circle and tangent to it. This case has been so constructed that we would then have the line drawn in our previous example. Third, having found the straight line, we find the corresponding value of $P$; call it $p_L$. The quickest way to do this is to read the vertical coordinate of the intersection of the line with the quarter-circle, which we see in Fig. 612 to be 20 percent, or 0.20. Hence $p_L = 0.20$, and we have now gone through in reverse essentially the procedure of Sec. 19.6.4.2.

EXAMPLE 613  TWO-SIDED 90 PERCENT CONFIDENCE LIMITS
FOR $n = 20$, $p = 0.15$

We now solve graphically a problem which was solved numerically in Sec. 14.7.2. Since $n = 20$ and $p = 0.15$, we have $X = 3$, $n - X = 17$. We first find the lower confidence limit by the procedure just described.

*First,* plot the point with coordinates $n - X + 1 = 18$, $X = 3$, as in Fig. 614A.

*Second,* around this point, draw a circle with radius given by $K = 1.64$, found from Table 391 (see Sec. 14.7.2). We take this length from the Full Scale: note the arc drawn on the Full Scale of Fig. 614A.

*Third,* draw the lower tangent between this circle and the origin.

*Fourth,* find the vertical coordinate of the intersection of the tangent line with the quarter-circle. This is seen from Fig. 614A to be about 4 percent, whence $p_L = 0.04$, as compared with the result 0.048 determined by the method of Sec. 14.7.2.

In practice, the upper limit would be determined from the same graph, though with a new circle, having a slightly different center; but for clarity we show this separately as Fig. 614B.

*First,* plot the point with coordinates $n - X = 17$, $X + 1 = 4$ (see Sec. 19.6.4.2).

*Second,* draw a circle with radius $K = 1.64$ around this point.

**Shortcuts**

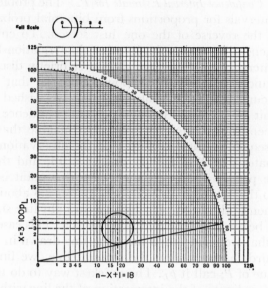

**FIG. 614A.** Determination of $p_L$ of 90 percent confidence interval.
($p = 0.15, n = 20.$)

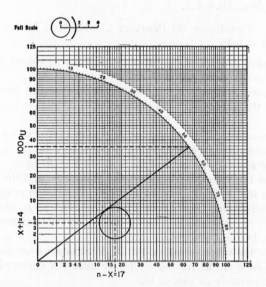

**FIG. 614B.** Determination of $p_U$ of 90 percent confidence interval.
($p = 0.15, n = 20.$)

*19.6 Binomial Probability Paper*

*Third,* draw the upper tangent between this circle and the origin.

*Fourth,* find the vertical coordinate of the intersection of the tangent line with the quarter-circle. This is seen in Fig. 614B to be about $p_U = 0.35$, as compared with our earlier result of 0.351.

Thus the confidence interval determined graphically extends from about 0.04 to 0.35—for most practical purposes an adequate approximation to the earlier result, 0.048 to 0.351.

### 19.6.4.4 Comparing Two Proportions.

EXAMPLE 615    COMPARING EMPLOYEE PERFORMANCE

Of 106 new female employees hired by a company, 67 attained or exceeded a certain level on a group of psychological tests, and 39 fell below

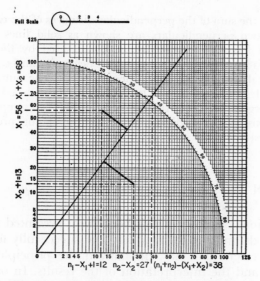

**FIG. 615.** Comparing $p_1 = 56/67$ and $p_2 = 12/39$.

this level. After a ninety-day probationary period, 56 of the 67 who exceeded the level, and 12 of the 39 who fell below the level, were retained.[5] Is the difference in the proportions retained more than might reasonably be expected between two independent samples from the same population? If so, it may be that the tests are useful in predicting success, and would be helpful in choosing employees.

Let $n_1 = 67$, $X_1 = 56$, $p_1 = 56/67$ and $n_2 = 39$, $X_2 = 12$, $p_2 = 12/39$. Note that if the null hypothesis is true (namely that the proportion retained

---

5. Thomas H. Wallace, "Pre-Employment Tests and Post-Employment Performance," *Journal of Business,* Vol. 28 (1955), pp. 73–74.

is the same for those above the test level and those below it),

$$p = \frac{X_1 + X_2}{n_1 + n_2}$$

is an estimate of the true proportion $P$ (see the first part of Sec. 13.3.2.1).

*First,* plot a population line for the value of $p$ (see Fig. 615) by taking $(n_1 + n_2) - (X_1 + X_2) = 38$ and $(X_1 + X_2) = 68$ as coordinates for a point and connecting this point by a straight line with the origin.

*Second,* plot a point for the sample whose proportion is above $p$, in this case the first sample. The coordinates of this point are $(n_1 - X_1 + 1, X_1)$, that is, (12, 56).

*Third,* plot a point for the sample whose proportion is below $p$, in this case the second sample. The coordinates of this point are $(n_2 - X_2, X_2 + 1)$, that is, (27, 13).

*Fourth,* find the sum of the perpendicular distances of the two points from the line (the two perpendiculars are shown as solid lines on Fig. 615). The combined lengths of these distances, measured by the full scale, is about 7.3. This must be multiplied by 0.7 before it can be regarded as a standard normal variable $K$. Here, then, $K = 5.1$. The numerical method of Sec. 13.3.2.1 gives $K = 5.26$. The probability of such a large $K$ arising by chance is negligible. Thus, the difference in the two sample proportions signifies the existence of a difference in the population proportions. Apparently the test has predictive value.

## 19.7
## CONCLUSION

The previous chapters of this book have presented standard statistical methods, most of them widely and successfully used for many years. The presentation has emphasized the principles underlying the methods and the interpretation of the results. In order to avoid introducing various special technical devices which, while invaluable to the practicing statistician, do not further illuminate the principles of analysis or the interpretation of results, we have sometimes introduced a little extra arithmetic which renders the normal distribution a sufficiently accurate sampling distribution. In this way, the need for tables and detailed explanations of Student's distribution, the chi-square distribution, the variance ratio distribution, the binomial distribution, and other special distributions has been obviated, and the discussion of these distributions has been confined to the principles underlying them and their role in reaching decisions from data.

Two other reasons, besides the desire to minimize technical detail, have led us to this treatment. First, the additional arithmetic necessary to use the normal distribution when other distributions would be directly applicable can often be circumvented by using special graphical methods. Some of these methods, which are often accurate enough for practical purposes, are presented in Sec. 19.6 of this chapter. Second, the standard or "classical" methods described in the earlier chapters can often be replaced satisfactorily by short-cut methods, usually based on substituting for each actual observation in a sample its rank—that is, an integer indicating whether it is the smallest (1), the next-to-smallest (2), and so on, to the largest (*n*). Not only are these ranking methods simpler to use, but frequently they are nearly as effective (have nearly as steep operating-characteristic curves or nearly as narrow confidence intervals) as the classical methods in situations to which the classical methods are appropriate; and ranking methods have the further advantage of being appropriate in some situations where the classical methods are not.

This chapter has shown how, by the use of ranks, to compute a confidence interval, or make a significance test for an average (specifically, a median); to compare two averages or to compare a set of averages, either from independent samples or from matched samples; and to test or measure the correlation between paired observations.

Graphical methods, based on double square-root graph paper (binomial probability paper) have been presented for finding probabilities for chi-square, variance ratio, and binomial probabilities. For binomial probabilities, the specific applications of testing a hypothesis about a population proportion, making a confidence interval estimate of a proportion, and comparing two proportions have been shown.

If in the future you occasionally want to carry out a statistical analysis, our advice is to turn to this chapter first. The appropriate earlier chapter should be referred to, however, for a general background of statistical considerations pertinent to the problem and the principles of analysis and interpretation. If you find yourself engaged seriously and frequently in statistical analysis, you will want to rely mainly on the earlier chapters, supplemented by a calculating machine or *Barlow's Tables* (better, by both) and perhaps supplemented also by one of the collections of statistical tables mentioned in the footnote of pages 591–592.

With this chapter, we do not close the subject of statistical reasoning. We hope, instead, that we have opened it for you.

## DO IT YOURSELF

EXAMPLE 618A

Find 95 percent confidence limits for the median from the sample of Table 206.

EXAMPLE 618B

Use the Wilcoxon two-sample test to decide whether or not there is a significant difference between the average expenses per trip of Tables 502A and 502B.

EXAMPLE 618C

Use the Wilcoxon signed-rank test to decide whether or not there is a significant difference in mean performance between the two tests of Example 201.

EXAMPLE 618D

Use the Kruskal-Wallis test to decide whether average daily bottle-cap production differs significantly among the three machines whose outputs[6] for selected days are shown below:

| Machine 1: | 340 | 345 | 330 | 342 | 338 |
| Machine 2: | 339 | 333 | 344 | | |
| Machine 3: | 347 | 343 | 349 | 355 | |

EXAMPLE 618E

Use the Friedman test in analyzing Table 277. What are your conclusions?

EXAMPLE 618F

Use the rank correlation test to decide whether there is any significant correlation between performance on the two tests of Example 201.

EXAMPLE 618G

Use the graphical method of Sec. 19.6.2 to obtain the probability of exceeding the value of $\chi^2$ obtained from Example 442C.

EXAMPLE 618H

Use the graphical method of Sec. 19.6.3 to obtain the probability of exceeding the value of $F$ obtained from applying the method of Sec. 13.2.3 to Table 600A.

---

6. Artificial data taken from William H. Kruskal and W. Allen Wallis, "Use of Ranks in One-Criterion Variance Analysis," *Journal of American Statistical Association*, Vol. 47 (1952), p. 588.

*Do It Yourself*

EXAMPLE 619A

Use the method of Sec. 19.6.4.2 in analyzing any of the examples at the end of Chap. 12, pages 409–411.

EXAMPLE 619B

Use the method of Sec. 19.6.4.3 to obtain 95 percent confidence limits for the observed results in Example 410E.

EXAMPLE 619C

Use the technique of Example 615 in analyzing Example 208C.

# APPENDIX
## SQUARES AND SQUARE ROOTS
## AND RANDOM DIGITS

# Squares and Square Roots and Random Digits

## SQUARE ROOTS

Table 626 shows the square root to three decimals for each integer, and for ten times each integer, from 0 to 1,000. Square roots of all other numbers can be obtained from these. Squares can be found inversely.

### Whole Numbers from 0 to 999

Look in the column corresponding with the last digit and the row corresponding with the preceding digit or digits. The first (plain type) figure is the square root required.

*Example* 1. To find $\sqrt{206}$, look in column 6 and row 20. The first (plain type) entry is 14.353, which is the required value. As an approximate check, note that 206 is between $14^2 = 196$ and $15^2 = 225$.

*Example* 2. To find $\sqrt{15}$, look in row 1, column 5. The plain-type entry, 3.873, is the value required. As an approximate check, note that 15 is between $3^2 = 9$ and $4^2 = 16$.

### Other Numbers

Move the decimal place to the left or the right until there are three digits before the decimal. The next step depends on whether the decimal was moved an even or an odd number of places.

**Decimal moved an even number of places.** Treat the three digits before the decimal exactly like a whole number from 100 to 999, taking the

**623**

**Appendix**

plain-type entry. Then move the decimal back, but *only half as many places.*

*Example* 3.   To find $\sqrt{20{,}600}$, first move the decimal two places to the left, giving $\sqrt{206.00}$. In row 20, column 6, the plain-type entry is 14.353. Move the decimal one place to the right, giving 143.53 as the required answer. As an approximate check note that 20,600 is between $140^2 = 19{,}600$ and $150^2 = 22{,}500$.

*Example* 4.   To find $\sqrt{0.000206}$, first move the decimal six places to the right, giving $\sqrt{206}$. This is shown by the plain-type entry in row 20, column 6, to be 14.353. Move the decimal three places to the left, giving 0.014353 as the required answer. As an approximate check, note that 0.000206 is between $0.014^2 = 0.000196$ and $0.015^2 = 0.000225$.

**Decimal moved an odd number of places.** Treat the three digits before the decimal as a whole number from 100 to 999 *but* use the *italic* entry. Then move the decimal back, but only about half as many places. It is impossible to return the decimal exactly half as many places, since half an odd number is not an integer; it should be put *half a place to the left of the position indicated by the rule "half as many places in the opposite direction."*

*Example* 5.   To find $\sqrt{2060}$, first move the decimal one place to the left, giving $\sqrt{206.0}$. In row 20, column 6, the italic entry is 45.387. Move the decimal 0 places to the right—that is, put it half a place to the left of half a place to the right. As an approximate check, note that 2060 is between $40^2 = 1600$ and $50^2 = 2500$.

*Example* 6.   To find $\sqrt{0.00206}$, first move the decimal five places to the right. In row 20, column 6, the italic entry is 45.387. Move the decimal in this three places to the left—that is, put it half a place to the left of two and one-half places to the left. Thus, the required value is 0.045387. As an approximate check, note that 0.00206 is between $0.040^2 = 0.001600$ and $0.050^2 = 0.002500$.

## Interpolation

*Example* 7.   To approximate $\sqrt{20617}$, take a value 0.17 of the way between $\sqrt{20600}$ and $\sqrt{20700}$, which are the nearest numbers with three significant figures (the final 00's are simply "spacers" to locate the decimal point, and signify nothing else about the quantities measured). Then $\sqrt{20600} = 143.53$ (Example 3). Similarly, $\sqrt{20700} = 143.87$. The value 0.17 of the way from 143.53 to 143.87 is

$$143.53 + 0.17(143.87 - 143.53) = 143.53 + 0.17 \times 0.34$$
$$= 143.53 + 0.06 = 143.59.$$

## SQUARES

Squares of integers up to 100 are easily found from the table, and squares of other numbers can be approximated.

*Example* 8.   To find $68^2$, note that 68 comes between the italicized entries *67.971* and *68.044*, in row 46, columns 2 and 3. Its square, therefore, is between 4620 and 4630, the final 0's being added because the entries near 68 are italicized. The final digit of the square of any number ending in 8 is $4(8^2 = 64)$, so 4624 is the required value.

*Example* 9.   To find $27.7^2$, note that 27.7 comes between the plain-type entries 27.695 and 27.713 in row 76, columns 7 and 8. Hence $27.7^2$ is between 767 and 768. Since 27.7 is $\frac{5}{18}$ or 0.28 of the way from 27.695 to 27.713, $27.7^2$ may be approximated as 767.28. Since the correct value must have two decimal places and end in $9(7^2 = 49)$, 767.29 may be taken as the required number.

## TABLE 626

### Square Roots of $N$ and $10N$

$N$ = 0 to 199

|    | 0 | 1 | 2 | 3 | 4 | 5 | 6 | 7 | 8 | 9 |
|----|-----|-----|-----|-----|-----|-----|-----|-----|-----|-----|
| 0  | 0 | 1.000 | 1.414 | 1.732 | 2.000 | 2.236 | 2.449 | 2.646 | 2.828 | 3.000 |
|    | *0* | *3.162* | *4.472* | *5.477* | *6.325* | *7.071* | *7.746* | *8.367* | *8.944* | *9.487* |
| 1  | 3.162 | 3.317 | 3.464 | 3.606 | 3.742 | 3.873 | 4.000 | 4.123 | 4.243 | 4.359 |
|    | *10.000* | *10.488* | *10.954* | *11.402* | *11.832* | *12.247* | *12.649* | *13.038* | *13.416* | *13.784* |
| 2  | 4.472 | 4.583 | 4.690 | 4.796 | 4.899 | 5.000 | 5.099 | 5.196 | 5.292 | 5.385 |
|    | *14.142* | *14.491* | *14.832* | *15.166* | *15.492* | *15.811* | *16.125* | *16.432* | *16.733* | *17.029* |
| 3  | 5.477 | 5.568 | 5.657 | 5.745 | 5.831 | 5.916 | 6.000 | 6.083 | 6.164 | 6.245 |
|    | *17.321* | *17.607* | *17.889* | *18.166* | *18.439* | *18.708* | *18.974* | *19.235* | *19.494* | *19.748* |
| 4  | 6.325 | 6.403 | 6.481 | 6.557 | 6.633 | 6.708 | 6.782 | 6.856 | 6.928 | 7.000 |
|    | *20.000* | *20.248* | *20.494* | *20.736* | *20.976* | *21.213* | *21.448* | *21.679* | *21.909* | *22.136* |
| 5  | 7.071 | 7.141 | 7.211 | 7.280 | 7.348 | 7.416 | 7.483 | 7.550 | 7.616 | 7.681 |
|    | *22.361* | *22.583* | *22.804* | *23.022* | *23.238* | *23.452* | *23.664* | *23.875* | *24.083* | *24.290* |
| 6  | 7.746 | 7.810 | 7.874 | 7.937 | 8.000 | 8.062 | 8.124 | 8.185 | 8.246 | 8.307 |
|    | *24.495* | *24.698* | *24.900* | *25.100* | *25.298* | *25.495* | *25.690* | *25.884* | *26.077* | *26.268* |
| 7  | 8.367 | 8.426 | 8.485 | 8.544 | 8.602 | 8.660 | 8.718 | 8.775 | 8.832 | 8.888 |
|    | *26.458* | *26.646* | *26.833* | *27.019* | *27.203* | *27.386* | *27.568* | *27.749* | *27.928* | *28.107* |
| 8  | 8.944 | 9.000 | 9.055 | 9.110 | 9.165 | 9.220 | 9.274 | 9.327 | 9.381 | 9.434 |
|    | *28.284* | *28.460* | *28.636* | *28.810* | *28.983* | *29.155* | *29.326* | *29.496* | *29.665* | *29.833* |
| 9  | 9.487 | 9.539 | 9.592 | 9.644 | 9.695 | 9.747 | 9.798 | 9.849 | 9.899 | 9.950 |
|    | *30.000* | *30.166* | *30.332* | *30.496* | *30.659* | *30.822* | *30.984* | *31.145* | *31.305* | *31.464* |
| 10 | 10.000 | 10.050 | 10.100 | 10.149 | 10.198 | 10.247 | 10.296 | 10.344 | 10.392 | 10.440 |
|    | *31.623* | *31.780* | *31.937* | *32.094* | *32.249* | *32.404* | *32.558* | *32.711* | *32.863* | *33.015* |
| 11 | 10.488 | 10.536 | 10.583 | 10.630 | 10.677 | 10.724 | 10.770 | 10.817 | 10.863 | 10.909 |
|    | *33.166* | *33.317* | *33.466* | *33.615* | *33.764* | *33.912* | *34.059* | *34.205* | *34.351* | *34.496* |
| 12 | 10.954 | 11.000 | 11.045 | 11.091 | 11.136 | 11.180 | 11.225 | 11.269 | 11.314 | 11.358 |
|    | *34.641* | *34.786* | *34.928* | *35.071* | *35.214* | *35.355* | *35.496* | *35.637* | *35.777* | *35.917* |
| 13 | 11.402 | 11.446 | 11.489 | 11.533 | 11.576 | 11.619 | 11.662 | 11.705 | 11.747 | 11.790 |
|    | *36.056* | *36.194* | *36.332* | *36.469* | *36.606* | *36.742* | *36.878* | *37.014* | *37.148* | *37.283* |
| 14 | 11.832 | 11.874 | 11.916 | 11.958 | 12.000 | 12.042 | 12.083 | 12.124 | 12.166 | 12.207 |
|    | *37.417* | *37.550* | *37.683* | *37.815* | *37.947* | *38.079* | *38.210* | *38.341* | *38.471* | *38.601* |
| 15 | 12.247 | 12.288 | 12.329 | 12.369 | 12.410 | 12.450 | 12.490 | 12.530 | 12.570 | 12.610 |
|    | *38.730* | *38.859* | *38.987* | *39.115* | *39.243* | *39.370* | *39.497* | *39.623* | *39.749* | *39.875* |
| 16 | 12.649 | 12.689 | 12.728 | 12.767 | 12.806 | 12.845 | 12.884 | 12.923 | 12.961 | 13.000 |
|    | *40.000* | *40.125* | *40.249* | *40.373* | *40.497* | *40.620* | *40.743* | *40.866* | *40.988* | *41.110* |
| 17 | 13.038 | 13.077 | 13.115 | 13.153 | 13.191 | 13.229 | 13.266 | 13.304 | 13.342 | 13.379 |
|    | *41.231* | *41.352* | *41.473* | *41.593* | *41.713* | *41.833* | *41.952* | *42.071* | *42.190* | *42.308* |
| 18 | 13.416 | 13.454 | 13.491 | 13.528 | 13.565 | 13.601 | 13.638 | 13.675 | 13.711 | 13.748 |
|    | *42.426* | *42.544* | *42.661* | *42.778* | *42.895* | *43.012* | *43.128* | *43.243* | *43.359* | *43.474* |
| 19 | 13.784 | 13.820 | 13.856 | 13.892 | 13.928 | 13.964 | 14.000 | 14.036 | 14.071 | 14.107 |
|    | *43.589* | *43.704* | *43.818* | *43.932* | *44.045* | *44.159* | *44.272* | *44.385* | *44.497* | *44.609* |

TABLE 626 (Continued)

SQUARE ROOTS OF $N$ AND $10N$

$N = 200$ to $399$

| | 0 | 1 | 2 | 3 | 4 | 5 | 6 | 7 | 8 | 9 |
|---|---|---|---|---|---|---|---|---|---|---|
| 20 | 14.142 | 14.177 | 14.213 | 14.248 | 14.283 | 14.318 | 14.353 | 14.387 | 14.422 | 14.457 |
| | 44.721 | 44.833 | 44.944 | 45.056 | 45.166 | 45.277 | 45.387 | 45.497 | 45.607 | 45.717 |
| 21 | 14.491 | 14.526 | 14.560 | 14.595 | 14.629 | 14.663 | 14.697 | 14.731 | 14.765 | 14.799 |
| | 45.826 | 45.935 | 46.043 | 46.152 | 46.260 | 46.368 | 46.476 | 46.583 | 46.690 | 46.797 |
| 22 | 14.832 | 14.866 | 14.900 | 14.933 | 14.967 | 15.000 | 15.033 | 15.067 | 15.100 | 15.133 |
| | 46.904 | 47.011 | 47.117 | 47.223 | 47.329 | 47.434 | 47.539 | 47.645 | 47.749 | 47.854 |
| 23 | 15.166 | 15.199 | 15.232 | 15.264 | 15.297 | 15.330 | 15.362 | 15.395 | 15.427 | 15.460 |
| | 47.958 | 48.062 | 48.166 | 48.270 | 48.374 | 48.477 | 48.580 | 48.683 | 48.785 | 48.888 |
| 24 | 15.492 | 15.524 | 15.556 | 15.588 | 15.620 | 15.652 | 15.684 | 15.716 | 15.748 | 15.780 |
| | 48.990 | 49.092 | 49.193 | 49.295 | 49.396 | 49.497 | 49.598 | 49.699 | 49.800 | 49.900 |
| 25 | 15.811 | 15.843 | 15.875 | 15.906 | 15.937 | 15.969 | 16.000 | 16.031 | 16.062 | 16.093 |
| | 50.000 | 50.100 | 50.200 | 50.299 | 50.398 | 50.498 | 50.596 | 50.695 | 50.794 | 50.892 |
| 26 | 16.125 | 16.155 | 16.186 | 16.217 | 16.248 | 16.279 | 16.310 | 16.340 | 16.371 | 16.401 |
| | 50.990 | 51.088 | 51.186 | 51.284 | 51.381 | 51.478 | 51.575 | 51.672 | 51.769 | 51.865 |
| 27 | 16.432 | 16.462 | 16.492 | 16.523 | 16.553 | 16.583 | 16.613 | 16.643 | 16.673 | 16.703 |
| | 51.962 | 52.058 | 52.154 | 52.249 | 52.345 | 52.440 | 52.536 | 52.631 | 52.726 | 52.820 |
| 28 | 16.733 | 16.763 | 16.793 | 16.823 | 16.852 | 16.882 | 16.912 | 16.941 | 16.971 | 17.000 |
| | 52.915 | 53.009 | 53.104 | 53.198 | 53.292 | 53.385 | 53.479 | 53.572 | 53.666 | 53.759 |
| 29 | 17.029 | 17.059 | 17.088 | 17.117 | 17.146 | 17.176 | 17.205 | 17.234 | 17.263 | 17.292 |
| | 53.852 | 53.944 | 54.037 | 54.129 | 54.222 | 54.314 | 54.406 | 54.498 | 54.589 | 54.681 |
| 30 | 17.321 | 17.349 | 17.378 | 17.407 | 17.436 | 17.464 | 17.493 | 17.521 | 17.550 | 17.578 |
| | 54.772 | 54.863 | 54.955 | 55.045 | 55.136 | 55.227 | 55.317 | 55.408 | 55.498 | 55.588 |
| 31 | 17.607 | 17.635 | 17.664 | 17.692 | 17.720 | 17.748 | 17.776 | 17.804 | 17.833 | 17.861 |
| | 55.678 | 55.767 | 55.857 | 55.946 | 56.036 | 56.125 | 56.214 | 56.303 | 56.391 | 56.480 |
| 32 | 17.889 | 17.916 | 17.944 | 17.972 | 18.000 | 18.028 | 18.055 | 18.083 | 18.111 | 18.138 |
| | 56.569 | 56.657 | 56.745 | 56.833 | 56.921 | 57.009 | 57.096 | 57.184 | 57.271 | 57.359 |
| 33 | 18.166 | 18.193 | 18.221 | 18.248 | 18.276 | 18.303 | 18.330 | 18.358 | 18.385 | 18.412 |
| | 57.446 | 57.533 | 57.619 | 57.706 | 57.793 | 57.879 | 57.966 | 58.052 | 58.138 | 58.224 |
| 34 | 18.439 | 18.466 | 18.493 | 18.520 | 18.547 | 18.574 | 18.601 | 18.628 | 18.655 | 18.682 |
| | 58.310 | 58.395 | 58.481 | 58.566 | 58.652 | 58.737 | 58.822 | 58.907 | 58.992 | 59.076 |
| 35 | 18.708 | 18.735 | 18.762 | 18.788 | 18.815 | 18.841 | 18.868 | 18.894 | 18.921 | 18.947 |
| | 59.161 | 59.245 | 59.330 | 59.414 | 59.498 | 59.582 | 59.666 | 59.749 | 59.833 | 59.917 |
| 36 | 18.974 | 19.000 | 19.026 | 19.053 | 19.079 | 19.105 | 19.131 | 19.157 | 19.183 | 19.209 |
| | 60.000 | 60.083 | 60.166 | 60.249 | 60.332 | 60.415 | 60.498 | 60.581 | 60.663 | 60.745 |
| 37 | 19.235 | 19.261 | 19.287 | 19.313 | 19.339 | 19.365 | 19.391 | 19.416 | 19.442 | 19.468 |
| | 60.828 | 60.910 | 60.992 | 61.074 | 61.156 | 61.237 | 61.319 | 61.400 | 61.482 | 61.563 |
| 38 | 19.494 | 19.519 | 19.545 | 19.570 | 19.596 | 19.621 | 19.647 | 19.672 | 19.698 | 19.723 |
| | 61.644 | 61.725 | 61.806 | 61.887 | 61.968 | 62.048 | 62.129 | 62.209 | 62.290 | 62.370 |
| 39 | 19.748 | 19.774 | 19.799 | 19.824 | 19.849 | 19.875 | 19.900 | 19.925 | 19.950 | 19.975 |
| | 62.450 | 62.530 | 62.610 | 62.690 | 62.769 | 62.849 | 62.929 | 63.008 | 63.087 | 63.166 |

**TABLE 626 (Continued)**

SQUARE ROOTS OF $N$ AND $10N$

$N = 400$ to $599$

| | 0 | 1 | 2 | 3 | 4 | 5 | 6 | 7 | 8 | 9 |
|---|---|---|---|---|---|---|---|---|---|---|
| 40 | 20.000 | 20.025 | 20.050 | 20.075 | 20.100 | 20.125 | 20.149 | 20.174 | 20.199 | 20.224 |
|    | *63.246* | *63.325* | *63.403* | *63.482* | *63.561* | *63.640* | *63.718* | *63.797* | *63.875* | *63.953* |
| 41 | 20.248 | 20.273 | 20.298 | 20.322 | 20.347 | 20.372 | 20.396 | 20.421 | 20.445 | 20.469 |
|    | *64.031* | *64.109* | *64.187* | *64.265* | *64.343* | *64.420* | *64.498* | *64.576* | *64.653* | *64.730* |
| 42 | 20.494 | 20.518 | 20.543 | 20.567 | 20.591 | 20.616 | 20.640 | 20.664 | 20.688 | 20.712 |
|    | *64.807* | *64.885* | *64.962* | *65.038* | *65.115* | *65.192* | *65.269* | *65.345* | *65.422* | *65.498* |
| 43 | 20.736 | 20.761 | 20.785 | 20.809 | 20.833 | 20.857 | 20.881 | 20.905 | 20.928 | 20.952 |
|    | *65.574* | *65.651* | *65.727* | *65.803* | *65.879* | *65.955* | *66.030* | *66.106* | *66.182* | *66.257* |
| 44 | 20.976 | 21.000 | 21.024 | 21.048 | 21.071 | 21.095 | 21.119 | 21.142 | 21.166 | 21.190 |
|    | *66.332* | *66.408* | *66.483* | *66.558* | *66.633* | *66.708* | *66.783* | *66.858* | *66.933* | *67.007* |
| 45 | 21.213 | 21.237 | 21.260 | 21.284 | 21.307 | 21.331 | 21.354 | 21.378 | 21.401 | 21.424 |
|    | *67.082* | *67.157* | *67.231* | *67.305* | *67.380* | *67.454* | *67.528* | *67.602* | *67.676* | *67.750* |
| 46 | 21.448 | 21.471 | 21.494 | 21.517 | 21.541 | 21.564 | 21.587 | 21.610 | 21.633 | 21.656 |
|    | *67.823* | *67.897* | *67.971* | *68.044* | *68.118* | *68.191* | *68.264* | *68.337* | *68.411* | *68.484* |
| 47 | 21.679 | 21.703 | 21.726 | 21.749 | 21.772 | 21.794 | 21.817 | 21.840 | 21.863 | 21.886 |
|    | *68.557* | *68.629* | *68.702* | *68.775* | *68.848* | *68.920* | *68.993* | *69.065* | *69.138* | *69.210* |
| 48 | 21.909 | 21.932 | 21.954 | 21.977 | 22.000 | 22.023 | 22.045 | 22.068 | 22.091 | 22.113 |
|    | *69.282* | *69.354* | *69.426* | *69.498* | *69.570* | *69.642* | *69.714* | *69.785* | *69.857* | *69.929* |
| 49 | 22.136 | 22.159 | 22.181 | 22.204 | 22.226 | 22.249 | 22.271 | 22.293 | 22.316 | 22.338 |
|    | *70.000* | *70.071* | *70.143* | *70.214* | *70.285* | *70.356* | *70.427* | *70.498* | *70.569* | *70.640* |
| 50 | 22.361 | 22.383 | 22.405 | 22.428 | 22.450 | 22.472 | 22.494 | 22.517 | 22.539 | 22.561 |
|    | *70.711* | *70.781* | *70.852* | *70.922* | *70.993* | *71.063* | *71.134* | *71.204* | *71.274* | *71.344* |
| 51 | 22.583 | 22.605 | 22.627 | 22.650 | 22.672 | 22.694 | 22.716 | 22.738 | 22.760 | 22.782 |
|    | *71.414* | *71.484* | *71.554* | *71.624* | *71.694* | *71.764* | *71.833* | *71.903* | *71.972* | *72.042* |
| 52 | 22.804 | 22.825 | 22.847 | 22.869 | 22.891 | 22.913 | 22.935 | 22.956 | 22.978 | 23.000 |
|    | *72.111* | *72.180* | *72.250* | *72.319* | *72.388* | *72.457* | *72.526* | *72.595* | *72.664* | *72.732* |
| 53 | 23.022 | 23.043 | 23.065 | 23.087 | 23.108 | 23.130 | 23.152 | 23.173 | 23.195 | 23.216 |
|    | *72.801* | *72.870* | *72.938* | *73.007* | *73.075* | *73.144* | *73.212* | *73.280* | *73.348* | *73.417* |
| 54 | 23.238 | 23.259 | 23.281 | 23.302 | 23.324 | 23.345 | 23.367 | 23.388 | 23.409 | 23.431 |
|    | *73.485* | *73.553* | *73.621* | *73.689* | *73.756* | *73.824* | *73.892* | *73.959* | *74.027* | *74.095* |
| 55 | 23.452 | 23.473 | 23.495 | 23.516 | 23.537 | 23.558 | 23.580 | 23.601 | 23.622 | 23.643 |
|    | *74.162* | *74.229* | *74.297* | *74.364* | *74.431* | *74.498* | *74.565* | *74.632* | *74.699* | *74.766* |
| 56 | 23.664 | 23.685 | 23.707 | 23.728 | 23.749 | 23.770 | 23.791 | 23.812 | 23.833 | 23.854 |
|    | *74.833* | *74.900* | *74.967* | *75.033* | *75.100* | *75.166* | *75.233* | *75.299* | *75.366* | *75.432* |
| 57 | 23.875 | 23.896 | 23.917 | 23.937 | 23.958 | 23.979 | 24.000 | 24.021 | 24.042 | 24.062 |
|    | *75.498* | *75.565* | *75.631* | *75.697* | *75.763* | *75.829* | *75.895* | *75.961* | *76.026* | *76.092* |
| 58 | 24.083 | 24.104 | 24.125 | 24.145 | 24.166 | 24.187 | 24.207 | 24.228 | 24.249 | 24.269 |
|    | *76.158* | *76.223* | *76.289* | *76.354* | *76.420* | *76.485* | *76.551* | *76.616* | *76.681* | *76.746* |
| 59 | 24.290 | 24.310 | 24.331 | 24.352 | 24.372 | 24.393 | 24.413 | 24.434 | 24.454 | 24.474 |
|    | *76.811* | *76.877* | *76.942* | *77.006* | *77.071* | *77.136* | *77.201* | *77.266* | *77.330* | *77.395* |

TABLE 626 (Continued)

SQUARE ROOTS OF $N$ AND $10N$

$N$ = 600 to 799

|    | 0 | 1 | 2 | 3 | 4 | 5 | 6 | 7 | 8 | 9 |
|----|---|---|---|---|---|---|---|---|---|---|
| 60 | 24.495 | 24.515 | 24.536 | 24.556 | 24.576 | 24.597 | 24.617 | 24.637 | 24.658 | 24.678 |
|    | 77.460 | 77.524 | 77.589 | 77.653 | 77.717 | 77.782 | 77.846 | 77.910 | 77.974 | 78.038 |
| 61 | 24.698 | 24.718 | 24.739 | 24.759 | 24.779 | 24.799 | 24.819 | 24.839 | 24.860 | 24.880 |
|    | 78.102 | 78.166 | 78.230 | 78.294 | 78.358 | 78.422 | 78.486 | 78.549 | 78.613 | 78.677 |
| 62 | 24.900 | 24.920 | 24.940 | 24.960 | 24.980 | 25.000 | 25.020 | 25.040 | 25.060 | 25.080 |
|    | 78.740 | 78.804 | 78.867 | 78.930 | 78.994 | 79.057 | 79.120 | 79.183 | 79.246 | 79.310 |
| 63 | 25.100 | 25.120 | 25.140 | 25.159 | 25.179 | 25.199 | 25.219 | 25.239 | 25.259 | 25.278 |
|    | 79.373 | 79.436 | 79.498 | 79.561 | 79.624 | 79.687 | 79.750 | 79.812 | 79.875 | 79.937 |
| 64 | 25.298 | 25.318 | 25.338 | 25.357 | 25.377 | 25.397 | 25.417 | 25.436 | 25.456 | 25.475 |
|    | 80.000 | 80.062 | 80.125 | 80.187 | 80.250 | 80.312 | 80.374 | 80.436 | 80.498 | 80.561 |
| 65 | 25.495 | 25.515 | 25.534 | 25.554 | 25.573 | 25.593 | 25.612 | 25.632 | 25.652 | 25.671 |
|    | 80.623 | 80.685 | 80.747 | 80.808 | 80.870 | 80.932 | 80.994 | 81.056 | 81.117 | 81.179 |
| 66 | 25.690 | 25.710 | 25.729 | 25.749 | 25.768 | 25.788 | 25.807 | 25.826 | 25.846 | 25.865 |
|    | 81.240 | 81.302 | 81.363 | 81.425 | 81.486 | 81.548 | 81.609 | 81.670 | 81.731 | 81.792 |
| 67 | 25.884 | 25.904 | 25.923 | 25.942 | 25.962 | 25.981 | 26.000 | 26.019 | 26.038 | 26.057 |
|    | 81.854 | 81.915 | 81.976 | 82.037 | 82.098 | 82.158 | 82.219 | 82.280 | 82.341 | 82.401 |
| 68 | 26.077 | 26.096 | 26.115 | 26.134 | 26.153 | 26.173 | 26.192 | 26.211 | 26.230 | 26.249 |
|    | 82.462 | 82.523 | 82.583 | 82.644 | 82.704 | 82.765 | 82.825 | 82.885 | 82.946 | 83.006 |
| 69 | 26.268 | 26.287 | 26.306 | 26.325 | 26.344 | 26.363 | 26.382 | 26.401 | 26.420 | 26.439 |
|    | 83.066 | 83.126 | 83.187 | 83.247 | 83.307 | 83.367 | 83.427 | 83.487 | 83.546 | 83.606 |
| 70 | 26.458 | 26.476 | 26.495 | 26.514 | 26.533 | 26.552 | 26.571 | 26.589 | 26.608 | 26.627 |
|    | 83.666 | 83.726 | 83.785 | 83.845 | 83.905 | 83.964 | 84.024 | 84.083 | 84.143 | 84.202 |
| 71 | 26.646 | 26.665 | 26.683 | 26.702 | 26.721 | 26.739 | 26.758 | 26.777 | 26.796 | 26.814 |
|    | 84.261 | 84.321 | 84.380 | 84.439 | 84.499 | 84.558 | 84.617 | 84.676 | 84.735 | 84.794 |
| 72 | 26.833 | 26.851 | 26.870 | 26.889 | 26.907 | 26.926 | 26.944 | 26.963 | 26.981 | 27.000 |
|    | 84.853 | 84.912 | 84.971 | 85.029 | 85.088 | 85.147 | 85.206 | 85.264 | 85.323 | 85.381 |
| 73 | 27.019 | 27.037 | 27.055 | 27.074 | 27.092 | 27.111 | 27.129 | 27.148 | 27.166 | 27.185 |
|    | 85.440 | 85.499 | 85.557 | 85.615 | 85.674 | 85.732 | 85.790 | 85.849 | 85.907 | 85.965 |
| 74 | 27.203 | 27.221 | 27.240 | 27.258 | 27.276 | 27.295 | 27.313 | 27.331 | 27.350 | 27.368 |
|    | 86.023 | 86.081 | 86.139 | 86.197 | 86.255 | 86.313 | 86.371 | 86.429 | 86.487 | 86.545 |
| 75 | 27.386 | 27.404 | 27.423 | 27.441 | 27.459 | 27.477 | 27.495 | 27.514 | 27.532 | 27.550 |
|    | 86.603 | 86.660 | 86.718 | 86.776 | 86.833 | 86.891 | 86.948 | 87.006 | 87.063 | 87.121 |
| 76 | 27.568 | 27.586 | 27.604 | 27.622 | 27.641 | 27.659 | 27.677 | 27.695 | 27.713 | 27.731 |
|    | 87.178 | 87.235 | 87.293 | 87.350 | 87.407 | 87.464 | 87.521 | 87.579 | 87.636 | 87.693 |
| 77 | 27.749 | 27.767 | 27.785 | 27.803 | 27.821 | 27.839 | 27.857 | 27.875 | 27.893 | 27.911 |
|    | 87.750 | 87.807 | 87.864 | 87.920 | 87.977 | 88.034 | 88.091 | 88.148 | 88.204 | 88.261 |
| 78 | 27.928 | 27.946 | 27.964 | 27.982 | 28.000 | 28.018 | 28.036 | 28.054 | 28.071 | 28.089 |
|    | 88.318 | 88.374 | 88.431 | 88.487 | 88.544 | 88.600 | 88.657 | 88.713 | 88.769 | 88.826 |
| 79 | 28.107 | 28.125 | 28.142 | 28.160 | 28.178 | 28.196 | 28.213 | 28.231 | 28.249 | 28.267 |
|    | 88.882 | 88.938 | 88.994 | 89.051 | 89.107 | 89.163 | 89.219 | 89.275 | 89.331 | 89.387 |

TABLE 626 (Continued)

SQUARE ROOTS OF $N$ AND $10N$

$N = 800$ to $1000$

| | 0 | 1 | 2 | 3 | 4 | 5 | 6 | 7 | 8 | 9 |
|---|---|---|---|---|---|---|---|---|---|---|
| 80 | 28.284 | 28.302 | 28.320 | 28.337 | 28.355 | 28.373 | 28.390 | 28.408 | 28.425 | 28.443 |
| | 89.443 | 89.499 | 89.554 | 89.610 | 89.666 | 89.722 | 89.778 | 89.833 | 89.889 | 89.944 |
| 81 | 28.460 | 28.478 | 28.496 | 28.513 | 28.531 | 28.548 | 28.566 | 28.583 | 28.601 | 28.618 |
| | 90.000 | 90.056 | 90.111 | 90.167 | 90.222 | 90.277 | 90.333 | 90.388 | 90.443 | 90.499 |
| 82 | 28.636 | 28.653 | 28.671 | 28.688 | 28.705 | 28.723 | 28.740 | 28.758 | 28.775 | 28.792 |
| | 90.554 | 90.609 | 90.664 | 90.719 | 90.774 | 90.830 | 90.885 | 90.940 | 90.995 | 91.049 |
| 83 | 28.810 | 28.827 | 28.844 | 28.862 | 28.879 | 28.896 | 28.914 | 28.931 | 28.948 | 28.965 |
| | 91.104 | 91.159 | 91.214 | 91.269 | 91.324 | 91.378 | 91.433 | 91.488 | 91.542 | 91.597 |
| 84 | 28.983 | 29.000 | 29.017 | 29.034 | 29.052 | 29.069 | 29.086 | 29.103 | 29.120 | 29.138 |
| | 91.652 | 91.706 | 91.761 | 91.815 | 91.869 | 91.924 | 91.978 | 92.033 | 92.087 | 92.141 |
| 85 | 29.155 | 29.172 | 29.189 | 29.206 | 29.223 | 29.240 | 29.257 | 29.275 | 29.292 | 29.309 |
| | 92.195 | 92.250 | 92.304 | 92.358 | 92.412 | 92.466 | 92.520 | 92.574 | 92.628 | 92.682 |
| 86 | 29.326 | 29.343 | 29.360 | 29.377 | 29.394 | 29.411 | 29.428 | 29.445 | 29.462 | 29.479 |
| | 92.736 | 92.790 | 92.844 | 92.898 | 92.952 | 93.005 | 93.059 | 93.113 | 93.167 | 93.220 |
| 87 | 29.496 | 29.513 | 29.530 | 29.547 | 29.563 | 29.580 | 29.597 | 29.614 | 29.631 | 29.648 |
| | 93.274 | 93.327 | 93.381 | 93.434 | 93.488 | 93.541 | 93.595 | 93.648 | 93.702 | 93.755 |
| 88 | 29.665 | 29.682 | 29.698 | 29.715 | 29.732 | 29.749 | 29.766 | 29.783 | 29.799 | 29.816 |
| | 93.808 | 93.862 | 93.915 | 93.968 | 94.021 | 94.074 | 94.128 | 94.181 | 94.234 | 94.287 |
| 89 | 29.833 | 29.850 | 29.866 | 29.883 | 29.900 | 29.917 | 29.933 | 29.950 | 29.967 | 29.983 |
| | 94.340 | 94.393 | 94.446 | 94.499 | 94.552 | 94.604 | 94.657 | 94.710 | 94.763 | 94.816 |
| 90 | 30.000 | 30.017 | 30.033 | 30.050 | 30.067 | 30.083 | 30.100 | 30.116 | 30.133 | 30.150 |
| | 94.868 | 94.921 | 94.974 | 95.026 | 95.079 | 95.131 | 95.184 | 95.237 | 95.289 | 95.341 |
| 91 | 30.166 | 30.183 | 30.199 | 30.216 | 30.232 | 30.249 | 30.265 | 30.282 | 30.299 | 30.315 |
| | 95.394 | 95.446 | 95.499 | 95.551 | 95.603 | 95.656 | 95.708 | 95.760 | 95.812 | 95.864 |
| 92 | 30.332 | 30.348 | 30.364 | 30.381 | 30.397 | 30.414 | 30.430 | 30.447 | 30.463 | 30.480 |
| | 95.917 | 95.969 | 96.021 | 96.073 | 96.125 | 96.177 | 96.229 | 96.281 | 96.333 | 96.385 |
| 93 | 30.496 | 30.512 | 30.529 | 30.545 | 30.561 | 30.578 | 30.594 | 30.610 | 30.627 | 30.643 |
| | 96.437 | 96.488 | 96.540 | 96.592 | 96.644 | 96.695 | 96.747 | 96.799 | 96.850 | 96.902 |
| 94 | 30.659 | 30.676 | 30.692 | 30.708 | 30.725 | 30.741 | 30.757 | 30.773 | 30.790 | 30.806 |
| | 96.954 | 97.005 | 97.057 | 97.108 | 97.160 | 97.211 | 97.263 | 97.314 | 97.365 | 97.417 |
| 95 | 30.822 | 30.838 | 30.854 | 30.871 | 30.887 | 30.903 | 30.919 | 30.935 | 30.952 | 30.968 |
| | 97.468 | 97.519 | 97.570 | 97.622 | 97.673 | 97.724 | 97.775 | 97.826 | 97.877 | 97.929 |
| 96 | 30.984 | 31.000 | 31.016 | 31.032 | 31.048 | 31.064 | 31.081 | 31.097 | 31.113 | 31.129 |
| | 97.980 | 98.031 | 98.082 | 98.133 | 98.184 | 98.234 | 98.285 | 98.336 | 98.387 | 98.438 |
| 97 | 31.145 | 31.161 | 31.177 | 31.193 | 31.209 | 31.225 | 31.241 | 31.257 | 31.273 | 31.289 |
| | 98.489 | 98.539 | 98.590 | 98.641 | 98.691 | 98.742 | 98.793 | 98.843 | 98.894 | 98.944 |
| 98 | 31.305 | 31.321 | 31.337 | 31.353 | 31.369 | 31.385 | 31.401 | 31.417 | 31.432 | 31.448 |
| | 98.995 | 99.045 | 99.096 | 99.146 | 99.197 | 99.247 | 99.298 | 99.348 | 99.398 | 99.448 |
| 99 | 31.464 | 31.480 | 31.496 | 31.512 | 31.528 | 31.544 | 31.559 | 31.575 | 31.591 | 31.607 |
| | 99.499 | 99.549 | 99.599 | 99.649 | 99.700 | 99.750 | 99.800 | 99.850 | 99.900 | 99.950 |

$$\sqrt{1,000} = 31.623 \qquad \sqrt{10,000} = 100$$

## RANDOM DIGITS

Table 632 reproduces the first 10,000 random decimal digits from The Rand Corporation, *A Million Random Digits with 100,000 Normal Deviates* (Glencoe, Illinois: Free Press, 1955). A method of using these to draw a random sample is described in Sec. 10.9.2.

A starting digit may be selected by placing the point of a pencil blindly on a page that has been chosen by tossing a coin twice, using page 632 for two heads, 633 for two tails, 634 for heads then tails, and 635 for tails then heads. Let the digit nearest the pencil point and the following two digits represent the line number. If these three digits are 000, 200, 400, 600, or 800, use line 0. If they are 001, 201, . . . , use line 1. In general, subtract the largest whole multiple of 200, and use the remainder as the line number. To find the starting digit within the line, take the two digits following the three used for the line number; if they exceed 50, subtract 50. Thus, the digits 27 or 77 lead to starting in column 27.

Having chosen a starting point, it is ordinarily satisfactory to read from left to right across lines or down columns, whichever is more convenient. An added refinement, helpful when this table is being used but a larger one is really needed, is to choose at random a direction in which to read the digits. Take the next digit, other than 0 or 9, after those used for finding the starting line and column. If it is 1, read to the right; if 2, to the left; if 3, up; if 4, down; if 5, diagonally up and right (northeast); if 6, diagonally up and left (northwest); if 7, diagonally down and left (southwest); if 8, diagonally down and right (southeast).

Suppose, for example, that the pencil point is nearest to line 42, column 8. The digits 406 lead to line 6 (that is, 406 − 400), and the digits 96 to column 46 (that is, 96 − 50). With the refinement, ordinarily unnecessary, of choosing a direction at random, the next digit, 3, leads to reading upward, so the sequence of random digits 0, 2, 7, 2, 3, 9, 7 is obtained. Here we come to the edge of the table, so for more digits we must select a new starting point and direction.

## TABLE 632

### 10,000 Random Digits

| Line No. | 1–5 | 6–10 | 11–15 | 16–20 | 21–25 | 26–30 | 31–35 | 36–40 | 41–45 | 46–50 |
|---|---|---|---|---|---|---|---|---|---|---|
| 0 | 10097 | 32533 | 76520 | 13586 | 34673 | 54876 | 80959 | 09117 | 39292 | 74945 |
| 1 | 37542 | 04805 | 64894 | 74296 | 24805 | 24037 | 20636 | 10402 | 00822 | 91665 |
| 2 | 08422 | 68953 | 19645 | 09303 | 23209 | 02560 | 15953 | 34764 | 35080 | 33606 |
| 3 | 99019 | 02529 | 09376 | 70715 | 38311 | 31165 | 88676 | 74397 | 04436 | 27659 |
| 4 | 12807 | 99970 | 80157 | 36147 | 64032 | 36653 | 98951 | 16877 | 12171 | 76833 |
| 5 | 66065 | 74717 | 34072 | 76850 | 36697 | 36170 | 65813 | 39885 | 11199 | 29170 |
| 6 | 31060 | 10805 | 45571 | 82406 | 35303 | 42614 | 86799 | 07439 | 23403 | 09732 |
| 7 | 85269 | 77602 | 02051 | 65692 | 68665 | 74818 | 73053 | 85247 | 18623 | 88579 |
| 8 | 63573 | 32135 | 05325 | 47048 | 90553 | 57548 | 28468 | 28709 | 83491 | 25624 |
| 9 | 73796 | 45753 | 03529 | 64778 | 35808 | 34282 | 60935 | 20344 | 35273 | 88435 |
| 10 | 98520 | 17767 | 14905 | 68607 | 22109 | 40558 | 60970 | 93433 | 50500 | 73998 |
| 11 | 11805 | 05431 | 39808 | 27732 | 50725 | 68248 | 29405 | 24201 | 52775 | 67851 |
| 12 | 83452 | 99634 | 06288 | 98083 | 13746 | 70078 | 18475 | 40610 | 68711 | 77817 |
| 13 | 88685 | 40200 | 86507 | 58401 | 36766 | 67951 | 90364 | 76493 | 29609 | 11062 |
| 14 | 99594 | 67348 | 87517 | 64969 | 91826 | 08928 | 93785 | 61368 | 23478 | 34113 |
| 15 | 65481 | 17674 | 17468 | 50950 | 58047 | 76974 | 73039 | 57186 | 40218 | 16544 |
| 16 | 80124 | 35635 | 17727 | 08015 | 45318 | 22374 | 21115 | 78253 | 14385 | 53763 |
| 17 | 74350 | 99817 | 77402 | 77214 | 43236 | 00210 | 45521 | 64237 | 96286 | 02655 |
| 18 | 69916 | 26803 | 66252 | 29148 | 36936 | 87203 | 76621 | 13990 | 94400 | 56418 |
| 19 | 09893 | 20505 | 14225 | 68514 | 46427 | 56788 | 96297 | 78822 | 54382 | 14598 |
| 20 | 91499 | 14523 | 68479 | 27686 | 46162 | 83554 | 94750 | 89923 | 37089 | 20048 |
| 21 | 80336 | 94598 | 26940 | 36858 | 70297 | 34135 | 53140 | 33340 | 42050 | 82341 |
| 22 | 44104 | 81949 | 85157 | 47954 | 32979 | 26575 | 57600 | 40881 | 22222 | 06413 |
| 23 | 12550 | 73742 | 11100 | 02040 | 12860 | 74697 | 96644 | 89439 | 28707 | 25815 |
| 24 | 63606 | 49329 | 16505 | 34484 | 40219 | 52563 | 43651 | 77082 | 07207 | 31790 |
| 25 | 61196 | 90446 | 26457 | 47774 | 51924 | 33729 | 65394 | 59593 | 42582 | 60527 |
| 26 | 15474 | 45266 | 95270 | 79953 | 59367 | 83848 | 82396 | 10118 | 33211 | 59466 |
| 27 | 94557 | 28573 | 67897 | 54387 | 54622 | 44431 | 91190 | 42592 | 92927 | 45973 |
| 28 | 42481 | 16213 | 97344 | 08721 | 16868 | 48767 | 03071 | 12059 | 25701 | 46670 |
| 29 | 23523 | 78317 | 73208 | 89837 | 68935 | 91416 | 26252 | 29663 | 05522 | 82562 |
| 30 | 04493 | 52494 | 75246 | 33824 | 45862 | 51025 | 61962 | 79335 | 65337 | 12472 |
| 31 | 00549 | 97654 | 64051 | 88159 | 96119 | 63896 | 54692 | 82391 | 23287 | 29529 |
| 32 | 35963 | 15307 | 26898 | 09354 | 33351 | 35462 | 77974 | 50024 | 90103 | 39333 |
| 33 | 59808 | 08391 | 45427 | 26842 | 83609 | 49700 | 13021 | 24892 | 78565 | 20106 |
| 34 | 46058 | 85236 | 01390 | 92286 | 77281 | 44077 | 93910 | 83647 | 70617 | 42941 |
| 35 | 32179 | 00597 | 87379 | 25241 | 05567 | 07007 | 86743 | 17157 | 85394 | 11838 |
| 36 | 69234 | 61406 | 20117 | 45204 | 15956 | 60000 | 18743 | 92423 | 97118 | 96338 |
| 37 | 19565 | 41430 | 01758 | 75379 | 40419 | 21585 | 66674 | 36806 | 84962 | 85207 |
| 38 | 45155 | 14938 | 19476 | 07246 | 43667 | 94543 | 59047 | 90033 | 20826 | 69541 |
| 39 | 94864 | 31994 | 36168 | 10851 | 34888 | 81553 | 01540 | 35456 | 05014 | 51176 |
| 40 | 98086 | 24826 | 45240 | 28404 | 44999 | 08896 | 39094 | 73407 | 35441 | 31880 |
| 41 | 33185 | 16232 | 41941 | 50949 | 89435 | 48581 | 88695 | 41994 | 37548 | 73043 |
| 42 | 80951 | 00406 | 96382 | 70774 | 20151 | 23387 | 25016 | 25298 | 94624 | 61171 |
| 43 | 79752 | 49140 | 71961 | 28296 | 69861 | 02591 | 74852 | 20539 | 00387 | 59579 |
| 44 | 18633 | 32537 | 98145 | 06571 | 31010 | 24674 | 05455 | 61427 | 77938 | 91936 |
| 45 | 74029 | 43902 | 77557 | 32270 | 97790 | 17119 | 52527 | 58021 | 80814 | 51748 |
| 46 | 54178 | 45611 | 80993 | 37143 | 05335 | 12969 | 56127 | 19255 | 36040 | 90324 |
| 47 | 11664 | 49883 | 52079 | 84827 | 59381 | 71539 | 09973 | 33440 | 88461 | 23356 |
| 48 | 48324 | 77928 | 31249 | 64710 | 02295 | 36870 | 32307 | 57546 | 15020 | 09994 |
| 49 | 69074 | 94138 | 87637 | 91976 | 35584 | 04401 | 10518 | 21615 | 01848 | 76938 |

TABLE 632 (Continued)

10,000 RANDOM DIGITS

| Line No. | 1–5 | 6–10 | 11–15 | 16–20 | Column Number 21–25 | 26–30 | 31–35 | 36–40 | 41–45 | 46–50 |
|---|---|---|---|---|---|---|---|---|---|---|
| 50 | 09188 | 20097 | 32825 | 39527 | 04220 | 86304 | 83389 | 87374 | 64278 | 58044 |
| 51 | 90045 | 85497 | 51981 | 50654 | 94938 | 81997 | 91870 | 76150 | 68476 | 64659 |
| 52 | 73189 | 50207 | 47677 | 26269 | 62290 | 64464 | 27124 | 67018 | 41361 | 82760 |
| 53 | 75768 | 76490 | 20971 | 87749 | 90429 | 12272 | 95375 | 05871 | 93823 | 43178 |
| 54 | 54016 | 44056 | 66281 | 31003 | 00682 | 27398 | 20714 | 53295 | 07706 | 17813 |
| 55 | 08358 | 69910 | 78542 | 42785 | 13661 | 58873 | 04618 | 97553 | 31223 | 08420 |
| 56 | 28306 | 03264 | 81333 | 10591 | 40510 | 07893 | 32604 | 60475 | 94119 | 01840 |
| 57 | 53840 | 86233 | 81594 | 13628 | 51215 | 90290 | 28466 | 68795 | 77762 | 20791 |
| 58 | 91757 | 53741 | 61613 | 62269 | 50263 | 90212 | 55781 | 76514 | 83483 | 47055 |
| 59 | 89415 | 92694 | 00397 | 58391 | 12607 | 17646 | 48949 | 72306 | 94541 | 37408 |
| 60 | 77513 | 03820 | 86864 | 29901 | 68414 | 82774 | 51908 | 13980 | 72893 | 55507 |
| 61 | 19502 | 37174 | 69979 | 20288 | 55210 | 29773 | 74287 | 75251 | 65344 | 67415 |
| 62 | 21818 | 59313 | 93278 | 81757 | 05686 | 73156 | 07082 | 85046 | 31853 | 38452 |
| 63 | 51474 | 66499 | 68107 | 23621 | 94049 | 91345 | 42836 | 09191 | 08007 | 45449 |
| 64 | 99559 | 68331 | 62535 | 24170 | 69777 | 12830 | 74819 | 78142 | 43860 | 72834 |
| 65 | 33713 | 48007 | 93584 | 72869 | 51926 | 64721 | 58303 | 29822 | 93174 | 93972 |
| 66 | 85274 | 86893 | 11303 | 22970 | 28834 | 34137 | 73515 | 90400 | 71148 | 43643 |
| 67 | 84133 | 89640 | 44035 | 52166 | 73852 | 70091 | 61222 | 60561 | 62327 | 18423 |
| 68 | 56732 | 16234 | 17395 | 96131 | 10123 | 91622 | 85496 | 57560 | 81604 | 18880 |
| 69 | 65138 | 56806 | 87648 | 85261 | 34313 | 65861 | 45875 | 21069 | 85644 | 47277 |
| 70 | 38001 | 02176 | 81719 | 11711 | 71602 | 92937 | 74219 | 64049 | 65584 | 49698 |
| 71 | 37402 | 96397 | 01304 | 77586 | 56271 | 10086 | 47324 | 62605 | 40030 | 37438 |
| 72 | 97125 | 40348 | 87083 | 31417 | 21815 | 39250 | 75237 | 62047 | 15501 | 29578 |
| 73 | 21826 | 41134 | 47143 | 34072 | 64638 | 85902 | 49139 | 06441 | 03856 | 54552 |
| 74 | 73135 | 42742 | 95719 | 09035 | 85794 | 74296 | 08789 | 88156 | 64691 | 19202 |
| 75 | 07638 | 77929 | 03061 | 18072 | 96207 | 44156 | 23821 | 99538 | 04713 | 66994 |
| 76 | 60528 | 83441 | 07954 | 19814 | 59175 | 20695 | 05533 | 52139 | 61212 | 06455 |
| 77 | 83596 | 35655 | 06958 | 92983 | 05128 | 09719 | 77433 | 53783 | 92301 | 50498 |
| 78 | 10850 | 62746 | 99599 | 10507 | 13499 | 06319 | 53075 | 71839 | 06410 | 19362 |
| 79 | 39820 | 98952 | 43622 | 63147 | 64421 | 80814 | 43800 | 09351 | 31024 | 73167 |
| 80 | 59580 | 06478 | 75569 | 78800 | 88835 | 54486 | 23768 | 06156 | 04111 | 08408 |
| 81 | 38508 | 07341 | 23793 | 48763 | 90822 | 97022 | 17719 | 04207 | 95954 | 49953 |
| 82 | 30692 | 70668 | 94688 | 16127 | 56196 | 80091 | 82067 | 63400 | 05462 | 69200 |
| 83 | 65443 | 95659 | 18288 | 27437 | 49632 | 24041 | 08337 | 65676 | 96299 | 90836 |
| 84 | 27267 | 50264 | 13192 | 72294 | 07477 | 44606 | 17985 | 48911 | 97341 | 30358 |
| 85 | 91307 | 06991 | 19072 | 24210 | 36699 | 53728 | 28825 | 35793 | 28976 | 66252 |
| 86 | 68434 | 94688 | 84473 | 13622 | 62126 | 98408 | 12843 | 82590 | 09815 | 93146 |
| 87 | 48908 | 15877 | 54745 | 24591 | 35700 | 04754 | 83824 | 52692 | 54130 | 55160 |
| 88 | 06913 | 45197 | 42672 | 78601 | 11883 | 09528 | 63011 | 98901 | 14974 | 40344 |
| 89 | 10455 | 16019 | 14210 | 33712 | 91342 | 37821 | 88325 | 80851 | 43667 | 70883 |
| 90 | 12883 | 97343 | 65027 | 61184 | 04285 | 01392 | 17974 | 15077 | 90712 | 26769 |
| 91 | 21778 | 30976 | 38807 | 36961 | 31649 | 42096 | 63281 | 02023 | 08816 | 47449 |
| 92 | 19523 | 59515 | 65122 | 59659 | 86283 | 68258 | 69572 | 13798 | 16435 | 91529 |
| 93 | 67245 | 52670 | 35583 | 16563 | 79246 | 86686 | 76463 | 34222 | 26655 | 90802 |
| 94 | 60584 | 47377 | 07500 | 37992 | 45134 | 26529 | 26760 | 83637 | 41326 | 44344 |
| 95 | 53853 | 41377 | 36066 | 94850 | 58838 | 73859 | 49364 | 73331 | 96240 | 43642 |
| 96 | 24637 | 38736 | 74384 | 89342 | 52623 | 07992 | 12369 | 18601 | 03742 | 83873 |
| 97 | 83080 | 12451 | 38992 | 22815 | 07759 | 51777 | 97377 | 27585 | 51972 | 37867 |
| 98 | 16444 | 24334 | 36151 | 99073 | 27493 | 70939 | 85130 | 32552 | 54846 | 54759 |
| 99 | 60790 | 18157 | 57178 | 65762 | 11161 | 78576 | 45819 | 52979 | 65130 | 04860 |

TABLE 632 (Continued)

10,000 RANDOM DIGITS

| Line No. | 1–5 | 6–10 | 11–15 | 16–20 | Column Number 21–25 | 26–30 | 31–35 | 36–40 | 41–45 | 46–50 |
|---|---|---|---|---|---|---|---|---|---|---|
| 100 | 03991 | 10461 | 93716 | 16894 | 66083 | 24653 | 84609 | 58232 | 88618 | 19161 |
| 101 | 38555 | 95554 | 32886 | 59780 | 08355 | 60860 | 29735 | 47762 | 71299 | 23853 |
| 102 | 17546 | 73704 | 92052 | 46215 | 55121 | 29281 | 59076 | 07936 | 27954 | 58909 |
| 103 | 32643 | 52861 | 95819 | 06831 | 00911 | 98936 | 76355 | 93779 | 80863 | 00514 |
| 104 | 69572 | 68777 | 39510 | 35905 | 14060 | 40619 | 29549 | 69616 | 33564 | 60780 |
| 105 | 24122 | 66591 | 27699 | 06494 | 14845 | 46672 | 61958 | 77100 | 90899 | 75754 |
| 106 | 61196 | 30231 | 92962 | 61773 | 41839 | 55382 | 17267 | 70943 | 78038 | 70267 |
| 107 | 30532 | 21704 | 10274 | 12202 | 39685 | 23309 | 10061 | 68829 | 55986 | 66485 |
| 108 | 03788 | 97599 | 75867 | 20717 | 74416 | 53166 | 35208 | 33374 | 87539 | 08823 |
| 109 | 48228 | 63379 | 85783 | 47619 | 53152 | 67433 | 35663 | 52972 | 16818 | 60311 |
| 110 | 60365 | 94653 | 35075 | 33949 | 42614 | 29297 | 01918 | 28316 | 98953 | 73231 |
| 111 | 83799 | 42402 | 56623 | 34442 | 34994 | 41374 | 70071 | 14736 | 09958 | 18065 |
| 112 | 32960 | 07405 | 36409 | 83232 | 99385 | 41600 | 11133 | 07586 | 15917 | 06253 |
| 113 | 19322 | 53845 | 57620 | 52606 | 66497 | 68646 | 78138 | 66559 | 19640 | 99413 |
| 114 | 11220 | 94747 | 07399 | 37408 | 48509 | 23929 | 27482 | 45476 | 85244 | 35159 |
| 115 | 31751 | 57260 | 68980 | 05339 | 15470 | 48355 | 88651 | 22596 | 03152 | 19121 |
| 116 | 88492 | 99382 | 14454 | 04504 | 20094 | 98977 | 74843 | 93413 | 22109 | 78508 |
| 117 | 30934 | 47744 | 07481 | 83828 | 73788 | 06533 | 28597 | 20405 | 94205 | 20380 |
| 118 | 22888 | 48893 | 27499 | 98748 | 60530 | 45128 | 74022 | 84617 | 82037 | 10268 |
| 119 | 78212 | 16993 | 35902 | 91386 | 44372 | 15486 | 65741 | 14014 | 87481 | 37220 |
| 120 | 41849 | 84547 | 46850 | 52326 | 34677 | 58300 | 74910 | 64345 | 19325 | 81549 |
| 121 | 46352 | 33049 | 69248 | 93460 | 45305 | 07521 | 61318 | 31855 | 14413 | 70951 |
| 122 | 11087 | 96294 | 14013 | 31792 | 59747 | 67277 | 76503 | 34513 | 39663 | 77544 |
| 123 | 52701 | 08337 | 56303 | 87315 | 16520 | 69676 | 11654 | 99893 | 02181 | 68161 |
| 124 | 57275 | 36898 | 81304 | 48585 | 68652 | 27376 | 92852 | 55866 | 88448 | 03584 |
| 125 | 20857 | 73156 | 70284 | 24326 | 79375 | 95220 | 01159 | 63267 | 10622 | 48391 |
| 126 | 15633 | 84924 | 90415 | 93614 | 33521 | 26665 | 55823 | 47641 | 86225 | 31704 |
| 127 | 92694 | 48297 | 39904 | 02115 | 59589 | 49067 | 66821 | 41575 | 49767 | 04037 |
| 128 | 77613 | 19019 | 88152 | 00080 | 20554 | 91409 | 96277 | 48257 | 50816 | 97616 |
| 129 | 38688 | 32486 | 45134 | 63545 | 59404 | 72059 | 43947 | 51680 | 43852 | 59693 |
| 130 | 25163 | 01889 | 70014 | 15021 | 41290 | 67312 | 71857 | 15957 | 68971 | 11403 |
| 131 | 65251 | 07629 | 37239 | 33295 | 05870 | 01119 | 92784 | 26340 | 18477 | 65622 |
| 132 | 36815 | 43625 | 18637 | 37509 | 82444 | 99005 | 04921 | 73701 | 14707 | 93997 |
| 133 | 64397 | 11692 | 05327 | 82162 | 20247 | 81759 | 45197 | 25332 | 83745 | 22567 |
| 134 | 04515 | 25624 | 95096 | 67946 | 48460 | 85558 | 15191 | 18782 | 16930 | 33361 |
| 135 | 83761 | 60873 | 43253 | 84145 | 60833 | 25983 | 01291 | 41349 | 20368 | 07126 |
| 136 | 14387 | 06345 | 80854 | 09279 | 43529 | 06318 | 38384 | 74761 | 41196 | 37480 |
| 137 | 51321 | 92246 | 80088 | 77074 | 88722 | 56736 | 66164 | 49431 | 66919 | 31678 |
| 138 | 72472 | 00008 | 80890 | 18002 | 94813 | 31900 | 54155 | 83436 | 35352 | 54131 |
| 139 | 05466 | 55306 | 93128 | 18464 | 74457 | 90561 | 72848 | 11834 | 79982 | 68416 |
| 140 | 39528 | 72484 | 82474 | 25593 | 48545 | 35247 | 18619 | 13674 | 18611 | 19241 |
| 141 | 81616 | 18711 | 53342 | 44276 | 75122 | 11724 | 74627 | 73707 | 58319 | 15997 |
| 142 | 07586 | 16120 | 82641 | 22820 | 92904 | 13141 | 32392 | 19763 | 61199 | 67940 |
| 143 | 90767 | 04235 | 13574 | 17200 | 69902 | 63742 | 78464 | 22501 | 18627 | 90872 |
| 144 | 40188 | 28193 | 29593 | 88627 | 94972 | 11598 | 62095 | 36787 | 00441 | 58997 |
| 145 | 34414 | 82157 | 86887 | 55087 | 19152 | 00023 | 12302 | 80783 | 32624 | 68691 |
| 146 | 63439 | 75363 | 44989 | 16822 | 36024 | 00867 | 76378 | 41605 | 65961 | 73488 |
| 147 | 67049 | 09070 | 93399 | 45547 | 94458 | 74284 | 05041 | 49807 | 20288 | 34060 |
| 148 | 79495 | 04146 | 52162 | 90286 | 54158 | 34243 | 46978 | 35482 | 59362 | 95938 |
| 149 | 91704 | 30552 | 04737 | 21031 | 75051 | 93029 | 47665 | 64382 | 99782 | 93478 |

TABLE 632 (Continued)

10,000 RANDOM DIGITS

| Line No. | 1–5 | 6–10 | 11–15 | 16–20 | 21–25 | 26–30 | 31–35 | 36–40 | 41–45 | 46–50 |
|---|---|---|---|---|---|---|---|---|---|---|
| | | | | | Column Number | | | | | |
| 150 | 94015 | 46874 | 32444 | 48277 | 59820 | 96163 | 64654 | 25843 | 41145 | 42820 |
| 151 | 74108 | 88222 | 88570 | 74015 | 25704 | 91035 | 01755 | 14750 | 48968 | 38603 |
| 152 | 62880 | 87873 | 95160 | 59221 | 22304 | 90314 | 72877 | 17334 | 39283 | 04149 |
| 153 | 11748 | 12102 | 80580 | 41867 | 17710 | 59621 | 06554 | 07850 | 73950 | 79552 |
| 154 | 17944 | 05600 | 60478 | 03343 | 25852 | 58905 | 57216 | 39618 | 49856 | 99326 |
| 155 | 66067 | 42792 | 95043 | 52680 | 46780 | 56487 | 09971 | 59481 | 37006 | 22186 |
| 156 | 54244 | 91030 | 45547 | 70818 | 59849 | 96169 | 61459 | 21647 | 87417 | 17198 |
| 157 | 30945 | 57589 | 31732 | 57260 | 47670 | 07654 | 46376 | 25366 | 94746 | 49580 |
| 158 | 69170 | 37403 | 86995 | 90307 | 94304 | 71803 | 26825 | 05511 | 12459 | 91314 |
| 159 | 08345 | 88975 | 35841 | 85771 | 08105 | 59987 | 87112 | 21476 | 14713 | 71181 |
| 160 | 27767 | 43584 | 85301 | 88977 | 29490 | 69714 | 73035 | 41207 | 74699 | 09310 |
| 161 | 13025 | 14338 | 54066 | 15243 | 47724 | 66733 | 47431 | 43905 | 31048 | 56699 |
| 162 | 80217 | 36292 | 98525 | 24335 | 24432 | 24896 | 43277 | 58874 | 11466 | 16082 |
| 163 | 10875 | 62004 | 90391 | 61105 | 57411 | 06368 | 53856 | 30743 | 08670 | 84741 |
| 164 | 54127 | 57326 | 26629 | 19087 | 24472 | 88779 | 30540 | 27886 | 61732 | 75454 |
| 165 | 60311 | 42824 | 37301 | 42678 | 45990 | 43242 | 17374 | 52003 | 70707 | 70214 |
| 166 | 49739 | 71484 | 92003 | 98086 | 76668 | 73209 | 59202 | 11973 | 02902 | 33250 |
| 167 | 78626 | 51594 | 16453 | 94614 | 39014 | 97066 | 83012 | 09832 | 25571 | 77628 |
| 168 | 66692 | 13986 | 99837 | 00582 | 81232 | 44987 | 09504 | 96412 | 90193 | 79568 |
| 169 | 44071 | 28091 | 07362 | 97703 | 76447 | 42537 | 98524 | 97831 | 65704 | 09514 |
| 170 | 41468 | 85149 | 49554 | 17994 | 14924 | 39650 | 95294 | 00556 | 70481 | 06905 |
| 171 | 94559 | 37559 | 49678 | 53119 | 70312 | 05682 | 66986 | 34099 | 74474 | 20740 |
| 172 | 41615 | 70360 | 64114 | 58660 | 90850 | 64618 | 80620 | 51790 | 11436 | 38072 |
| 173 | 50273 | 93113 | 41794 | 86861 | 24781 | 89683 | 55411 | 85667 | 77535 | 99892 |
| 174 | 41396 | 80504 | 90670 | 08289 | 40902 | 05069 | 95083 | 06783 | 28102 | 57816 |
| 175 | 25807 | 24260 | 71529 | 78920 | 72682 | 07385 | 90726 | 57166 | 98884 | 08583 |
| 176 | 06170 | 97965 | 88302 | 98041 | 21443 | 41808 | 68984 | 83620 | 89747 | 98882 |
| 177 | 60808 | 54444 | 74412 | 81105 | 01176 | 28838 | 36421 | 16489 | 18059 | 51061 |
| 178 | 80940 | 44893 | 10408 | 36222 | 80582 | 71944 | 92638 | 40333 | 67054 | 16067 |
| 179 | 19516 | 90120 | 46759 | 71643 | 13177 | 55292 | 21036 | 82808 | 77501 | 97427 |
| 180 | 49386 | 54480 | 23604 | 23554 | 21785 | 41101 | 91178 | 10174 | 29420 | 90438 |
| 181 | 06312 | 88940 | 15995 | 69321 | 47458 | 64809 | 98189 | 81851 | 29651 | 84215 |
| 182 | 60942 | 00307 | 11897 | 92674 | 40405 | 68032 | 96717 | 54244 | 10701 | 41393 |
| 183 | 92329 | 98932 | 78284 | 46347 | 71209 | 92061 | 39448 | 93136 | 25722 | 08564 |
| 184 | 77936 | 63574 | 31384 | 51924 | 85561 | 29671 | 58137 | 17820 | 22751 | 36518 |
| 185 | 38101 | 77756 | 11657 | 13897 | 95889 | 57067 | 47648 | 13885 | 70669 | 93406 |
| 186 | 39641 | 69457 | 91339 | 22502 | 92613 | 89719 | 11947 | 56203 | 19324 | 20504 |
| 187 | 84054 | 40455 | 99396 | 63680 | 67667 | 60631 | 69181 | 96845 | 38525 | 11600 |
| 188 | 47468 | 03577 | 57649 | 63266 | 24700 | 71594 | 14004 | 23153 | 69249 | 05747 |
| 189 | 43321 | 31370 | 28977 | 23896 | 76479 | 68562 | 62342 | 07589 | 08899 | 05985 |
| 190 | 64281 | 61826 | 18555 | 64937 | 13173 | 33365 | 78851 | 16499 | 87064 | 13075 |
| 191 | 66847 | 70495 | 32350 | 02985 | 86716 | 38746 | 26313 | 77463 | 55387 | 72681 |
| 192 | 72461 | 33230 | 21529 | 53424 | 92581 | 02262 | 78438 | 66276 | 18396 | 73538 |
| 193 | 21032 | 91050 | 13058 | 16218 | 12470 | 56500 | 15292 | 76139 | 59526 | 52113 |
| 194 | 95362 | 67011 | 06651 | 16136 | 01016 | 00857 | 55018 | 56374 | 35824 | 71708 |
| 195 | 49712 | 97380 | 10404 | 55452 | 34030 | 60726 | 75211 | 10271 | 36633 | 68424 |
| 196 | 58275 | 61764 | 97586 | 54716 | 50259 | 46345 | 87195 | 46092 | 26787 | 60939 |
| 197 | 89514 | 11788 | 68224 | 23417 | 73959 | 76145 | 30342 | 40277 | 11049 | 72049 |
| 198 | 15472 | 50669 | 48139 | 36732 | 46874 | 37088 | 73465 | 09819 | 58869 | 35220 |
| 199 | 12120 | 86124 | 51247 | 44302 | 60883 | 52109 | 21437 | 36786 | 49226 | 77837 |

# Index

*Index*

# Index

*Index*

# Index

# Index

## Index